The Quality Toolbox

Second Edition

Also available from ASQ Quality Press:

The Quality Improvement Glossary
Don Siebels

The Quality Improvement Handbook
ASQ Quality Management Division and John E. Bauer, Grace L. Duffy, Russell T. Westcott, editors

Quality's Greatest Hits: Classic Wisdom from the Leaders of Quality
Zigmund Bluvband

The Executive Guide to Improvement and Change
G. Dennis Beecroft, Grace L. Duffy, John W. Moran

Business Process Improvement Toolbox
Bjørn Andersen

From Quality to Business Excellence: A Systems Approach to Management
Charles Cobb

The Change Agents' Handbook: A Survival Guide for Quality Improvement Champions
David W. Hutton

Making Change Work: Practical Tools for Overcoming Human Resistance to Change
Brien Palmer

Principles and Practices of Organizational Performance Excellence
Thomas J. Cartin

Customer Centered Six Sigma: Linking Customers, Process Improvement, and Financial Results
Earl Naumann and Steven H. Hoisington

The Certified Quality Manager Handbook, Second Edition
Duke Okes and Russell T. Westcott, editors

To request a complimentary catalog of ASQ Quality Press publications, call 800-248-1946, or visit our Web site at http://qualitypress.asq.org.

The Quality Toolbox

Second Edition

Nancy R. Tague

ASQ Quality Press
Milwaukee, Wisconsin

American Society for Quality, Quality Press, Milwaukee 53203

Printed in the United States of America

19 20 21 5 4

Library of Congress Cataloging-in-Publication Data
Application in progress

Tague, Nancy R., 1955–
The quality toolbox / Nancy R. Tague.—2nd ed.
p. cm.
Includes bibliographical references and index.
ISBN 978-0-87389-871-3 (hard cover, perfect case bind : alk. paper)
1. Group problem solving. 2. Self-directed work teams. I. Title.

ISBN 978-0-87389-871-3

Acquisitions Editor: Matt Meinholz
Managing Editor: Paul O'Mara
Production Administrator: Randall Benson

ASQ Mission: The American Society for Quality advances individual, organizational, and community excellence worldwide through learning, quality improvement, and knowledge exchange.

Printed on acid-free paper

ASQ

Quality Press
600 N. Plankinton Ave.
Milwaukee, WI 53203-2914
E-mail: authors@asq.org
The Global Voice of Quality™

For my parents

who taught me the most important lessons about quality
and who have always believed in me

Table of Contents

List of Figures and Tables

Preface to the First Edition

The idea for this book originated when a group of facilitators in my company, well down the road of quality improvement, asked me to teach them new tools to use with their quality teams. They were stuck in a rut of using just a few familiar standards: brainstorming, multivoting, fishbone and Pareto diagrams. Their knowledge of the wide choice of methods and techniques that can be used in quality improvement was limited. Frustrated at being able to teach so few of the available tools in a training session, I decided to create a reference that they could use to locate and learn new tools on their own.

The question they asked after, "What tools exist?" was, "When do we use them?" The facilitators knew far more tools than they commonly used, but they did not know how to choose and apply tools at appropriate times during the process of quality improvement. So woven through the reference book was guidance on fitting the tools into the quality improvement process.

Since then, the book has been used with groups just getting started with quality improvement. It gives them more confidence with the basic tools and the quality improvement process they have learned. It also gives them a way to continue learning *just-in-time,* as they encounter needs for new methods. Team members, as well as facilitators and team leaders, have copies of the *Toolbox* on their shelves and refer to it between or during meetings.

Sometimes anything labeled "quality" is considered separate from day-to-day activities, but quality improvement extends into many areas that are not labeled "quality." Anyone planning strategy, solving a problem, developing a project plan, seeking ideas or agreement from other people, or trying to understand the customer better can use these tools to produce higher quality outcome more easily. By whatever name we call it, quality improvement should be a significant part of everything that every one of us does.

The Quality Toolbox is a comprehensive reference to a variety of methods and techniques: those most commonly used for quality improvement, many less commonly used, and a half dozen created by the author—not available elsewhere. The reader will find the widely used seven basic quality control tools (for example, fishbone diagram and Pareto chart) as well as the newer management and planning tools, sometimes called the seven new QC tools (for example, affinity diagram and arrow diagram). Tools are included for generating and organizing ideas, evaluating ideas, analyzing processes, determining root causes, planning, and basic data-handling and statistics.

Most reference books of statistical techniques do not include other quality improvement tools. Yet, those improving the quality of their work will need both kinds of tools at different times. This is true in both manufacturing and service organizations. In service organizations, and business and support functions of all organizations, people often fear statistical tools. They do not understand when and how to call upon their power. By combining both types of tools and providing guidance for when to use them, this book should open up the wide range of methods available for improvement.

The book is written and organized to be as simple as possible to use so that anyone can find and learn new tools without a teacher. Above all, *The Quality Toolbox* is an instruction book. The reader can learn new tools or, for familiar tools, discover new variations or applications. It also is a reference book. It is organized so that a half-remembered tool can be found and reviewed easily and so that the reader can quickly identify the right tool to solve a particular problem or achieve a specific goal.

With this book close at hand, a quality improvement team becomes capable of more efficient and effective work with less assistance from a trained quality consultant. I hope that quality and training professionals also will find the *Toolbox* a handy reference and quick way to expand their repertoire of tools, techniques, applications, and tricks.

Preface to the Second Edition

In the ten years since the first edition of this book was published, much has changed and much has stayed the same in quality improvement. The fundamental tools continue to be essential, but the value of other tools, especially statistical ones, has become more widely acknowledged. Thanks to Six Sigma, statistical tools such as hypothesis testing, regression analysis, and design of experiments, which have always been powerful tools for understanding and improving processes, are being used more regularly within quality improvement projects. A variety of previously lesser-known nonstatistical tools has also been taught by Six Sigma, lean, and other methodologies that have become widespread over the last ten years.

In updating this book, I have added 34 tools and 18 variations. Many of these tools existed when the first edition was published but were not yet used widely, or by the typical quality improvement team, or (I must confess) by me and my organization. Some of these new tools were used in other fields, such as the social sciences, but had not yet been adopted into quality improvement. As I wrote this edition, I discovered that I could spend years discovering the many varieties of quality improvement methodologies and the many creative applications of tools for improving work processes. At some point, however, I had to say "Enough!" and send the book to print.

Some of the added tools—for example, design of experiments and benchmarking—are too complex to be learned from the description in this book, but an overview is provided so that the reader will know when the tool is appropriate or even essential to the improvement process and be encouraged to use it, with expert assistance. Many of the new tools, however, are not difficult and can be learned easily from this book. That has always been the intent: to help team members and facilitators find and use the right tool at the right time in their improvement processes.

Another decade of Baldrige Award winners has continued to show us that there are many paths to excellence and that the tools and methods of quality improvement apply well to nonindustrial areas such as education and healthcare. This edition includes examples from a wider range of applications. Readers of the first edition asked for more case studies. The "Quality Improvement Stories" chapter has been expanded to include detailed case studies from three Baldrige Award winners.

A new chapter, "Mega-Tools: Quality Management Systems," puts the tools into two contexts: the historical evolution of quality improvement and the quality management systems within which the tools are used.

The last ten years have also seen increasing use of computers, especially the prevalence of software for all sorts of quality improvement tasks and the explosion of the Internet. This edition recognizes the computer as a valuable assistant to remove drudgery from many of the tools' procedures. However, it is critical to know the tool well enough to set up the computer's task and to understand and act on the results the computer provides, and this edition gives the reader that knowledge.

Computers have taught us all how valuable icons can be for instant understanding. This edition liberally uses icons with each tool description to reinforce for the reader what kind of tool it is and where it is used within the improvement process.

Before beginning work on this second edition, I used a basic quality principle: I asked my customers (readers) what the first edition did well and what could be improved. The needs of both seasoned quality practitioners and those just beginning to learn about quality were kept in mind as the book was written. I hope the improvements to *The Quality Toolbox* delight you by exceeding your expectations!

Acknowledgments

The tools of quality improvement have been developed by many people over a long time. Some of the toolmakers are well known, but many of the tools have been talked and written about so often that their origins are lost. I have been able to identify originators of many of the tools and have given credit to their inventiveness in the Resources. I am equally grateful to those whose names I do not know. Everyone who has contributed to the body of quality knowledge has helped us all find the satisfaction of learning, improving, and becoming just a bit more excellent.

Creating this book required the guidance and help of many people. My first teachers, mentors, and colleagues in quality improvement were Tom Dominick and Mark Rushing, and I am grateful for the experience of working with both of them. Tom introduced me to quality improvement as a discipline and shared his enthusiasm for the value of quality methods in an organization. He taught me to borrow, adapt, and customize from many sources. Mark's depth of understanding, innovativeness, and integrative thinking make him a source for a wealth of ideas and insight. I have learned much from working beside him and from our conversations about new ideas and applications.

Too many people to name individually have helped me understand the concepts and methods of quality improvement, especially my colleagues in quality. My thanks to each of you who have shared your ideas, successes, and failures with me. I am grateful to all the people of Ethyl and Albemarle Corporations, whom I have taught and learned from, as we labored together to apply these methods to improve our work and organizations.

I am indebted to my colleague Dave Zimmerman, who created initial drafts of several tools when this book was first expanded beyond a thin compendium of the most basic ones. He first brought brainwriting and list reduction to my attention. The entertaining example for the importance–performance analysis also was devised by Dave.

For the second edition, many people were generous with their time and knowledge. My thanks to Romesh Juneja of Albemarle for sharing his perspective on Six Sigma and to Sharron Manassa formerly of ASQ for invaluable research assistance. When I searched for applications outside my experience, many people generously stepped forward with experiences and insight that made the examples and case studies come to life. Thanks to Lance Casler, Anne Papinchak, Traci Shaw, and Lyn Tinnemeyer-Daniels of Medrad;

Patty Gerstenberger and Alan Huxman of St. Luke's Hospital of Kansas City; Sandy Cokeley-Pedersen and Susan Grosz of Pearl River School District; Pam Misenheimer, Cheryl Niquette, Marsha Plante, and Lesley Steiner.

I am indebted to Roger Berger and Davis Bothe for their painstaking reviews of the manuscript, their awesome attention to detail, and the depth of their knowledge. Their comments improved the book greatly.

Deep appreciation goes to Dianne Muscarello, who provided computer expertise and first manuscript review for both editions, as well as constant support and encouragement. In addition, she created most of the new graphics for the second edition. I couldn't have done this without her help.

Finally, my gratitude to the readers of the first edition, whose use of the book confirmed that it fills a need and whose comments have helped shape this second edition.

1

How to Use This Book

A carpenter with only a hammer, a screwdriver, a pair of pliers, and a straight-blade saw can build cabinets that are functional, but plain and crude. The carpenter with many different tools at hand will be able to create unique and well-crafted items and solve problem situations.

Like a carpenter's toolbox, *The Quality Toolbox* provides you with a choice of many tools appropriate to the wide variety of situations that occur on the road to continuous improvement. In fact, 148 different tools and variations are described with step-by-step instructions.

What is a quality tool? Webster defines a tool as: "Any implement, instrument, or utensil held in the hand and used to form, shape, fasten, add to, take away from, or otherwise change something . . . . Any similar instrument that is the working part of a power-driven machine . . . . Anything that serves in the manner of a tool; a means."[1] So tools are relatively small, often parts of a larger unit; they do something; each is designed for a very specific purpose. Thus, concepts are not tools, because they don't do anything, and methodologies or systems are not tools, because they are large and do too many things, although each of these have been called tools. Quality tools are the diagrams, charts, techniques and methods that, step by step, accomplish the work of quality improvement. They are the means to accomplish change.

If the *Toolbox* were only a step-by-step guide to many tools, it would be difficult to use. No one wants to read such a book cover to cover. How can you know a tool will be useful if you don't already know the tool? Several aids help guide you to the right tool for the situation.

THE TOOL MATRIX

The Tool Matrix (Table 1.1, page 8) lists all the tools in the book and categorizes them in three different ways to help you find the right one. To search for a tool, ask yourself three questions:

1. *What do we want to do with this tool?* A carpenter who wants to cut something will look for some type of saw, not for a screwdriver. Quality improvement tools also can be grouped according to how they are used.

Project planning and implementing tools: When you are managing your improvement project.

Idea creation tools: When you want to come up with new ideas or organize many ideas.

Process analysis tools: When you want to understand a work process or some part of a process. Processes start with inputs coming from suppliers, change those inputs, and end with outputs going to customers.

Data collection and analysis tools: When you want to collect data or analyze data you have already collected.

Cause analysis tools: When you want to discover the cause of a problem or situation.

Evaluation and decision-making tools: When you want to narrow a group of choices to the best one, or when you want to evaluate how well you have done something. This includes evaluating project results.

The tools in the Tool Matrix are grouped according to these categories. Notice that some tools show up in several categories. These versatile tools can be used in a variety of ways.

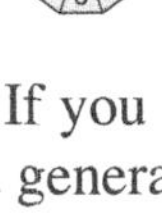

2. *Where are we in our quality improvement process?* A carpenter would use fine sandpaper only when the cabinet is almost done. Some tools are useful only at certain steps in the quality improvement process.

If you are not sure what this question means, read chapter 2. It describes ten steps of a general process for quality improvement. This process was deliberately written in ordinary, commonsense language. A translation to standard quality terminology is shown beside it. Your organization's process probably is written differently and has more or fewer steps. However, you should be able to find all the elements of your process in the ten-step process.

In the Tool Matrix, the columns list the ten steps. Each step of the process in which a tool can be used is marked with an X. The versatile tools that appear in several categories often have different steps marked from category to category, as their use changes.

3. *Do we need to expand or to focus our thinking?* The process of quality improvement goes through alternating periods of expanding our thinking to many different ideas and focusing our ideas to specifics. The expanding period is creative and can generate new and innovative ideas. The focusing period is analytical and action oriented. To obtain results, you eventually have to stop considering options, decide what to do, and do it!

See Figure 1.1 for an illustration of how the expand–focus sequence works. To choose the most worthwhile problem to attack, first expand your thinking to many different problems—big, small, annoying, and expensive problems—by analyzing the process and collecting data. Next, focus your thinking: with evaluation tools, use a set of criteria to choose one well-defined problem to solve.

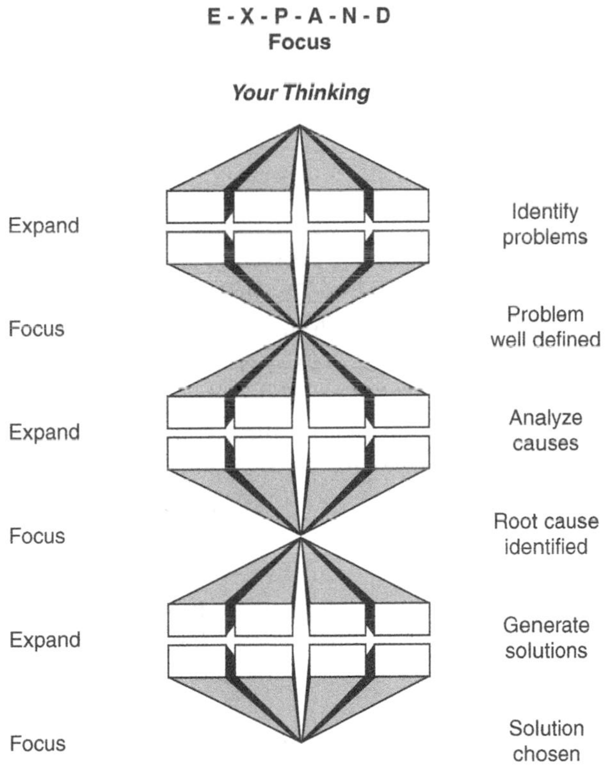

Figure 1.1 Expand–focus sequence.

Now, expand your thinking to many possible causes of the problem using tools like the fishbone diagram or is–is not matrix. *Could it be this? Could it be that? Maybe what's happening is. . . .* After getting lots of ideas, use methods such as data collection, analysis, and logical reasoning to narrow all possible causes to the few that really are the culprits.

Finally, expand your thinking once again to many ways to solve the problem, using tools like idea creation and data analysis. From a variety of solutions, use evaluation tools to choose the one most likely to work in your unique circumstances.

Some tools are designed specifically to help you expand your thinking. Others are designed solely to help you focus. A few encompass both modes: the first few steps of the tool expand your thinking and the final steps lead you through focusing. Some tools can either expand or focus your thinking, depending on how and when they are used. For example, flowcharts can be used to expand your thinking to all possible problems in a process, or they can guide a group to focus on the one way everyone has agreed a process will operate from this time forward.

The third column of the Tool Matrix shows an E for *expansion* or an F for *focusing*. Tools that encompass both modes or that can be used for either purpose are indicated by E/F.

Example

Let's look at an example of the Tool Matrix in use. Suppose your team has tested a solution; it worked, and you are ready to install it throughout your organization. Suppose that as you are beginning to plan how to do that, your team wants to consider what might go wrong. How do you find potential tools to help you?

First ask, "What do we want to do with this tool?" You need to plan and implement, so look at the Tool Matrix in the group labeled "Project planning and implementing tools." There are 28 tools in that group.

Then ask, "Where are we in our quality improvement process?" You are at step 9: "If it worked, how can we do it every time?" or, in quality jargon, "Standardize." On the Tool Matrix, under the column for that step, you find 24 tools marked with an X.

Your third question is, "Do we need to expand or focus our thinking?" By considering everything that might go wrong, you are expanding your thinking, so you eliminate the 11 tools marked F. That leaves 13 possible tools.

What next? Now you are ready to turn to chapter five, the main part of the book, and browse through the tools.

THE TOOLS

The tools are listed in alphabetical order, rather than by categories, so that whenever you know the name of the tool, you know exactly where to find it. Each tool has six sections.

- *Description.* A few sentences explain what the tool is and what it does. In addition, the icons you saw on pages 2 and 3 appear again to remind you to what category the tool belongs, at what steps of the process the tool can be used, and whether it is an expanding or focusing tool.
- *When to Use.* This section describes the situations in which you would want to use this tool. A situation might be a particular stage of the quality improvement process, a certain kind of problem, or after another tool has been used. If two or more situations should be true when the tool is used, "and . . ." links the statements. Otherwise, "or . . ." shows that the tool can be used if any of the statements are true. "Especially" means that the statement following is a situation where the tool is particularly useful.
- *Procedure.* A step-by-step numbered procedure guides you through using the tool. This section is very basic, so you can always use it as a quick reference.
- *Example.* You are introduced to a situation when the tool was appropriate for a team's situation, and the tool's use is explained. Calculations, the thinking behind various steps, and the conclusions that could be drawn also are explained. Some of these examples are fictional; others are based on actual situations. Whenever a tool involves a chart or diagram, an example or drawing is shown.
- *Variations.* When the tool can have several different appearances or methods, the step-by-step procedure for each variation is written out. Often, examples are provided for the variation. Occasionally, separate "description," "when to use," or "considerations" sections are also necessary. In most cases, the variation has a unique name. Occasionally it is simply called "Variation."
- *Considerations.* This section includes tips, tricks, and warnings—notes to help you use the tool more easily, avoid problems, or add additional flair or sophistication. Thus, this section adds all the detail and color that were omitted from the basic procedure.

Example

Let's return to the example, with your team ready to spread a solution throughout the organization. What happens after using the Tool Matrix to narrow the list of tools to 13? Browse through those 13 tools, reading just the "Description" and "When to Use." When you flip to the *contingency diagram,* you will read, "The contingency diagram uses brainstorming and a negative thinking process to identify how process problems occur or what might go wrong in a plan. Then the negative thinking is reversed to generate solutions or preventive measures." When do you use this? "When planning implementation of a phase of a project, especially the solution," and "Before launching a change." That sounds like exactly what you need.

But you should continue browsing through all 13 tools. You will discover that *potential problem analysis* and *process decision program chart* are designed for similar purposes and applications. Then read the sections for each of the three tools carefully in order to decide which one is most appropriate for your team's situation.

GENERIC TOOLS

If you walked into a hardware store and asked for "a saw," the clerk would respond, "What *kind* of saw?" "Saw" is a name for a broad group of tools; there are many specific saws designed for particular purposes. Similarly, there are several generic kinds of quality tools that can be customized for particular purposes.

For example, "graph" is a general name for a generic tool. There are hundreds of types of graphs—line graphs, bar charts, pie charts, box plots, histograms, control charts, and so on—which have been developed over the years to respond to specific needs or uses for graphs. People have named and described these graph variations so that others won't have to keep reinventing the wheel, or the graph, in this case.

Anyone familiar with the general definition of graph—"a visual display of numerical data to achieve deeper or quicker understanding of the meaning of the numbers"—might devise a new way of displaying data to suit a particular need. That type of graph might be used only once, or it might become another named graph that others with the same need could use.

Quality improvement practitioners are creative and inventive, so many tool variations have been devised. This book does not include many that are straightforward applications of the generic tool, which is especially common with check sheets, matrices and tables. It does include ones that are in some way unique, where there was a creative or conceptual jump between the generic tool and the variation. Separate entries are allotted for tools that have complex procedures (such as the control chart) or a narrower application (such as the decision matrix). Otherwise, the variation is described under the generic tool's listing. Each of the generic tools has its own entry.

So what are these generic tools? Here is a list, with examples of unique variations:

Check sheet and checklist: defect concentration diagram, project charter checklist, guidance questions

Flowchart: deployment flowchart, macro flowchart, process mapping, top-down flowchart, work flow diagram

Graph: control chart, histogram, Pareto chart, run chart, scatter diagram

Matrix Diagram: decision matrix, prioritization matrix, house of quality

Table: contingency table, check sheet, stakeholder analysis table, voice of the customer table

Tree Diagram: decision tree, fault tree analysis, process decision program chart, why–why diagram

Two-Dimensional Chart: effective–achievable chart, Kano model, plan–results chart

Learn the generic tools first, and then the variations will be easier.

Once you are familiar with the generic tools, you too might improvise to develop a tool variation that fits a particular need. If it is widely applicable, share your new tool with others.

Now you know how to use this book to identify and learn the tools that are most useful in your specific situation. You might be wondering, "How do all these tools fit together to create improvements?" The next chapter, "Mega-Tools: Quality Management Systems," discusses the organizationwide processes into which the tools fit. Chapter 3, "The Quality Improvement Process," outlines the improvement model and types of tools used each step of the way, and Chapter 4, "Quality Improvement Stories," tells how four teams actually used the tools to improve their work.

ENDNOTE

1. Michael Agnes, ed., *Webster's New World College Dictionary,* 4th edition (Foster City, CA: IDG Books Worldwide, 2000).

Table 1.1 Tool matrix.

	Tool	E/F	1 Charter & Plans	2 Customer Needs	3 Current State	4 Oppor-tunities	5 Root Causes	6 Changes	7 Do It	8 Monitor	9 Stan-dardize	10 Learnings
Project Planning and Implementing Tools	ACORN test	F	X									
	Arrow diagram	F	X					X	X		X	
	Balanced scorecard	F	X	X	X			X	X	X		
	Barriers and benefits exercise	E						X	X		X	X
	Checklist (generic)	F	X						X	X	X	
	Contingency diagram	E							X		X	
	Continuum of team goals	F	X							X		X
	Flowchart (generic)	E/F	X	X	X	X	X	X	X	X	X	X
	Force-field analysis	E						X	X		X	X
	Gantt chart	F	X					X	X		X	
	Matrix diagram (generic)	F	X	X				X	X		X	X
	Meeting evaluation	F	X	X	X	X	X	X	X	X	X	X
	Mind map	E	X	X	X	X	X	X	X		X	X
	Operational definitions	F	X	X	X	X	X	X	X	X	X	X
	Plan–do–study–act cycle	F	X	X	X	X	X	X	X	X	X	X
	Plan–results chart	F								X	X	X
	Potential problem analysis	E/F						X	X		X	
	Presentation	F	X	X	X		X		X	X	X	X
	Process decision program chart	E/F						X	X		X	
	Project charter	F	X	X	X	X	X	X	X	X	X	X
	Project charter checklist	F	X									
	Relations diagram	E/F	X	X		X		X			X	

Continued

	Tool	E/F	1 Charter & Plans	2 Customer Needs	3 Current State	4 Oppor-tunities	5 Root Causes	6 Changes	7 Do It	8 Monitor	9 Stan-dardize	10 Learnings
Project Planning and Implementing Tools	Stakeholder analysis	E/F	X	X				X	X		X	
	Storyboard	E/F	X	X	X	X	X	X	X	X	X	X
	Table (generic)	E/F	X	X	X	X	X	X	X	X	X	X
	Tree diagram (generic)	E	X	X	X	X	X	X	X	X	X	X
	Two-dimensional chart (generic)	F	X	X	X	X		X	X	X	X	X
	Wordsmithing	E/F	X	X	X		X	X			X	X
Idea Creation Tools	Affinity diagram	E/F	X	X	X	X	X	X			X	X
	Benchmarking	E/F	X			X		X				
	Brainstorming	E	X	X		X	X	X			X	X
	Brainwriting	E	X	X		X	X	X			X	X
	Desired-result fishbone	E						X				
	5W2H	E	X					X		X	X	X
	Mind map	E	X	X	X	X	X	X	X		X	X
	NGT	E	X	X		X	X	X			X	X
	Relations diagram	E/F	X	X	X	X	X	X			X	X
	Storyboard	E/F	X	X	X	X	X	X			X	X
Process Analysis Tools	Benchmarking	E/F	X			X		X				
	Cause-and-effect matrix	F	X	X								
	Cost-of-poor-quality analysis	E			X	X	X					
	Critical-to-quality analysis	E		X	X	X	X					
	Critial-to-quality tree	F		X	X	X						
	Failure modes and effects analysis	E				X	X				X	
	5W2H	E			X	X					X	

Continued

	Tool	E/F	1 Charter & Plans	2 Customer Needs	3 Current State	4 Oppor-tunities	5 Root Causes	6 Changes	7 Do It	8 Monitor	9 Stan-dardize	10 Learnings
Process Analysis Tools	Flowchart (generic)	E/F	X	X	X	X	X	X	X	X	X	
	House of quality	F		X	X	X						
	Matrix diagram (generic)	F		X	X	X			X	X	X	X
	Mistake-proofing	F						X	X		X	X
	Relations diagram	E/F	X	X	X	X	X	X			X	
	Requirements table	E		X	X	X				X	X	
	Requirements-and-measures tree	E		X	X	X				X	X	
	SIPOC diagram	E/F		X	X							
	Storyboard	E/F		X	X			X	X		X	X
	Tree diagram (generic)	E/F		X	X	X		X	X		X	X
	Value-added analysis	E			X	X	X					
	Work flow diagram	E/F			X	X	X	X	X		X	
Data Collection and Analysis Tools	Balanced scorecard	F	X	X	X			X	X	X		
	Benchmarking	E/F	X			X		X				
	Box plot	F			X	X	X	X	X	X	X	
	Check sheet (generic)	F			X	X	X		X	X	X	
	Contingency table	F			X	X	X			X		
	Control chart	F			X	X	X		X	X	X	
	Correlation analysis	F			X	X	X			X		
	Cycle time chart	F			X	X	X			X		
	Design of experiments	F					X	X				
	Graph (generic)	F	X	X	X	X	X	X	X	X	X	X
	Histogram	F			X	X	X	X	X	X	X	

Continued

	Tool	E/F	1 Charter & Plans	2 Customer Needs	3 Current State	4 Oppor-tunities	5 Root Causes	6 Changes	7 Do It	8 Monitor	9 Stan-dardize	10 Learnings
Data Collection and Analysis Tools	Hypothesis testing	F			X	X	X	X		X		
	Importance–performance analysis	F		X	X	X				X	X	
	Normal probability plot	F			X	X	X		X	X	X	
	Operational definitions	F		X	X	X	X	X	X	X	X	
	Pareto chart	F		X	X	X	X			X	X	
	Performance index	F	X		X	X				X	X	
	PGCV index	F		X	X	X				X	X	
	Process capability study	F			X	X				X	X	
	Radar chart	F			X	X			X	X	X	
	Regression analysis	F					X	X		X	X	
	Repeatability and reproducibility study	F			X	X	X		X	X	X	
	Run chart	F			X	X	X		X	X	X	
	Sampling	F		X	X	X	X		X	X	X	
	Scatter diagram	F				X	X			X		
	Stratification	F	X	X	X	X	X	X		X	X	
	Survey	E/F	X	X	X	X	X	X		X	X	
	Table (generic)	E/F	X	X	X	X	X	X	X	X	X	X
	Voice of the customer table	E		X	X							
Cause Analysis Tools	Contingency diagram	E					X	X				
	Failure modes and effects analysis	E/F					X	X				
	Fault tree analysis	E					X					
	Fishbone diagram	E					X					

Continued

Continued

	Tool	E/F	1 Charter & Plans	2 Customer Needs	3 Current State	4 Oppor-tunities	5 Root Causes	6 Changes	7 Do It	8 Monitor	9 Stan-dardize	10 Learnings
Cause Analysis Tools	Force-field analysis	E					X	X				
	Is–is not matrix	F					X					
	Matrix diagram (generic)	F					X	X				
	Pareto chart	F					X					
	Relations diagram	E/F					X	X				
	Scatter plot	F					X					
	Stratification	F					X					
	Tree diagram (generic)	E					X	X				
	Why–why diagram	F					X					
Evaluation and Decision-Making Tools	Criteria filtering	F	X	X		X		X				
	Decision matrix	F	X			X		X				
	Decision tree	F			X	X		X	X		X	
	Effective–achievable chart	F	X			X		X				
	Reverse fishbone diagram	F						X				
	List reduction	F	X	X		X		X				
	Matrix diagram (generic)	F	X	X				X			X	X
	Multivoting	F	X	X		X		X				
	Paired comparisons	F	X	X		X		X				
	PMI	F	X			X		X			X	
	Prioritization matrix	F	X	X		X		X				
	Tree diagram (generic)	F	X	X	X	X		X	X	X	X	
	Two-dimensional chart (generic)	F	X	X	X	X		X	X	X	X	X

2

Mega-Tools: Quality Management Systems

Discussions of quality tools often include huge mechanisms with cryptic names such as QFD, ISO, Six Sigma, and lean. Although sometimes called "tools," these are really systems for organizing and managing improvement across an organization. They involve philosophical concepts and methodologies as well as collections of smaller tools.

While a blueprint might be called a tool, a carpenter would consider it in a different category from screwdrivers and hammers. Similarly, these quality mega-tools are different from the specific tools that are the primary focus of this book. This chapter will provide overviews of the major quality improvement systems to help clarify what they are and how the tools of this book are used within them. First, however, we must set the context with a brief history of quality improvement.

THE EVOLUTION OF QUALITY

The story actually begins in the middle ages with craftsmen's guilds, but we'll start with Walter A. Shewhart, a Bell Laboratory statistician. In the 1920s, based on earlier English work in agriculture, he developed control charts and the principles of modern statistical process control. Shewhart's statistical principles were applied in American industry in the 1930s but lost favor after World War II as the booming market gave American manufacturing easy primacy.

Dr. W. Edwards Deming, a statistician who worked for the USDA and the Census Bureau, learned statistical process control from Shewhart and taught it to engineers and statisticians in the early 1940s. He became frustrated that managers did not understand the benefits of these methods and therefore did not support them. After World War II, he went to Japan to advise on census issues and in the early 1950s was invited to lecture to the Union of Japanese Scientists and Engineers (JUSE) on quality control. At the time, "Made in Japan" was a synonym for low-quality junk. Deming taught Japanese industrialists statistical and managerial concepts and told them that by applying these concepts, they could have the world asking for their products.

Dr. Joseph M. Juran was an electrical engineer trained in industrial statistics at Western Electric. Like Deming, he applied his knowledge in Washington, D.C., during World War II. Like Deming, he was invited to lecture to the JUSE, focusing on planning and management's responsibilities for quality. Drs. Deming and Juran were both decorated by Emperor Hirohito.

Deming had been right. By the 1970s, American auto and electronics industries were reeling from Japanese high-quality competition. In 1980, a TV documentary titled, "If Japan Can, Why Can't We?" got the attention of American companies. Teams went to Japan to study what Toyota, Mitsubishi, Nissan, and others were doing, and Drs. Deming and Juran were suddenly in demand as consultants to American CEOs. Major corporations, including the Big Three automakers—Ford, General Motors, and Chrysler—began programs of quality management and statistical quality control. The new quality philosophy taught that the quality of incoming materials was important, so these companies pressed their suppliers to begin quality efforts as well. Those suppliers turned to their suppliers, and quality programs cascaded through American industry.

TOTAL QUALITY MANAGEMENT

Total Quality Management (TQM) is any quality management system that addresses all areas of an organization, emphasizes customer satisfaction, and uses continuous improvement methods and tools. TQM is based on the concepts taught by quality management gurus Deming, Juran, Crosby, Ishikawa, and others.

Listening to Deming and Juran and observing the methods that had yielded such success in Japan, American quality programs emphasized far more than just statistics. Approaches that embraced the entire organization, not just the production area, and that included a change in management style, not just statistical tools, came to be called Total Quality Management (TQM).

TQM was the name used in 1985 by the Naval Air Systems Command for its program. Since then, the term has been widely adopted and does not refer to a specific program or system. Practitioners of TQM might follow a program based primarily on Deming's fourteen management points, the Juran Trilogy (quality planning, quality control, and quality improvement), Philip Crosby's Four Absolutes of Quality Management, or some customized composite. Regardless of the flavor, TQM programs include three components: management philosophy, an improvement process or model, and a set of tools that include the seven quality control (QC) tools.

All of the quality gurus agree that a fundamental cause of quality problems in any organization is management. The leaders of organizations adopting TQM usually need to make fundamental changes in their management philosophy and methods. Common elements of any TQM program include senior management leadership of quality, employee involvement and empowerment, customer-defined quality and a focus on customer satisfaction, a view of work as process, and continuous improvement.

An improvement process or model provides the "how-to" for specific improvements. It is a framework for teams or individuals to follow each time they tackle a specific issue. Chapter 3 discusses improvement processes and defines the generic ten-step model used in this book.

Tools are the means for action, as discussed in Chapter 1. The most fundamental tools and the first ones developed are the seven quality control tools.

Tools: The Seven QC Tools and the Seven MP Tools

The seven QC tools were first emphasized by Kaoru Ishikawa, professor of engineering at Tokyo University and father of quality circles. His original seven tools were: cause-and-effect diagram (also called Ishikawa or fishbone chart), check sheet, Shewhart's control charts, histogram, Pareto chart, scatter diagram, and stratification. Some lists replace stratification with flowchart or run chart. They are variously called the seven quality control tools, the seven basic tools, or the seven old tools. Regardless of the name, a set of seven simple yet powerful tools are used in every system of quality improvement.

In 1976, the JUSE saw the need for tools to promote innovation, communicate information, and successfully plan major projects. A team researched and developed the seven new QC tools, often called the seven management and planning (MP) tools or simply the seven management tools. Not all the tools were new, but their collection and promotion were. The seven MP tools are: affinity diagram, relations diagram, tree diagram, matrix diagram, matrix data analysis, arrow diagram, and process decision program chart (PDPC). The order listed moves from abstract analysis to detailed planning. All of the old and new tools are included in this book, with the exception of matrix data analysis. That tool, a complex mathematical technique for analyzing matrices, is often replaced in the list by the similar prioritization matrix, which is included here.

The seven new tools were introduced in the United States in the mid-1980s with *hoshin planning,* a breakthrough strategic planning process that links visionary goals to work plans. Through the 1980s and 1990s, many people provided innovative additions to the concepts, methods, and tools of quality improvement. Genichi Taguchi developed new methods of applying experimental design to quality control. Masaaki Imai popularized the term and concept *kaizen,* which means small, continuous improvements, often using the PDSA cycle. Quality function deployment (QFD), benchmarking, ISO 9000 and ISO 14000, the Baldrige Award, Six Sigma, theory of constraints, and lean manufacturing are all either new developments or revitalization and repackaging of prior concepts and methods.

Problems and Benefits

TQM doesn't always generate the hoped-for results. In the last decade, it was sometimes considered a fad that had flashed and was dying. However, enough organizations have used it with outstanding success that, to paraphrase Mark Twain, the reports of its death have been greatly exaggerated. Applying any system of quality management requires such tremendous change to an organization's culture that it is very difficult to accomplish. Picking out just the pieces that are appealing or easy won't work. Imitating a successful organization won't work either, because their starting point and their organization were different than yours. Quality management can only be successful with a great deal of learning, intense analysis, hard work, and focused attention over an extended period of time.

The next sections describe some of the most recent innovations to quality management. These systems are the mega-tools we discussed at the beginning of this chapter. Each is an evolutionary step beyond the foundations laid by Deming, Juran, and the early Japanese practitioners of quality.

QUALITY FUNCTION DEPLOYMENT

Quality function deployment (QFD) is a structured process for planning the design of a new product or service or for redesigning an existing one. QFD first emphasizes thoroughly understanding what the customer wants or needs. Then those customer wants are translated into characteristics of the product or service. Finally, those characteristics are translated into details about the processes within the organization that will generate the product or service.

History and Applications

QFD was developed by Dr. Yoji Akao for Japanese tire and ship manufacturing in the late 1960s and early 1970s. Previous quality methods had addressed only problems that arose during production. With QFD, customer satisfaction dictates product and process design, so that customers are happy with the first and every product rolling out of production.

QFD was introduced to the United States in 1983 and has since spread around the world. Originally applied to manufacturing, it has been used to design services as diverse as police work, healthcare, law, kindergarten and college curricula, and a realistic, interactive dinosaur for a theme park.

Benefits and Problems

QFD shortens the design time and reduces the costs of achieving product or service introduction. The planning stage may take longer than without QFD, but expensive corrections and redesigns are eliminated. Eventually, fewer customer complaints, greater customer satisfaction, increased market share, and higher profits are achieved.

QFD requires cross-functional teams. It establishes a focus on the customer throughout the organization. And in its final steps, quality control measures are generated that ensure customer satisfaction after process start-up. As a result, QFD can be much more than a simple tool or planning process. It can be a key element of an organization's quality system.

Because of the emphasis on cross-functional teams and customer focus, introducing QFD into an organization may clash with the existing culture. This is a double-edged sword. Done thoughtfully, QFD can lead to the additional benefits of increased teamwork and customer focus. Done without considering the cultural conflicts, QFD can fail.

If you want to introduce QFD into your organization, learn more through reading and training, and enlist experienced assistance. Like any large-scale endeavor using new methods and requiring cultural change, the first efforts must be carefully planned, and benefit greatly from the insights of someone who has "been there, done that." QFD can be a powerful system to take an organization beyond preventing problems to truly pleasing the customer.

Tools and Methods

The house of quality is a key tool in QFD. The house of quality starts with customer requirements, stated in the voice of the customer, and relates them to quality characteristics of the product or service. From the house, decisions are made about which characteristics to emphasize and specific numerical targets to aim for in their design. See *house of quality* in Chapter 5 for more information about this tool.

Sometimes the house of quality is considered synonymous with QFD. But QFD is much more than one diagram. In the full QFD process, after you determine targets for critical characteristics, those characteristics must be translated into details about parts or components (for manufacturing applications) or functions (for service applications). In turn, those details must be translated into process designs and finally into specific task descriptions and procedures. Diagrams structured like the house of quality are typically used for those steps. Other tools, such as matrices, tables, and tree diagrams, are used within the QFD process to analyze customers, reliability, safety, or cost and to deploy the plans into action.

Satisfying Customers and the Kano Model

Customer satisfaction is the primary goal in QFD. A model of customer satisfaction developed by Noriaki Kano is often considered when QFD is applied. The model says that customer requirements fall into three groups: satisfiers, dissatisfiers, and delighters or exciters. (Kano called the groups one-dimensional attributes, must-be attributes, and attractive attributes.)

Satisfiers are requirements customers usually state if asked about their requirements. As satisfiers are increased, customer satisfaction increases. An example might be a computer's memory capacity or speed.

Dissatisfiers are requirements customers don't even think to mention, yet without them they would be very upset. They're expected or taken for granted. For example, a power cord for a computer is a dissatisfier, because that unstated requirement has the potential to create great customer dissatisfaction if it is not met.

Delighters or *exciters* are extras customers haven't even imagined, yet if they were offered, the customer would be thrilled. They're the whiz-bang features that make a customer say, "Wow!" An example might be a computer that doesn't need passwords because it recognizes fingerprints or voice waves. Without the delighter, the customer isn't dissatisfied. After all, the customer has never even dreamed of that feature! But when it is provided, the customer's satisfaction increases dramatically.

One unusual feature of delighters is that today's delighter becomes tomorrow's satisfier or dissatisfier. CD drives in computers were delighters when first introduced. Now they're expected to be present, and their features are satisfiers.

Kano's model is drawn as a two-dimensional chart (Figure 2.1). Moving from left to right, more or better requirements are provided. Moving from bottom to top, satisfaction increases.

The three lines represent the three types of requirements. Satisfiers follow the straight line through the intersection of the two axes. Satisfaction increases linearly as more or better satisfiers are provided.

Dissatisfiers follow the curved line at the bottom of the graph. When dissatisfiers are missing, satisfaction plummets. (Remember, dissatisfiers are features customers expect.) As dissatisfiers are added, satisfaction stabilizes.

Delighters follow the curved line at the top of the graph. Satisfaction isn't affected when they're missing, but when they are added, satisfaction skyrockets.

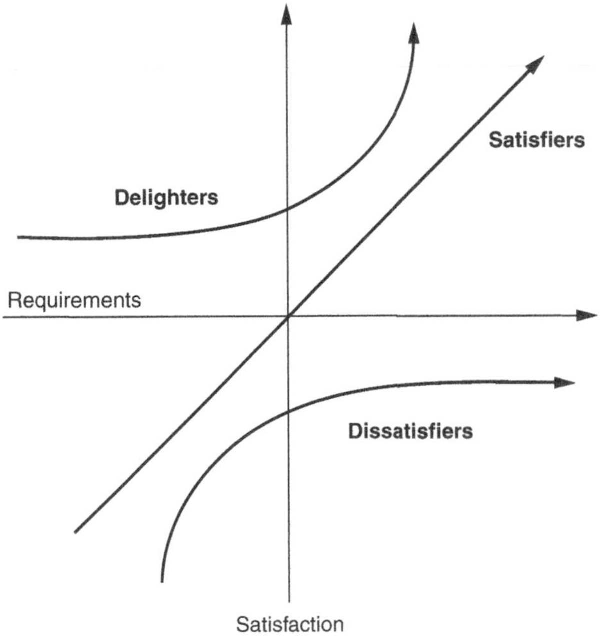

Figure 2.1 Kano model.

How does the Kano model affect QFD? When you are gathering customer requirements, find creative ways to gather delighters and dissatisfiers, which customers will not normally mention. Also, the numerical methods applied in the house of quality should be modified if the requirements are delighters or dissatisfiers, which do not follow linear relationships with satisfaction. Various methods developed to do this can be found in recent books and papers on QFD.

ISO 9000

ISO 9000 is a set of international standards for quality. Organizations are audited and certified against the requirements of the standard. The standard includes elements that are considered important in a quality management system, from senior management responsibility to documentation to continuous improvement. Customers of a certified organization can be assured that their supplier has minimum acceptable practices in place.

History and Applications

Quality standards have their roots in the 12th century practice of hallmarking silver. To protect customers, an item was tested at an assay office, and if it met minimum standards of silver content, it was stamped with a hallmark.

Starting in 1959, the US and UK defense departments issued standards for quality control to ensure that materials, parts, and equipment provided by suppliers were of suitable quality. No one wanted bombs that wouldn't explode—or exploded prematurely! Government inspectors visited suppliers to ensure compliance to the standards. Other standards proliferated through the 1960s: by NASA's space program, NATO, Canadian and British electrical utilities, the UK's Ministry of Defence. All these standards were based on a philosophy of inspecting the product for quality after production and required auditing by the customer's inspectors. In Britain alone, 17,000 inspectors were employed by the government.

During the 1970s, the approach changed from customers' inspection of their suppliers to independent, third-party inspectors performing this function. The term "quality assurance" also became prevalent, rather than "quality control." In 1979, the British standards office issued BS 5750, covering quality assurance in nonmilitary applications. Other countries followed suit. Finally, in 1987, the International Organization for Standardization (known as ISO, from the Greek word meaning "equal"), an alliance of the standards bodies of 91 countries worldwide, issued the ISO 9000 series of standards to facilitate international trade. It was essentially identical to BS 5750.

Use of ISO 9000 spread rapidly, as customers required it of their suppliers in order to be assured of a minimum level of quality practice. In 1994, ISO issued minor revisions to the standard. However, concern was widespread about whether the standards promoted quality approaches that reflected post–World War II knowledge about managing quality. In 2000, ISO issued a major revision of the standard, reducing the number of documents in the series and placing more emphasis on concepts such as work processes and customer satisfaction. Transition to the new standard was required by December 2003.

For suppliers to the automotive industry, ISO/TS 16949:2002 follows the format of ISO 9001:2000 but adds additional requirements. This standard will supplant QS-9000, a standard published by Ford, General Motors, and DaimlerChrysler. ISO 14001, a standard for environmental management systems, is aligned with ISO 9001:2000 for easy integration of environmental and quality management systems.

What ISO 9000 Requires

The ISO 9000 series consists of three documents: *Fundamentals and vocabulary* (ISO 9000), *Requirements* (ISO 9001), and *Guidelines for performance improvements* (ISO 9004). (The numerical jump between 9001 and 9004 is because the original series had two other documents that were eliminated.) There are other supplemental documents in the ISO 9000 family.

The major sections of ISO 9001:2000 are:

- Quality management system and documentation requirements
- Management responsibility
- Resource management
- Product realization, including customer-related processes, design and development, purchasing, and production and service provision
- Measurement, analysis and improvement

Organizations wishing to be certified against ISO 9000 must study the standard, assess where they are in compliance and where practices are lacking, then make changes to their quality management system. Training, standardization of procedures, documentation, and internal auditing are usually required. Then accredited auditors are hired to audit the organization, and if the organization passes, it is certified. Periodic surveillance and recertification audits are required to maintain certification.

Problems and Potential

The first (1987 and 1994) version of ISO 9000 was criticized for its emphasis on inspection, control of nonconforming product, and documentation. It reflected a traditional quality control approach of "inspect-in quality." It could work with modern total quality management concepts, especially if used within the final standardization step of an improvement process, but it had to be creatively applied by people who had learned elsewhere the principles of TQM. By itself, it led to an outmoded management system.

ISO 9001:2000 was intended to incorporate current quality management principles, as well as to be more user-friendly. Its changes included greater emphasis on processes, customer satisfaction, the role of top management, data analysis, and continual improvement. Eight quality management principles are included, although their use is not required for certification: customer focus, leadership, involvement of people, process approach, system approach to management, continual improvement, factual approach to decision making, and mutually beneficial supplier relationships.

One major criticism of any standard-based approach is that it guarantees only minimum levels of system development and performance. This is understandable, since standards were born out of a need to guarantee a minimum level of quality to customers. Proponents argue that the recent changes will lead an organization through cycles of improvement that will take it beyond minimum requirements.

Past experience has shown that when applied well, ISO 9000 standards can benefit the organization by ensuring consistency and sustainability of its quality management system. As of this writing, the changes to ISO 9000 are too new to assess whether the standard alone can form the foundation of an effective quality management system.

MALCOLM BALDRIGE NATIONAL QUALITY AWARD

The Malcolm Baldrige National Quality Award (MBNQA) is an annual national award given to high-performing organizations (or divisions). It was created to help US businesses focus on the systems and processes that would lead to excellent performance and improve their global competitiveness. The MBNQA criteria provide a guide for developing management systems that can achieve high levels of quality, productivity, customer satisfaction, and market success.

History and Applications

In 1950, the Union of Japanese Scientists and Engineers instituted the annual Deming Prize to reward companies showing excellence in product quality. The prize and its criteria, based on Dr. Deming's teachings, are believed to be a significant motivating factor in the "Japanese miracle"—the transformation of the Japanese economy following World War II.

In 1987, when such a transformation was desired for American businesses, a similar award was established by an act of Congress, named after a recently-deceased Secretary of Commerce who believed strongly in the importance of quality management to America's prosperity and strength. The National Institute of Standards and Technology (NIST) administers the award jointly with the American Society for Quality.

Originally, awards could be given in three categories: manufacturing, service, and small business. Since 1999, awards also can be given in the categories of education and healthcare. Not all categories are awarded every year, and some years there are multiple winners in a category.

Dozens of other quality awards—local, state, and in other countries—have been modeled on the MBNQA. Even Japan has created a new award—the Japan Quality Award—that is similar to the Baldrige.

Criteria, Process, and Benefits

The award criteria were not permanently fixed. Instead, they are continually reviewed and modified based on proven practices of the best-performing companies in the world. Factors considered are how widespread and broadly applicable a practice is and evidence of links to performance results. Therefore, the MBNQA criteria are an up-to-date guide to the latest consensus on management systems and practices that lead to performance excellence.

The criteria are built around seven major categories, based on key organizational processes. (See Figure 2.2.) Leaders drive the organization through leadership, strategic planning, and customer and market focus. The work of the organization is accomplished through human resource focus and process management. All these processes are supported by measurement, analysis, and knowledge management. Finally, business results show the outcomes of the other processes and provide feedback. The results category includes not only financial results, but also results in areas of customer focus, product and service, human resources, organizational effectiveness, and governance and social responsibility.

The criteria do not require any particular organizational structure, systems, improvement methods or tools. Instead they state necessary elements of a successful quality management system. The scoring guidelines indicate the relative importance of the elements.

Many managers and organizations around the world use the criteria to help them build high-performing management systems, without any intention of applying for the award. Although the number of applicants in any year is usually less than a hundred, millions of

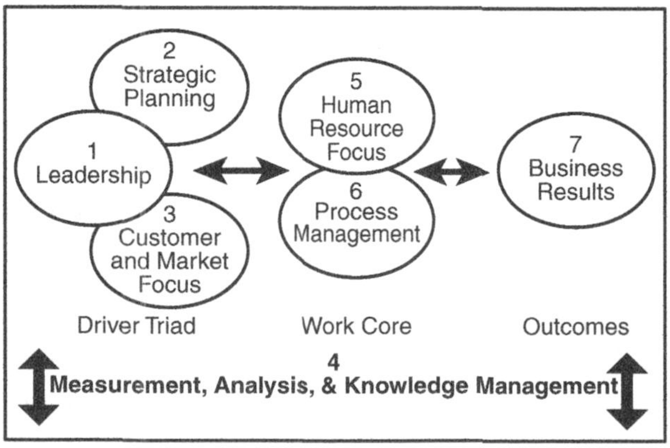

Figure 2.2 Baldrige Award categories.

copies of the criteria have been distributed. In addition, the criteria are not copyrighted, can be copied freely and are available at the NIST Web site. (See Resources.)

An organization applies for the award by submitting a 50-page written application. Each application is reviewed by at least eight members of the board of examiners, which is composed of several hundred volunteer quality experts. High-scoring applicants receive site visits from a team of reviewers. All applicants receive detailed written feedback on their strengths and improvement opportunities.

Many organizations have found that the self-scrutiny required in applying for the award and the feedback from reviewers have been valuable guides to further improvement. Other organizations use the criteria, write an application, but then do their own review and feedback. The MBNQA criteria have become a diagnostic tool to assess an organization's status compared to best-in-class and to identify areas for further work.

Not only do the criteria identify critical elements for creating a high-performing organization, but also winners of the award must share detailed information on how they were able to change their cultures and achieve success. One example of the results of this sharing is the Six Sigma process, developed by Motorola, recipient of the MBNQA in 1988, and now used widely around the world.

BENCHMARKING

Benchmarking is a structured process for comparing your organization's work practices to the best similar practices you can identify in other organizations and then incorporating these best ideas into your own processes.

In surveying, a benchmark is a permanent mark with known position and altitude. It is used as a reference point when other positions around it are measured. In business, to "benchmark" has for many years meant comparing your own products, services, or financial results to those of your competitors. In quality improvement, benchmarking has attained a new, very specific meaning.

Robert C. Camp, author of the book that introduced benchmarking to the world, defined it in his book's title: *Benchmarking: The Search for Industry Best Practices that Lead to Superior Performance.* Jack Grayson, founder of American Productivity and Quality Center (APQC), has said, "Benchmarking is the practice of being humble enough to admit that someone else is better at something, and being wise enough to learn how to match them and even surpass them at it."

History

Xerox Corporation developed benchmarking in the early 1980s. The patent protection on their copying process had expired, and suddenly they were faced with competitors selling product below Xerox's costs. Xerox's manufacturing organization studied the processes of their Japanese subsidiary and their competitors, applied the best ideas to their own processes, and turned the business around. Soon after, benchmarking became a companywide effort, one of three pillars of the effort to attain "leadership through quality." (The other two were employee involvement and their quality process.)

Camp was one of the Xerox benchmarking pioneers. In his book, he credits the first use of benchmarking to the Japanese manufacturers who in the 1960s and 1970s challenged the supremacy of American manufacturing with Japanese quality improvements. They visited American factories and offices in droves, asking questions and taking pictures. A Japanese word, *dantotsu,* means "best of the best," and that is what they were seeking. Wherever and whenever they found a good idea, they adopted it. For example, they saw bar coding in American grocery stores and began using it to control manufacturing processes.

Xerox, however, developed, formalized and promoted benchmarking. They used the method in support and service operations as well as manufacturing. Camp's book spread the word. When the Baldrige guidelines came out in 1987, they required organizations to seek "world-class" operations for comparison, and in 1991 the guidelines began using the word "benchmarks." Organizations began banding together into consortiums or clearinghouses to conduct benchmarking studies together, since a large study can be very expensive. The APQC's International Benchmarking Clearinghouse is the oldest such consortium, begun in 1992. Others have been established in Hong Kong, by government agencies such as NASA and the Department of Energy for their divisions and contractors, and by industry groups. Widespread use of the Internet and other computer technology has made gathering and sharing information easier. Today benchmarking remains an important "mega-tool" for any organization serious about improving its competitiveness.

What's Different About Benchmarking

For years, companies have analyzed their competitors' products or services, studying what is different from their own. Sometimes they use the word benchmarking to refer to this effort. Managers talk about "benchmarking performance," meaning comparing operating measurements or results within their industry. However, in the kind of benchmarking we're talking about, it's more important to know *how* something was done. That's why the definition of benchmarking refers to processes or practices—things people do. Any practice can be benchmarked, from customer satisfaction measurement to warehouse handling.

Benchmarking also looks for *enablers,* elements of the organization's culture, environment, structure, skills, and experience that help make the practice possible. A computer-based practice may work well in an organization with a "computer culture" but be hard to adapt to an organization lacking computer training and support groups.

The definition also says "best." Benchmarking is not interested in keeping up with the Joneses, or even inching ahead. Benchmarking is about leap-frogging to clear superiority. This is the difference between benchmarking and most other quality improvement techniques, which focus on incremental improvement. The words "step-change," "quantum leap," and "breakthrough" are often used with benchmarking.

Another difference with this kind of benchmarking is where you look for best practices. In the past, companies have had a tunnel-like focus on their competitors: trying to catch up, stay even, or get ahead. With benchmarking, you look anywhere for the best practices you can find.

Since the focus is on practices, it doesn't matter whether the organization uses that practice for the same purpose you do. You might look in organizations in your own industry but competing in different markets, or you might look in very different companies. For example, drug and candy manufacturers both use handling and packaging processes involving small objects and high standards of cleanliness. If a hospital wants to benchmark the process for delivering charts, lab samples, and medicines, they might study pizza or package delivery. The most innovative ideas come from cross-industry benchmarking—like grocery-store bar codes moving to factories and hospitals.

Besides looking outside, benchmarkers do *internal* benchmarking. Especially in large organizations, the same practice is often done differently in different locations. By standardizing across the organization the best internal practice, immediate gains are achieved.

Finally, benchmarking involves an unusual amount of cooperation between organizations. Instead of secret reverse-engineering of your competitor's product or secret-shopping in your competitor's store, benchmarking is characterized by openness and sharing. Partners exchange information and data about their practices and results. Certain topics such as pricing strategies legally must not be revealed, but a wide range of other types of information can be shared, even with competitors.

The Benchmarking Process and Tools

Xerox developed a ten-step process. Other organizations modified the process to fit their cultures, so you may see as few as four or as many as nine steps. Fortunately, all the processes have the same essential components. They are just grouped and described in different ways. For example, some processes put "identify partner organizations" in the planning phase and other processes put it in the data collection phase.

See *benchmarking* in Chapter 5 for a simplified description of the process. When you are ready to begin benchmarking, you will need many more details than can be provided in this book.

Often, the benchmarking process is drawn as a circle. After changes have been put in place and superior performance achieved, you recycle and do it again. Not only must you select other processes to benchmark, you must periodically revisit ones you have already studied. The world does not stand still; "best" is a moving target. You must recalibrate your benchmarks periodically in order to maintain superior performance.

The benchmarking process is a complete improvement model in itself. However, benchmarking often is listed as a tool to be used within an improvement or reengineering process to identify stretch goals or unique solutions. See *benchmarking* in Chapter 5 for more information.

Since benchmarking is a complete improvement methodology, many of the tools in this book can be used during the process. The most important ones are: prioritization tools for selecting a topic and partners, flowcharts for understanding your process and your partners', surveys and questionnaires, and data analysis tools.

Benefits . . .

The biggest benefit of benchmarking is that it can create huge leaps in performance. Instead of tweaking the process and creating gradual improvement, you can jump to another level. The Z chart (Figure 2.3) shows the difference between incremental quality improvement and benchmarking breakthroughs. (The chart also shows that often the best-practice organization is better at incremental improvement, too.)

Benchmarking establishes goals that are ambitious yet realistic—after all, someone else is already doing it. Complacency can be shaken up by seeing what others are achieving. Benchmarking encourages creativity and innovation and promotes an attitude of learning.

Benchmarking also keeps you from reinventing wheels, which wastes energy and resources. Because you are not starting from scratch, changes can be developed and implemented more quickly. Getting better gets faster.

. . . and Problems

Benchmarking doesn't always bring such shining results. Some organizations have tried and discarded it as the latest fad that doesn't work. Usually, those organizations made a mistake in how they approached it. Some of these mistakes are common to all quality

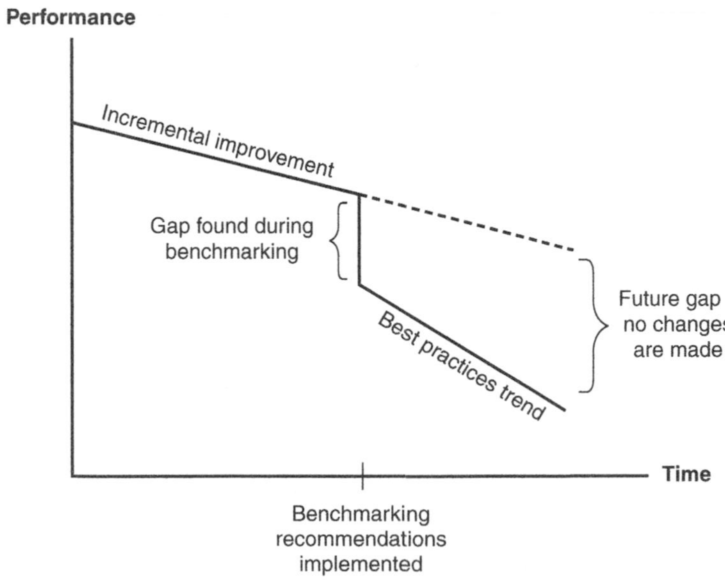

Figure 2.3 Z chart.

management systems: inadequate management support, inappropriate scope, inadequate resources, and giving up after only one attempt. After all, the first time is always the hardest. Other mistakes unique to benchmarking are the "not invented here" syndrome, skipping the planning or adapting phases, and benchmarking before one's own processes are clearly understood and under control. See *benchmarking* in Chapter 5 for more details.

Benchmarking is an ongoing process. After the first study, an organization should identify the next critical success factor and benchmark that . . . then the next . . . then the next . . . .

SIX SIGMA

Six Sigma is an organization-wide approach used to achieve breakthrough improvements tied to significant bottom-line results. Unlike previous TQM approaches, Six Sigma specifies exactly how the organization's managers should set up and lead the effort. Key features are the use of data and statistical analysis, highly trained project leaders known as Black Belts and Green Belts, project selection based on estimated bottom-line results, and the dramatic goal of reducing errors to about three per million opportunities.*

History and Applications

Traditional statistical process control (SPC) standards called for improving defect levels measured in percentages (parts per hundred). In the 1980s, Motorola realized that this was nowhere near adequate to meet their competition. They needed to measure defects in parts per million. The company also realized that its quality problems were caused by the way it managed the organization.

Motorola developed and implemented *Six Sigma Quality*, a unique approach to dramatically improving quality. In 1988, Motorola became one of the first winners of the new Malcolm Baldrige National Quality Award. Because winners are required to share their methods, Six Sigma Quality became public knowledge. Other companies began using and improving on it, especially GE and AlliedSignal. In the 1990s and 2000s, it spread to thousands of other organizations.

Six Sigma programs have reported huge savings when applied to complex, even organizationwide problems that need breakthrough solutions. It is best focused on reducing variation in any major process from the production floor to headquarters offices. Customer requirements, defect prevention, and waste and cycle time reduction are the kinds of issues that Six Sigma addresses. Small or local problems do not need the major investment of a Six Sigma project and are better handled by local or departmental teams.

---

* Six Sigma is a registered trademark and service mark of Motorola, Inc.

What Does Six Sigma Mean?

Sigma—written with the Greek letter σ—is a measure of a process's variation or spread. The process is improved by making that spread smaller, producing output that is more consistent and has fewer defects or errors. Under traditional quality standards, the spread is reduced until the specification limit is 3σ away from the process mean. (See Figure 2.4.) With this standard, 0.135 percent of output still would be outside the specifications. This unacceptable output generates *cost of poor quality:* lost time, money, and customers.

With Six Sigma Quality, process variation is squeezed even more, reducing σ until the specification limit is 6σ away from the mean. (See Figure 2.5.) Notice that the specification limit has not moved but the measuring stick—σ—is smaller. An assumption is made that over time the mean might shift as much as 1.5σ. With these conditions, unacceptable output would be only 3.4 per million, or 0.00034 percent. This is *six sigma* performance.

Common measures of performance in Six Sigma programs are *sigma levels* and *defects per million opportunities.* The assumptions and mathematics involved in calculating these metrics have been questioned by statisticians.[1] However, the goal of achieving process capability of 2.0, which corresponds to six sigma, is highly appropriate for today's demanding world. Organizations undertaking Six Sigma strive for these near-perfect error rates in their processes.

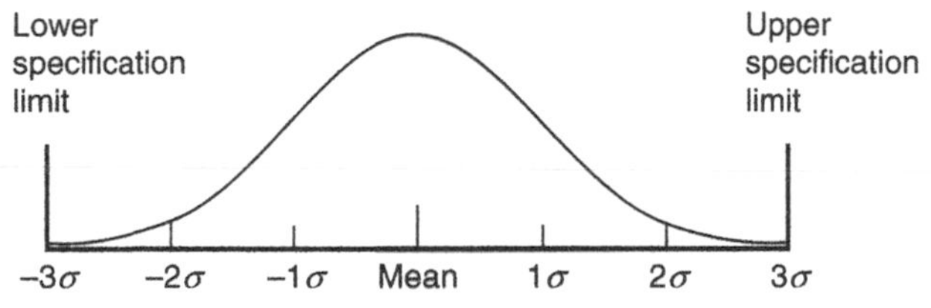

Figure 2.4 "3σ" process.

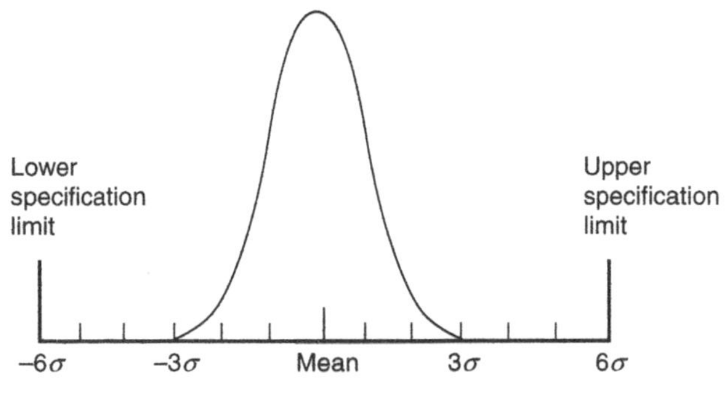

Figure 2.5 "6σ" process.

Methods and Tools

Six Sigma prescribes an improvement process known as DMAIC: define, measure, analyze, improve, control.

- *Define* the improvement project's goals, deriving them from customer needs or wants.
- *Measure* the current process and establish metrics to monitor progress toward goals.
- *Analyze* the current process to understand problems and their causes.
- *Improve* the process by identifying and piloting solutions to the problems.
- *Control* the improved process with standardization and ongoing monitoring.

For developing a new product or process or for processes that need total overhaul, there is a modified version called Design for Six Sigma (DFSS). A process often used in DFSS is called DMADV: define, measure, analyze, design, verify.

After each phase of DMAIC or DMADV, a tollgate review is held. Standardized checklists and questions ensure that all the necessary work of that phase has been done.

Most of the tools of Six Sigma are familiar from TQM, SPC, and other improvement methodologies. All the seven basic QC tools are used. Sampling, measurement system analysis, hypothesis testing, correlation, and regression analysis are important because of an emphasis on using data to make decisions. Some tools have been modified from more general tools and are applied in specific ways within Six Sigma, such as project charter, SIPOC diagram, CTQ trees, and cause-and-effect matrix.

Other mega-tools often are used. During the *define* phase, gathering and hearing the voice of the customer (VOC) is an important concept from QFD. Sometimes called a tool, VOC is really an umbrella that includes specific tools such as surveys, e-surveys, focus groups, interviews, customer complaints, and so on. During the *analyze* phase, the power of design of experiments is often applied. Benchmarking, SPC, and even ISO 9000 have been used within the Six Sigma methodology.

What makes Six Sigma different is not the tools but the focused way in which they are applied. Six Sigma demands intelligent use of data, emphasizing statistical analysis and designed experiments. It also takes advantage of the widespread use of personal computers and statistical software, making once-onerous calculations quick and simple.

Six Sigma in the Organization

The methodology of Six Sigma goes beyond the improvement process and tools. Improvement is driven by projects carefully selected by the organization's leaders to ensure impact on key stakeholders (customers, shareholders, and/or employees), integration with the organization's goals, and bottom-line results. Senior managers are champions or sponsors of improvement projects. Perhaps most noticeable in Six Sigma companies is the cadre of highly trained project leaders known as Green Belts and Black Belts.

Green Belts spend 5 to 10 percent of their time managing a Six Sigma project, working with a cross-functional team. Black Belts work full time with multiple Six Sigma projects, leading teams and advising Green Belts. Generally, after several years working in Six Sigma, Black Belts move to different jobs and other employees are trained to fill their roles. Eventually the organization is leavened with employees well-versed in the philosophy and methods of Six Sigma.

One of the most common reasons for failure of quality improvement efforts has been lack of guidelines for establishing and leading the effort and, as a result, uneven support from top management. Six Sigma uses the most powerful methods and tools of previous quality improvement efforts and adds to them a blueprint for how the organization's leaders should lead the effort. In that sense, Six Sigma may be viewed as a maturation of the learning and work about quality improvement that took place from the 1920s through the 1980s.

LEAN MANUFACTURING

Also called *lean production, lean enterprise,* or simply *lean*

> *Lean manufacturing refers to a system of methods that emphasize identifying and eliminating all non-value-adding activities—waste—from a manufacturing or manufacturing support organization. Processes become faster and less expensive. Lean manufacturing is characterized by fast cycle times, just-in-time methods, pull systems, little or no inventory, continuous flow or small lot sizes, production leveling, and reliable quality. Lean organizations are efficient, flexible, and highly responsive to customer needs.*

History and Applications

Although the principles, concepts, and tools of lean manufacturing are often credited to Taiichi Ohno, an engineer at Toyota, Ohno himself said he learned many of his ideas from Henry Ford and Ford Motor Company. As with Shewhart's statistical process control, American industry forgot the concepts it had developed and relearned them from the Japanese in the 1980s. The term *lean* was coined in 1990 by an MIT research team.

The concepts of constraint management (theory of constraints or TOC) and synchronous flow manufacturing (SFM) were developed by Dr. Eliyahu Goldratt and popularized through the 1992 book *The Goal.* These concepts are complementary to lean manufacturing.

The concepts of lean require changes in support functions such as product and process design, purchasing, shipping, and, indeed, throughout the entire supply chain. An organization implementing lean concepts in all these ways is called a *lean enterprise.*

By addressing waste reduction, lean manufacturing efforts solve problems of cycle time reduction, standardization, flexibility, and quick response to customer needs.

While the concepts of waste reduction could apply anywhere, for waste is certainly present in service industries and office functions as well as in manufacturing, many of the techniques of lean are specific to manufacturing and have to be translated to apply outside manufacturing environments.

Although lean manufacturing has the potential to revitalize manufacturing, its principles, while widely known, are not yet widely applied. Like every other system of organizational improvement, implementation requires difficult organizational and cultural change.

Identifying Waste

Waste is much more than scrap and defects. Ohno defined seven forms of waste: defective product (costs of poor quality, including final inspections, scrap, and rework), inventory (both finished product and work-in-process), overproduction, non-value-added processing, transportation of goods (both within and outside the factory), motion (of people, tools, or product), and waiting.

Even value-added activities often have components of waste. How many turns of the wrench are required before the nut is seated? Is the product rotated to be able to reach the bolt? How many times is the tool picked up and put down?

Masaaki Imai says that value is added at brief moments—“Bang!”—and everything else in the process step is waste. Think about the process of firing a gun 300 years ago, when powder, wad, and ball were separately dropped into the gun barrel and rammed with a rod before the gun could finally be fired. Firing the gun was the value-added step; everything else was waste. Although that process was taken for granted then, today we recognize how wasteful it was of time, motion, materials, and even human lives. Once non-value-added components of any process are recognized, we can begin to think of ways to eliminate them.

Constraint management and synchronous flow manufacturing are approaches that come from Goldratt’s Theory of Constraints but fit beautifully with lean concepts. A *constraint* is the slowest process step: the bottleneck. Under constraint management, improvement efforts are focused at the constraint, because increasing throughput anywhere else will not increase overall throughput. Time lost at other process steps can be made up, but time lost at the constraint is lost forever. Synchronous flow manufacturing modifies just-in-time and pull system approaches so that the constraint sets the pace for the rest of the process. Under the Theory of Constraints, even traditional cost accounting methods used to plan production are turned upside down, so that the true costs of wasteful processes become clear.

Methods and Tools

Because lean manufacturing starts with identifying waste, tools such as value-added analysis and value stream mapping are used along with the seven QC tools and an improvement process such as kaizen. Mistake-proofing is widely used in lean. Because of the emphasis on adding value, linkage of customer needs to process activities is often accomplished through QFD. Lean concepts find their way into product and process design, which is called *design for manufacturing* (DFM) or *design for assembly* (DFA).

An important tool unique to lean is 5S, named after five Japanese words that roughly translate to sort (seiri), set (seiton), shine (seiso), standardize (seiketsu), and sustain (shitsuke). Sometimes the 5Ss are translated into CANDO: clearing up, arranging, neatness, discipline, and ongoing improvement. One basic idea behind 5S is that waste hides in clutter and dirt. An oil leak is noticeable on an uncluttered, clean, white floor. The other basic idea of 5S is that rooting out waste requires continual, focused attention and commitment.

Many of the tools of lean are techniques or systems that solve common waste problems in manufacturing. Pyzdek[2] has described lean as offering "a proven, pre-packaged set of solutions" to waste. These include:

- Autonomation, or machines that sense conditions and adjust automatically
- Factory layout variations, such as cellular manufacturing (machines organized around the part), and flexible processes (maneuverable tools that can be quickly reconfigured)
- Just-in-time, in which materials are received and operations performed just before they are needed by downstream processes
- Level loading, which aims to create a consistent, market-sensitive production schedule
- Preventive maintenance, which reduces unplanned downtime through well-functioning equipment
- Pull systems, in which the pace of downstream activities "pulls" work into the beginning of the process
- Quick changeover methods or setup reduction, such as single-minute exchange of die (SMED), which eliminate long machine downtimes between tasks
- Single-unit processing, or at least small lot processing, to eliminate the waste of batch-and-queue operation
- Standardization, to keep waste from creeping back into the process
- Visual controls, such as warning lights or lines on the floor to mark reorder levels

COMBINING MEGA-TOOLS

Lean manufacturing focuses on speed and efficiency: eliminating waste.

Six Sigma focuses on quality: eliminating defects through reduced variation.

MBNQA focuses on leadership: implementing management systems for performance excellence.

Benchmarking focuses on best practices: seeking great ideas for breakthrough improvements.

ISO 9000 focuses on consistency: eliminating unpleasant surprises through standardization and discipline.

Quality function deployment focuses on customers: creating products and services they want.

These approaches are compatible with one another. Many organizations are successfully combining mega-tools. For example, a foundation of ISO 9000 will bring order to an undisciplined organization and create a base on which to build improvements through Six Sigma and lean. Six Sigma plus lean equals a system that can address complex problems of process variation as well as the very different problems of cycle time. MBNQA criteria can guide senior management to establish the organizational systems and culture that will support improvement. QFD and benchmarking often are explicitly listed as tools needed for Six Sigma, lean, or MBNQA.

Other mega-tools from outside the quality arena can be integrated with your quality management system. For example, TRIZ (pronounced "trees") is a system for analyzing difficult problems and developing innovative ideas. It was developed quite separately from the Japanese and American evolution of quality improvement, beginning in communist USSR during the 1950s. (TRIZ is the acronym for a Russian phrase meaning "theory of inventive problem solving.") TRIZ's originator, Genrich Altshuller, had the idea that great innovations contain basic principles which can be extracted and used to create new innovations faster and more predictably. TRIZ's principles, methodology, tools, and knowledge base emerged from the study of several million historical innovations. First used for commercial applications in the 1980s, it was introduced to the rest of the world in 1991 and has been applied not only to technology but also to business, social systems, arts, and other nontechnical areas. Quality practitioners saw how valuable it is for generating innovative ideas and have combined it with QFD, Six Sigma, and other quality mega-tools. TRIZ is too complex to be covered in this book, but it is a valuable addition to the toolbox of anyone who is looking for innovative product or process improvements or who must solve problems containing inherent contradictions.

Each of these mega-tools should be chosen based on the needs of the organization. Don't pick up a hammer, then look around for things you can pound. Decide what you want to build and what starting materials you have, then choose the tools that will accomplish your goals.

ENDNOTES

1. See, for example, Donald J. Wheeler, "The Six-Sigma Zone," *SPC Ink* (Knoxville, TN: SPC Press, 2002). ww.spcpress.com/ink_pdfs/The%20Final%206%20Sigma%20Zone.pdf (accessed March 5, 2004).
2. Thomas Pyzdek, *The Six Sigma Handbook* (New York: McGraw-Hill, 2003).

3

The Quality Improvement Process

An improvement process or model is one of the fundamental elements in any quality management system. It is a guide, a framework, a road map. In our carpentry analogy, it's the step-by-step process of building cabinets: first, obtain a plan; second, assemble materials; next, construct the base, and so on to the final steps, attach hardware and install in location. Regardless of what the cabinets will look like or what function they will serve, the basic method for doing the work will be the same. Similarly, in quality programs, an improvement process provides a consistent method for doing the work of improvement.

THE BENEFITS OF AN IMPROVEMENT PROCESS

A quality improvement process presents a series of steps to think about and work through. These steps help you ask questions, gather information, and take actions effectively and efficiently. Thus, a quality improvement process provides a framework that guides you from the initial improvement challenge to successful completion of the effort.

A quality improvement process's biggest benefit is to prevent you from skipping important steps along the way. For example, groups might not think about their customers, or they jump to a solution without first understanding root causes. Following a quality improvement process will keep you from making these mistakes.

A quality improvement process also helps a group work together and communicate their progress to others. Everyone knows what you are trying to accomplish at any point and where you are headed next.

A quality improvement process can be used in any time frame. While it often takes months to work a difficult problem through the entire process, it's also useful when improvement ideas must be generated quickly. In an hour or two, the process can guide your thinking through various aspects of a situation to a well-founded plan.

Also, a quality improvement process can be used by anyone. While improvement teams most often employ the process, it can be used by any group or individual, from plant site to executive offices.

A GENERIC QUALITY IMPROVEMENT PROCESS

Different organizations use different improvement processes or models. The most basic is the plan–do–study–act cycle. Six Sigma uses DMAIC—design, measure, analyze, implement, control. This book uses a detailed, generic ten-step process. In this chapter and throughout the book, this model will demonstrate how the process of improvement proceeds and how tools are used within the process.

Figure 3.1 shows the ten-step quality improvement process. It uses ordinary, common-sense language. Each step is also rewritten in terminology commonly used in quality programs. Both are included so one or both phrases will communicate clearly to you.

A third way of understanding the quality improvement process is shown in Figure 3.2. This flowchart shows that the sequence of working through the process might not

Common Terminology	Quality Terminology
1. What do I or we want to accomplish?	Identify charter and make initial plans.
2. Who cares and what do they care about?	Identify customers and requirements.
3. What are we doing now and how well are we doing it?	Assess current state.
4. What can we do better?	Define preferred state, gaps between current and preferred state, and improvement opportunities.
5. What prevents us from doing better? (What are the underlying problems?)	Identify barriers and root causes.
6. What changes could we make to do better?	Develop improvement solutions and plans.
7. Do it.	Implement plans.
8. How did we do? If it didn't work, try again.	Monitor results; recycle if necessary.
9. If it worked, how can we do it every time?	Standardize.
10. What did we learn? Let's celebrate!	Conclude project, identify lessons learned, and recognize accomplishment.

Figure 3.1 Ten-step quality improvement process.

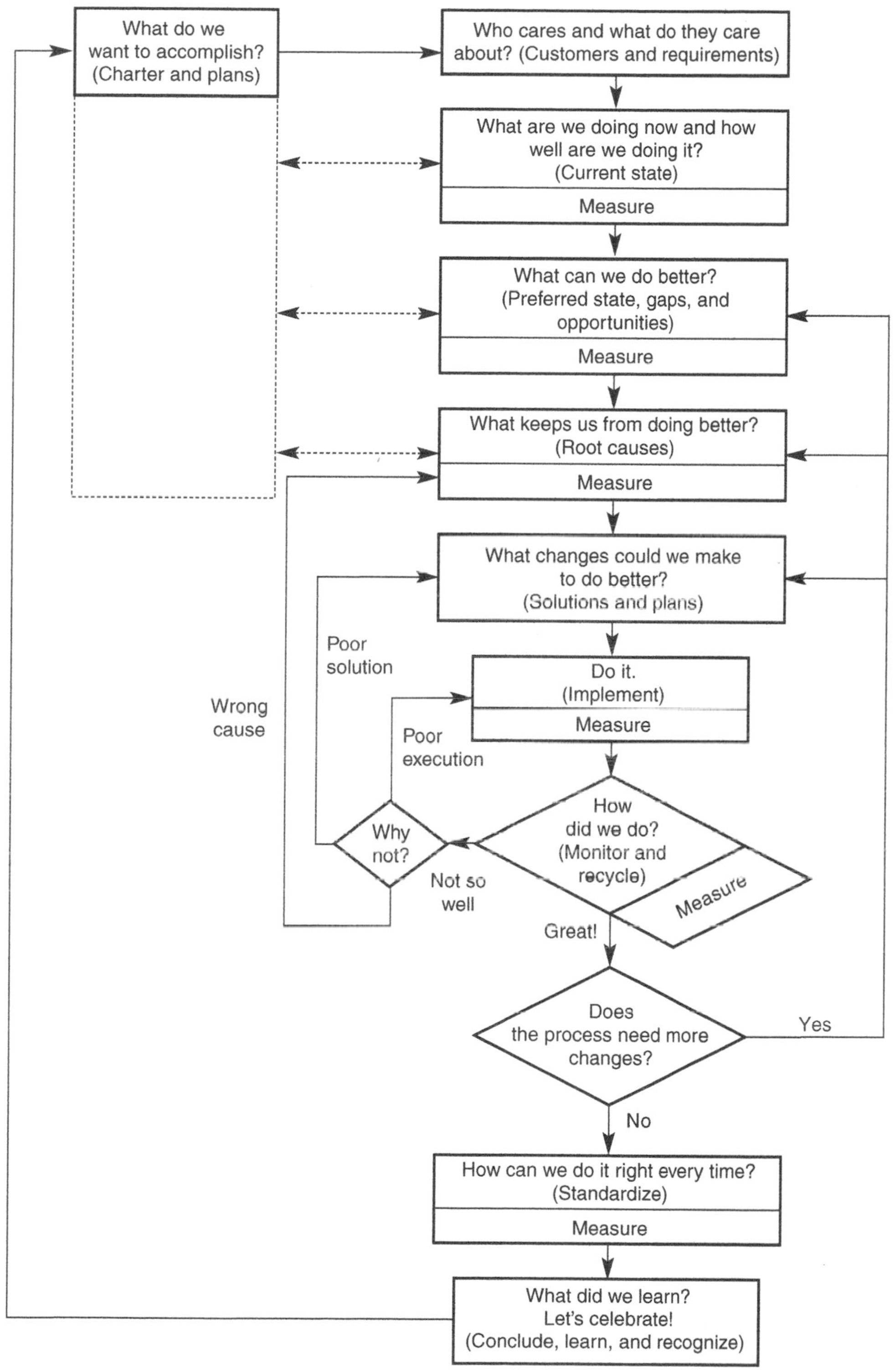

Figure 3.2 Ten-step quality improvement process flowchart.

proceed smoothly from first step to last. One might need to go back and adjust the charter, or return to an earlier step when a change does not work, or recycle to work on a second change after implementing a first. Also, because measurement is significant in any quality improvement process, the flowchart shows which steps of the process have measurement as a key element.

This process was described as "detailed" because it has more steps than most, and as "generic" because all ten steps can be found within any other improvement model. It was written with more steps in order to separate the different perspectives and questions necessary in an improvement project. Because the point of this book is to help you figure out when to use which tool, a detailed model allows better matching of tools with their appropriate uses.

If your organization has an official improvement process, it will be written differently from this one and have a different number of steps. You should, however, be able to find all the elements of the ten-step process in your process. For example, Six Sigma's DMAIC might be roughly matched to the ten-step process like this:

- "Define" includes steps 1, 2, and some aspects of 3.
- "Measure" completes step 3.
- "Analyze" includes steps 4 and 5.
- "Improve" combines steps 6, 7, and 8.
- "Control" concludes with steps 9 and 10.

The rest of this chapter discusses each step of this quality improvement process in detail, offering tips to make the quality improvement process itself a powerful tool for improvement.

THE TEN-STEP QUALITY IMPROVEMENT PROCESS

1. What do we want to accomplish?

Identify charter and make initial plans.

Identifying Your Charter

- The charter statement should be provided by the steering committee when the team is established.
- It is essential to spend time and effort in group discussions and in communication with the steering committee to make sure that a good charter statement is agreed upon and understood by everyone.
- Agreement among the team members and steering committee about the scope of the project (how broad or narrow) is important.
- Could you explain the importance of your project to external customers?

- If you started with a fuzzy charter that has since been clarified, consider whether the team still has the right members to address the new, focused charter.
- Idea creation, high-level process analysis, and decision-making tools are especially useful here.

Making Initial Plans

- Although you may not know much detail yet, lay out a general plan including key steps. See Figure 5.62, page 259, a top-down flowchart, for an example of a plan.
- Try setting milestones backward. Do you have a deadline for finishing? When do you want to have a solution ready to put in place? When will you have finished studying how things are now? How much time will you allow for gathering customer input?
- Do you have review points (sometimes called tollgates) established with management or a steering committee? Such reviews can establish timing for your key milestones.
- Filling in bars and diamonds on a Gantt chart will reflect the team's accomplishments and sustain motivation.
- As you move through your quality improvement process, revise and update your plan. You will be able to add more specifics later.
- The project planning and implementing tools are especially useful here.

2. Who cares and what do they care about?

Identify customers and their requirements.

- Products and customers can be identified for each output on the flowchart. Supplies and suppliers can be identified for each flowchart input.
- Talk with each of your key customers and suppliers about what they need.
- Invite key customers and suppliers to sit on teams or involve them through interviews.
- Concepts from Kano's model of customer satisfaction (page 18) are useful in understanding how to ask customers about their requirements and how to understand what they tell you.
- Never try to go past this step without getting direct input from real, live customers.
- Several process analysis tools are designed especially for this step, as well as data collection and analysis tools that collect and analyze voice-of-the-customer data.

3. What are we doing now and how well are we doing it?

Assess current state.

Defining the Process

- For most situations, you will begin this step by drawing a flowchart.
- Include cost-of-quality steps: inspection, rework, delays, what happens when things go wrong.
- Start with a simple flowchart that shows the big picture, then draw one with more detail. For example, you could start with a top-down flowchart and add detail for critical steps. Convert a detailed flowchart into a deployment flowchart to show who is involved when.
- Need more detail? Pick a key step in your process and draw a flowchart showing that step's substeps. Trying to handle both explicit detail and a large scope on your flowchart at the same time could overwhelm your team.
- As much as possible, involve people who *do* the process.
- Carefully identify inputs and outputs to capture interfaces and cross-functional activities.
- In some situations, a flowchart isn't applicable. It is still important to understand the current state before you proceed.
- Try considering the current state from your customer's point of view.
- Based on your new knowledge of the process, is the charter statement still valid? Do you have all the right people on the team?
- Not surprisingly, process analysis tools are especially useful here. Don't overlook several idea creation tools that help you work with the "big picture."

Collecting and Analyzing Data

- To answer the question "How well are we doing?" you usually must start taking measurements in this step.
- How will you know when you have improved? That is one of the questions you are trying to answer with measurement. Develop measurements that will answer that question for you later.
- Study customer requirements and flowcharts to help determine appropriate elements to measure.
- Consider both process and product measures. Process measures tell how things are running, while they are running. Product measures tell you results, after the process is finished. You need both kinds of indicators.

- When planning data collection, think about how the data will be analyzed and used. This affects what kind of data you will want to collect and how you will collect it. Sometimes the format that will be used for analysis can also be used for collection.
- A written data collection plan provides a guide throughout the rest of the improvement process. It should include what will be measured; how often; how the data will be obtained, including sampling instructions; how the data will be recorded; control limits (if control charts are used); specification or action limits; action guidelines; and who is responsible. See *operational definitions* for more information on creating clear data collection plans.
- The measurement system should be studied here to ensure that measurements are producing reliable results.
- Almost all of the data collection and analysis tools are useful here.

4. What can we do better?

Define preferred state, gaps between current and preferred state, and improvement opportunities.

Defining the Preferred State

- First, expand your thinking and identify many possibilities. What would the *best* look like? What are all the things you could do better?
- This is an expansion point; get as many ideas as the group can generate.
- Many teams have an almost irresistible urge to jump to solutions at this point. Often, a problem statement is really a premature solution in disguise: "We don't have enough people." It is an important role of the facilitator to keep the team focused on problems and causes until both are well understood.
- Draw an ideal flowchart and then determine the gap between what is really happening and the ideal. The ideal should eliminate all rework, checking, approvals, delays, fixing of errors—anything that finds or solves problems.
- Think boldly and creatively when envisioning the ideal state. Do not take anything for granted; question all assumptions. The methods of lateral thinking, developed by Edward de Bono, can be very powerful at this step. (See Resources at the end of this book.)
- If there are no problem areas in your flowchart, include more detail.
- The other tools listed on the Tool Matrix for this step can be combined with the flowchart of what is happening today to create the ideal flowchart. Or, those tools alone may make it clear where opportunities for improvement lie.

- For processes where time is important, use a cycle time flowchart to identify how long each step should take and where slowdowns and waits now occur.
- Compare your measurements to customer needs.
- Idea creation tools and the expanding process analysis tools are especially useful here.

Determining Gaps and Improvement Opportunities

- After expanding your thinking to lots of improvement opportunities, focus on the most significant.
- Whenever possible, use data to confirm problems.
- Often, using several data analysis tools gives more insight than just one.
- It is always preferable to prioritize with data, rather than with opinions.
- Use list reduction or multivoting as filters to reduce a long list of ideas to a manageable number.
- Focus on the most significant ways to improve. What will really make a difference to your customers? What will make a long-term difference to your organization?
- Exception to the previous statement: If your team is new, it may be a good idea to go for a quick success with an easier objective in order to learn. Identify a short-term, easily achievable opportunity. These are often called *low-hanging fruit.* You can later go back and work on more difficult problems.
- Set challenging targets for improvement. Challenging targets generate breakthroughs. They motivate people to stretch themselves and to find ways to achieve dramatic improvements.
- Most of the data collection and analysis tools can be used here, as well as the focusing process analysis tools plus the evaluation and decision-making tools.

5. What prevents us from doing better? (What are the underlying problems?)

Identify barriers and root causes.

Identifying Possible Causes

- Again, first expand your thinking to many potential causes.
- The root cause is the fundamental, underlying reason for a problem. The root cause is what causes a problem to happen repeatedly. If you do not identify the root cause, you are just putting a bandage on the problem. It will probably happen again. When you remove the root cause, you have fixed it permanently.

- Explore many ideas before zeroing in on one root cause to solve.
- Thorough probing to root causes will lead you to broad, fundamental issues. Often management policies will be involved—such as training or the design of a system.
- Solving a problem at a deep, fundamental level often will solve other related issues. Remember "inch-wide, mile-deep": choose narrow scopes, but dig deeply into them for best results.
- Idea generation tools as well as the expanding process analysis and cause analysis tools are useful here.

Focusing on the Root Cause

- When you have identified many potential causes, focus on the true root cause.
- Use data to be sure you have the right cause.
- When planning data collection, think about how the data will be analyzed and used. This affects what kind of data you will want to collect and how.
- Some root causes can be verified only by trying out a solution.
- You can return later to work on causes not chosen to be tackled the first time.
- The focusing cause analysis tools and many data collection and analysis tools are useful here.

6. What changes could we make to do better?

Develop improvement solutions and plans.

Generating Potential Solutions

- Once again, generate lots of possible solutions before settling on one.
- All too often, groups start planning to carry out their first idea. Play with additional ideas and you will probably be able to improve on your first one.
- Get as many solutions as possible, realistic or crazy.
- This is another place where it is important to think boldly and creatively. Lateral thinking methods can help you break out of conventional thinking to original ideas. (See Resources.)
- Consider combining two merely *OK* solutions into one hybrid *great* solution.
- Ask your customers and suppliers to help develop solutions to problems that affect them.
- Use idea creation tools as well as expanding process analysis tools.

Choosing the Best Solution

- When you have several workable solutions, evaluate them carefully to focus on the best one.
- Ask two types of questions when evaluating potential solutions: How well will this solution achieve the desired results? (How effective will it be?) How successful are we likely to be in carrying out this solution? (How achievable is it?)
- Brainstorm evaluation criteria. What factors need to be considered under the broad headings of "effective" and "achievable"? These become headings of a decision matrix.
- How could a solution be changed to eliminate problems that show up on the decision matrix?
- After a decision matrix has been completed, summarize with an effective–achievable chart.
- Discussion about consensus may be important here. What does consensus mean? How can we be sure we have achieved it?
- If possible, involve in the decision process those who will participate in carrying out the solution, including customers and suppliers. You will increase their buy-in.
- Before choosing a solution, the team may need to ask, "Do we have enough information to choose a solution?"
- The evaluation and decision-making tools are especially useful here. Data analysis tools are also invaluable to determine optimum process settings.

Plan the Change

- When a solution is chosen, plan carefully how you will carry it out.
- The more carefully developed your plan, the easier implementation will be.
- A good plan includes what, who, when.
- Be sure to include contingency planning. What can go wrong? Who or what will stand in the way? Identify potential obstacles or barriers and decide how to avoid or overcome them. Plan and put in writing how you will react in the worst-case scenario.
- Have you considered the human elements that will affect your success? Who will feel threatened? What new procedures will be hard to remember?
- Be sure to plan measurement, both to determine the results of the change and to monitor how well your plan was implemented.
- Project planning and implementing tools are valuable here. Cause analysis tools are useful for contingency planning.

7. Do it!

Implement solution.

- If possible, test the chosen solution on a small scale. Choose a subset of the eventual full scope of your plan, such as one region, one unit, or one class. Choose a typical subset, but avoid one with big or unusual obstacles. Be sure customers are involved. Plan to continue your test long enough to observe problems. If you are successful, in step 9 you can expand your solution to the full scope.
- As you roll out your plan, be alert for unexpected observations and problems with data collection.
- Does your solution appear to be creating other problems?
- Be patient.
- Project planning and implementing tools and data collection tools are most useful here. Process analysis tools developed in previous steps are also helpful for communicating the changed process.

8. How did we do? Try again if necessary.

Monitor results. Recycle if necessary.

- Did you accomplish your implementation plan? Did you accomplish your objectives and original charter? The answers to these two questions will indicate whether you need to recycle, and to where.
- If you didn't accomplish what you had hoped, three reasons are possible. Your plan may have been poorly executed. You may have developed a poor solution. Or, you may have attacked the wrong cause. You will recycle to step 7 for better execution, to step 6 for a better solution, or to step 5 to find the right cause.
- When you review results, ask whether your targets have been achieved, not just whether there was a change.
- If you plan to test solutions to other root causes or other problems in the same process, it may be better to return to step 4 or 5 now. When all your tests are done, you can go on to step 9 and expand all your changes to the entire process at one time.
- Process analysis tools developed in previous steps continue to be used now. Data collection and analysis tools are critical for monitoring improvement. Some of the project planning and implementing tools such as the plan–results chart and the continuum of team goals are important for determining whether the overall project goals were met.

9. If it worked, how can we do it every time?

Standardize.

- In this step, you introduce to everyone changes that you tested on a trial basis. Also, you make sure that whatever changes have been made become routine. You prevent any slipping back into the same old problems. There are several aspects to standardizing an improved process.
- First, "new and improved" won't do much good unless it is used consistently throughout the organization. Plan to expand your test solution to everywhere it applies.
- Second, everyone involved in the new process must know what to do. Formal and informal training is essential to spread a new process throughout the organization.
- Document new procedures. This will make training and ongoing control easier. Flowcharts and checklists are great ways to document procedures.
- Make sure documentation and checklists are easily available in the workplace.
- Third, humans don't like change. Advertise the benefits of the new process so people will want the change. Use dramatic transitional aids to help people understand and remember new ways of doing things. Make the change as easy as possible for the people involved.
- Fourth, plan for problems. Ask, "What is most likely to go wrong?"
- Set up indicators that will flag a problem about to occur. Also try to set up indicators that things are running well.
- A control plan documents critical variables, how they are monitored, desired conditions, signals that indicate problems, and appropriate response to those signals.
- Consider all elements of the new process: people, methods, machines, materials, and measurement. Have they all been standardized?
- Finally, it is critical to identify who owns the improved process, including responsibility for continued tracking of measurements. A process owner will hold the gain the team has made and set new targets for continuous improvement.
- Your standardization plan also should have a way to monitor customers' changing needs.
- Idea creation and expanding process analysis tools help you be creative in extending improvements elsewhere. Focusing process analysis tools and project planning and implementing tools are used to lock in improvements throughout the organization. Measurements must be standardized with data collection and analysis tools.

10. What did we learn? Let's celebrate!

Conclude project, identify lessons learned, and recognize accomplishment.

- Don't quit yet! Take some time to reflect, learn, and feel good about what you have accomplished.
- Use your charter statement, implementation plans, and team minutes to remind you of the process your team has followed.
- Identify and record what you have learned about teamwork, improvements, the improvement process you followed, gathering data, and your organization. What went well? What would you do differently next time?
- Use your organization's structure (steering committee, quality advisors, best-practice database, newsletters, conferences, and so on) to share your learning.
- Do you have ideas about how to improve your organization's improvement process? Use this project's successes and missteps to improve how other teams function.
- You have worked, struggled, learned, and improved together; that is important to recognize. Plan an event to celebrate the completion of your project.
- Endings often cause painful feelings; this may affect your last few meetings. Be aware of this and acknowledge it if it occurs. Planning your celebration can help deal with this.
- If yours is a permanent team, celebrate the end of each project before you go on to a new one.
- If your team is scheduled to disband, ask yourselves, "Is there some other aspect of this project, or some other project, that this team is ideally suited to tackle?" Your momentum as a successful team may carry you to even greater improvements.
- Idea creation tools are used to extract learnings from your experiences. Project planning and implementing tools are used to organize and document learnings and to transfer them elsewhere in the organization.
- Congratulations!

TOOLS FOR TEAM EFFECTIVENESS

While individuals can improve their own work, it takes a group to improve a process. And let's face it: it's hard to get a group of diverse people to mesh well enough to perform smoothly the many sequential tasks of the quality improvement process. A team might have incredible knowledge and creativity but still flounder because team members don't work well together. To be successful, a quality improvement team must master not only the quality improvement process and quality tools but also the challenges of group effectiveness.

The broad topics of group dynamics and team management are beyond the scope of this book. However, here are some basic facilitation tools, including team roles, that help any group function more effectively and efficiently.

Facilitator

Quality improvement teams are most effective and efficient when they have a facilitator. This role may be filled by the team leader or by another individual, from either within the work group or a resource group. The more critical or highly visible the project, the more important it is that the team be guided by someone specially trained and experienced in facilitation skills.

Regardless who acts as facilitator, his or her most important function is focusing on meeting and project process, not content. *Content* means ideas and suggestions, problems and solutions, data and analysis. *Process* means how work is done: keeping meetings on track, enabling contributions from all team members, dealing with group problems, following the quality improvement process, and using appropriate quality tools correctly. The first three process topics are well covered in books on group dynamics and team management. The latter two aspects of process directly relate to the subject of this book. Of course, the team leader (if that is a separate role) and team members also can and should help with these tasks. But until a team has experience, members rely on the facilitator to guide them through the quality improvement process, to suggest tools, and to lead the group through the tools' procedures.

It's worth repeating: having a facilitator who is focused on process is the most valuable tool for getting more out of meetings and making the project move forward rapidly.

Planning and Agenda

The project leader and facilitator (if these are two separate roles) should plan the meeting together, with input from team members. These questions can help with the planning:

- What is the purpose of the meeting?*
- At what step are we in our quality improvement process?*
- What results are desired from the meeting?*
- Who should be there—or need not be there?*
- What pre-meeting work should be completed? What information should be brought to the meeting?*
- What methods and tools will be used?*
- What alternate methods and tools could be used if this plan doesn't work?
- Which of these tools is the group familiar with? Which will need to be taught?

- What materials are needed?
- What is the best sequence of activities?*
- How long should each topic take?*
- What issues or problems might arise? How will they be handled?

Items marked with an asterisk should be included on an agenda sent to meeting attendees in advance. This allows everyone to come to the meeting prepared to contribute.

Flipchart

Nothing has yet taken the place of the basic flipchart for capturing and displaying a group's ideas. Once ideas are captured on flipchart pages, the group can see them, reflect on them, build on them, and make progress. Flipchart pages can be posted around the meeting room as a constant display, unlike transparencies. Flipcharts also generate a record that can be taken from the meeting room and transcribed, unlike most whiteboards. (Some whiteboards do print a copy, but it is too small to display for the group.) Following are some tips on using flipcharts:

- Use markers that don't bleed through the paper.
- Have a variety of different colored markers. Black, blue, purple, and green can be seen best, but red is good for adding emphasis. Avoid yellow and orange—they become invisible at a distance. Use different colors to add meaning.
- Whenever you use flipcharts, have tape with you—strong enough to hold the pages but gentle enough not to peel paint. (Or substitute tacks if meeting room walls have tack strips or pinboard.) As a page is completed, tear it off and post it on a wall where everyone can see it. Don't just flip the page over (despite the name of this tool) or the ideas will be unavailable to the group.
- See the brainstorming considerations (page 131) for other suggestions about capturing ideas on flipcharts.

Whiteboard

- Whiteboards are useful for creating large diagrams, such as a fishbone or tree diagram. The kind that can make copies are even more useful for such tasks.
- Whiteboards are also useful for tasks that don't need to be preserved, such as tallying paired comparisons.
- Use a whiteboard instead of a flipchart if it's big enough for all the anticipated ideas and a permanent record isn't necessary (are you sure?) or if you absolutely have no choice—but then be prepared for someone to spend time after the meeting transcribing the information.

- Plan the size and layout of your writing carefully so that you won't run out of room on the whiteboard before the group runs out of ideas.

Parking Lot

Title a flipchart page with those words and post it on the wall before the meeting. When an idea arises that is off-subject, write it on the parking lot without discussing it. As the meeting proceeds, if a parking lot item becomes relevant, bring it into the discussion and cross it off the parking lot. At the end of the meeting, review the items left in the parking lot. Assign action items, if appropriate; retain items for future meetings (put the idea in the meeting record); or drop the item if the group agrees.

Scribe

In fast-paced or large meetings, have someone other than the facilitator capture ideas on the flipchart or whiteboard. This frees the facilitator to call upon contributors, manage discussion, help summarize complicated ideas, monitor time, and otherwise run the meeting.

- Rotate the scribe job. Otherwise the person acting as scribe never gets to participate as fully as the others.
- In a really hectic meeting, such as a fast and furious brainstorming session, two simultaneous scribes may be necessary, writing alternate contributions.
- Scribing can help involve someone who seems remote from the discussion.
- The scribe can contribute ideas to the discussion too. However, if you are scribing, you should never just write your idea. Instead, switch to a participant role for a moment, get recognition from the facilitator, stop scribing, and state your idea to the group. Then go back to scribing by writing your idea.

Other Roles

- A timekeeper keeps track of how much time has been spent on a topic and how much time is left. The group agrees in advance how much time is appropriate; having a timekeeper then frees the rest of the group (including the facilitator) to focus on the subject, methods, or tools instead of thinking about time.
- Use a timekeeper to meet the group's needs. A timekeeper may be necessary only during one exercise. Or if time has been a problem for the group, the timekeeper may be needed throughout the meeting.
- A note taker keeps track of major discussion points, decisions made, action items, assignments, deadlines, and items set aside for future work. The note taker later prepares the meeting record.

- The note taker is not the scribe. The scribe writes ideas on the flipchart or whiteboard for all to see. The note taker takes notes for the record.
- Roles are often combined. For example, the facilitator may act as timekeeper and/or note taker. However, knowing they are separate functions can help your team share responsibilities.

Action List

An ongoing list of action items keeps track of what the team agreed should be done, who agreed to do it, when it should be completed, and status. This list should be reviewed at the end of every meeting. Meeting records should include the updated action list. See Medrad's corrective action tracker (page 63) for an example.

Meeting Record

After the meeting, the note taker combines his or her notes plus key information from flipchart work into a meeting record sent to participants, managers, sponsors, and any other stakeholders who need to stay informed of the project's progress.

Meeting Evaluation

Evaluating the meeting and the team's progress gets everyone thinking about process. See *meeting evaluation* in Chapter 5 (page 345) for more information.

GUIDANCE QUESTIONS CHECKLIST

A team's management, steering committee, or quality advisor often is not sure how best to provide guidance and support to the team. The following questions can be asked to probe what the team is doing and help guide it through the process. Team members also could ask themselves these questions.

Some of the questions are general and can be asked throughout the quality improvement process. Others relate to particular steps of the process.

The following questions can be asked throughout the improvement process:

- Have you revisited your charter to see if it is still appropriate?
- What is the purpose of what you are doing?
- Show me your data.
- How do you know?
- Where are you in the quality improvement process? What step?
- What will you do next?

- What tools did you use? What tools do you plan to use?
- Are you on schedule?
- What are your major milestones over the next couple of months?
- Do you have the resources you need?
- Do you have the knowledge and skills you need?
- What obstacles are you running into?
- What can I or we do to help remove obstacles?
- What people issues have you encountered?
- What have you learned?

The following questions are related to specific steps of the process:

1. Charter and plans
   - Where does your charter lie on the continuum of team goals?
   - What is your goal?
   - How is your goal linked to organizational goals and objectives?
   - What is the scope of your project? Is it manageable?
   - Do you have the right people on your team from all the functions involved?
   - How will your team meet? How will you communicate outside meetings?
2. Customers and needs
   - Who are your customers?
   - How have you obtained information about your customers' needs?
   - What are your customers' needs?
   - Have you translated your customers' needs into specific, measurable requirements?
   - How are you involving your customers in your quality improvement process?
3. Current state
   - Do you have an "as-is" flowchart of your process? How was it developed and verified?
   - What have you done to analyze your flowchart for problems?
   - What are the key measures of current performance?

- Have you verified the reliability of the measurement system?
- How will you measure improvement?
- How will you know you are improving?

4. Preferred state, gaps, and improvement opportunities
   - What is your vision of the ideal process?
   - In a perfect world, how would this process be done?
   - What are the gaps between customers' requirements and current performance?
   - Did you identify any quick fixes or low-hanging fruit?
   - Has your project goal changed as a result of studying the current process?
5. Root causes
   - Why? Why? Why? Why? Why?
   - What is the root cause of this problem?
   - What other causes were studied and eliminated?
   - How did you verify this cause?
6. Solutions and plans
   - What alternative solutions have you evaluated?
   - What methods did you use to encourage creative solutions?
   - What criteria did you use to choose the best solution?
   - What constraints prevented you from choosing different solutions?
   - How does this solution address the root cause(s)?
   - Did you do a test or pilot? What did you learn from it?
   - What is your implementation plan?
   - What are the possible problems? What are your contingency plans?
   - How will you communicate your plan?
   - How will you monitor your plan?
   - How will we know if the plan is successful?
7. Implement
   - What problem are you trying to address with this action?
   - Are things going according to plan? On schedule?

- How are the changes being received?
- What are you doing to deal with problems you are encountering?

8. Monitor and recycle
   - What results did you get?
   - Does the process performance meet the customers' requirements?
   - What were the gaps between actual results and your goal?
   - If it did not work, was it because of the wrong solution or the wrong root cause?
   - If it did not work, how do you plan to try again?
9. Standardize
   - Have the process changes been standardized? How?
   - What ongoing measurement is in place?
   - Is there new documentation? Work instructions or procedures?
   - Has routine training been modified?
   - What indicators or measures will tell us if the process starts to regress?
   - Are appropriate actions documented for foreseeable problems?
   - Is a plan for continual improvement in place?
   - Who is responsible for this process? Is this person prepared to take over monitoring and improvement?
   - How will a change in customer requirements be identified and met?
   - Are additional improvements needed in this process?
   - Is there someplace else this solution or improvement could be applied?
10. Conclude, learn, and recognize
    - What did you learn from this experience with the quality improvement process?
    - How can the organization take advantage of what you have learned?
    - What changes should the organization make to improve the improvement process?
    - How can your knowledge and experience continue to be applied? Is there something else you want to work on?
    - How would you like to be recognized for your accomplishments?

4

Quality Improvement Stories

Stories and examples often are the best teachers. This chapter presents case studies that show how groups actually use the tools within the quality improvement process. Since there is a tendency to think of statistical tools as used just in manufacturing and other kinds of tools just in offices, these stories also illustrate how both kinds of tools have value in both settings.

This chapter concentrates on showing how the tools are used together to create improvement. To understand how this book can help you find tools that fit a particular need, refer to Chapter 1.

Three of these case studies are from organizations that have earned the Malcolm Baldrige National Quality Award (MBNQA) for their achievements. The last story is a composite of several teams' experiences.

MEDRAD, INC.: FREIGHT PROCESSING TEAM

Medrad designs and manufactures products used in medical imaging procedures such as MRI, CT scanning, and cardiovascular imaging. This 1300-employee company, headquartered near Pittsburgh, Pennsylvania, distributes and services its products worldwide. Medrad started its total quality management process in 1988 and six years later began using MBNQA criteria to assess and improve its business management system. In 1995, the company achieved certification to ISO 9000. Medrad first applied for the MBNQA in 1996, using feedback from the examiners to create cycles of improvement. After five applications and four site visits, in 2003 Medrad was the sole manufacturing recipient of the Malcolm Baldrige National Quality Award.

Medrad has used a **balanced scorecard** since 1997 to set organizationwide goals. Figure 4.1 shows their five long-term goals, which are supported by short- and long-term

Corporate Goal	Target	Corporate Objective	Priority
Exceed financials	CMB (profit) growth > revenue growth		1
		C	2
Grow the company	Revenue growth > 15% per year	***O***	3
		N	4
		F	6
		I	7
		D	8
		E	12
Improve quality and productivity	Grow CMB (profit) per employee > 10% per year	***N***	5
		T	10
Improve customer satisfaction	Continuous improvement in Top Box ratings	***I***	9
		A	11
Improve employee growth and satisfaction	Continuous improvement in employee satisfaction above best-in-class Hay benchmark	***L***	See # 5

Figure 4.1 Medrad: balanced scorecard.

targets and "Top 12" corporate objectives. In turn, departments, teams, and individuals develop actions to help accomplish these goals. (See *balanced scorecard* for more details.) The following story describes a project whose three objectives addressed four of the scorecard goals. For their "best practice" use of improvement processes and tools and their results, this team received a top honor, Gold Team, in Medrad's President's Team Award recognition program.

Define

Managers in the accounting department wanted to reduce the work required to process freight invoices in order to free up people to do more value-added work, to eliminate a temporary position, and also to prepare the department to handle future company growth. Although the goal was clear, how to accomplish it was not. With the help of a performance improvement facilitator from Medrad's Performance Excellence Center, they drafted goals and objectives and identified several areas of possible improvement. When they constructed a high-level **Pareto analysis,** one area jumped out: freight payables. This process of paying freight bills involves verifying the rates charged, determining which cost center within Medrad should be charged for each shipment, putting data into computer systems, and finally paying the bill. The facilitator chose to use the Six Sigma DMAIC methodology, because project conditions matched that methodology well: existing process, unknown causes, and unknown solutions.

A project plan and **charter** were developed. Figure 4.2 shows the charter after an update midway through the project. (Names have been changed here and throughout the story for privacy. See *project charter* for more discussion of this charter.) Three specific

Project Charter	
Project Title: Freight Processing Improvement	**Business Unit:** Accounting Operations
Project Leader: John Doe **Team Members & Responsibilities:** Meg O'Toole (Accounting), Deanne Zigler (Supply Chain) **Facilitator:** Steve White **Black Belt:** Steve White **Sponsor:** Maria Martinez **Process Owner:** George Gregory, Red Morris	**Business Unit Description:** Daily processing of accounting transactions: mailing invoices, paying vendors, processing invoices, etc. **Unit Manager:** John Doe

Problem Statement: A large number of hours are spent processing the payment of inbound and outbound freight.

Project Objective: Improve the efficiency of the freight payables process through better utilization of company resources and improvements in the service provided to the department's customers.

Business Case/Financial Impact: The improvements will save $xxM annually in labor. These improvements will enable the process to support future growth of the company.

Project SMART Goals:		**Baseline**	**Current**	**Goal**
• Reduce the labor needed to process freight bills.	CCID	8–18	10–27	
• Make processes more robust to handle more capacity.	**Defects** **DPM**	xx% xxx,xxx	xx% xx,xxx	x% xx,xxx
• Increase awareness of process improvement tools.	**Process σ**	x.xx	x.xx	x.xx

Project Scope Is: The current process of freight payables, including minor improvements to systems that provide inputs to this process.
Project Scope Is Not: Major system enhancements; other accounting operations functions.

Deliverables:
- Identification of business elements critical in importance to our customers.
- Identification of baseline measurements and performance goals.
- Process plan and budget required for implementation of the improvements.
- Upon approval, successful development, implementation, and control of the improvements.
- List of identified opportunities for future projects.

Support Required: IT—Suzanne Smith to adjust ERP (estimate 8 days). Shipping software consultant to incorporate entry screen changes. Planning—Mark Savoy (4 hours).

Schedule: (key milestones and dates)	**Target**	**Actual**	**Status**
Project Start	7-07-03		
D—Define: Confirm scope and purpose statements. Document findings from site visits.	7-31-03	7-31-03	complete
M—Measure: Determine appropriate measurements. Calculate baseline and performance goals.	8-15-03	8-20-03	complete
A—Analyze: Perform process analysis and identify drivers of baseline performance. Document analysis of opportunity areas. Gain approval for I & C.	9-19-03	10-02-03	complete
I—Innovative improvement: Improve and implement new process.	11-10-03		planning
C—Control: Verify that performance goals were met. Standardize the process and list future improvement opportunities.	12-01-03		

Realized Financial Impact:		**Validated by:**
Prepared by: Steve White	**Date:** 7-7-03	**Revision date:** 10-27-03
Approvals	**Manager:**	**Sponsor:**

Figure 4.2 Medrad: project charter.

	Exceed financials	Grow the company	Improve quality and productivity	Improve employee satisfaction
Reduce internal labor	◎		◎	◎
Make processes more robust		◎	◎	◎
Process improvement tools			◎	◎

Figure 4.3 Medrad: project objectives matrix.

goals were defined. A **matrix** (Figure 4.3) shows how the three project goals relate to the company's **balanced scorecard** goals. This project was listed in team members' annual objectives, which created connection and alignment between individual activities and corporate scorecard goals. The third project goal, "increase awareness of process improvement tools," was included because project sponsors agreed that it was important to spend extra project time to understand and learn quality tools, rather than simply use them under the facilitator's guidance. This empowers the team to tackle new opportunities in the future. A team member said, "The charter initially set direction and got everybody on board, but it also kept us on the right track as we went through the project." The team points to clearly defined goals as a best practice that helped their project succeed.

The charter made it clear that the focus of the project was low-hanging fruit, improvements easily achievable within a short time frame and low investment. Sections 4 and 5 define what is and is not within the scope. While everyone acknowledged that major system enhancements were possible, they were not within this project's scope. On a **continuum of team goals,** this project would not be located at "improve a service" or "improve a process" but rather at "solve a problem." The problem was high manpower use.

A small team was selected, composed of one person each from accounting and supply chain, whose regular jobs included the freight payable process, and an accounting manager who co-led the team along with the performance improvement facilitator. Later, Information Technology (IT) people were called in, but they were not asked to attend meetings until their expertise was needed—a best practice that helped the project use resources efficiently.

During the design stage, a **SIPOC diagram** was made to identify stakeholders of the process. Figure 4.4 shows a portion of the diagram, including a **macro flowchart** of the entire shipping process; the full diagram also included a **detailed flowchart** of the freight payables process. The process overlaps two functional areas, accounting and supply chain, and involves several computer systems. External customers are the freight carriers (vendors) who receive payment, and internal customers are the individuals or departments (cost centers) who ship items.

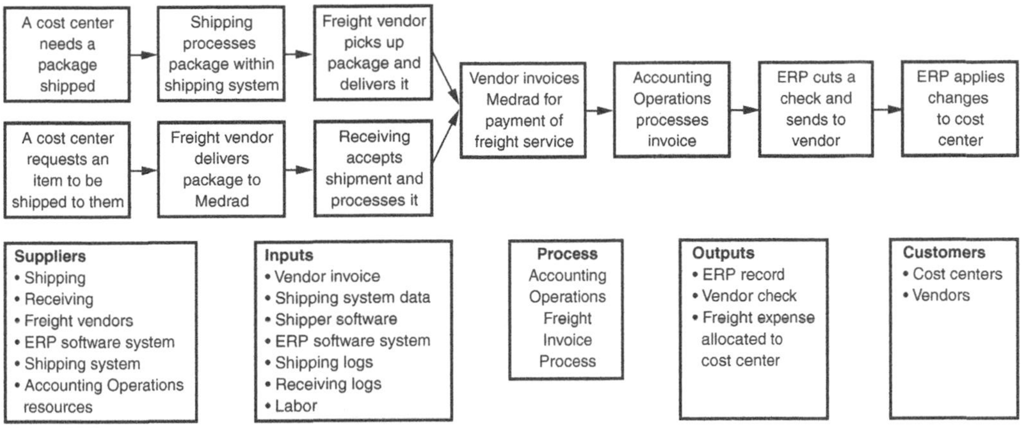

Figure 4.4 Medrad: SIPOC diagram.

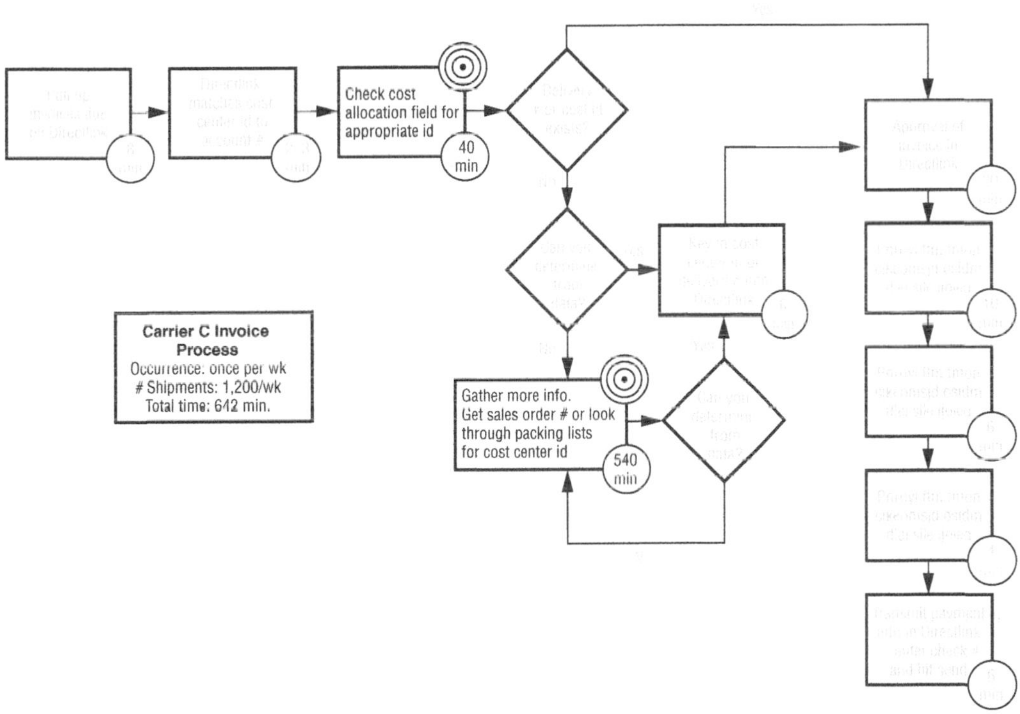

Figure 4.5 Medrad: as-is process map.

The next step was to **interview** stakeholders to understand the perspectives of shipping department and cost center personnel. Then the team created as-is **process maps.** Eight flowcharts were required, because the process is different for each carrier and sometimes for inbound and outbound shipments. Figure 4.5 shows the map created for carrier C. Most of the process steps are deliberately blurred in this reproduction of the map, except for two that the team identified as contributing most to excessive labor. (The circles with times were added later in the project.)

According to team members, creating the process maps was another best practice that contributed to their project's success. They identified waste such as redundant work, complexity, missing handoffs. Those team members whose work processes touch gained a side benefit: they now understand better what the other does and how their work affects the other.

Measurement

When the project began, no measurement tools existed for determining how much time the processes required or how many errors were encountered. The team developed a **table** (Figure 4.6) in a spreadsheet to capture this information. The team members from accounting and supply chain tracked how long each step of each process took. One time-consuming step was researching missing cost center numbers (CCIDs), so they also tracked how often those numbers were missing. Columns to the left of the double line show collected data. Columns to the right are calculated values. Tracking continued throughout the project.

Times for each step of the processes were added to the process flowcharts. Carrier C's flowchart shows 540 minutes per week spent researching missing CCID numbers. The metrics were used to create **control charts:** a **chart of individuals** for processing times and a ***p*-chart** for percentages of missing CCIDs. Each batch was a day's processed invoices, so batch size varied, making the *p* chart appropriate. Figure 4.7 shows the *p*-chart near the end of the project.

Analysis

A simplified **decision matrix** was used with a two-phase procedure to prioritize improvement opportunities. Figure 4.8 shows the first five rows of the matrix. Process steps that contributed to excessive time are listed in the rows as potential improvement opportunities. The second column from the right shows the minutes spent annually on each step, the first criteria. For only the most time-consuming steps, the team assessed the second criteria, difficulty of improvement, on a scale of 1 to 5. An opportunity with

Date	# Shipments	# Invoices	Vendor Type Carrier A	Carrier B	Carrier C	Start Time	End Time	# CCID Missing	Total Time	# CCID Missing/ Shipment	Min/ Shipment
19 Aug	714	20			x	1445	1635	169	110	0.24	0.15
22 Aug	818	84			x	0845	1042	159	127	0.19	0.16

Figure 4.6 Medrad: project metrics table.

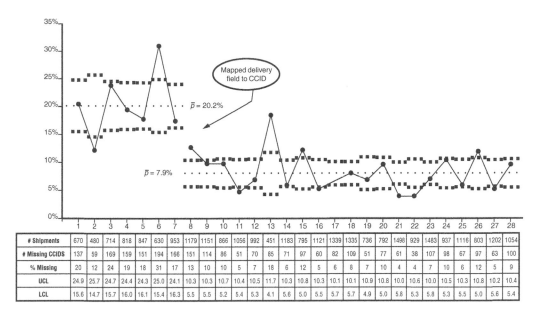

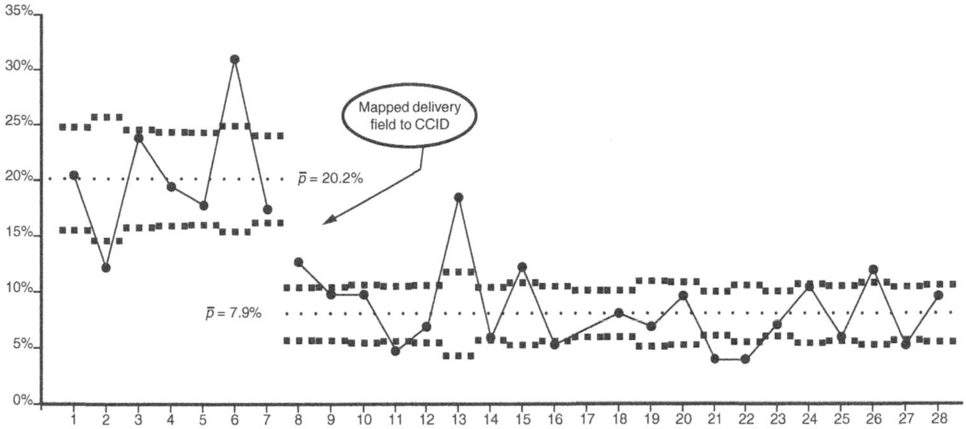

# Shipments	670	480	714	818	847	630	953	1179	1151	866	1056	992	451	1183	795	1121	1339	1335	736	792	1498	929	1483	937	1116	803	1202	1054
# Missing CCIDS	137	59	169	159	151	194	166	151	114	86	51	70	85	71	97	60	82	109	51	77	61	38	107	98	67	97	63	100
% Missing	20	12	24	19	18	31	17	13	10	10	5	7	18	6	12	5	6	8	7	10	4	4	7	10	6	12	5	9
UCL	24.9	25.7	24.7	24.4	24.3	25.0	24.1	10.3	10.3	10.7	10.4	10.5	11.7	10.3	10.8	10.3	10.1	10.1	10.9	10.8	10.0	10.6	10.0	10.5	10.3	10.8	10.2	10.4
LCL	15.6	14.7	15.7	16.0	16.1	15.4	16.3	5.5	5.5	5.2	5.4	5.3	4.1	5.6	5.0	5.5	5.7	5.7	4.9	5.0	5.8	5.3	5.8	5.3	5.5	5.0	5.6	5.4

Figure 4.7 Medrad: *p* chart.

Rank	Ref #	Type	Process Step	Weekly Shipments	Annual Shipments	Average Minutes per Shipment	Annual Impact of Step (minutes)	Difficulty of Improvement 1 = Easy 5 = Difficult
1	1	Carrier C	Check cost allocation field ID	1200	62,400	0.19	11,856	2
2	11	Carrier B outbound	Look up delivery # in ERP to get CCID	70	3,640	2	7,280	2
3	4A	Carrier B outbound sales	Post w/clearing	70	3,640	3.54	12,886	5
4	5	Carrier B outbound and inbound	Delivery # not found/ PA process	50	2,600	1.76	4,576	
5	9	Carrier B inbound	Research to find the PO of non-inventory shipments	25	1,300	3	3,900	

Figure 4.8 Medrad: decision matrix.

huge potential time savings scored 5 in difficulty, because improvement would require major changes in companywide procedures. Because the project's scope was restricted to low-hanging fruit, that opportunity was eliminated from consideration.

Each of the opportunities chosen in the prioritization step was analyzed for root causes, using a variation of the **fishbone diagram** analysis. Instead of drawing the traditional fish skeleton, flipchart pages were headed with the categories of causes—people, environment, technology, procedures, and so on—and as the team brainstormed, causes were placed onto appropriate pages. The facilitator explains, "We didn't introduce the fishbone because sometimes just its structure makes it seem more complex than it really is." One cause of missing CCIDs was incorrect numbers keyed in by shipping personnel. Another cause was no number keyed in, and a cause for that was the CCID missing on the packing slip turned in by the shipper.

Brainstorming was used again to identify solutions for the causes. Some causes required additional research. IT people were brought in to develop solutions to computer issues. The causes of missing CCIDs were first labeled "skill and knowledge issues" under the "people" category, and training solutions were discussed. As the conversation went deeper, the team realized that training would work until turnover occurred and training didn't happen or a different message was communicated. The team asked themselves how technology could ensure that the CCID is entered. Their **mistake-proofing** solution had two parts. CCID became a required data entry field, to prevent omitting the number; and a drop-down list showed all cost center numbers, to reduce data entry errors. Mistake-proofing the CCID entry, another best practice, required the team to take their thought process to another level to find a systemic solution.

Some of the solutions changed the process maps. To-be **flowcharts** were drawn for the new processes. Figure 4.9 shows as-is and to-be **flowcharts** for one carrier. Again, the steps of the process are intentionally blurred here; what is important is how much simpler the new process became. Some process maps did not change. All these process maps can be used in the future for training.

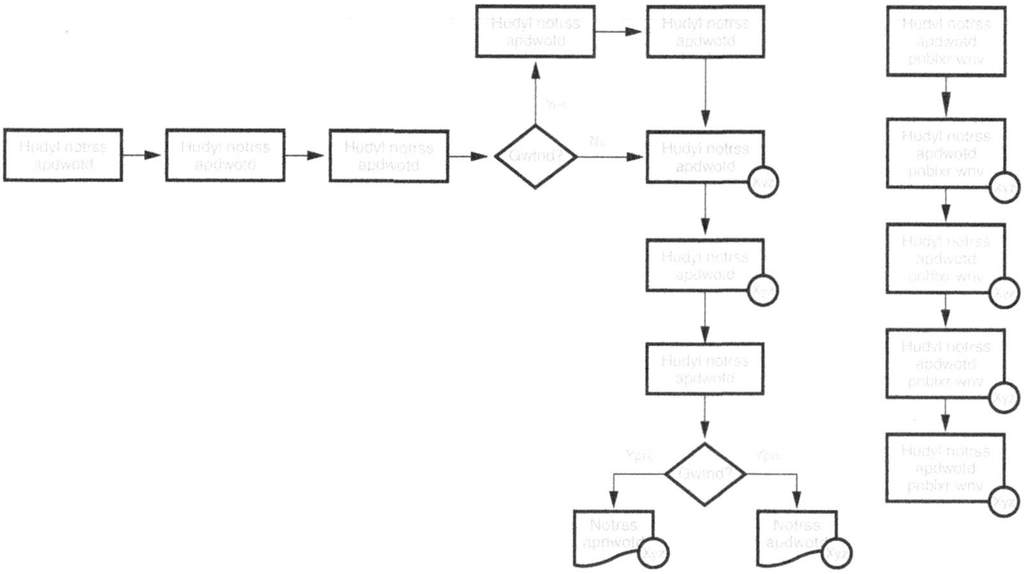

Figure 4.9 Medrad: as-is and to-be process maps.

Innovative Improvement

People who would be affected by the solutions were consulted to understand their perspective and improve the plans. Then the plans were carried out by the team and IT. The team continued tracking metrics, but because improvement opportunities focused on a few carriers' processes, only their charts were maintained.

Many changes were implemented. One solution corrected the interfaces between computer systems, so those working the freight payables process had access to information they needed. Non-value-added steps were eliminated. For example, paper invoices had been manually sorted by carrier, a time-consuming chore, but the team realized that only once in several years was it necessary to retrieve a paid invoice. Another time-consuming step, reviewing rates on every paper invoice and signing each one, was modified so that invoices under a certain dollar amount have blanket authorization. That change speeded about 90 percent of all paper invoices. Carriers, who are both suppliers and customers, were involved in some solutions. A carrier who used to provide a paper invoice for every shipment was asked to provide consolidated billing. To solve the problem of personnel in field offices not providing cost center information, the team worked with carriers to make the information mandatory for shipments charged to Medrad.

To manage project implementation, Medrad uses a tool they call a corrective action tracker (Figure 4.10). Another variation of the generic **table,** it not only lists actions, responsible parties, targeted completion dates, and status, but also relates actions to steps on the process map and to actions that must happen before or after.

Action #	Ref #	Ref Step	Action	Owner	Target Date	Current Status	Linked Actions	Closed
1	#11	Look up CCID	Set up meeting with IT (Suzanne S.) to discuss the requirements of this step	Meg	9/16	Waiting for Suzanne to confirm that will take about 1 hr to complete	/10	/
2	#4a	Post w/ clearing	Discuss concept with finance and sales to understand their points of view	John	9/16	• Turned down idea. Need accountability for freight. • Need to investigate other opportunities	/2	x/
3	5	Delivery # not found	Get more details about duration of manual processing for carriers C and D	Deanne	9/16	This is on Mark S.'s list, but not high priority. Deanne will follow up with Mark and provide support to raise priority. Met with Mark and this is higher priority now.	/5	/

Current status: Provide details to identify any risk of missing the target date.
Linked actions: Preceding action # / Subsequent action #
Closed: Closing of current action item / Closing of all related items

Figure 4.10 Medrad: corrective action tracking table.

When the CCID data entry was changed in the computer system, the results were immediate. On the **control charts,** the percentage of missing CCIDs and the time required to process invoices both dropped dramatically when that change was made and stayed at that lower level. (The percentage of missing CCIDs is not zero because of shipments dropped off at carriers' facilities outside Medrad.) Overall, significant labor savings were achieved. For example, processing time for one carrier's invoices was reduced by more than six hours per week.

The team considers ongoing measurement to be another best practice. The facilitator explains, "Often people can feel the results, but they don't have a quantitative measure for the impact of the project. Using tools like these, it was very easy to communicate to people what our impact was, what the savings were. And it gives the team a feeling of accomplishment because it's visual; they can see the impact they had by participating in this project."

Control

Responsibility for the improved freight payables process was handed off to the functional groups. Because these groups were involved throughout the project, the disbanding of the team and hand-off to the groups was seamless. The measurement tools were passed to accounting so they could continue to monitor performance.

"We have a lot of ideas for things that we see us working on in the future," says one team member, "but for the purpose of our team, it had to have an end point." The next step is "benchmarking, benchmarking, benchmarking": learning from best-practice companies in the area of freight processing. They also plan to extend this project's improvements to other types of shipments and other carriers.

The team highlighted the five best practices that had contributed to their success. They also identified lessons they had learned from challenges they encountered. For example, they learned that during the planning stage, they should consider competing projects and commitments, not only current ones but also ones that will occur throughout the expected time frame of the project. Their project had to be lengthened and team members had to juggle too many balls because of other important activities in their regular jobs.

Another important lesson was about getting started. Despite Medrad's strong continuous improvement culture, "sometimes projects seem more complex, maybe require more project management experience or different tools that people don't feel they have expertise using," says this project's facilitator. "They brought me into the picture. 'How would we go about this? What tools would we use?' Take small bites. You're sitting there looking at a gigantic cookie, and it's a little intimidating. Small bites out of the project have large impact."

Another team member said, "Sometimes you're concentrated so much on what you do every day, and you know all these different issues. So on a project you're not really sure where to start. You get bogged down with talking about all these things, and you don't really solve anything. You don't know what direction to go in and don't know how you're supposed to lay it out. Even something as simple as a process map—you don't even think to do that sometimes. By having him [the facilitator] on our team, it helped us step back and realize how not to get bogged down with all the other issues and to find ways to get moving in the right direction. Actually, it was a joy, and it was one of the best projects I've been on, because we were so focused and because of the quality tools we used as a group."

PEARL RIVER SCHOOL DISTRICT

Pearl River School District (PRSD), located 25 miles north of New York City, has creatively adapted the tools and methods of quality to fit its needs. They explain, "We deal with people. Our students are our customers, and the product we deliver is to allow them to achieve to their highest ability." Pearl River has applied the principles of continuous improvement so well that results over a twelve-year period earned them the Malcolm Baldrige National Quality Award in 2001.

Pearl River's "Golden Thread" is a **tree diagram** (Figure 4.11) that communicates how districtwide goals are translated into subgoals and individual projects. **Surveys** are used extensively to gather input from students, faculty, staff, and community. (See *tree diagram* and *survey* for details.)

Pearl River uses data extensively to monitor performance on everything they do. A computerized data warehouse automatically pulls in data from a variety of sources and makes it easily available in **tables** or **graphs, stratified** in whatever way is needed for analysis. Root cause analysis is applied to any areas needing improvement. They view their work as processes and have studied and **flowcharted** dozens of key processes, from teacher recruitment and hiring to maintenance work orders. They use the **plan–do–study–act (PDSA) cycle** extensively. Figure 4.12 shows how they have

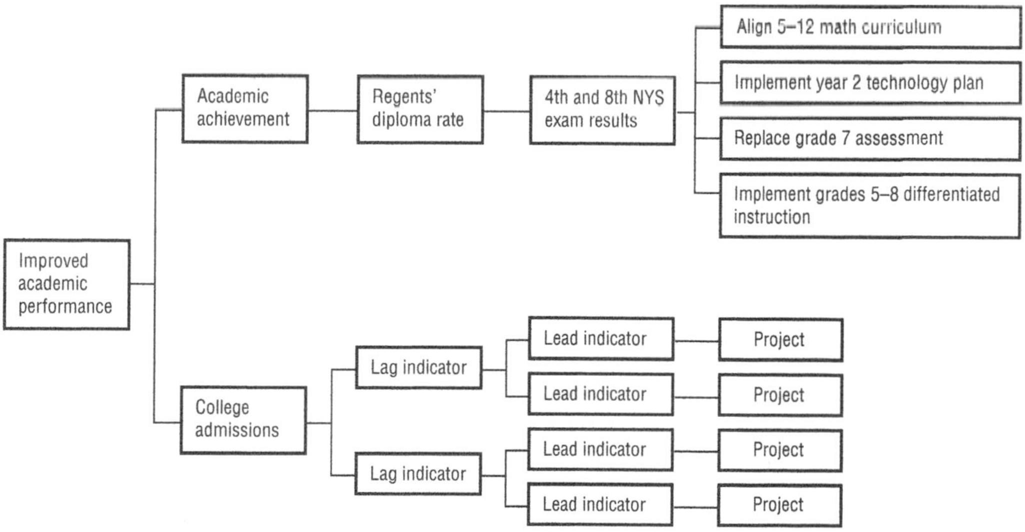

Figure 4.11 Pearl River: tree diagram.

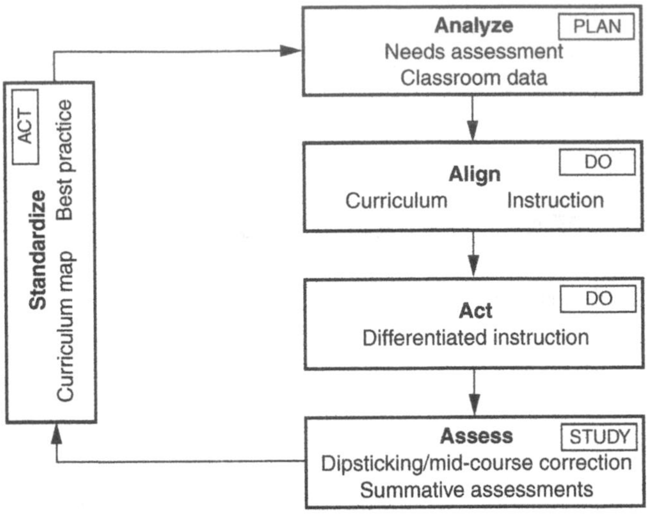

Figure 4.12 Pearl River: PDSA approach.

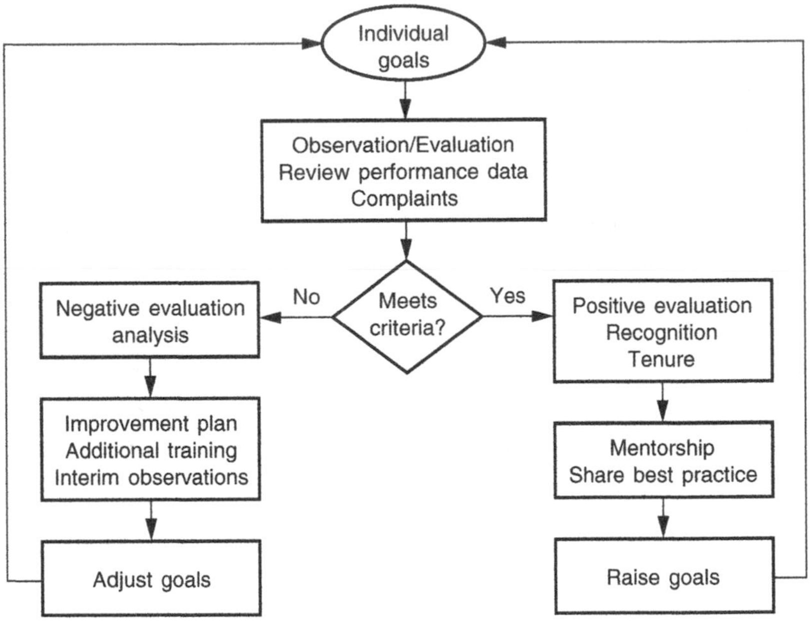

Figure 4.13 Pearl River: performance evaluation process.

transformed it into the A+ Approach for curriculum planning. See the PDSA example on page 391 for more details.

Pearl River believes education is all about what happens in the classroom. The PRSD performance evaluation process, shown in Figure 4.13, supports teachers in setting and

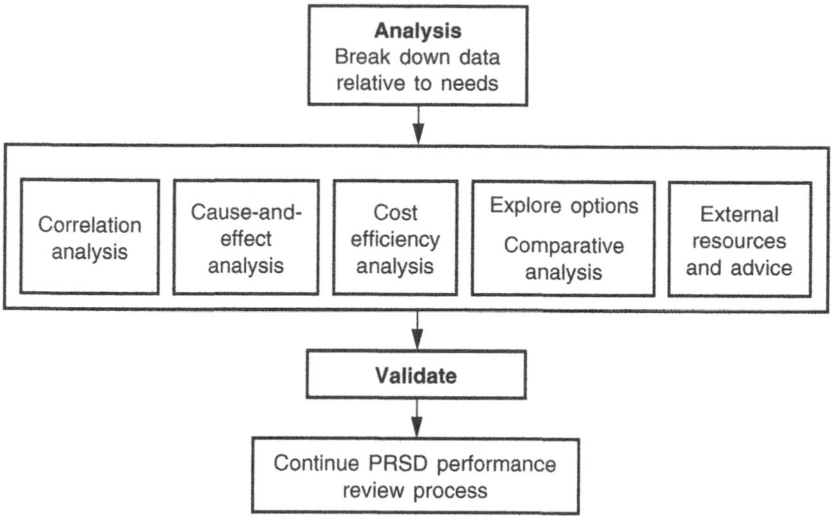

Figure 4.14 Pearl River: analysis process.

achieving goals for improved teaching, aligned with district goals, and evaluates teacher performance, identifying and correcting problems when they appear. The steps of this process encompass all but step 2 of this book's generic 10-step improvement process. Step 2 is built into the design of the data warehouse, which collects data from and for significant stakeholders. Their data analysis process, Figure 4.14, forms the core of their improvement cycle and is used during every iteration before deciding what steps to take next.

The following stories from Pearl River School District illustrate how the improvement process—and especially using data as a tool for problem analysis—applies even to education's fundamental activities of teaching and learning.

Closing the Gap

A high school teacher met with her principal in September to review performance and set improvement goals, as all teachers do at the start of the school year. An experienced teacher of a core academic subject, she had taught in the district for only one year. Data from the warehouse showed that the average of her students' scores on state exams was ten percent lower than the average of their classroom work. The principal, who had observed this teacher's classroom many times, was perplexed by the huge gap.

"I don't get it," the principal said. "I know you're a great teacher. Why do you think this happened?"

As the principal and teacher explored the problem, they realized what was going on. The teacher had given her students too many extra-credit assignments. Because their grades were high, they thought they knew the material and didn't study as hard as they should have. Inaccurate data feedback was affecting the students' study process.

The teacher changed her approach. She also continued to reflect on what was happening in her classroom. In addition to reducing extra-credit assignments, she was quicker to recommend struggling students for the math lab, and she started reviews for the state exam earlier.

The following year, the gap had completely disappeared. Her students' classroom average remained high, and the average of their state exam scores matched their classroom work.

Pearl River High School's principal says she has used data to study all kinds of problems, now that she understands its power. "The power of data lies not only in studying figures but also in knowing what figures to study." State exam scores alone would not have indicated how the teacher's approach needed changing. The cause could have been anything from teaching the wrong material to poor classroom management skills. However, comparing two sets of data gave additional information about the current state that led to the correct cause of the problem.

Focusing on the Cause

In another instance, the performance of a brand-new teacher in an academic subject was not satisfactory. A huge percentage of his students failed the midterm exam. Again, the cause could have been anything from a group of kids with poor study skills to poor teaching to a poorly designed exam. Before a quarterly performance review, the principal did several data comparisons.

First, she compared the midterm exam scores of this teacher's students to other teachers' students. These kids did significantly worse. The cause was not the exam.

Second, she compared grades in these students' other classes to their grades in this subject. They were doing well elsewhere. The cause was not the student's study skills or learning capacity.

Third, she compared the results on this midterm exam, which is a departmentwide exam, to results on classroom quarterly exams, which are prepared by the teacher. The students had performed well on the quarterly exams. Poor teaching was probably not the cause, unless the teacher's tests were badly designed.

These comparisons pointed to a different cause: the teacher was not teaching the same material as the other teachers. The teacher and principal could now begin to seek confirmation of this cause and plan changes.

When data is used for problem solving, judgmental issues like personalities can be avoided. "The discussion is cooler and cleaner," says Pearl River's principal. "Not necessarily more pleasant, but on a different, less emotional plane." While education's results are dependent on what happens inside classrooms between teachers and students, the field has no monopoly on emotionally charged issues. Every organization functions through people and therefore is full of potentially explosive emotional land mines. Thoughtful use of data can help us weave through those land mines and focus on true problem causes.

ST. LUKE'S HOSPITAL: CHARGING STANDARDS

St. Luke's Hospital is the largest hospital in Kansas City, Missouri, with 450 doctors, over 3000 employees, and almost 600 beds. For many years it has been committed to total quality and continuous improvement. In 1995, the hospital began using the MBNQA criteria to help it strive for excellence. By 2003, St. Luke's success in improving healthcare, patient satisfaction, and bottom-line results earned it the Malcolm Baldrige National Quality Award.

Although it is a not-for-profit hospital, like any organization it must manage its financial resources so that it can continue to fulfill its three-part mission of outstanding health care, medical education, and research. St. Luke's uses a planning process based on an organizationwide **balanced scorecard** (Figure 4.15), deployed throughout the organization with 90-day action plans in all departments.

The following story illustrates how the principles and tools of quality improvement were applied to a hospital process that overlapped medical care and financial accounting. It also demonstrates that the steps of the improvement process need not be addressed in strict sequential order, if the project suggests a different organization. St. Luke's uses a five-step improvement process, but the ten-step process is indicated here as another illustration of those steps.

The Problem

Hospitals itemize patient charges to include supplies used and services that are above and beyond normal nursing care on the unit. For example, if a patient receives a cardiac catheterization, the catheter and dressings should be itemized. After the procedure, the

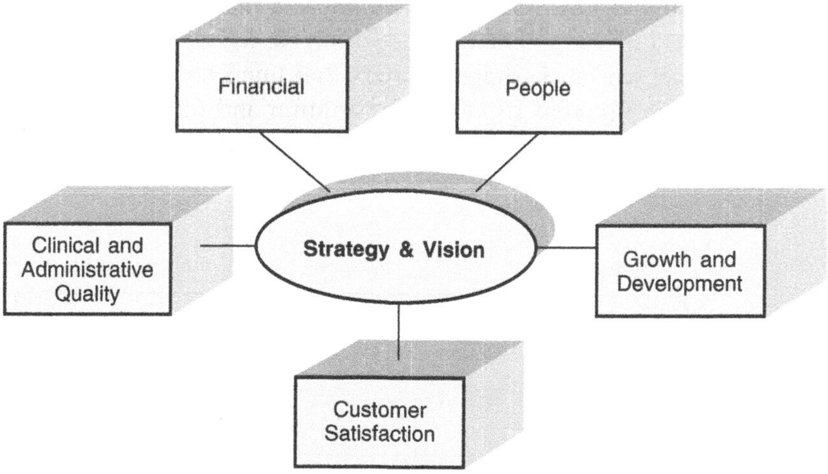

Figure 4.15 St. Luke's: balanced scorecard.

nurse on the unit closely monitors the patient for about four hours, watching for complications, and that special nursing care also is charged.

But in the fast-paced, high-pressure nursing units at St. Luke's, these charges often were not captured. A **mistake-proofing** system had been devised to make it easy to capture charges for supplies. Each item had a yellow sticker, and when the item was used for a patient, the sticker was removed from the item and placed on the patient's card. But many charges were missed anyway. The hospital's charge audit department reviews records when requested by insurance companies, looking for invalid charges, but the auditors discovered the opposite problem: an overwhelming percentage of patient records were missing charges. Because not every record is audited, the true cost to the hospital's bottom line was unknown, but St. Luke's knew they were losing a lot of money.

Team Formation

To solve this problem, the Performance Improvement (PI) Steering Committee **chartered** a team to develop new charging standards. Eleven team members represented every nursing unit, different positions from nursing director to staff nurses to unit secretaries, different functions such as information services (IS) and accounting, a facilitator from quality resources, and a director-level team leader. (Step 1: What do we want to accomplish?)

Because of the high priority of this issue, the steering committee and team leadership decided to accelerate the project with an approach that was unusual at the time, although St. Luke's has since adopted it for other projects. The team met only four times—but those meetings lasted all day long. With meetings every Friday and work assignments between meetings, the project kicked off at the beginning of April and rolled out a new process in July. Many people complain that quality improvement projects take too long. St. Luke's has proved it doesn't have to be so!

The most-used tool was an **L-shaped matrix** that hung on the wall throughout all the team meetings. (See Figure 4.16.) Only the column and row labels were filled in. Brainchild of the team leader, it visually conveyed two intersecting dimensions of the charging problem. (Step 1.) The charging process has three phases: knowing what to

	Knowledge	Resources	Accountability
What to charge			
Entering the charge			
Auditing the charge			

Figure 4.16 St. Luke's: L-shaped matrix.

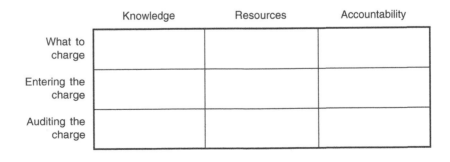

charge (and not to charge), entering the charge so it can be captured on the patient's record, and auditing charges later. For each of these three phases, three categories of requirements have to be met: knowledge of the charging process; resources available, such as access to the computer system; and accountability, so that people know for what part of the process they are responsible. After a few meetings, one of the team members commented, "You know, the more I stare at that matrix, the more it's starting to make sense." A tool's structure, without detail, can be used to convey complex concepts or relationships.

First Meeting

The first team meeting, held in April, was devoted to the first phase of the charging process, knowing what to charge. The team **brainstormed** requirements under the three broad categories—knowledge, resources, and accountability—for each staff position involved in the process. (Step 2: Who cares and what do they care about?) The brainstorming began by asking questions like, "What do you need to know? What resources do you need in order to do this job?" A **table** derived from the charging concepts matrix was used to organize the ideas generated by the brainstorming. (See Figure 4.17.) This table became a work-in-progress and was used throughout all the team meetings.

Flowcharting was used to understand how items are added or changed on the master list of charges. Figure 4.18 shows a portion of the **detailed flowchart** developed during this first Friday.

Accountability	Knowledge	Resources
Director	What procedures and supplies used by each area B94 process (what is a B94, when to use) What items are "outside" How to monitor for changes in cost/prices Respond to the CPT change e-mails	Access to list of CDM Form on N drive Internet Intranet resources B94 committee Monthly material management report REVEAL training on reports
Administrative Assistant	How to maintain the unit charge sheets How to communicate changes Monitor invoices of outside items How to post the charges	Access to CDM inquiry Access to PMM
Patient Care Provider	Charge sheets and how to use Charging process Why they can charge for items Importance (results) of entering charges	Training on charge sheets Access to charge sheets More terminals Initial training and training process Who to ask questions of
Information Associate	Usual charges for different types of patients and diagnosis or procedure	Initial training and continued training People resources Charge sheet

Figure 4.17 St. Luke's: table.

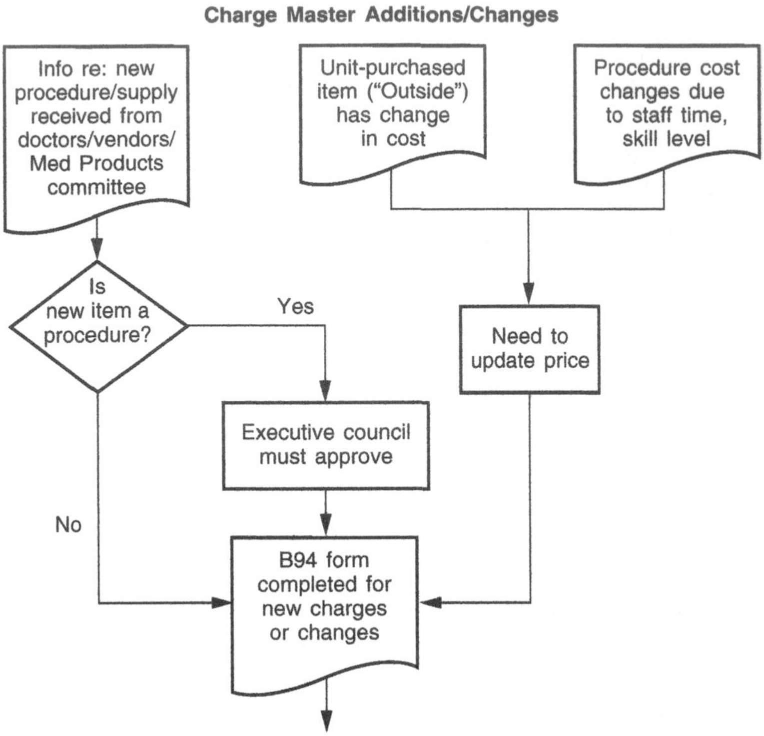

Figure 4.18 St. Luke's: detailed flowchart.

Between the first and second meetings, team members had assignments to visit the hospital units, **interview** people doing the work, and find out exactly how the charging process was currently being done. (Step 3: What are we doing now and how well are we doing it?)

Second Meeting

The second Friday was devoted to phase two of the charging process, entering the charge. The day began with individuals reporting what they had learned from the units. The meeting room's walls were hung with many different **flowcharts,** showing that each unit had its own variation of the process. Another member had gathered all the charging forms used in the units. Again, they were all different. The team discussed the problems and opportunities represented by the variations. (Step 4: What can we do better?)

The team continued to work on the charging standards table (Step 2). They also began working on a **fishbone diagram** (Figure 4.19) to identify the causes of inaccurate charging. (Step 5: What are the underlying problems?) The fishbone revealed to the team how complex the process really is. Then they developed a **detailed flowchart** (Figure 4.20) of a generic, ideal process for collecting and capturing charges. "Day one" and "day two" of the to-be flowchart were done during this meeting. (Step 6: What changes could we make to do better?)

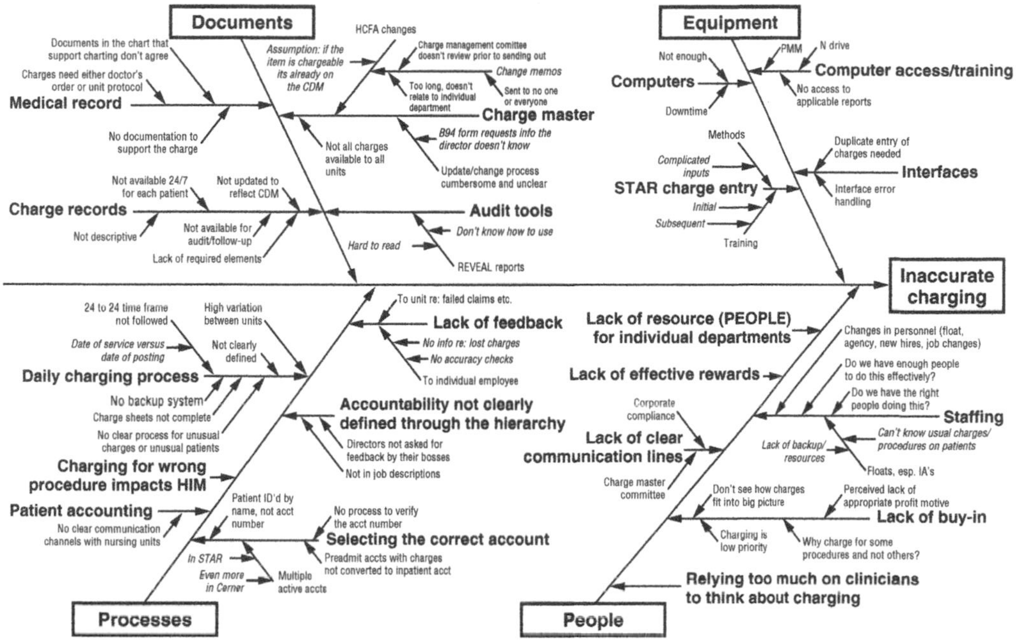

Figure 4.19 St. Luke's: fishbone diagram.

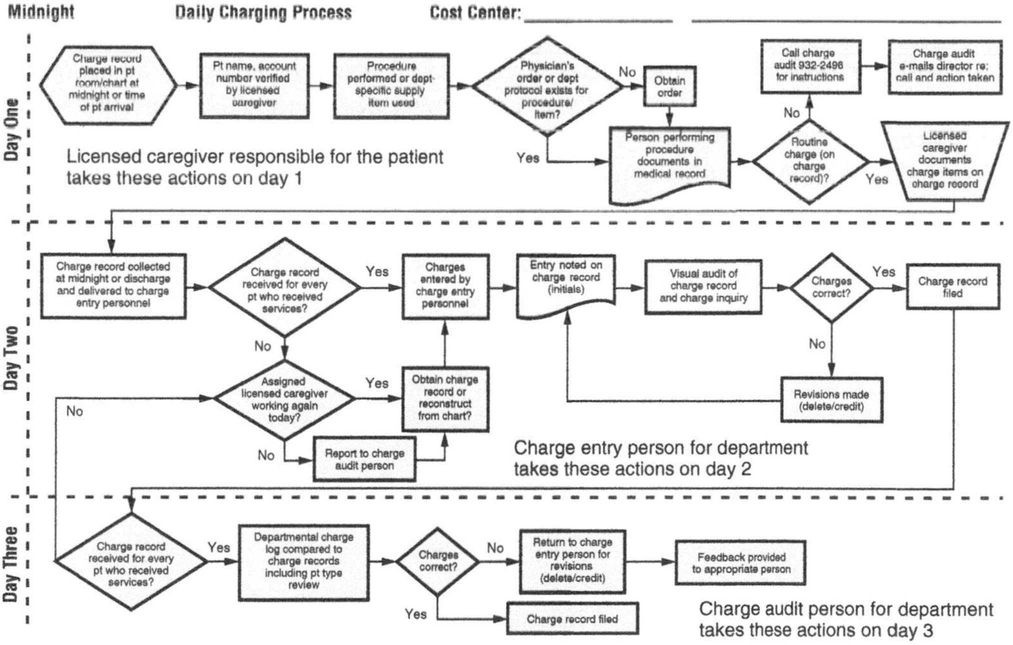

Figure 4.20 St. Luke's: to-be detailed flowchart.

In the week between meetings two and three, a subteam designed a new charging form, with standard information plus areas that could be customized by each unit. Because the charging code depends on location as well as the item—a bandage in ER has a different charge code than a bandage in ICU—each unit had to be able to list its own unique codes. (Step 6.)

Third Meeting

The third meeting was devoted to the third phase of the charging process: auditing the charge. Additional representatives from IS attended this meeting. The team reviewed the reports that were available from the computer system (Step 3) and redesigned them (Step 4). In one case, a new report was needed. For charging it is important to know what patients were on a unit the previous day, but no existing report gave that information. One of the IS people volunteered to develop a new "patients who were here" report. **Brainstorming** was used to plan new and redesigned reports, asking questions such as, "What do we want on the report? What do we not want on the report? What's on there now that isn't helpful? What do we need in place of that?" (Step 4.)

The team also continued work on the charging standards **table** (Step 2) and the **fishbone diagram** (Step 5), adding information related to auditing. "Day three" was added to the daily charging process **flowchart,** showing audit activities. (Step 6.)

Fourth Meeting

The fourth Friday was devoted to planning implementation. (Step 6: What changes could we make to do better?) During previous meetings, the group had begun to realize that a team could never be responsible for the new process. Instead, it had to be embedded into the organization's job responsibilities and accountabilities. Fortunately, St. Luke's has a well-developed performance management system, with clear job descriptions and a goal-setting process for each individual that relates to the organization's core values. One of the core values is resource management, which clearly ties to the charging process.

The team developed a process and time line for rolling out the new process, summarized in a **Gantt chart.** (See Figure 4.21.) Communications, training, and documentation were included.

During the afternoon of this last meeting, the quality resources director led the team members in a short course on skills of team leadership. This was considered both a reward for the team members, who enjoyed the training, and a way of capitalizing on the experience the team members had gained by increasing their skills for use on future projects. (Step 10: What did we learn? Let's celebrate!)

Implementation

The PI steering committee approved the new process, and the rollout began. (Step 7: Do it!) A meeting was held at the end of May for all the unit directors, attended by senior management, for a **presentation** communicating the new standards, processes, and forms.

Charging Standards Implementation Timeline

ID	Task Name	Start Date	End Date	Duration
1	Director's meeting to introduce standards and documents	5/28/01	5/29/01	1d
2	Directors complete documents	5/29/01	6/4/01	5d
3	Directors turn in documents	6/5/01	6/5/01	0d
4	Charge Std Leadership prepares materials for department training	6/5/01	6/8/01	4d
5	Department meetings of Charge Resource Persons	6/11/01	6/15/01	5d
6	IS training as needed	6/18/01	6/29/01	10d
7	Department orientation	6/25/01	6/29/01	5d
8	GO LIVE	7/1/01	7/2/01	1d

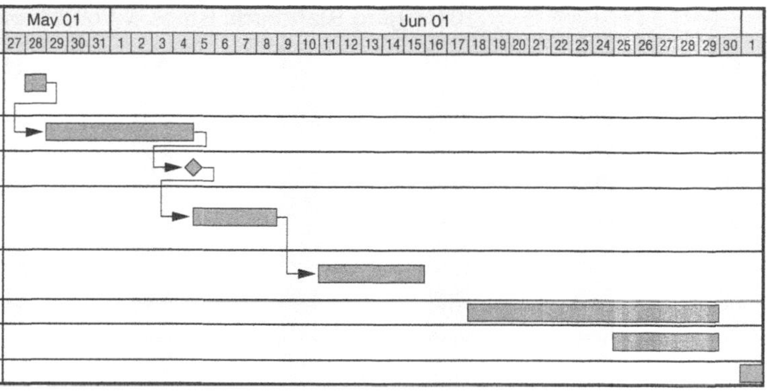

Figure 4.21 St. Luke's: Gantt chart.

Section 2: Entering the Charge

Directors will be accountable for the following for each cost center assigned to the director:

- Designating the position(s) responsible for completing the Charge Record (charge sheet or medical record document used to enter charges)
- Designating the position(s) responsible for charge entry

The Charge Entry Person will be accountable for the following:

- Validating that Charge Records were completed for all patients receiving services within that revenue center for the identified 24-hour period, even if there were no charges
- Validating the patient type, date of service, and account number
- Ensuring that all charges are entered within five days of discharge (date of discharge is day one)

The organization will be held accountable for providing the following resources and training:

- Training on charge entry and charge inquiry in STAR and any other software used to post charges for that revenue center
- How to identify the patient type and correct account number
- Clearly defined written process for charging
- Communicating the importance of accurate charging and its impact on the organization

Figure 4.22 St. Luke's: policy document.

Each director received a packet of materials, containing baseline data for that unit, examples of new and redesigned forms and reports, and the timeline for implementation.

The blank charging concepts matrix used throughout the project had now been transformed into documents capturing process details. The directors' packets included a document organized by the three phases of the process, describing what happens and who does it. (See Figure 4.22.) But the team went a step further and gave the directors

Charging Standards Roles & Positions

Cost Center(s) Number Name of person completing form:

_____________ _______ _________________________ Ext: _______

Responsibility: Charge Entry Person

Accountabilities:

1. Validating that charge records were completed for all patients receiving services within that cost center for the identified 24-hour period, even if there were no charges
2. Validating the patient type, date of service, and account number
3. Ensuring that all charges are entered within five days of discharge (date of discharge is day one)

Computer Access Needed:

- Access to charge entry and charge inquiry in STAR

Training Needed:

- What are customary charges for patients on that unit
- Charge entry and charge inquiry in STAR
- How to identify the patient type and appropriate account number
- Revenue center charging process
- The importance of accurate charging and its impact on the organization

Position of this person: _______________________________

Figure 4.23 St. Luke's: roles document.

a second document. It contained the same information, but it was organized by job role. For each role, it listed what the person does in each phase of the charging process, what knowledge and skills are needed, and the computer access required for the tasks. (See Figure 4.23.) A third document, based on the second, was a **checklist** for the director to list employees and check off training completed and computer access available. (See Figure 4.24 for a portion of this checklist.) The team saw that it was important to provide multiple tools that give information in the way people will need to use it.

The directors also received the generic daily charging process flowchart. At the bottom were seven questions, such as "Where is the charge record placed in your unit?" and "What are the titles/names of charge entry persons?" After the meeting, the directors provided answers to the questions, and the team prepared a customized flowchart for each unit.

Team members provided computer training as needed to unit employees. The week before the transition, unit directors held orientations with their staff to prepare them for the rollout. On the "go-live" day, everyone began using the new process.

Place a check in the box if the training is COMPLETE or the person has access.

Charge Entry Person	**Training**					**Computer Access**		
Name	Customary Charges	Charge Entry	ID Patient	Charge Process	Importance	Charge Entry		

Figure 4.24 St. Luke's: checklist.

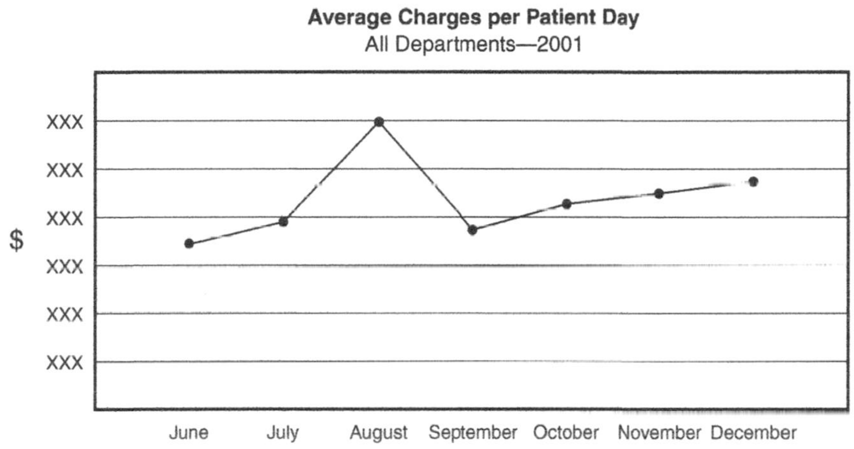

Figure 4.25 St. Luke's: line graph.

How Did We Do?

One of the directors commented after the directors' meeting that it was the first he had attended where it was very clear at the end what he needed to do. The team's care with defining responsibilities and resources paid off with a smooth, successful implementation.

Figure 4.25 shows a **line graph** of average charges per patient per day (CPPD), starting in June before the process was implemented and continuing through the rest of the year. (Step 8: How did we do?) (For confidentiality, the actual data has been masked.) August shows a spike, as everyone paid extra attention to the new process. Charges dropped in September, then gradually began increasing.

The team had recognized that continuing responsibility for maintaining charging standards had to be embedded in the organization. One indication of continuing attention to the new process was that departments chose to include charging standards goals in their 90-day action plans for the **balanced scorecard's** financial goal. Each unit developed its own metrics for monitoring accuracy. (Step 9: How can we do it every time?)

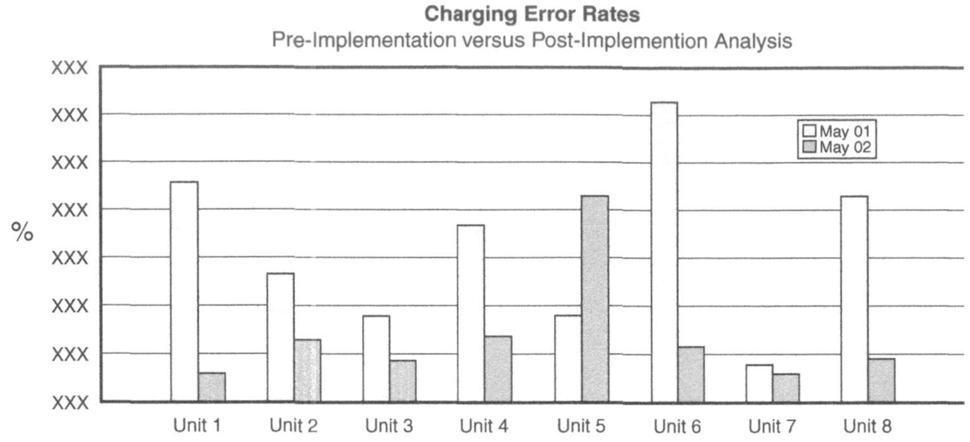

Figure 4.26 St. Luke's: bar graph.

A labor-intensive analysis of charging error rates had been done when the team first formed. A year later, the analysis was repeated. Figure 4.26 shows a **bar graph** comparing error rates in eight different units before and after the changes. (Step 8.) (Again, actual data has been masked.) In all cases but one, the improvement is obvious and dramatic. The one exception in unit 5 occurred because the regular person was on vacation most of that month; the following month, error rates dropped to a very low number. Unit 7's improvement appears small; however, that unit has extremely large charges, so even a small improvement contributes greatly to the bottom line.

ZZ-400 MANUFACTURING UNIT

The story of ZZ-400, a chemical manufacturing unit, is a composite of the experiences of several real teams, plus some fiction to tie them all together. Each of the figures shown here is repeated in Chapter 5, with more explanation, as an example of that tool.

ZZ-400 is the company's nickname for one of its products. The workers in the ZZ-400 unit, one of many within a large chemical plant, operate the continuous manufacturing process around the clock. They are supported by central facilities that service all units in the plant: analytical laboratory, maintenance shop, shipping and receiving, engineering services, and administrative functions such as accounting and human resources.

ZZ-400 began its quality improvement process at the same time as the rest of the plant, after training in quality concepts, methods, and tools. A team volunteered to work together to plan initial projects.

Step 1: What do we want to accomplish?

A sign had hung on the control room wall for over a year, designed by several of the unit operators:

Best-Run Unit in the World

Making the Best ZZ-400 in the World!

The team agreed that the sign was more a dream than a reality, so it was a statement of their vision. But they also realized that it was vague. What does "best-run unit" mean? What does "best ZZ-400" mean? They had no **operational definitions** for "best" in either case.

The team decided that their initial objective should be to lay groundwork for all other improvement activities in the unit by creating unit-wide understanding of customers' needs, current process performance, and metrics for measuring "best." They created a **top-down flowchart** (Figure 4.27) showing a plan of these initial steps and set target dates for each activity.

Step 2: Who cares and what do they care about?

The team created a **SIPOC diagram** to identify key customers. They quickly listed the companies that purchased ZZ-400 but were astonished by how many other outputs they had and how many groups interacted with them.

The unit prided itself on close contact with its customers. But when the team started pulling together all the customer specification sheets to keep in one place, they realized that many were several years old and some of the specifications were unfamiliar. In addition, they had never talked with internal groups about the information they needed.

They arranged for one or two team members to visit each external customer's plant to find out exactly how ZZ-400 was handled and used. Before these **interviews**, they developed a **checklist** of questions to be asked and information to be given. Back in the

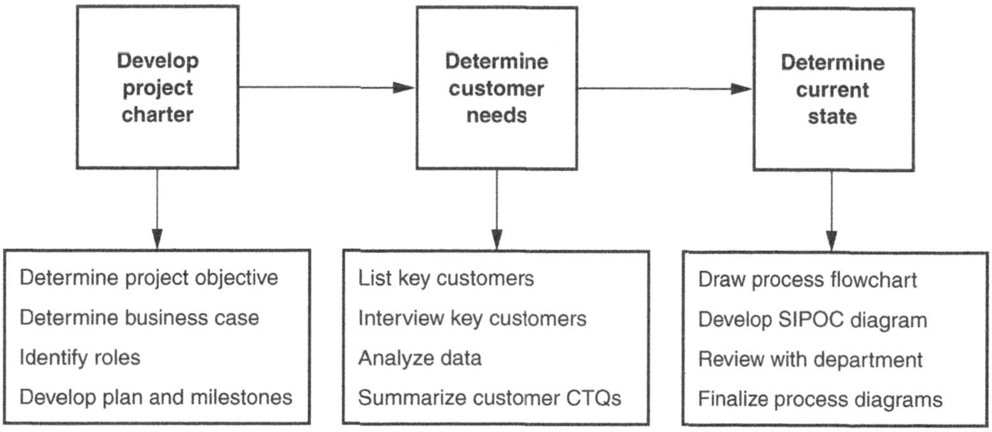

Figure 4.27 ZZ-400: top-down flowchart.

Customer Requirements

	Customer *D*	Customer *M*	Customer *R*	Customer *T*
Purity %	> 99.2	> 99.2	> 99.4	> 99.0
Trace metals (ppm)	< 5	—	< 10	< 25
Water (ppm)	< 10	< 5	< 10	—
Viscosity (cp)	20–35	20–30	10–50	15–35
Color	< 10	< 10	< 15	< 10
Drum		✔		
Truck	✔			✔
Railcar			✔	

Figure 4.28 ZZ-400: L-shaped matrix.

unit, a task team developed an **L-shaped matrix** (Figure 4.28) that showed each customer's product specifications and choice of packaging and delivery options.

They also **interviewed** representatives of the internal groups. That information was summarized in a **requirements table.** Because maintenance and analytical lab play a crucial day-to-day role in keeping the unit running, the team asked for a representative from each group to join the team.

Step 3: What are we doing now and how well are we doing it?

The team began to focus on its processes. A **detailed flowchart** of the manufacturing process was already a daily tool. To decide which process variables to monitor, everyone reviewed the newly collected customer specifications as well as their own expert knowledge of the most sensitive variables in the process. Collecting all this input, they set up **control charts** for each critical process characteristic (Figure 4.29) with historical data used to establish control limits. The entire unit began monitoring its manufacturing process with the control charts.

Step 4: What can we do better?

In order to know where they could improve, the team had to define "best-run unit" and "best ZZ-400." First, team members **brainstormed** (Figure 4.30) a list of over 30 indicators of "best." To make some sense out of all these ideas, they used an **affinity diagram.** They modified the procedure slightly to allow for the fact that the entire team was on shiftwork and wanted to minimize meetings. On a rarely used door in the control room, members set up sticky notes with the brainstormed ideas on them and passed along a request for everyone to work with the notes whenever they had time. After several days, a natural grouping and some headings had emerged (Figure 4.31).

$\overline{X}$ and R or Moving Average–Moving Range Chart

Process: ZZ-400	Variable: Purity	Units: %	Limits set by: PW	Date: 3/29/05
	$UCL_{\overline{X}}$: 99.88 $LCL_{\overline{X}}$: 99.20	$\overline{\overline{X}}$: 99.54	UCL_R: 0.59	$\overline{R}$: 0.18

Date	3/4	3/5	3/6	3/7	3/8	3/9	3/10	3/11	3/12	3/13	3/14	3/15	3/16	3/17	3/18	3/19	3/20	3/21	3/22	3/23
#1	99.7	99.7	99.4	99.5	99.7	99.6	99.3	99.6	99.5	99.7	99.6	99.5	99.7	99.7	99.4	99.6	99.2	99.7	99.4	99.5
#2	99.6	99.4	99.3	99.8	99.4	99.9	99.2	99.3	99.6	99.6	99.8	99.4	99.8	99.3	99.5	99.7	99.5	99.6	99.3	99.6
#3																				
#4																				
#5																				
#6																				
Sum																				
Avg.	99.65	99.55	99.35	99.65	99.55	99.75	99.25	99.45	99.55	99.65	99.70	99.45	99.75	99.50	99.45	99.65	99.35	99.65	99.35	99.55
Range	0.1	0.3	0.1	0.3	0.3	0.3	0.1	0.3	0.1	0.1	0.2	0.1	0.1	0.4	0.1	0.1	0.3	0.1	0.1	0.1

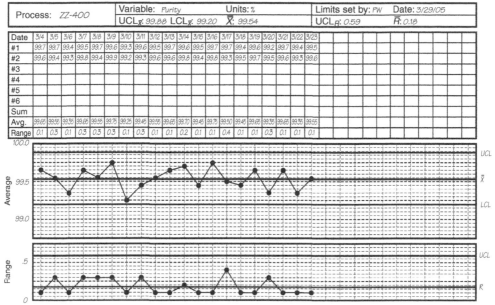

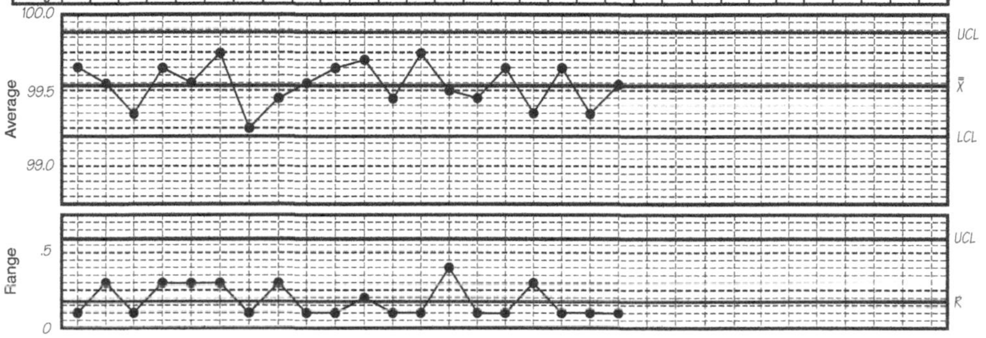

Figure 4.29 ZZ-400: $\overline{X}$ and R chart.

Possible Performance Measures	
% purity	# of OSHA recordables
% trace metals	# of customer returns
Maintenance costs	Customer complaints
# of emergency jobs	Overtime/total hours worked
lbs. produced	$/lb. produced
Environmental accidents	Raw material utilization
Material costs	Yield
Overtime costs	Utility cost
# of pump seal failures	ppm water
Viscosity	Color
Cp_k values	Service factor
Safety	Time between turnarounds
Days since last lost-time	Hours worked/employee
% rework or reject	lbs. waste
Hours downtime	Housekeeping score
% uptime	% capacity filled

Figure 4.30 ZZ-400: brainstorming.

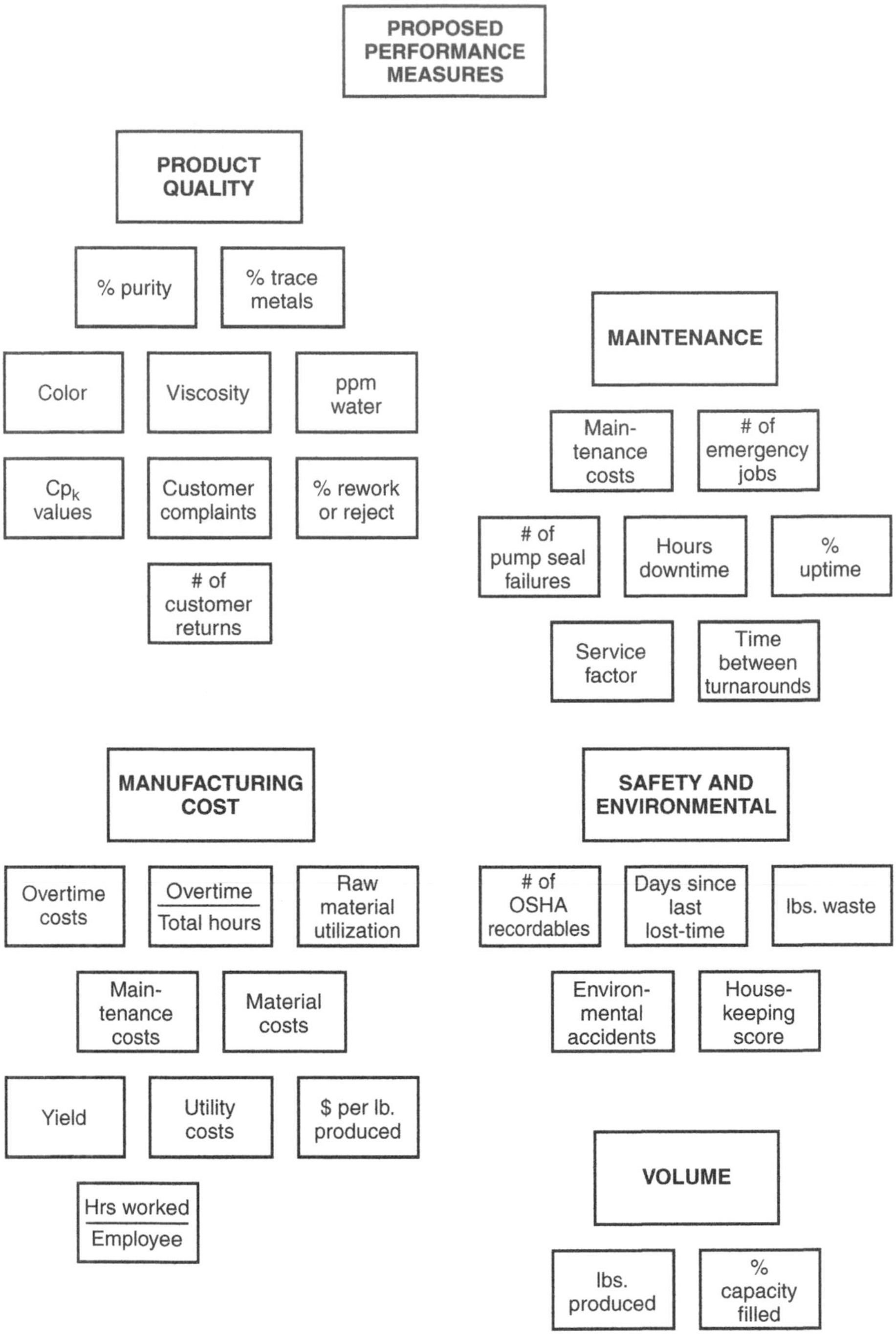

Figure 4.31 ZZ-400: affinity diagram.

The affinity diagram defined five broad areas of performance—product quality, equipment maintenance, manufacturing cost, production volume, and safety and environmental. The team used **list reduction** to select one overall measure for each area. Those five measures became the basis for a **performance index** (Figure 4.32).

The team had data on current performance from their work in step 3, but they needed to establish goals for "best" performance. It was easy to set zero as the target for rework, unplanned downtime, and safety and environmental warnings. For volume and cost goals, they went back to data from their customer and business group interviews.

Those interviews also helped them prioritize their work in the five performance areas. Product quality, most heavily weighted on the performance index and most important to external customers, was one of the first areas for improvement. New volunteers formed a team and were given a project charter, recommended by the first team and supported by plant management.

ZZ-400 Purity Team

Step 1: What do we want to accomplish?

This team was given a very specific objective: increase ZZ-400 purity process average and reduce its variation so that all product meets the tightest specification of 99.4 percent without rework. They went quickly through steps 2 through 4, since the first team had covered much of the groundwork.

Step 2: Who cares and what do they care about?

The team reviewed the customer interview information to learn more about the problems encountered by customers when low or variable-purity product entered their processes. From the business group interviews, they learned that not only did the expense of rework increase product cost, but also the lost capacity prevented the business group from expanding sales.

Step 3: What are we doing now and how well are we doing it?

They studied the process flowchart, process control charts, and other historical production data. While the process was usually in control, as shown in Figure 4.29, occasionally it went out of control, dipping below 99.0 percent purity.

Step 4: What can we do better?

They talked to R&D chemists and engineers to learn that it was possible for purity to consistently achieve 99.95 percent with the equipment they had. They decided that their goals would be first to eliminate the out-of-control periods, second to decrease variation around the average, and third to raise the average.

Step 5: What prevents us from doing better? (What are the underlying problems?)

As they discussed possible causes of the periodic drop in purity, one of the ideas volunteered was a traditional but unproven belief around the unit that overall purity dropped whenever traces of iron showed up. The subteam decided to test this belief.

Date: *June, 2005*

% Rework	Hours Downtime	Pounds Made	Safety & Environmental Flags	$ / Pound			Criteria
26	*28*	*42*	*9*	*278*			Performance
0	0	≥ 70	0	≤ 143			10
2	5	69	1	164			9
6	10	67	2	185			8
11	15	65	4	206			7
16	20	60	6	227			6
21	25	55	8	248			5
24	30	50	10	269			4
28	40	45	12	290			3
31	50	40	13	311			2
34	60	35	14	332			1
≥ 36	≥ 70	≤ 30	≥ 15	≥ 353			0
3	*4*	*2*	*4*	*3*			Score
25	15	15	20	20			Weight
75	*60*	*30*	*100*	*60*			Value

Index = *325*

Jan	Feb	Mar	Apr	May	Jun	Jul	Aug	Sep	Oct	Nov	Dec
			300	*315*	*325*						

Figure 4.32 ZZ-400: performance index.

Source: James Riggs and Glenn Felix, *Productivity by Objectives,* ©1983, pp. 225, 233. Adapted by permission of Prentice Hall, Englewood Cliffs, N.J.

A **scatter diagram** (Figure 4.33) of purity and iron content showed no relationship between the two variables. Then the mechanic reminded everyone that there were three reactors. Maybe the reactors were different enough that their data should not be combined. The team modified the scatter diagram, this time using **stratification** to separate the three reactors' data. (Figure 4.34) Sure enough, for two of the reactors, increased

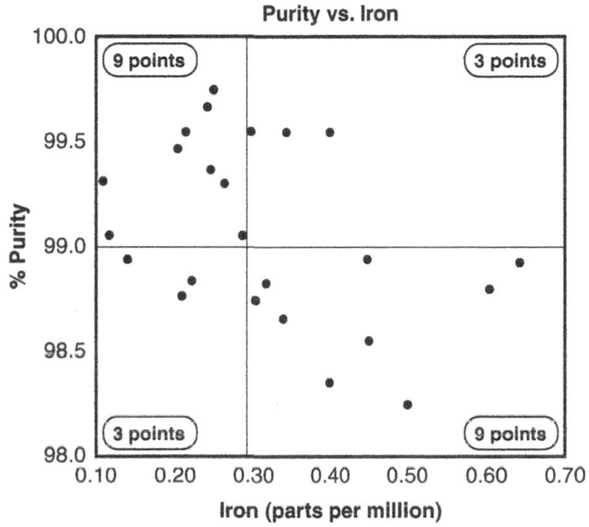

Figure 4.33 ZZ-400: scatter diagram.

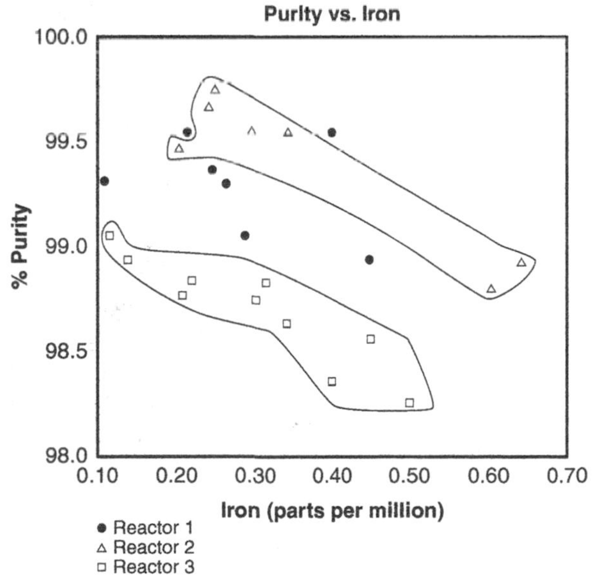

Figure 4.34 ZZ-400: stratified scatter diagram.

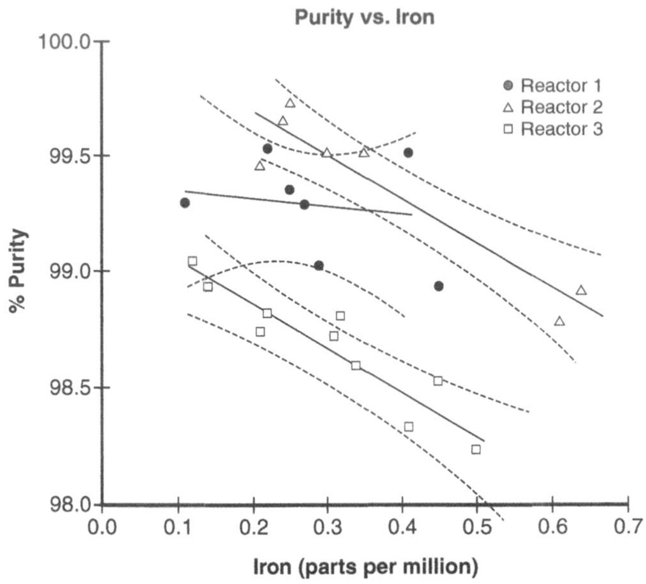

Figure 4.35 ZZ-400: regression analysis.

iron was related to decreased purity. **Regression analysis** (Figure 4.35) confirmed the relationship.

The team now knew that iron was one cause of lower purity. But why was iron in the process? Where was it coming from? All members of the unit worked on a **time-delay fishbone** (Figure 4.36), to identify possible sources of iron. The **flowchart** was examined to determine what was different about reactor 1, which did not show increased iron. They examined **control charts** and log sheets to try to understand what else was happening when the problem occurred. An **is–is not diagram** (Figure 4.37) was used to bring together everything they knew about the situation.

The diagram revealed that the problem occurred at the same time as pump preventive maintenance, and only when one particular spare pump was operated. That pump was used for reactors 2 and 3; another was used for reactor 1.

They asked "Why?" again. Why did that pump cause problems? Pump data sheets revealed that the pump was made of carbon steel, not stainless steel like the other two pumps and necessary for the fluid it pumped. That fluid rusts carbon steel, producing iron.

Why was that pump different? Maintenance records revealed the root cause. The spare pump for reactors 2 and 3 had been installed during a midnight emergency and never replaced. Maintenance and unit crews weren't sufficiently aware of materials of construction needed for different fluids.

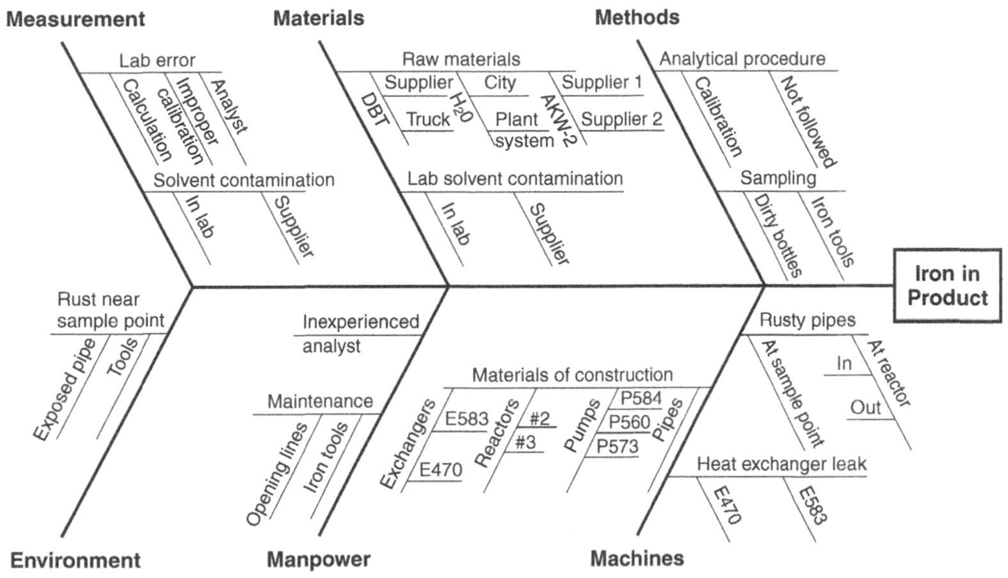

Figure 4.36 ZZ-400: fishbone diagram.

Problem Statement: Iron contamination in ZZ-400	**Is** Describe what does occur	**Is Not** Describe what does not occur, though it reasonably might	**Distinctions** What stands out as odd?
What objects are affected? What occurs?	Purity drops Iron increases	Other metal impurities Increase in moisture Change in color	Only iron increases
Where does the problem occur? • Geographical • Physical • On an object	Reactors 2 and 3	Reactor 1	Not reactor 1
When does the problem occur? When first? When since? How long? What patterns? Before, during, after other events?	Lasts about one shift About once a month Often during same week as pump PM When P-584C runs	Continuous Any particular shift Any other pump than P-584C	Just P-584C
Extent of problem How many problems? How many objects or situations have problems? How serious is the problem?	Purity drops from 99% range to 98% range Puts us out of spec Iron up to 0.65 (3 × normal)	Purity below 98.2% Pluggage problem Other specification problems	
Who is involved? (Do not use this question to blame.) To whom, by whom, near whom does this occur?	Mechanics doing pump PM		Mechanics usually around

Figure 4.37 ZZ-400: is–is not diagram.

Source: © Copyright, Kepner-Tregoe, Inc. Reprinted with permission.

Step 6: What changes could we make to do better?
Step 7: Do it.

A stainless steel pump was quickly ordered and installed.

Step 8: How did we do? If it didn't work, try again.

During the next preventive maintenance cycle, the process stayed in control.

The team knew they weren't finished. Variation and process average were still unacceptable. They recycled back to step 5 to search for root causes of those problems.

Step 9: If it worked, how can we do it every time?

In the meantime, the team members realized that if the wrong equipment could be installed once, it could happen again. They set out to educate themselves about material of construction standards. The team studied information provided by engineering then summarized the facts in a **T-shaped matrix** showing material of construction related both to specific pieces of equipment and to the various liquids that the unit handled. **Presentations** were made to each shift and to the maintenance crew, and the written documents were placed in unit maintenance guides.

Step 10: What did we learn? Let's celebrate!

Plant management knew that material-of-construction errors could happen throughout the plant, so they asked the team to find a way to share their work. The team created a large **storyboard** (Figure 4.38) that traveled around the plant. Other units adapted the materials matrix for their processes.

OTHER IMPROVEMENT STORIES

Woven through the examples in this book are other quality improvement stories, fictitious and real. To help you compare tools or see how tools link together, those stories are listed here along with the tools they illustrate.

Catalog Retailer

Contingency table page 148
House of quality page 308
Hypothesis test page 318
SIPOC diagram page 475
Voice of the customer table page 511

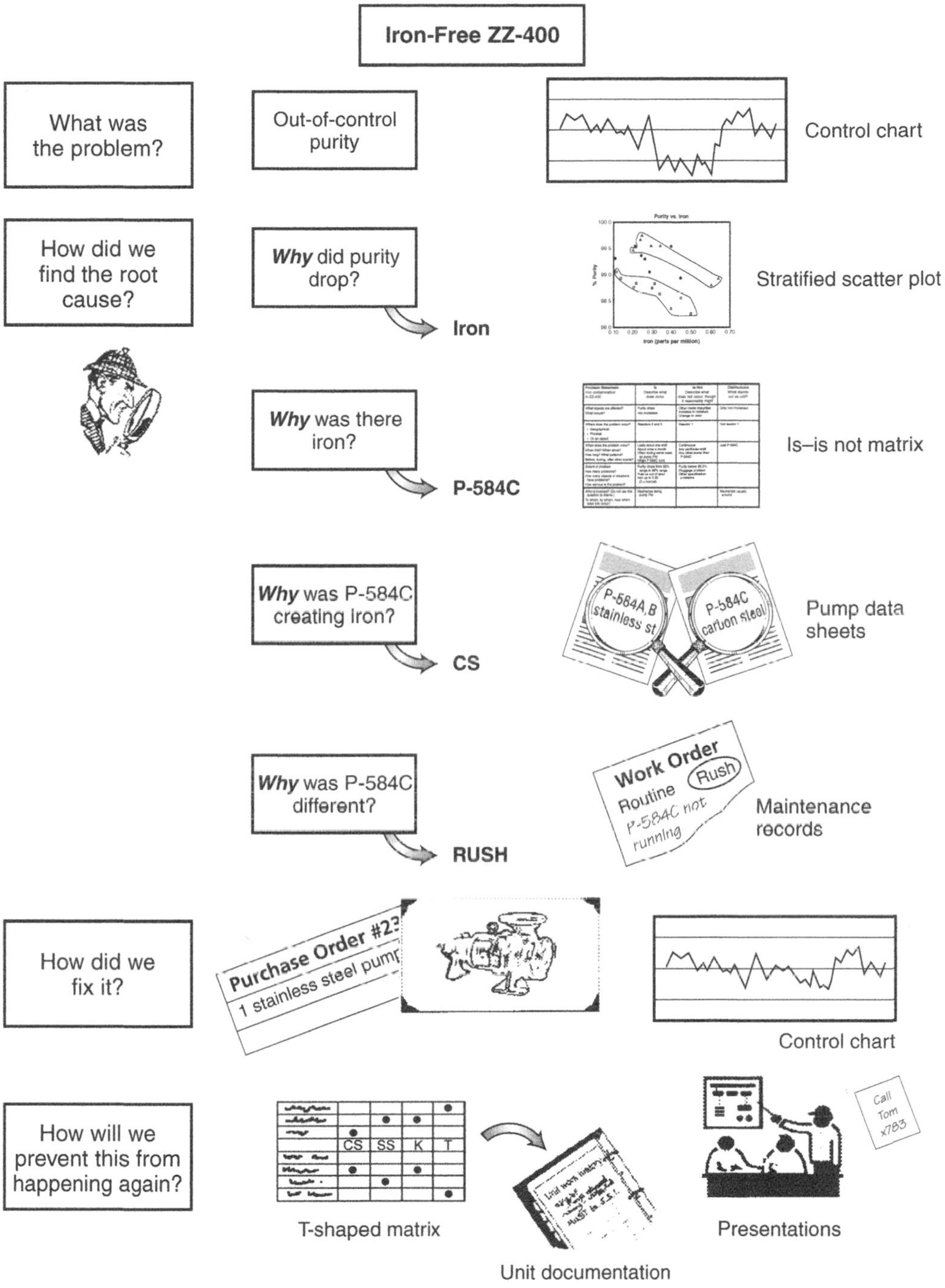

Figure 4.38 ZZ-400: storyboard.

Education

Manufacturing: Filling an Order

Medical Group: Chronic Illness Management Program

Medical: Other Examples

Restaurant: Parisian Experience

Restaurant: Ben-Hur's Pizza

Another Restaurant

Survey Development Team

5

The Tools

ACORN test

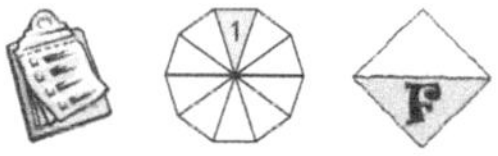

Description

The ACORN test is a check on a mission or project charter goal statement to determine whether it is well-defined.

When to Use

- When writing the mission of either an organization or a project team, or . . .
- While drafting a project charter goal, to guide its development, or . . .
- When the project charter goal has been written, as a check on its quality, or . . .
- When writing subgoals

Procedure

Ask the following questions about the statement of the mission or project charter goal:

A—Accomplishment. Does the goal describe results rather than behaviors? Could fulfillment of the goal be verified if the people responsible for it were not there?

C—Control. Do the group's actions determine whether the goal is accomplished? If it depends on actions of others, it is not within the group's control.

O—Only Objective. If this and only this were accomplished, would it be enough? If something else also is required, then subgoals have been identified, not the true goal. (Subgoals do not have to pass the *O* part of the test.)

R—Reconciliation. Will accomplishing this goal prevent another group within the organization from accomplishing its goal? Or, does another group share the same goal? Two groups should not be addressing either the same goal or conflicting goals. (Subgoals do not have to pass the *R* part of the test.)

N—Numbers. Can the goal be measured? It must be possible to generate practical, cost-effective data to measure the goal. Measurement will indicate whether or not the goal has been achieved.

Examples

Engineering Department

The engineering department of a manufacturing company has drafted this mission statement:

> The Engineering Department serves the business units, manufacturing department, and our customers by being the best at designing and constructing safe, environmentally responsible, easily maintained and operated facilities—better, faster, and at lower total cost than anyone else.

The department checks the statement against the ACORN test.

A—Engineering has a tangible *accomplishment* when the mission is fulfilled: the steel and concrete of a manufacturing plant.

C—Engineering has broad *control* over its projects, although R&D, manufacturing, and business units provide input and have veto power.

O—Engineering's *only* function is designing and managing the construction of new facilities.

R—No one else has this function, so the mission is *reconcilable.*

N—*Numbers* are routinely available for project schedule and cost as well as safety, environmental, and operational performance. However, the group decided it must develop an operational definition and measures for "better," which everyone intuitively understands to mean best technology and methods.

Parisian Experience Restaurant

The Parisian Experience Restaurant Customer Satisfaction (PERCS) team wrote this charter statement as it began its work:

> The PERCS team will improve the customer satisfaction ratings of our restaurant by 20%.

This draft failed the ACORN test. Team members realized they do not have control over customer satisfaction ratings: every waiter, chef, cashier, and manager plays a part. So they tried again with the following statement:

> The PERCS team will direct and support the continuing improvement of customer satisfaction with our restaurant.

That statement failed the ACORN test, too. This time, the statement described behaviors ("direct and support"), not accomplishments. What is the team really going to do? Back to the drawing board for draft number three, which passed the test:

> The PERCS team will identify opportunities to improve customer satisfaction with our restaurant. It will develop and recommend plans that transform those opportunities into value for our customers and our business.

Considerations

- If a goal statement does not pass the ACORN test, it may actually be in conflict with the goals of the entire company or institution.
- An overall mission statement must pass all five parts of the ACORN test. A subgoal will never pass the *O* (only objective) part of the test. It may not pass the *R* (reconcilable) part either, if this goal needs to be balanced against another. But it must pass the *A* (accomplishment), *C* (control), and *N* (numbers) tests.

  Example: Two subgoals of the Engineering Department, to build plants quickly and to build them so they are easily maintained, may not be completely reconcilable with each other. It may add time to the design phase of a project to obtain input from maintenance. These two subgoals must be balanced in order for the department to achieve its mission.
- When the mission of an organization is well defined according to the ACORN test, one can proceed to evaluate performance against that mission in a meaningful way.

affinity diagram

Also called: affinity chart, K-J method
Variation: thematic analysis

Description

The affinity diagram organizes a large number of ideas into their natural relationships. This method taps the team's creativity and intuition. It was created in the 1960s by Japanese anthropologist Jiro Kawakita.

When to Use

- When you are confronted with many facts or ideas in apparent chaos, or . . .
- When issues seem too large and complex to grasp
- And when group consensus is necessary

Typical situations are:

- After a brainstorming exercise, or . . .
- When analyzing verbal data, such as survey results, or . . .
- Before creating a tree diagram or storyboard

Procedure

Materials needed: sticky notes or cards, marking pens, large work surface (wall, table, or floor).

1. Record each idea with marking pens on a separate sticky note or card. (During a brainstorming session, write directly onto sticky notes or cards if you suspect you will be following the brainstorm with an affinity diagram.) Randomly spread notes on a large work surface so all notes are visible to everyone. The entire team gathers around the notes and participates in the next steps.
2. *It is very important that no one talk during this step.* Look for ideas that seem to be related in some way. Place them side by side. Repeat until all notes are grouped. It's okay to have "loners" that don't seem to fit a group. It's all right to move a note someone else has already moved. If a note seems to belong in two groups, make a second note.

3. *You can talk now.* Participants can discuss the shape of the chart, any surprising patterns, and especially reasons for moving controversial notes. A few more changes may be made. When ideas are grouped, select a heading for each group. Look for a note in each grouping that captures the meaning of the group. Place it at the top of the group. If there is no such note, write one. Often it is useful to write or highlight this note in a different color.
4. Combine groups into supergroups if appropriate.

Example

The ZZ-400 manufacturing team used an affinity diagram to organize its list of potential performance indicators. Figure 5.1 shows the list team members brainstormed. Because the team works a shift schedule and members could not meet to do the affinity diagram together, they modified the procedure.

They wrote each idea on a sticky note and put all the notes randomly on a rarely used door. Over several days, everyone reviewed the notes in their spare time and moved the notes into related groups. Some people reviewed the evolving pattern several times. After a few days, the natural grouping shown in Figure 5.2 had emerged.

Notice that one of the notes, "Safety," has become part of the heading for its group. The rest of the headings were added after the grouping emerged. Five broad areas of

Possible Performance Measures	
% purity	# of OSHA recordables
% trace metals	# of customer returns
Maintenance costs	Customer complaints
# of emergency jobs	Overtime/total hours worked
lbs. produced	$/lb. produced
Environmental accidents	Raw material utilization
Material costs	Yield
Overtime costs	Utility cost
# of pump seal failures	ppm water
Viscosity	Color
Cp_k values	Service factor
Safety	Time between turnarounds
Days since last lost-time	Hours worked/employee
% rework or reject	lbs. waste
Hours downtime	Housekeeping score
% uptime	% capacity filled

Figure 5.1 Brainstorming for affinity diagram example.

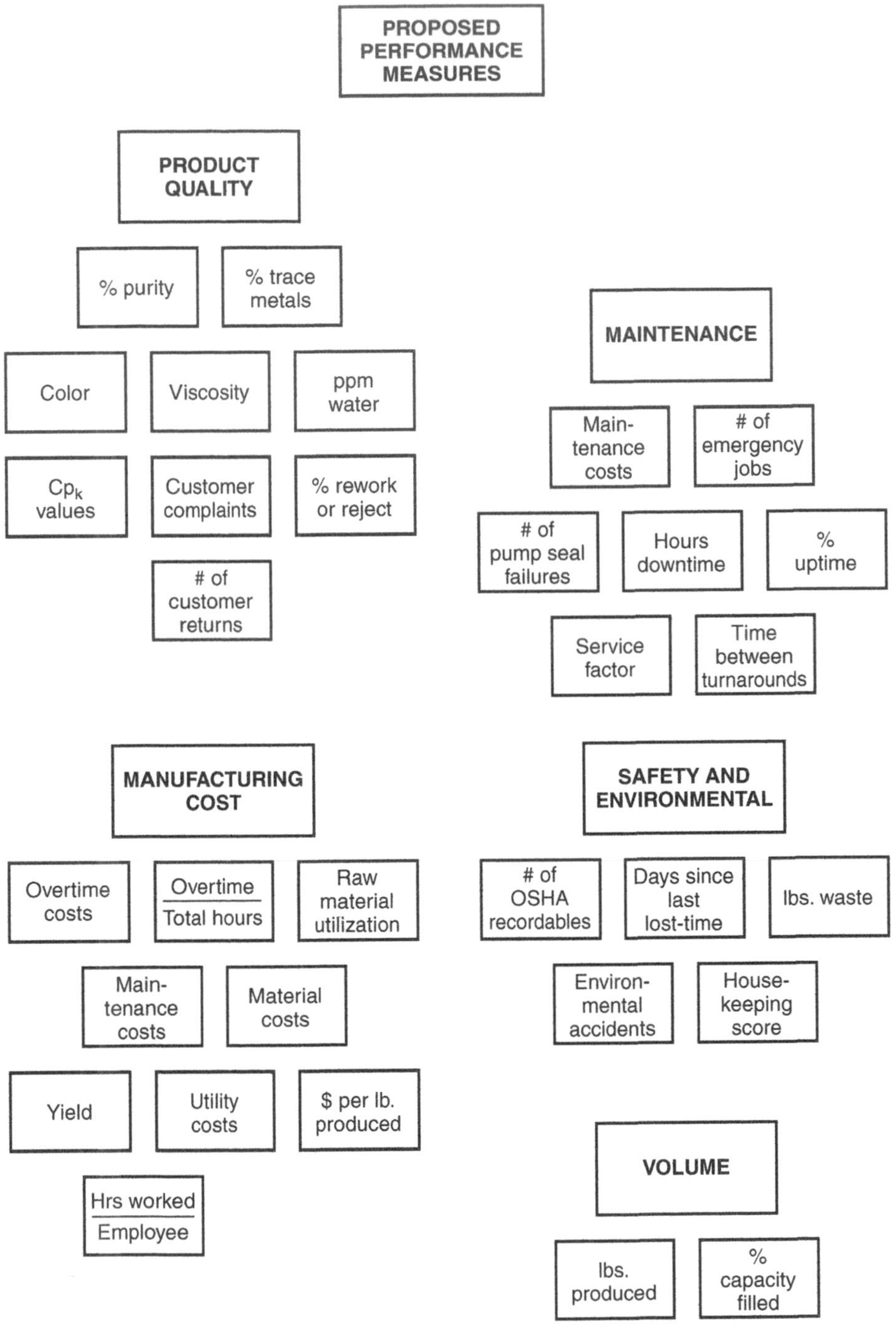

Figure 5.2 Affinity diagram example.

performance were identified: product quality, equipment maintenance, manufacturing cost, production volume, and safety and environmental.

This example is part of the ZZ-400 improvement story in Chapter 4.

Considerations

- The affinity diagram process lets a group move beyond its habitual thinking and preconceived categories. This technique accesses the great knowledge and understanding residing untapped in our intuition.
- Very important "Do nots": Do not place the notes in any order. Do not determine categories or headings in advance. Do not talk during step 2. (This is hard for some individuals!)
- Allow plenty of time for step 2. You can post the randomly-arranged notes in a public place and allow grouping to happen over several days.
- Most groups that use this technique are amazed at how powerful and valuable a tool it is. Try it once with an open mind and you will be another convert.
- Use markers. With regular pens, it is hard to read ideas from any distance.

thematic analysis

Description

Thematic analysis is a method for analyzing open-ended questions or any nonnumerical information by an iterative process of sorting and looking for relationships.

When to Use

- When analyzing responses to open-ended questions on surveys (questionnaires, interviews, or focus groups) or . . .
- When analyzing results of benchmarking, or . . .
- When analyzing any nonnumeric information amassed during research

Procedure

1. From each response, extract key ideas or themes. Record each theme separately, with a reference linking it to its source. If using a computer, spreadsheet cells are ideal for this. Or you can write each theme on a sticky note or slip of paper.

2. Physically rearrange the records into a small number of categories—four to eight is ideal—with similar themes. Name each category. Code them, such as 1, 2, 3, 4 . . . or A, B, C, D . . . , and label each theme with its category code.
3. Working with one category at a time, look for patterns in the themes and rearrange the records into subcategories with similar themes. Again code each subcategory, such as 1.1, 1.2 . . . or A1, A2 . . . and label each theme's record.
4. Continue this process until no new relationships suggest themselves. Different numbers of subcategories and levels can exist side by side.

Considerations

- Thematic analysis is related to processes used in social science research, such as ethnography.
- It takes advantage of the power of personal computers to easily sort text, although it can be done manually with pieces of paper.
- Other instructions for this analysis require defining categories first and then rearranging the records. The method given here uses the power of the affinity diagram process to find patterns and relationships intuitively without language imposing structures.
- This method differs from the affinity diagram in that one level is sorted and identified before the next level is addressed. The process alternates between intuitive and language-based thinking.

arrow diagram

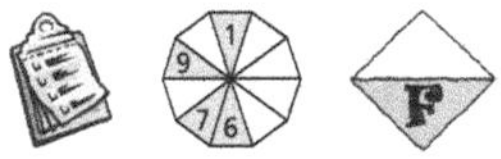

Also called: activity network diagram, network diagram, activity chart, node diagram, CPM (critical path method) chart
Variation: PERT (program evaluation and review technique) chart

Description

The arrow diagram shows the required order of tasks in a project or process, the best schedule for the entire project, and potential scheduling and resource problems and their solutions. The arrow diagram lets you calculate the *critical path* of the project. This is the flow of critical steps where delays will affect the timing of the entire project and where addition of resources can speed up the project.

When to Use

- When scheduling and monitoring tasks within a complex project or process with interrelated tasks and resources, and . . .
- When you know the steps of the project or process, their sequence, and how long each step takes, and . . .
- When project schedule is critical, with serious consequences for completing the project late or significant advantage to completing the project early

Procedure

Materials needed: sticky notes or cards, marking pens, large writing surface (newsprint or flipchart pages)

Drawing the Network

1. List all the necessary tasks in the project or process. One convenient method is to write each task on the top half of a card or sticky note. Across the middle of the card, draw a horizontal arrow pointing right.
2. Determine the correct sequence of the tasks. Do this by asking three questions for each task:
   - Which tasks must happen before this one can begin?
   - Which tasks can be done at the same time as this one?
   - Which tasks should happen immediately after this one?

   It can be useful to create a table with four columns—prior tasks, this task, simultaneous tasks, following tasks.
3. Diagram the network of tasks. If you are using notes or cards, arrange them in sequence on a large piece of paper. Time should flow from left to right and concurrent tasks should be vertically aligned. Leave space between the cards.
4. Between each two tasks, draw circles for events. An *event* marks the beginning or end of a task. Thus, events are nodes that separate tasks.
5. Look for three common problem situations and redraw them using dummies or extra events. A *dummy* is an arrow drawn with dotted lines used to separate tasks that would otherwise start and stop with the same events or to show logical sequence. Dummies are not real tasks.

   Problem situations:

   - Two simultaneous tasks start and end at the same events. *Solution:* Use a dummy and an extra event to separate them. In Figure 5.3, event 2 and the dummy between 2 and 3 have been added to separate tasks A and B.

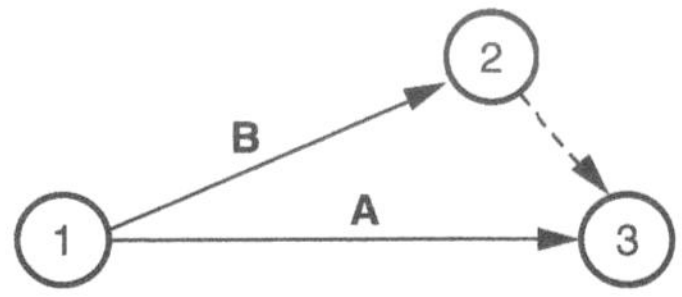

Figure 5.3 Dummy separating simultaneous tasks.

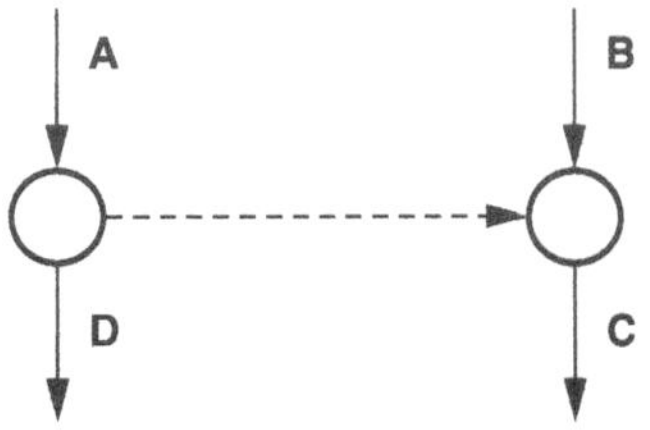

Figure 5.4 Dummy keeping sequence correct.

- Task C cannot start until both tasks A and B are complete; a fourth task, D, cannot start until A is complete, but need not wait for B. (See Figure 5.4.) *Solution:* Use a dummy between the end of task A and the beginning of task C.
- A second task can be started before part of a first task is done. *Solution:* Add an extra event where the second task can begin and use multiple arrows to break the first task into two subtasks. In Figure 5.5, event 2 was added, splitting task A.

6. When the network is correct, label all events in sequence with event numbers in the circles. It can be useful to label all tasks in sequence, using letters.

Scheduling: Critical Path Method (CPM)

7. Determine task times. The *task time* is the best estimate of the time that each task should require. Use one measuring unit (hours, days, or weeks) throughout, for consistency. Write the time on each task's arrow.
8. Determine the critical path. The *critical path* is the longest path from the beginning to the end of the project. Mark the critical path with a heavy line or color. Calculate the length of the critical path: the sum of all the task times on the path.

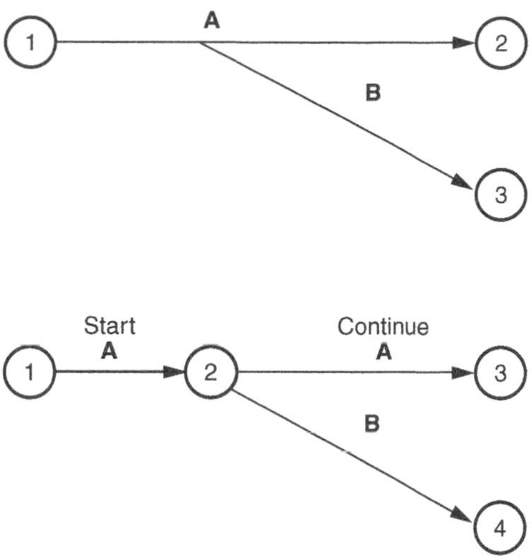

Figure 5.5 Using an extra event.

Figure 5.6 Arrow diagram time box.

9. Calculate the earliest times each task can start and finish, based on how long preceding tasks take. These are called earliest start (ES) and earliest finish (EF). Start with the first task, where ES = 0, and work forward. Draw a square divided into four quadrants, as in Figure 5.6. Write the ES in the top left box and the EF in the top right. For each task:

   Earliest start (ES) = the largest EF of the tasks leading into this one

   Earliest finish (EF) = ES + task time for this task

10. Calculate the latest times each task can start and finish without upsetting the project schedule, based on how long later tasks will take. These are called latest start (LS) and latest finish (LF). Start from the last task, where the latest finish is the project deadline, and work backwards. Write the LS in the lower left box and the LF in the lower right box.

Latest finish (LF) = the smallest LS of all tasks immediately following this one

Latest start (LS) = LF – task time for this task

11. Calculate slack times for each task and for the entire project.

Total slack is the time a job could be postponed without delaying the project schedule.

Total slack = *LS* – *ES* = *LF* – *EF*

Free slack is the time a task could be postponed without affecting the early start of any job following it.

Free slack = the earliest *ES* of all tasks immediately following this one – *EF*

Figure 5.7 shows a schematic way to remember which numbers to subtract.

Example

Tasks. Figure 5.8 shows an arrow diagram for a benchmarking project. There are 14 tasks, shown by the solid arrows. The number on the arrow is the task time in days, based on the experience and judgment of the group planning the project.

Events. There are 15 events, represented by the circled numbers. The events mark the beginning and ending times for the tasks. They will serve as milestones for monitoring progress throughout the project.

Dummies. Tasks E and F show the first kind of problem situation, where two tasks can occur simultaneously. Developing a detailed plan and conducting preliminary research can be done at the same time. Dummy 6–7 separates them. Task 10–11 is also a dummy to separate the simultaneous tasks J (develop questions) and K (identify current state).

The second kind of problem situation is illustrated around tasks H, I, J, K, and L. Task L (analyzing public data) cannot start until both tasks H (identifying key practices

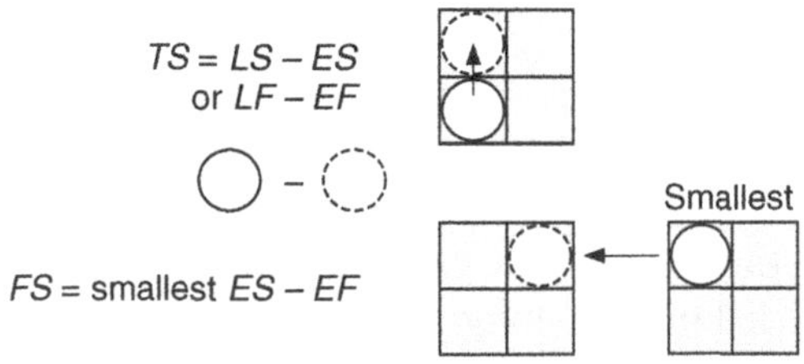

Figure 5.7 Remembering slack calculations.

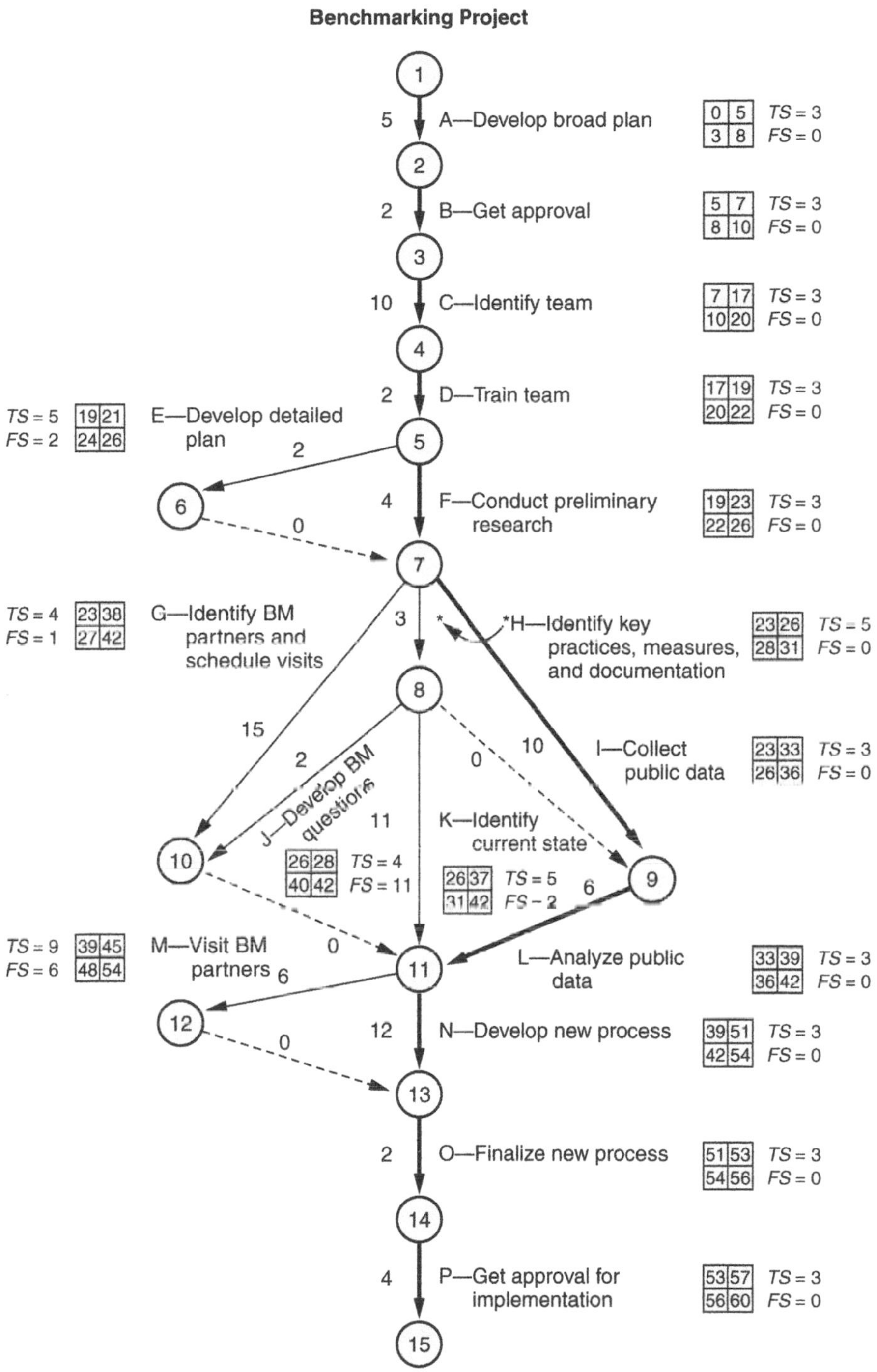

Figure 5.8 Arrow diagram example.

and measures) and I (collecting public data) are done. Task J (develop questions) and task K (identify current state) can start as soon as task H is complete. Task 8–9 is a dummy to separate the starting time of tasks J and K from that of task L.

The third kind of problem situation is illustrated around tasks M, N, and O. Originally, tasks N and O were one task—develop new process. Task M (visiting benchmark partners) can begin at the same time, but the visits need to be completed before final decisions can be made about the new process. To show this, subtask O (finalize new process) and extra event 13 were created. A dummy was also needed because tasks M and N are simultaneous. Now the network shows that the visits must be completed before the process can be finalized.

Critical Path. The longest path, marked in bold, is the critical path. Its tasks and their times are:

A	B	C	D	F	I	L	N	O	P
5	2	10	2	4	10	6	12	2	4 = 57

The team was surprised that the external visits, which members assumed would be a scheduling bottleneck, were not on the critical path—if they can schedule visits with a three week lead time, as task G assumes. The entire project will require 57 days if all tasks on the critical path proceed as scheduled.

Earliest and Latest Start and Finish Times. The earliest times are calculated forward from the start. The *ES* is the largest *EF* of the tasks just before the one being calculated. The *ES* for the first task is zero.

Task	ES	+	Task Time	=	EF
A	0		5		5
B	5		2		7
C	7		10		17
D	17		2		19
E	19		2		21
F	19		4		23
G	23		15		38

and so on.

The latest times are calculated backward from the end. *LF* is the smallest *LS* of the tasks just after the one being calculated. The project has a 12-week deadline, or 60 working days, so 60 is used as the *LF* for the last task.

Task	LF	–	Task Time	=	LS
P	60		4		56
O	56		2		54
N	54		12		42
M	54		6		48
L	42		6		36

and so on. All these times are shown beside each task, in the four-box grid.

Slack Times. The slack times for each task are shown beside each grid. Because 60 days have been allotted for the project and the critical path adds up to 57 days, there is slack time for all the tasks. For each of the tasks on the critical path, the total slack is three, the slack time for the entire project. The tasks not on the critical path have up to nine days total slack.

Surprisingly, there is free slack of six days for visiting the benchmark partners, so six days of scheduling problems with the partners can be tolerated without delaying later tasks. There is total slack of nine days for the visits, so if the project is on schedule up to that task, nine days' delay in making visits will not affect the project completion date.

Compare this diagram with the Gantt chart example (Figure 5.67, page 272). Both show the same project. Relationships and dependencies between tasks are clearer on the arrow diagram, but the Gantt chart makes it easier to visualize progress.

program evaluation and review technique (PERT) chart

This variation allows for uncertain time estimates for individual tasks. Draw the network as in steps 1 through 6 (pages 101–2) of the basic procedure. For scheduling, replace step 7 with:

7. Make three estimates of the time required for each task.
   - Minimum *(a)*—if everything went right
   - Most likely *(m)*—the normal time required; an average of the times if the task were done repeatedly
   - Maximum *(b)*—if everything went wrong

   Show all three times on the diagram, separated by dashes: 3–5–8. Next calculate the expected time (t_E) and the variance (σ^2) for each task.

$$t_E = \frac{a + 4m + b}{6} \qquad \sigma^2 = \left(\frac{b - a}{6}\right)^2$$

Go on to steps 8 through 11 (pages 102–104). Use expected time t_E instead of task time. Then add this step:

12. Determine the probability (P) that the project will be finished by its deadline, T_D. Calculate three numbers.

    T_E = total expected time = the sum of the expected times t_E for all the critical path tasks

    $\sigma^2_{T_E}$ = total variance = the sum of the variances for all the critical path tasks

$$Z = \frac{T_D - T_E}{\sqrt{\sigma^2_{T_E}}}$$

Look up the value for Z on the table of area under the normal curve (Table A.1). The number read from the center of the table, P, is the probability that the project will be finished by its deadline.

Example

To demonstrate these calculations, we will use a critical path with only four tasks. Table 5.1 shows the minimum, most likely, and maximum time estimates (a, m, and b), the expected time (t_E) and the variance (σ^2). If the deadline for the project is 25 days, then

Table 5.1 PERT example.

Task	$a-m-b$	t_E	σ^2
A	2 – 5 – 10	$\frac{2+20+10}{6}=5.3$	$\left(\frac{10-2}{6}\right)^2=1.8$
B	5 – 8 – 12	$\frac{5+32+12}{6}=8.2$	$\left(\frac{12-5}{6}\right)^2=2.3$
C	3 – 6 – 8	$\frac{3+24+8}{6}=5.8$	$\left(\frac{8-3}{6}\right)^2=0.7$
D	2 – 4 – 6	$\frac{2+16+6}{6}=4.0$	$\left(\frac{6-2}{6}\right)^2=0.4$
		$T_E = 23.3$	$\sigma^2_{T_E} = 5.2$
	$T_D = 25$	$Z=\frac{25-23.3}{\sqrt{5.2}}=0.75$ $P=0.77$	

$Z = 0.75$. Looking up 0.75 on Table A.1 shows that $P = 0.77$. The probability of completing the project in 25 days or less is 77 percent.

It may seem surprising that there is a 23 percent chance of not meeting the deadline, since the total expected time (T_E) for the project is 23.3 days, less than the deadline. Remember, however, that each task's time estimate had a worst case. If everything went wrong and all the worst-case estimates came true, the project would require 36 days.

Considerations

- If any task on the critical path is delayed, all tasks on the critical path will be pushed back. If there is no slack in the overall project, the project will be delayed.
- The timing of tasks with some free slack can be left to the judgment of those handling the task, as long as they do not delay the task more than the amount of free slack. *Example:* The team making benchmark visits can spread its schedule over 12 days instead of the planned six days without consulting with the rest of the project team.
- When tasks without free slack are delayed, the times and slack for all following jobs must be recalculated.
- Slack time can be negative. *Example:* If the time allotted for the benchmarking project was less than 57 days, or if the critical path tasks become delayed by more than three days, the slack time would become negative, and ways to make up that time would have to be found.
- To speed up a project schedule, find ways to increase resources or reduce scope only for those tasks on the critical path. Speeding up tasks not on the critical path will have no effect on the overall project time. It may be possible to move resources from noncritical tasks to critical ones. *Example:* To complete the benchmarking project in 10 weeks instead of 12, ways might be found to identify the team faster (task C), to get public data more quickly (task I), to get faster approvals (tasks B and P), or to speed up any of the other six tasks on the critical path. Collapsing the visit schedule (tasks G or M) will not complete the project sooner.
- However, when critical path tasks are shortened, the entire network must be recalculated. New tasks may now be on the critical path, and they can be examined for opportunities to shorten the schedule. *Example:* In the benchmark project, if a way were found to collect public data (task I) in five days instead of 10, scheduling visits (task G) would now lie on the critical path.
- Another way to shorten the project time is to rethink the sequence of tasks on the critical path. If some of them can be done simultaneously, total project time can be reduced.

- In the arrow diagram process, involve a team of people who have broad knowledge about the project or process.
- The easiest way to construct the diagram when first laying out the sequence is to find the path with the most tasks. Lay out that path first, then add other parallel paths.
- No loops are allowed in the network of tasks. A loop would be a sequence where task A is followed by task B, followed by task C, followed by task A.
- The length of an arrow is not related to the amount of time the task takes. Arrow lengths depend simply on the way you have chosen to depict the network of tasks.
- A common notation labels tasks with their starting and ending events. A task that starts at event 4 and ends at event 7 would be labeled task 4–7.
- For complex networks of tasks, computer software can make constructing and updating the chart quick. Software can also easily construct a Gantt chart and arrow diagram from the same data, ensuring that the two are always consistent with each other as the project progresses and updates are made.
- This chart fills some of the same functions as the Gantt chart. It analyzes the project's schedule more thoroughly, revealing the critical path and dependencies between tasks. However, the Gantt chart is easier to construct, can be understood at a glance, and can be used for monitoring progress.
- An alternative way of drawing the network is to represent tasks by circles or rectangles (nodes) and connect them with arrows showing sequence. That format is essentially a flowchart. Task time is written within the node, ES and LS are written to the left, separated by a colon (for example, 7:10 for task C), and EF and LF are similarly written to the right. For PERT analysis, the three time estimates are usually written on the arrow leaving the node. This alternative drawing method is easier to draw, but the one described in this book's procedure shows milestones more clearly.
- A variety of names have been used for this diagram. In general, they are all considered *activity network diagrams.* The drawing procedure detailed in this book is called *activities on the arc* or *activities on the arrow* and the resulting diagram is named *arrow diagram.* Arrow diagram and activity network diagram are names used in various descriptions of the seven MP tools. The alternative drawing method described above results in *activities on the node* and the resulting chart is named *node diagram.* Both methods of scheduling, PERT and CPM, can be used with either form of diagram. *Activity chart* is another name sometimes used. However, that name has been used for the Gantt chart as well, potentially leading to confusion. Since several names exist for each of these diagrams, it seems simpler to avoid using the name activity chart.

balanced scorecard

Description

The balanced scorecard is a set of measures that gives a quick overall view of the performance of an organization or business unit. Measures are grouped into four perspectives: customer, internal business, innovation and learning, and financial. Within each perspective, measures chosen by the organization reflect its business strategy. At every level of the organization, measures, targets, and actions are chosen that support the overall organization scorecard. Thus, the balanced scorecard allows everyone to plan for and monitor improvement on the issues most important to the organization's success.

When to Use

For senior managers:

- When developing the organization's mission and strategy, or . . .
- When communicating the mission and strategy throughout the organization, or . . .
- When planning programs and initiatives to achieve the organization's goals, or . . .
- Later, when routinely reviewing the organization's performance

For everyone in the organization:

- Only after a balanced scorecard has been created for the entire organization or business unit
- When planning what areas to address for improvement, or . . .
- When assessing how well current processes are meeting the needs of the customer or the organization, or . . .
- When setting improvement goals and planning how to achieve them, or . . .
- When planning what measures to use to monitor your improvement efforts, either during initial trial of a new solution or during standardization

Procedure

The entire process of developing a balanced scorecard is beyond the scope of this book. Here is a simplified outline to show the general concepts.

For senior management:

1. Reach a consensus on the vision and strategy of the organization.
2. Consider what each of these four perspectives means for the organization:
   - Customer
   - Internal business
   - Innovation and learning
   - Financial

   Decide whether the names should be changed to make the perspectives more meaningful for the organization, or even if a perspective should be added.
3. For each perspective, choose no more than five measures that would indicate progress toward achieving the organization's strategy.
4. Choose ambitious targets for each measure that, when achieved, will bring the organization closer to its vision of excellence.

At every level of the organization—group or individual—the process is very similar:

1. Review the balanced scorecards for the organization and for the unit to which you belong. Make sure you understand what will be measured and what those measures say about what is important to the organization.
2. Within each of the perspectives, choose no more than five measures at your level that indicate your contribution to the organizational measures.
3. Choose ambitious targets for your measures. This becomes your level's balanced scorecard.

Everyone in the organization puts the balanced scorecard into action with these steps:

4. Identify actions that will achieve the targets. For each action, identify which measures will indicate success of the actions.
5. Set intermediate milestones for improvement in the measures. Include when the milestone will be reached as well as the numeric target.
6. Review the scorecard measures regularly. Adjust actions, milestones, targets, or even measures or strategy when necessary.

Example

Medrad, Inc., recipient of the 2003 Malcolm Baldrige National Quality Award, uses a balanced scorecard approach to ensure that the whole organization focuses on its strategy and vision and to guide making decisions and prioritizing resources. They have renamed the perspectives using terms more meaningful to their organization and added an extra one to the standard four. Their perspectives are: exceed the financials (financial), grow the company (innovation and learning), improve quality and productivity (internal business), improve customer satisfaction (customer), and improve employee growth and satisfaction.

Figure 5.9 shows the organization's scorecard. Each of the five goals has a defined measurement method and a target. These long-term goals rarely change from year to year. Each year during the strategic planning process, senior managers confirm the scorecard goals and targets as well as 12 corporate objectives called the Top 12. Two of the objectives—achieve financials and achieve scorecard objectives—never change; the rest are one- to three-year programs critical to achieving the scorecard goals. Each month, the CEO communicates to all employees on the company's performance on the scorecard goals, and the finance department publishes monthly scorecard goal results on the intranet.

Departments, teams, and employees create their own objectives supporting the scorecard goals and the Top 12 objectives. This process is called waterfalling. Many departments have their own scorecards to monitor their waterfalled goals. When cross-functional projects are selected, one of the criteria is the impact on scorecard goals. Figure 5.10 shows how the objectives of the freight processing improvement team supported Medrad's corporate scorecard goals. Team members' individual annual objectives included this project, thus linking their performance assessment and evaluation to corporate priorities.

Corporate Goal	Target	Corporate Objective	Priority
Exceed financials	CMB (profit) growth > revenue growth		1
		C	2
Grow the company	Revenue growth > 15% per year	***O***	3
		N	4
		F	6
		I	7
		D	8
		E	12
Improve quality and productivity	Grow CMB (profit) per employee > 10% per year	***N***	5
		T	10
Improve customer satisfaction	Continuous improvement in Top Box ratings	***I***	9
		A	11
Improve employee growth and satisfaction	Continuous improvement in employee satisfaction above best-in-class Hay benchmark	***L***	See # 5

Figure 5.9 Balanced scorecard example.

	Exceed financials	Grow the company	Improve quality and productivity	Improve employee satisfaction
Reduce internal labor	◎		◎	◎
Make processes more robust		◎	◎	◎
Process improvement tools			◎	◎

Figure 5.10 Project objectives support balanced scorecard goals.

See the Medrad story in Chapter 4 (page 55) for more information about their improvement process. Also, see the St. Luke's Hospital story on page 69 for another example of a balanced scorecard.

Considerations

- The four perspectives were chosen to reflect the four most important aspects of organizational success:
  - Customer perspective: What do our customers think is important? How do they think we are doing?
  - Internal business perspective: In what business processes must we excel to satisfy our customers and shareholders?
  - Innovation and learning perspective: Do we know how to change and improve, to keep up with a changing world?
  - Financial perspective: Are the changes we make affecting the bottom line?
- Sometimes organizations using the scorecard change the names or add a perspective to reflect their unique circumstances. For example, 2003 Baldrige winner St. Luke's Hospital of Kansas City uses for its perspectives customer satisfaction, clinical and administrative quality, growth and development, financial, and people.
- Traditionally, organizations tracked only financial measures. The other three operational measures provide a more complete picture of the sustainability of any financial results.
- The balanced scorecard is most successful when the measures reflect factors that are truly critical to the success of the organization. Then the actions chosen

to affect these measures will be ones that really matter. Because time and people-power are limited in any organization, the scorecard should show only a small set of the most critical measures and actions.

- The scorecard is balanced because it prevents an organization from creating improvements in one area that hurt the organization in another. For example, focusing only on financial measures could lead an organization to undermine product quality or timely delivery, which would upset customers and ultimately lose sales.
- The balanced scorecard also helps a group think about which of several ways of accomplishing a goal would be best for the organization overall. What effects will each solution have on all four perspectives of the scorecard?
- Because "what you measure is what you get," the balanced scorecard is used more broadly as a management system. The thinking required to develop the measures makes it a powerful tool for focusing strategy. Senior managers clarify their vision and strategy, set a limited number of critical goals for the organization, and set priorities for actions to accomplish the goals. Organizations that use it as a management system get much greater results than those who adopt it only for performance measurement.
- Measures must be cascaded throughout the organization. At every level, a group must select measures, targets, and actions that will contribute to the changes desired by the organization.
- The balanced scorecard approach is a top-down way of setting goals that gets everyone pulling in the same direction. Goal-setting that works in the opposite direction—starting with frontline workers and moving toward upper management—creates an organization with mismatched, even contradictory, goals.
- Everyone must agree what the words of the scorecard mean. See *operational definitions.*
- Communication throughout the organization about the purpose, meaning, and data of the scorecard is critical.
- The assumptions and decisions that went into creating the scorecard should be reviewed frequently. Is the data available and accurate? Do the measures really capture all aspects of the vision the organization wants to achieve? Is the strategy leading us to our vision? Do changed circumstances call for changes in our measures, actions, or targets? The balanced scorecard is not cast in stone. It should be modified as the organization learns what measures are linked best to its strategy and what strategies lead to success.
- This is just an overview of a complex subject. See Resources for more information.

benchmarking

Description

Benchmarking is a structured process for comparing your organization's work practices to the best similar practices you can identify in other organizations and then incorporating the best ideas into your own processes.

When to Use

- When you want breakthrough improvements, and . . .
- When you need fresh ideas from outside your organization, and . . .
- After your own processes are well understood and under control

Procedure

Plan

1. Define a tightly focused subject of the benchmarking study. Choose an issue critical to the organization's success.
2. Form a cross-functional team. During the first step and this one, management's goals and support for the study must be firmly established.
3. Study your own process. Know how the work is done and measurements of the output.
4. Identify partner organizations that may have best practices.

Collect

5. Collect information directly from partner organizations. Collect both process descriptions and numeric data, using questionnaires, telephone interviews, and/or site visits.

Analyze

6. Compare the collected data, both numeric and descriptive.
7. Determine gaps between your performance measurements and those of your partners.
8. Determine the differences in practices that cause the gaps.

Adapt

9. Develop goals for your organization's process.
10. Develop action plans to achieve those goals.
11. Implement and monitor plans.

Example

The example for the Gantt chart, Figure 5.67 on page 272, describes a typical plan for a benchmarking study. The arrow diagram example, Figure 5.8 on page 105, discusses the same study.

Considerations

- Various benchmarking processes have been published that range from four to ten steps. The procedure here is intended to simplify a complex process. Before beginning to benchmark, study the references listed in the Resources.
- This process is a quality improvement model. The major difference between it and other models is that benchmarking gets ideas from outside the organization, whereas other improvement methods get ideas from within. Also, the benchmarking process is usually focused on a subject where dramatic changes, not incremental improvements, are needed.
- If you are using benchmarking as a tool within another improvement process, steps 1 through 3 and 9 through 11 are probably included in the other process.
- Before an organization can achieve the full benefits of benchmarking, its own processes must be clearly understood and under control. Before jumping into benchmarking, an organization should establish a quality improvement process that teaches it how to improve and how to work together across functions.
- Benchmarking studies require significant investments of manpower and time—up to a third of the team's time for five or more months—so management must champion the process all the way through. They also must be ready and willing to make changes based on what is learned.
- Too broad a scope dooms the project to failure. A subject that is not critical to the organization's success won't return enough benefits to make the study worthwhile.
- Inadequate resources can also doom a benchmarking study. This problem can be caused by taking on too large a subject or underestimating the effort involved. But it also can be caused by inadequate planning. The better you prepare, the more efficient site visits will be.

- Sometimes site visits are not necessary if information can be transferred in other ways. Using technology well also reduces costs.
- It's tempting to say, "Let's just go see what Platinum Star Corporation does. We don't need all that up-front work." Don't. The planning phase sets the stage for maximum learning from your partners.
- A phrase has been coined for benchmarking that skips the planning and adapting phases: industrial tourism. If you don't transform what you've learned into changes to your organization, you've wasted a lot of time, money, and effort.
- Sometimes changes are rejected because "that won't work here," "we already tried that," or "we're different"—the notorious "not invented here" syndrome. This problem can be avoided by establishing strong management support up front, including on the benchmarking team people who actually do the process, and communicating well to the organization the benchmarking process, the specific findings, and proposed changes.
- During the first project, the organization is learning the benchmarking process while doing it. The first time doing something usually is difficult. Hang in there, then go benchmark something else.
- See *benchmarking* in Chapter 2 for more information about this mega-tool.

benefits and barriers exercise

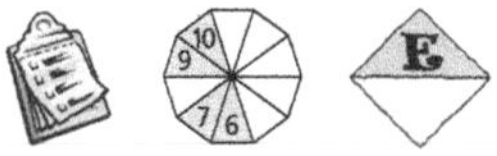

Description

The benefits and barriers exercise helps individuals see both personal and organizational benefits of a proposed change. It also identifies perceived obstacles to accomplishing the change so they can be addressed in design. Most importantly, it generates individual and group buy-in to the change.

When to Use

- When trying to decide whether to proceed with a change, or . . .
- When trying to generate buy-in and support for a change, or . . .
- After a concept has been developed, but before detailed design of a plan, to identify obstacles that need to be considered in the design

- Especially for major changes, such as launching a quality effort or implementing a recognition program

Procedure

Materials needed: flipchart paper, marking pen, and masking tape for each group of five to seven people; paper and pen or pencil for each individual.

1. Explain the purpose of the exercise and how it will be done. Emphasize that everyone's active involvement is important. Divide the participants into groups of five to seven each and assign breakout rooms and leaders, who have been coached in advance on their role.
2. Do benefits first. Show the group this statement, written on flipchart paper and posted where all can see:

   Assume that it is now two years in the future and we have been successful in implementing [name of concept or change]. What benefits do you see for yourself as an individual, for your work group, and for the company as a whole?
3. Within each small group, brainstorm benefits using the nominal group technique (NGT) method. (See *nominal group technique* for details on this structured brainstorming method.) Collect ideas by category in either of two ways:
   - Conduct three separate NGT sessions, first collecting benefits for the individual, then for the work group, then for the company.
   - Or, conduct one NGT, and mark each idea when collected with a code (such as *I, W, C*) indicating who benefits.
4. Within each small group, use multivoting to choose the top three benefits in each of the three categories. (See *multivoting* for more detail on this prioritization method.) Let each participant vote for his or her top five in each category. Each group selects a spokesperson. Allow 1 to 1½ hours for steps 3 and 4.
5. Reassemble the entire group. Each spokesperson reports on:
   - Briefly, how the exercise went, such as the extent of participation and how much agreement there was
   - The top three benefits in each category (a total of nine benefits)
6. Follow with barriers. Show the group this statement:

   What are the barriers that we as an organization are going to have to overcome in order to make [name of concept or change] a success and thereby achieve the benefits?

7. Within each small group, brainstorm barriers using the NGT method. Do not separate individual, work group, and company barriers.
8. Within each small group, identify the top three barriers, again using multivoting. A different spokesperson should be selected. This breakout session should last 45 minutes to an hour.
9. Reassemble the entire group. Each spokesperson reports the group's top three barriers.
10. Someone with respected leadership involved in the change, such as the project champion, should acknowledge the benefits and barriers and state that they will be considered in the design of the change.
11. After the meeting, transcribe the flipchart pages from all the groups and send them to the participants with the meeting notes.

Variation for a small group

When the group is too small to form several breakout groups, spokespersons are not needed. At step 5, the facilitator should ask the group to discuss briefly how the exercise went. Step 9, of course, is omitted.

Example

A work team looking for ways to improve safety performance decided to adopt a new approach, safety circles. They had given everyone in the unit information about safety circles and details about their plan. Because they knew that doubts and resistance were likely, they also planned a benefits and barriers exercise. The unit worked different shifts around the clock, so the team conducted four separate sessions. The first shift was not able to see the other groups' ideas until later, but the remaining shifts had the opportunity to compare their ideas immediately with ideas from the groups that had already done the exercise. After the exercise, some members of the unit were enthusiastic and others had a wait-and-see attitude, but all were willing to try the approach. Using the barriers identified during the exercise, the team modified and launched its plan.

Considerations

- When the proposed change is major and involves a lot of emotion and potential resistance, it is valuable to do this exercise with enough people to form several breakout groups. Fifteen to 28 individuals are ideal. Hearing other groups generate similar ideas validates each individual's ideas and feelings and reinforces each individual's commitment.

- An outside facilitator can be valuable if the change is significant and participants are high-ranking.
- Individuals must be familiar with the proposed change before the exercise can be done.
- During report-backs, the facilitator should listen for and comment on similarities between the small groups' ideas.
- All perceived barriers must be taken seriously and addressed in the design of the plan.

Conducting the NGT

- The small-group leaders should be experienced with using NGT and multivoting.
- Avoid discussing ideas until after the lists are completed. Also avoid responding to barriers until the end.
- The brainstorming rule of "no criticism, no censoring" applies.
- Whoever is writing the ideas should write what the speaker says as closely as possible. If the idea must be shortened, ask the speaker if the summary captures the idea accurately.
- Write large enough that everyone can read the ideas. Generally, that means letters about two inches high.
- As a page is filled, tear it off and tape it to the wall where it can be seen by the entire group.

box plot

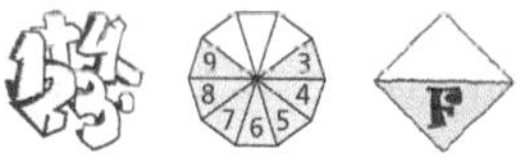

Also called: box-and-whisker plot

Description

The box plot is a graph that summarizes the most important statistical characteristics of a frequency distribution for easy understanding and comparison. Information about where the data falls and how far it spreads can be seen on the plot. The box plot is a powerful tool because it is simple to construct yet yields a lot of information.

When to Use

- When analyzing or communicating the most important characteristics of a batch of data, rather than the detail, and . . .
- When comparing two or more sets of data, or . . .
- When there is not enough data for a histogram, or . . .
- When summarizing the data shown on another graph, such as a control chart or run chart

Procedure

1. List all the data values in order from smallest to largest. We will refer to the total number of values, the count, as *n*. We will refer to the numbers in order like this: X_1 is the smallest number; X_2 is the next smallest number; up to X_n, which is the largest number.
2. *Medians.* Cut the data in half. Find the median—the point where half the values are larger and half are smaller.
   - If the total number of values (n) is odd: the median is the middle one. Count $(n + 1)/2$ from either end.

     $$\text{median} = X_{(n+1)/2}$$

   - If the total number of values (n) is even: the median is the average of the two middle ones. Count $n/2$ and $n/2 + 1$ from either end. Average those two numbers:

     $$\text{median} = (X_{n/2} + X_{n/2+1})/2$$

3. *Hinges.* Cut the data in quarters. Find the hinges—the medians of each half.
   - If n is even, the median is the average of $X_{n/2}$ and $X_{n/2+1}$. Take the values from 1 to $X_{n/2}$ and find their median just as in step 2. This is the lower hinge.
   - If the total number of values is odd, the median is $X_{(n+1)/2}$. Take the values from 1 to the median and find their median, just as in step 2. This is the lower hinge.

   Do the same with the values at the upper end to find the upper hinge.
4. *H-spread.* Calculate the distance between the hinges, or H-spread:

   $$\text{H-spread} = \text{upper hinge} - \text{lower hinge}$$

5. *Inner fences.* These are values separating data that are probably a predictable part of the distribution from data that are outside the distribution. Inner fences are located beyond each hinge at 1½ times the H-spread, a distance called a *step*.

upper inner fence = upper hinge + 1.5 × H-spread

lower inner fence = lower hinge – 1.5 × H-spread

6. *Outer fences.* Data beyond these values are far outside the distribution and deserving of special attention. Outer fences are located one step beyond the inner fences.

   upper outer fence = upper inner fence + 1.5 × H-spread

   lower outer fence = lower inner fence – 1.5 × H-spread

7. To draw the box plot, first draw one horizontal axis. Scale it appropriately for the range of data.
   - Draw a box with ends at the hinge values.
   - Draw a line across the middle of the box at the median value.
   - Draw a line at each inner fence value.
   - Draw a dashed crossbar at the *adjacent value*, the first value inside the inner fences.
   - Draw *whiskers*, dashed lines from the ends of the box to the adjacent values.
   - Draw small circles representing any *outside* data points: beyond the inner fences but inside the outer fences.
   - Draw double circles to represent *far out* data points: beyond the outer fences.
8. If you are comparing several data sets, repeat the procedure for each set of data.
9. Analyze the plot. Look for:
   - Location of the median
   - Spread of the data: how far the hinges and fences are from the median
   - Symmetry of the distribution
   - Existence of outside points

Example

Suppose two bowling teams, the Avengers and the Bulldogs, have the scores shown in Figure 5.11. Which team is better? We will draw a box plot of each team's scores and compare the two plots.

1. The scores are already in order from smallest to largest. There are 14 scores for each team, so $n = 14$.

Bowling Scores

The Avengers

126	134	137	142	145	148	149	150	155	157	160	165	170	198

The Bulldogs

103	139	147	152	153	154	155	159	161	163	163	165	176	183
			↑ Hinge			↑ Median				↑ Hinge			

Figure 5.11 Data for box plot example.

2. *Median.* There is an even number of scores, so the median is the average of the two middle ones. We must count $n/2$ and $n/2 + 1$ from one end.

$$n/2 = 14/2 = 7 \text{ and } n/2 + 1 = 8$$

Count to the seventh and eighth scores in each group and average them.

$$\text{Median A} = \frac{149+150}{2} = 149.5$$

$$\text{Median B} = \frac{155+159}{2} = 157$$

3. *Hinges.* We must find two medians, first of values 1 through 7 and then of values 8 through 14. There are seven values in each half, an odd number, so we count $(7 + 1)/2 = 4$ from either end.

lower hinge A = 142 upper hinge A = 160

lower hinge B = 152 upper hinge B = 163

4. *H-Spread.* The distance between hinges is

$$\text{H-spread} = \text{upper hinge} - \text{lower hinge}$$

$$\text{H-spread A} = 160 - 142 = 18$$

$$\text{H-spread B} = 163 - 152 = 11$$

5. *Inner fences.*

$$\text{upper inner fence} = \text{upper hinge} + 1.5 \times \text{H-spread}$$

$$\text{upper inner fence A} = 160 + 1.5 \times 18 = 160 + 27 = 187$$

$$\text{upper inner fence B} = 163 + 1.5 \times 11 = 163 + 16.5 = 179.5$$

$$\text{lower inner fence} = \text{lower hinge} - 1.5 \times \text{H-spread}$$

$$\text{lower inner fence A} = 142 - 27 = 115$$

$$\text{lower inner fence B} = 152 - 16.5 = 135.5$$

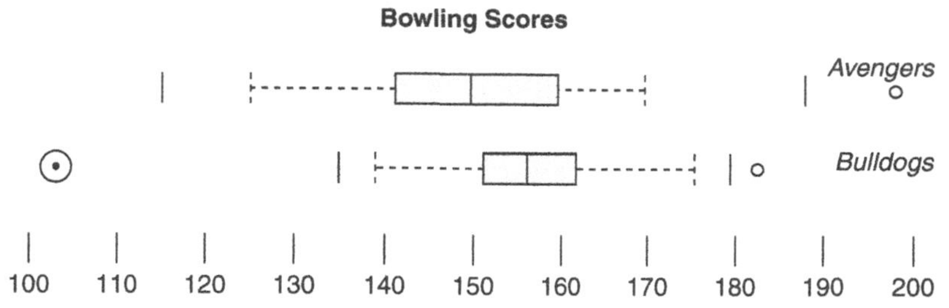

Figure 5.12 Box plot example.

6. *Outer fences.*

upper outer fence	= upper inner fence + 1.5 × H-spread				
upper outer fence A =	187	+	27	= 214
upper outer fence B =	179.5	+	16.5	= 196
lower outer fence	= lower inner fence – 1.5 × H-spread				
lower outer fence A =	115	–	27	= 88
lower outer fence B =	135.5		16.5	= 119

Figure 5.12 is the box plot of the two teams' scores. While the Avengers have a star and the Bulldogs have a poor player, overall the Bulldogs tend to score higher than the Avengers. The Bulldogs' smaller spread also indicates they score more consistently.

Variations

The box plot was created by John W. Tukey. Many variations have been proposed for calculating, drawing, and using box plots. Whenever you use a variation on the basic box plot, draw solid lines beyond the hinges to indicate that you are not conforming to Tukey's rules. Some variations are:

- *Simple box plot.* Instead of calculating and drawing fences and outliers, draw lines from the ends of the box (hinge values) to the highest and lowest data values.
- *Modified box plot.* Calculate the arithmetic average of all the data values and show it with a dot on the box plot. The closer the average is to the median, the more symmetrical the distribution.
- *Modified-width box plot.* When using two or more box plots to compare several data sets, the widths of the boxes can be drawn proportional to the sample size of the data sets.

- Parentheses can be drawn on the plot to represent 95% confidence limits.
- *Ghost box plot* or *box-plot control chart.* A box plot can be drawn with dotted lines directly on a control chart or other graph of individual data points to show a summary of the data. This variation is especially useful if several plots are drawn showing sequential subgroups of the data. For example, draw one ghost box plot in the middle of a set of 15 data points prior to a process change and another in the middle of the next set of 15 data points after the change.

brainstorming

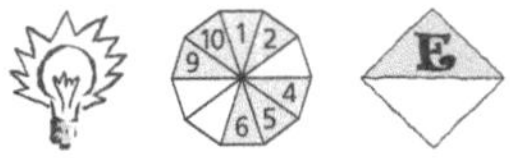

Variations: round robin brainstorming, wildest idea brainstorming, double reversal, starbursting, charette procedure
See also: nominal group technique, brainwriting, contingency diagram, and fishbone diagram

Description

Brainstorming is a method for generating a large number of creative ideas in a short period of time.

When to Use

- When a broad range of options is desired, and . . .
- When creative, original ideas are desired, and . . .
- When participation of the entire group is desired

Procedure

Materials needed: flipchart, marking pens, tape, and blank wall space.

1. Review the rules of brainstorming with the entire group:
   - No criticism, no evaluation, no discussion of ideas.
   - There are no stupid ideas. The wilder the better.
   - All ideas are recorded.
   - Piggybacking is encouraged: combining, modifying, expanding others' ideas.

2. Review the topic or problem to be discussed. Often it is best phrased as a *why, how,* or *what* question. Make sure everyone understands the subject of the brainstorm.
3. Allow a minute or two of silence for everyone to think about the question.
4. Invite people to call out their ideas. Record all ideas, in words as close as possible to those used by the contributor. No discussion or evaluation of any kind is permitted.
5. Continue to generate and record ideas until several minutes' silence produces no more.

Example

Read the Medrad, St. Luke's Hospital, and ZZ-400 improvement stories in Chapter 4 for examples of brainstorming used within the quality improvement process.

Variations

There are many versions of brainstorming. The basic one above is sometimes called free-form, freewheeling, or unstructured brainstorming. In addition to the variations described below, see *brainwriting* and *nominal group technique.*

round-robin brainstorming

When to Use

- When dominant group members stifle other members' ideas. However, in a group without that problem, this method can stifle creativity or cause impatience.

Procedure

At step 4, go around the group and have each person in turn say one idea. If they have no ideas on that turn, they may pass. Stop the brainstorming when everyone passes.

wildest idea brainstorming

When to Use

- To rekindle ideas when people are running out, or . . .
- To spark creativity when the ideas previously generated are unimaginative

Procedure

In step 4, allow only outrageous and unrealistic ideas. After step 5, ask the group to look at the ideas and see if they can be modified into realistic ones.

double reversal

When to Use

- When the purpose of your brainstorming is to develop problem solutions, and . . .
- When people are running out of ideas, to rekindle their imaginations, or . . .
- When the ideas previously generated are unimaginative, to spark creativity

Procedure

1. Review the rules of brainstorming with the entire group.
2. Review the topic or problem to be discussed. Make sure everyone understands the subject of the brainstorm.
3. Then, reverse the topic statement so that you're thinking about its opposite: how to make the problem worse or how to cause the opposite of the desired state.
4. Allow a minute or two of silence for everyone to think about the question.
5. Invite people to call out their ideas for making the problem worse. Record all ideas on a flipchart, in words as close as possible to those used by the contributor. No discussion or evaluation of any kind is permitted.
6. Continue to generate and record ideas until several minutes' silence produces no more.
7. Look at each idea and reverse it. Does the reversed statement lead to any new ideas? On a separate flipchart page, capture the reversal and any new ideas.

Example

A group of teachers was brainstorming solutions to the problem, "How to encourage parents to participate more in their children's education." They reversed the statement: "How to discourage parents from participating in their children's education." Here are some of their ideas, how they were double-reversed, and new ideas those led to.

Idea	Reversal/New Ideas
• Keep school doors locked	• Keep school doors unlocked • Open school for evening and weekend activities • Open school library to public
• Make teachers and principal seem remote and forbidding	• Make teachers and principal seem real and human • Share staff's personal background with parents
• Never tell parents what children are studying	• Always tell parents what children are studying • Send parents advance information about major projects or units of study
• Never communicate with parents	• Always communicate with parents • Collect parents' e-mail addresses • Provide teachers' e-mail addresses

starbursting

When to Use

- To identify potential problems with previously brainstormed ideas, or . . .
- To identify issues that must be considered before implementing an idea, or . . .
- To identify concerns about various ideas when trying to narrow a list of possibilities

Procedure

Follow the basic procedure above, with these changes:

2. Either display a list of previously brainstormed ideas or identify one idea that will be the focus.

4. Have people call out questions they think need to be answered. All questions are recorded. If a previously brainstormed list of ideas is the focus of starbursting, it helps to number ideas on the first list and write the appropriate number next to each question.

charette procedure

When to Use

- When a large group must develop ideas about several facets of a problem or project

Procedure

1. Before the meeting, determine how many people will be present and plan for breaking into small groups of five to eight people each. Appoint as many facilitators as there will be small groups.
2. Determine in advance, perhaps with a small planning group, what issues need to be brainstormed. Assign an issue to each facilitator. If there are more small groups than issues, some issues will be assigned to more than one facilitator. Do not assign more than one issue to a facilitator.
3. In the meeting, divide the participants into the small groups. Direct each group to sit together in a circle, with a facilitator and a flipchart. Announce a time limit for this activity.
4. Each group brainstorms ideas related to the facilitator's topic, following the basic procedure above. The facilitator records all ideas on a flipchart.
5. When the allotted time is up, each facilitator rotates to another group, taking the flipchart with him or her.
6. Each facilitator reviews with the new group the topic and the ideas already generated. The group brainstorms on that topic, adding ideas to the flipchart.
7. Repeat steps 5 and 6 until all groups have discussed every issue. If some groups discuss an issue more than once, that's okay. The flipchart will contain additional ideas on which they can piggyback.
8. During the last rotation, the group should identify the most significant ideas.
9. Everyone comes back together in a large group. Each facilitator displays his or her flipchart and summarizes the ideas generated.

Example

The two dozen faculty members of a school met to plan a program to encourage parent participation. Four issues were identified for discussion:

What after-school activities can the faculty promote to get parents into the school?

How can the faculty seem more real and human to parents?

How can parents be involved in student projects and curriculum units?

What are the best ways to communicate with parents?

Four teachers had been designated facilitators. The rest of the faculty divided into four groups of five teachers each. Each group discussed each topic for 15 minutes. After an hour, the group reconvened to review all ideas and plan the next steps.

Considerations

- Judgment and creativity are two functions that cannot occur simultaneously. That's the reason for the rules about no criticism and no evaluation.
- Laughter and groans are criticism. When there is criticism, people begin to evaluate their ideas before stating them. Fewer ideas are generated and creative ideas are lost.
- Evaluation includes positive comments such as "Great idea!" That implies that another idea that did not receive praise was mediocre.
- The more the better. Studies have shown that there is a direct relationship between the total number of ideas and the number of good, creative ideas.
- The crazier the better. Be unconventional in your thinking. Don't hold back any ideas. Crazy ideas are creative. They often come from a different perspective. Crazy ideas often lead to wonderful, unique solutions, through modification or by sparking someone else's imagination.
- Hitchhike. Piggyback. Build on someone else's idea.
- Unless you're using the round-robin variation, encourage people to rapidly speak whatever ideas pop into their heads. The recording of ideas must not slow down the idea-generation process. If ideas are coming too fast to keep up with, have several people record ideas.
- When brainstorming with a large group, someone other than the facilitator should be the recorder. The facilitator should act as a buffer between the group and the recorder(s), keeping the flow of ideas going and ensuring that no ideas get lost before being recorded.

- The recorder should try not to rephrase ideas. If an idea is not clear, ask for a rephrasing that everyone can understand. If the idea is too long to record, work with the person who suggested the idea to come up with a concise rephrasing. The person suggesting the idea must always approve what is recorded.
- Keep all ideas visible. When ideas overflow to additional flipchart pages, post previous pages around the room so all ideas are still visible to everyone.
- Brainstorming is not the only way to generate ideas. Many more creativity tools have been developed than can be squeezed into this book. For example, powerful tools called *lateral thinking* have been developed by Dr. Edward de Bono based on how the brain organizes information. These proven techniques do not depend on spontaneous inspiration but use specific methods to change concepts and perceptions and generate ideas. These tools have been used for purposes as varied as new product development and transforming the 1984 Olympic games. The Resources lists sources of creativity tools.
- See *affinity diagram, list reduction, multivoting,* and *starbursting* (above) for ways to work with ideas generated by brainstorming.

brainwriting

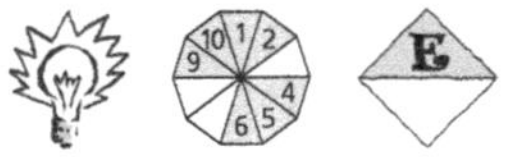

Variations: 6–3–5 method, Crawford slip method, pin cards technique, gallery method

Description

Brainwriting is a nonverbal form of brainstorming. Team members write down their ideas individually. Ideas are shared through an exchange of papers, then additional ideas are written down.

When to Use

Just as in brainstorming, brainwriting is used . . .

- When a broad range of options is desired, and . . .
- When creative, original ideas are desired, and . . .
- When participation of the entire group is desired

But try brainwriting instead of brainstorming if any of the following are true:

- When a topic is too controversial or emotionally charged for a verbal brainstorming session, or . . .

- When participants might feel safer contributing ideas anonymously, or . . .
- To encourage equal participation, when verbal brainstorming sessions are typically dominated by a few members, or . . .
- When some group members think better in silence, or . . .
- When ideas are likely to be complex and require detailed explanation

Procedure

Materials needed: a sheet of paper and pen or pencil for each individual.

1. Team members sit around a table. The facilitator reviews the topic or problem to be discussed. Often it is best phrased as a *why, how,* or *what* question. Make sure everyone understands the topic.
2. Each team member writes up to four ideas on a sheet of paper. He or she places the paper in the center of the table and selects another sheet.
3. Up to four new ideas are added to the list already on the sheet. That sheet goes back in the center and another sheet is chosen.
4. Continue this way for a predetermined time (usually 15 to 30 minutes) or until no one is generating more ideas. The sheets are collected for consolidation and discussion.

6–3–5 method

1. Six people sit around a table. (That is the 6.) The facilitator reviews the topic or problem to be discussed. Often it is best phrased as a *why, how,* or *what* question. Make sure everyone understands the topic.
2. Each person writes down three ideas. (That is the 3.)
3. After five minutes, sheets are passed to the left. (That is the 5.)
4. Continue the process in five-minute blocks until all participants have their own sheets back.
5. Collect sheets for consolidation and discussion.

Crawford slip method

When to Use

- When preserving anonymity is so important that papers cannot be shared because handwriting might be recognized, and . . .

- When the need for anonymity outweighs the value of sharing ideas to foster creativity

Procedure

1. Review the topic or problem to be discussed. Often it is best phrased as a *why, how,* or *what* question. Make sure everyone understands the topic.
2. Every participant individually brainstorms ideas on his or her own sheet of paper until the allotted time is up.
3. Papers are collected. Later, the facilitator privately consolidates the ideas into one list.
4. The consolidated list of ideas is presented to the group for discussion and selection.

pin cards technique

Materials needed: sticky notes, index cards, or slips of paper; marking pen for each individual; large work surface (wall, table, or floor).

Instead of using sheets of paper in the basic or 6–3–5 methods, use index cards, slips of paper, or sticky notes. Write one idea per card, slip, or note. Place completed cards on a large working surface for review and discussion.

As an additional variation, each participant is given paper of a unique color. His or her new ideas are always written on that color, to allow easy identification of the source of ideas.

gallery method

When to Use

- When a group has been sitting a long time and needs activity, or . . .
- When a large group is involved, or . . .
- When ideas are needed about several clearly defined aspects of a topic

Procedure

Materials needed: flipchart paper, marking pen for each individual

1. Post blank flipchart pages around the room. Be generous with paper. For a small group, post as many pages as there are participants.
2. (Optional) If the topic has several issues or questions that must be considered, label the pages with those issues or questions.

3. Review with the group the topic or problem to be discussed. If pages have been labeled in step 2, review them. Make sure everyone understands the topic.
4. Every participant goes to a flipchart page and writes up to four ideas or until a predetermined time is reached (often five minutes). Then he or she moves to another position.
5. The process continues until everyone has reviewed all the pages, and no more ideas are being generated. All flipchart pages are collected for consolidation and discussion.

Considerations

- Use a nonverbal tool when the team needs quiet time or when quieter members might not contribute fully.
- Passing sheets allows team members to build on each other's ideas without crushing them in the process.
- The 6–3–5 method works well for developing one or a few ideas. The emphasis can be placed on adding to the existing ideas on the page, rather than creating a list of scattered ideas. By the way, the method still works if you have five or eight people!
- Writing ideas individually on index cards, paper slips, or sticky notes, as in the pin cards technique, makes it easy to follow up with an organizing tool like the affinity diagram.
- See *brainstorming* for more information about basic principles of idea generation.
- See *nominal group technique* for a tool that combines brainstorming and brainwriting.

cause-and-effect matrix

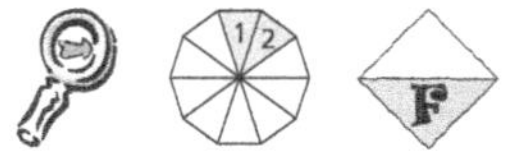

Also called: C&E matrix

Description

A cause-and-effect matrix links steps of a process with customer requirements. Rating requirements for importance and rating the relevance of each process step to each requirement helps you understand which aspects of the process are most significant in satisfying the customer.

When to Use

- After identifying critical-to-quality (CTQ) characteristics through QFD or a CTQ tree and . . .
- When determining which aspects of a process have the greatest impact on customer requirements, and . . .
- When determining where to focus improvement efforts

Procedure

1. Obtain or develop a list of key process output variables, usually customer requirements. These are often called critical-to-quality (CTQ) characteristics. A house of quality or a critical-to-quality tree can be used. These should be at a level of detail appropriate for the analysis you need. Write the output variables as column headings of an L-shaped matrix.
2. Assign a relative weight to each output variable, based on its importance to the customer. Write it with the column heading.
3. Obtain or develop a list of key process input variables. These may be process steps from a macro flowchart. They should be at a level of detail comparable to that of the output variables and appropriate for the analysis. Write them as row labels of the matrix.
4. Evaluate each input variable against each output variable, looking for whether the input influences the output. Ask, "If we changed this variable, would we see a resulting change in the output?" Use a rating scale where a low value means little or no change, a medium value means some change, and a high value means great change. Common rating scales are 1, 2, 3 or 1, 3, 5 or 1, 4, 9. Write the rating in the intersection cell.
5. Multiply the rating by the importance weight, and add the resulting products across each row. The rows with the highest totals are the input variables that have the greatest influence on important outputs.
6. As a check, add the products from step 5 down each column. The column totals should reflect the output variables' relative importance. If they do not, you may have errors in the ratings, missing input variables, output variables that cannot be affected by the process, or process linkages that have not been recognized.

Example

Ben-Hur's Pizza wishes to add home delivery to their services in order to expand their business. They have surveyed current and potential customers to determine what would make them order Ben-Hur pizza instead of a competitors' or instead of a different kind of food.

Summarized VOC data told them that when customers order-in pizza, they want "hot pizza, now, with my choice of toppings and crusts, at a reasonable cost." They also determined the relative importance of each requirement. A high-level process map showed five main steps of their process: developing the menu and recipes, buying and storing supplies and ingredients, making and baking the pizza, packaging, and delivery. This information was placed in a cause-and-effect matrix (Figure 5.13).

Each cell of the matrix was considered, asking "If we changed this process step, could it affect this customer requirement?" For example, menu and recipes have no effect on whether the pizza arrives hot. However, a more complicated recipe or a pizza with lots of toppings could take longer to prepare than a simple one, so they rated a moderate influence on "now." Menu and recipes have a great deal of influence on the customer's choices and on how much they have to charge for the pizza to cover their costs. The other large influence on "reasonable price" is their buying and storing process, because high-priced suppliers will drive up their costs, as will ordering more ingredients than they can use before they go bad or storing them so they do not keep.

The rest of the ratings were filled in, products written below the diagonal lines, and totals calculated. Delivery is the process step with the greatest influence on customer satisfaction, followed by menu and recipe development. Notice that the column totals wouldn't suggest that "choice" ties for first in importance, or that "price" is least important. Studying the matrix reveals why: Only one process step strongly affects "choice" and "now," while two strongly affect "hot" and "price." Ben-Hur Pizza should keep this in mind when they plan where to focus their efforts.

CTQs	Hot pizza	Now	Choice of toppings and crusts	Reasonable price	**Totals**
Process steps	8	6	8	5	
Menu and recipes	1 / 8	4 / 24	9 / 72	9 / 45	113
Buying and storing	1 / 8	1 / 6	4 / 32	9 / 45	91
Making and baking	1 / 8	4 / 24	1 / 8	1 / 5	45
Packaging	9 / 72	1 / 6	1 / 8	1 / 5	91
Delivery	9 / 72	9 / 54	1 / 8	4 / 20	154
	168	114	128	120	

Figure 5.13 Cause-and-effect matrix example.

Considerations

- This tool can be used at a high level, using process steps from a macro flowchart or SIPOC diagram and customer requirements in their own words. As process analysis continues, it can be used again at lower levels, with steps from a detailed process map and CTQs developed with a CTQ tree.
- Customers can be internal or external. Be sure to obtain true customer input instead of guessing about requirements and importance.
- This is called a "cause-and-effect" matrix because you are searching for inputs in the rows that cause the results, or effects, represented by the output variables or CTQs, in the columns. In Six Sigma shorthand, the causes are called X's and the effects are called Y's. You want to be able to make changes to X's in order to make the Y's match customer needs. In shorthand again, you are looking for a relationship Y = f(X) where the CTQ values (Y) are a function of how you set process variables (X).
- When you look for potential changes in step 4, either positive or negative changes get the same rating. Do not use negative scores for negative changes. Magnitude of change is all you are looking for.
- If members of the team disagree on the ratings, do not average the results or take a vote. Instead, discuss the ratings to learn what each person is thinking about. One person may be aware of an influence that the others are overlooking.
- This tool does not take into account the customer's view of current performance on this or similar products or services. This is important information to consider before deciding what areas of the process need improvement. Importance–performance analysis is one tool that can be used for this purpose.
- Sometimes ratings in this matrix are adjusted to include current performance. If you choose to do this, the interpretation of results must change. Results no longer show pure cause and effect. The rows with the highest totals are the process steps that have the greatest influence on important customer requirements relative to how well they are currently being satisfied. When performance changes, such as after improvements are made, the matrix will change. As more factors are rolled into one assessment, the interpretation becomes more difficult.
- A full-blown quality function deployment includes this type of analysis when parts or function characteristics are translated into process designs.
- While it's not necessary to write down the rating × importance products, it makes the calculations easier and errors less likely.
- The results of this tool are often used to develop a control plan or as input to FMEA, a multi-vari chart, correlation or regression analysis, or design of experiments.
- This tool is a variation of the decision matrix but used for a different purpose. See *decision matrix* for more information.

checklist

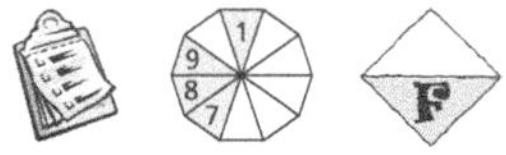

Also called: work procedure, confirmation check sheet

Description

A checklist is a predetermined list of items, actions, or questions that are checked off as they are considered or completed. A type of check sheet, this is a generic tool.

When to Use

- When a process or procedure with many steps must be performed over and over, or . . .
- When a process or procedure is done infrequently and might be forgotten, or . . .
- When you are beginning an activity that will have multiple steps or lots of detail, especially if it is something you've never done before, or . . .
- When you are beginning a project that will not be completed for a long time, or . . .
- When the same set of questions or issues can be applied repeatedly to different cases, or . . .
- When actions or steps of a process must be done in a certain order, or . . .
- When it is important that no item on the list is omitted

Specific situations that often use checklists:

- Work procedures
- Safety checks
- Audits
- Project reviews
- Event planning
- Meeting planning

Procedure

If a checklist does not already exist:

1. Decide what the purpose and scope of the checklist should be.
2. Research and/or brainstorm items that should be on the checklist.
3. Write the list of items. Be as clear and concise as possible.
4. Decide whether the order of the items is important. If so, determine the proper order. Also determine what must be done if a step is missed.
5. If the checklist will be used repeatedly, prepare a form listing the items and providing a place for checking off or dating completed items. If the order of steps is important, be sure to include that instruction on the form and directions for what to do if a step is skipped.
6. If the checklist will be used repeatedly, test it with people who did not help prepare it. Look for errors, omissions, illogical order, or unclear language.

If a checklist already exists, or after you have prepared your own checklist:

7. Keep the checklist where it will be easily available during the process, procedure, or activity.
8. Refer to the checklist every time the process, procedure, or activity is performed.
9. Check off or write the date as each item is done.
10. When the checklist is completed, make sure that every item has been checked off.
11. If the order of steps in a process or procedure is important and if an item is discovered unchecked when later items are done, follow the checklist's instructions for skipped steps.

Example

The ten-step quality improvement process in this book, or the process your organization follows, is a checklist. Every set of procedures in this book is also a checklist. Prepared checklists in this book include the project charter checklist and 5W2H. Checklists encountered in daily life include meeting agendas, to-do lists, grocery lists, and toy assembly instructions. See the St. Luke's Hospital and ZZ-400 stories in Chapter 4 for examples of using checklists within the quality improvement process.

Considerations

- Prepared checklists are widely available. For example, checklists that will guide the work of a project team can be found in books about teams or quality management. Such checklists are a condensation of other people's hard-won experience. Benefit from them!

- When you use a prepared checklist, feel free to review it in light of your particular situation and revise it as necessary. But before eliminating an item, think hard about why it's on the checklist and whether it's applicable to your situation.
- Before you jump into a new phase of a project, prepare a checklist of critical steps to take and perhaps another one of issues you don't want to overlook. The time spent will be well repaid.
- When you are creating a checklist for steps of a process, prepare a flowchart first to determine what the steps are and their sequence.
- At step 3, you might give the preliminary list of items from step 2 to the person in the group who is best with words. Writing by committee is difficult. Alternatively, if correct wording is a sensitive issue, consider using wordsmithing to include everyone's ideas.
- Review the operational definitions tool and be sure terms in your checklist are unambiguous.
- In step 7, be creative. Consider putting up a poster of the checklist on the wall near the work area. Or hang from a hook a clipboard with the checklist. If you use a checklist in every team meeting, keep it in the file folder you bring to the meetings. Or print it on the reverse side of the team agenda. For checklists that will be frequently used by a large number of people, consider laminated pocket-size cards.
- In step 8, never assume you remember the items on the checklist. The whole point of a checklist is to keep you from forgetting something, and if you've forgotten, you don't *know* you've forgotten! Look at the checklist to make sure.
- If items absolutely must be done in a certain order, use mistake-proofing so that an incorrect order is impossible.
- A checklist is a specialized form of the generic check sheet. See *check sheet* for more information.

check sheet

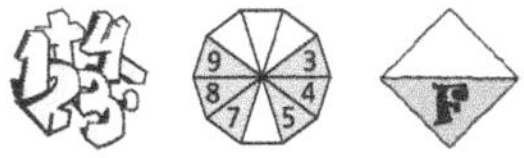

Variation: defect concentration diagram
See also: checklist

Description

A check sheet is a structured, prepared form for collecting and analyzing data. This is a generic tool that can be adapted for a wide variety of purposes.

When to Use

- When data can be observed and collected repeatedly by the same person or at the same location, and . . .
- When collecting data on the frequency or patterns of events, problems, defects, defect location, defect causes, and so forth, or . . .
- When collecting data from a production process

Procedure

1. Decide what event or problem will be observed. Develop operational definitions.
2. Decide when data will be collected and for how long.
3. Design the form. Set it up so that data can be recorded simply by making check marks or Xs or similar symbols and so that data does not have to be recopied for analysis.
4. Label all spaces on the form.
5. Test the check sheet for a short trial period to be sure it collects the appropriate data and is easy to use.
6. Each time the targeted event or problem occurs, record data on the check sheet.

Example

Figure 5.14 shows a check sheet used to collect data on telephone interruptions. The tick marks were added as data was collected over several weeks' time. What days are worst for interruptions? Which interruptions are most frequent? This check sheet was designed in the same format as a contingency table, so that the data can be analyzed with chi-square hypothesis tests without recopying into a different format.

Telephone Interruptions

Reason	Day					
	Mon	Tues	Wed	Thurs	Fri	Total
Wrong number	𝍸	𝍷𝍷	𝍷	𝍸	𝍸 𝍷𝍷	20
Info request	𝍷𝍷	𝍷𝍷	𝍷𝍷	𝍷𝍷	𝍷𝍷	10
Boss	𝍸	𝍷𝍷	𝍸 𝍷𝍷	𝍷	𝍷𝍷𝍷𝍷	19
Total	12	6	10	8	13	49

Figure 5.14 Check sheet example.

See also Figure 5.181, page 492 a check sheet for tracking telephone interviews. Read the ZZ-400 story in Chapter 4 for an example of using a check sheet within the quality improvement process.

defect concentration diagram

Also called: concentration diagram, defect map, defect location check sheet, location diagram

Description

A defect concentration diagram is a check sheet using an actual image of the object or area where problems could occur. During data collection, observed defects, mistakes, or problems are marked on the picture where they occurred.

When to Use

- When you are investigating defects, mistakes, or problems that occur on a physical object or within an area
- Especially when you suspect that the physical pattern of occurrence might provide clues to the cause of the problem

Procedure

1. Decide what object or area will be observed. Develop operational definitions.
2. Decide when data will be collected and for how long.
3. Design the form. Draw a clear picture of the object or a map of the area, showing all significant features. Make it large and uncluttered enough that data can be collected directly on the picture.
4. If different kinds of defects, mistakes, or problems will be observed, decide on symbols for each kind. Put a legend on the form showing the symbols and what they mean.
5. Test the diagram for a short trial period to be sure it collects the appropriate data and is easy to use.
6. Each time an observation occurs, record it on the diagram.
7. When the data collection period is over, analyze the data for patterns and trends that might indicate causes. If a visual analysis is inconclusive, you can measure distances between occurrences and analyze those measurements statistically.

Example

The most common use of this tool is to mark defects on a picture of a manufactured object coming off an assembly line. In another example, insurance adjustors and vehicle repair facilities use pictures of cars to clearly show damaged locations. Here's an example from a nonmanufacturing application.

A drugstore team trying to reduce shoplifting created a defect concentration diagram (Figure 5.15) to study how store layout contributed to the problem. Using a map of the store, they marked locations where shoplifted merchandise had been displayed. When they analyzed the diagram, they observed clustering around the edges of the store, but they couldn't relate the pattern to aisle length, type of merchandise, or clerk location. Finally, a team member suggested adding to the diagram the locations of surveillance cameras. A pattern became clear based on sight lines from the cameras.

Considerations

- A common way to lay out a check sheet is to create a table with descriptions of what you are observing (events, problems, types of errors, and so on) in a column down the left side (or a row across the top). Divide the remainder of the page into columns (or rows) for easy data collection. These might represent dates, times, locations of defects, or any other category that you wish to use to analyze the data later.

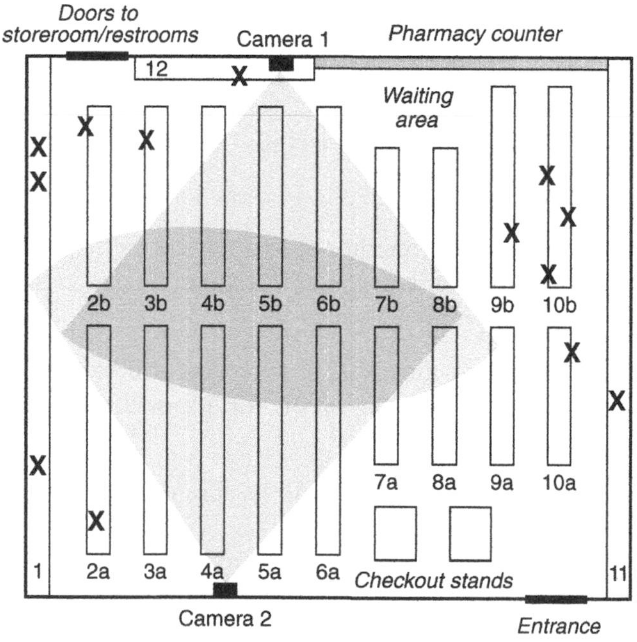

Figure 5.15 Defect concentration diagram example.

- Consider what information about the sources of the data might be important in analysis. Set up the check sheet so that data is captured. It's easier to capture information when the data is generated than to try to reconstruct it later. Sometimes it's impossible to capture additional information later. See *stratification* for ideas about information you might need to capture.
- When designing your form, use illustrations whenever possible. This makes the check sheet easier to use and can reveal patterns during your later analysis. For example, to collect data on damaged packages, include a sketch of the package and have data collectors put an X where they see the damage. See *defect concentration diagram* for more information about this tool, sometimes called a defect location check sheet.
- Think about how data is received or how you will want to analyze it later, and consider keeping separate sheets for different aspects of the collection or analysis. For example, if different people handle international and domestic shipments, use separate check sheets for international and domestic shipment errors.
- Keep the check sheet near the point where the data to be recorded will occur. For example, if you are monitoring types of telephone interruptions, keep the check sheet next to the telephone. This will help you collect the data consistently.
- If observations occur frequently, you may choose to record samples rather than every observation. In step 2, decide when an observation will be recorded. You may decide to use a time interval (every 30 minutes) or a frequency (every fifth phone call). See *sampling* for more information.
- A checklist (page 139) is a variation sometimes called a confirmation check sheet.

Defect Concentration Diagram

- A physical object might be a manufactured item rolling off an assembly line, an item that might have been damaged in shipment or handling, or a form that workers or customers must fill out.
- An area might be an office building, a laboratory, a warehouse, or a store (retail and/or storage areas).
- To study errors made in filling out a form, use a copy of the form as your defect concentration diagram.
- Include just enough detail on your picture so that the location of the problem is clear, but not enough that the picture is too cluttered to mark on.
- Use colored pens or pencils to mark the defects, so they are easily seen on the black-and-white picture. (But don't do this if you will need to photocopy the completed diagram for analysis or presentation.)

- When you are analyzing the data, try to look at it with a fresh eye if patterns aren't obvious. For example, turn the diagram upside down. Or have someone unfamiliar with the object or area look at the diagram. Alternatively, add more detail to the diagram—detail that was originally left out to unclutter the picture—and see if a pattern becomes obvious.

contingency diagram

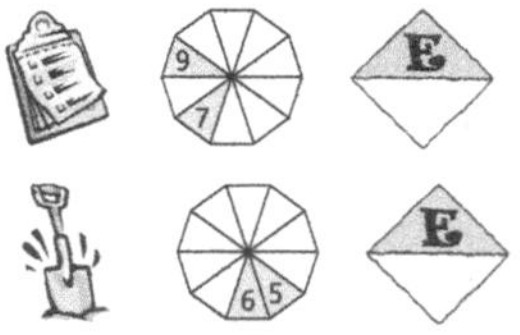

Description

The contingency diagram uses brainstorming and a negative thinking process to identify how process problems occur or what might go wrong in a plan. Then the negative thinking is reversed to generate solutions or preventive measures.

When to Use

- When identifying problem causes, or . . .
- When developing solutions to problems, or . . .
- When planning implementation of a phase of a project, especially the solution, or . . .
- Before launching a change
- Especially when negative thinking is preventing the group from generating ideas

Procedure

1. Identify your topic—the problem or the proposed action plan—and write it prominently on a flipchart.
2. Brainstorm how to make things go wrong. For a problem:
   - How can we make the problem happen?
   - How could we make it worse?

   For a plan or action:
   - How can our plan be made to fail?
   - What assumptions are we making that could turn out to be wrong?

Write each idea on the flipchart in words as close as possible to those used by the contributor.

3. When no more ideas are being generated, reverse your thinking. For each idea on the flipchart, describe actions that would prevent it. Write these beside or under the problem actions, in a different color.
4. When each negative idea has been reversed, think more broadly: modifying ideas, combining ideas, extending patterns, and other creative thinking techniques.

Example

A state agency used a contingency diagram (Figure 5.16) to identify potential problems with implementing their long-range plan. Their first four ideas are shown above the lines, with countermeasures below.

"Forbid transfers!" is obviously unrealistic, but in brainstorming no idea is rejected, and humor often inspires creativity. Later, that idea was combined with another to generate a workable way to create smooth transitions when key personnel leave.

Considerations

- This tool is useful when the group is locked into negative thinking that is preventing creativity. Channel that negative energy with this tool.
- Be sure to use the rules and techniques of brainstorming.
- First, generate many possibilities of things that could go wrong until the group runs out of ideas. Then go back and identify possible preventions. By sequencing the ideas this way, the brain is functioning in one mode at a time.
- If several possible preventions are thought of, write them all down.
- This tool is similar to the double reversal variation of brainstorming.

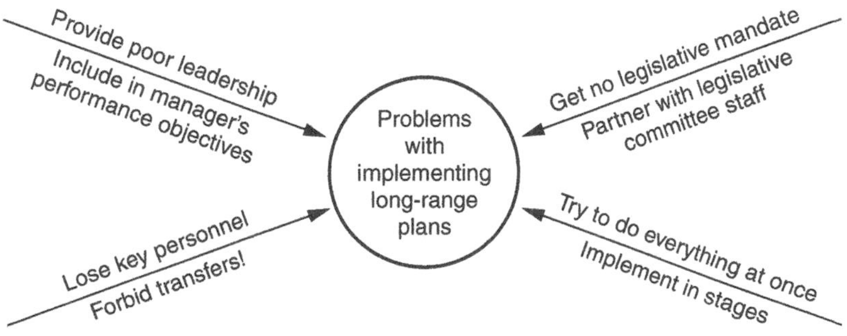

Figure 5.16 Contingency diagram example.

- For a more rigorous analysis of what could go wrong in a project, see *process decision program chart* and *potential problem analysis.*

contingency table

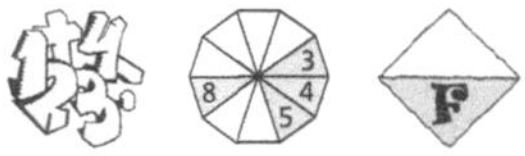

Description

A contingency table organizes categorical data for better analysis. It is composed of columns and rows that represent two different ways of analyzing the same set of data. Create a contingency table before doing a chi-square hypothesis test for variable independence or for comparing several groups' distributions across several categories.

When to Use

- When the data are counts of various categories, and . . .
- When the data are cross-classified, that is, classified one way into a set of categories and classified another way into a different set of categories
- Especially before doing a chi-square hypothesis test

Procedure

1. Determine the categories into which the data are classified. If the categories are ordinal, determine the appropriate order. Decide which categories will be the rows and which will be the columns. If only one categorization represents groups into which the respondents are divided, such as male and female, it is customary to put it in the rows.
2. Create a table with two more columns and two more rows than you need for the categories. The extra columns are for category labels and totals. Label the columns and rows.
3. Write in each cell (a box of the table) the number of occurrences for the category represented by the row *and* the category represented by the column. Fill in the totals. The lower right cell of the table is the total number of occurrences. (If you are using a spreadsheet, enter summation formulas in the cells of the row and column for totals.)

Example

A catalog clothing retailer wanted to know whether proposed changes to its product line would be similarly received in different parts of the country. They conducted a random survey of 750 customers. After describing the proposed new products, they asked the customers to estimate their likelihood of buying, using an ordinal scale from very unlikely to very likely.

The resulting data was categorized in two ways: by the customer's estimate of buying likelihood and by geographical region. Because the geographical regions represent groups of customers, those categories are placed in the rows. The contingency table looked like Figure 5.17.

See the second example of *hypothesis testing* to learn how the contingency table was used to answer the company's question.

Another Example

The same company planned to change the format and style of their catalog. They wanted to know if the new format would be effective at increasing orders. As a test, they sent 200,000 copies of the spring catalog in the new version to randomly selected customers. The other 1,800,000 catalogs were in the traditional version.

The data are shown in a 2 × 2 contingency table, Figure 5.18. This table is useful for analyzing proportions between two groups when one of the classifications' values

	Very Unlikely	Somewhat Unlikely	Somewhat Likely	Very Likely	Total
W	23	34	48	31	136
SW	43	58	41	37	179
SE	39	56	82	49	226
MW	17	26	19	16	78
NE	33	29	51	18	131
Total	155	203	241	151	750

Figure 5.17 Contingency table example.

	Standard Catalog	Test Catalog	Total
Did buy	34,852	4,972	39,824
Didn't buy	1,765,148	195,028	1,960,176
Total	1,800,000	200,000	2,000,000

Figure 5.18 2 × 2 contingency table example.

are yes/no, pass/fail, accept/reject, or something similar. In this format, the data can be entered into hypothesis testing software or an online calculator to determine whether the proportions are significantly different. (They are. The new catalog is effective at increasing orders.)

Considerations

- The total number in the lower right cell of the table can be used as a check for errors. It should equal both the sum of all the row totals and the sum of all the column totals.
- Computer spreadsheets are often useful for doing the math associated with contingency tables. The procedure here can be used for setting up a contingency table in a spreadsheet. In step 2, "draw" by placing labels in the appropriate spreadsheet cells.
- Check sheets can be designed so that they look just like the contingency table that will be required for analyzing the data later. See the example for *check sheets* and notice the similarity to a contingency table.

continuum of team goals

Description

The continuum of team goals shows how general or specific a goal or mission statement is. Used to foster communication, it helps everyone involved clarify their understanding of a team's purpose.

When to Use

- To stimulate and structure discussion about a team's goal or mission statement; especially
- When writing or improving a project goal or mission statement, or . . .
- In a steering committee or other management group discussion, to clarify the scope of a task to be assigned to a project team, or . . .
- When a team is first discussing their assignment and team members may have different ideas about its scope, or . . .
- In a meeting between the team and management, to achieve common understanding of the project scope, either at the beginning of the project or when it has stalled

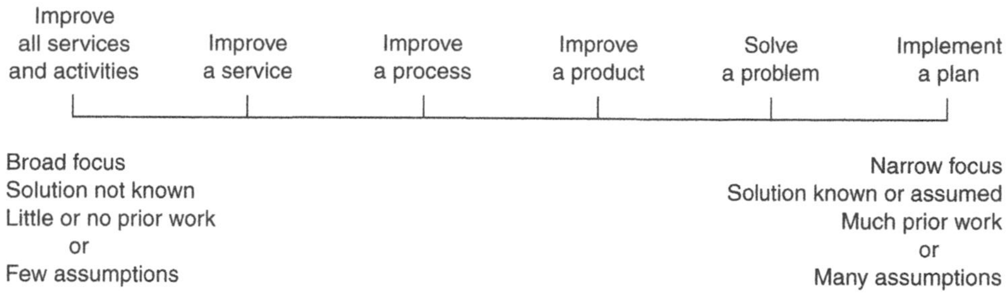

Figure 5.19 Continuum of team goals.

Procedure

1. If a goal or mission statement already exists, write it on a flipchart or overhead transparency so that all participants can see it.
2. Draw a continuum scale (Figure 5.19) on a flipchart or overhead transparency so that all participants can see it. Have each participant decide where along the continuum scale he or she would place the group's assignment, based on his or her understanding of what is expected of the project team. This can be done orally or in writing. If a goal or mission statement already exists, each participant also should decide where the statement falls along the continuum.
3. Record all participants' responses. Discuss the group chart until consensus is reached on the project scope. Be sure to consider whether this scope is appropriate.
4. If necessary, rewrite the goal or mission statement to reflect the consensus understanding of the project scope.

Examples

Below are examples of possible goals for the Parisian Experience restaurant. None are right or wrong. The best goal depends on what a group needs to accomplish and prior work that has been done by this group or others.

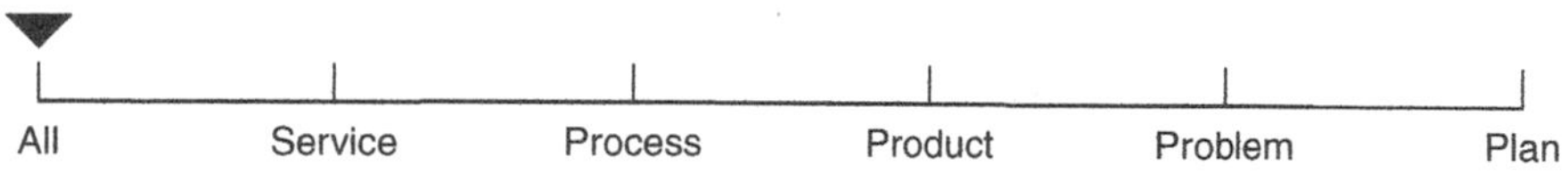

Our customers choose and return to the Parisian Experience restaurant because we offer the most authentic and tastiest French cuisine in the state, provided with unobtrusive and thoughtful service in an atmosphere of relaxed luxury that captures the feeling of cosmopolitan France.

Any mission statement written to describe an organization's purpose should be at the far left of the continuum because it includes all aspects of the organization.

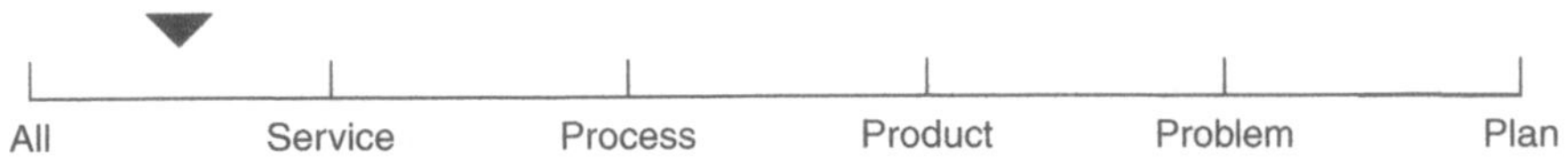

The customer satisfaction team will identify opportunities to improve customer satisfaction with our restaurant and will develop and recommend plans that transform those opportunities into value for our customers and our business.

This statement is located between "Improve all services and activities" and "Improve a service." Since the scale is continuous, goals may fall between the tick marks. While many services will be examined by the team (for example, providing gourmet food and luxurious atmosphere), activities such as buying ingredients or hiring employees will not be considered because they do not directly affect customer satisfaction.

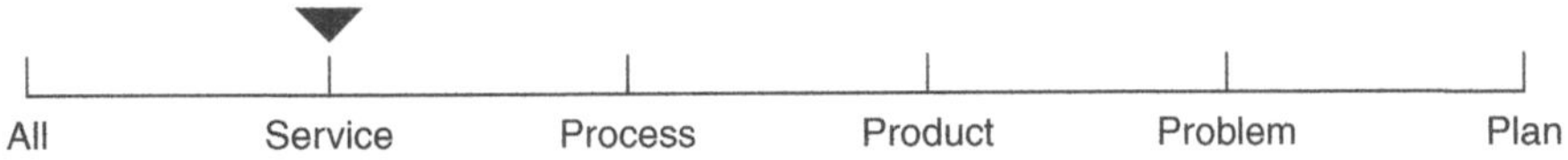

Enable quick, convenient, and unobtrusive payment.

It may seem odd to call "enabling payment" a service, but helping that necessary exchange to happen painlessly is indeed a service. This goal focuses the team on improving that service by whatever means it can create. The team is free to improve the current process or to devise a completely new one, such as restaurant membership with monthly billing.

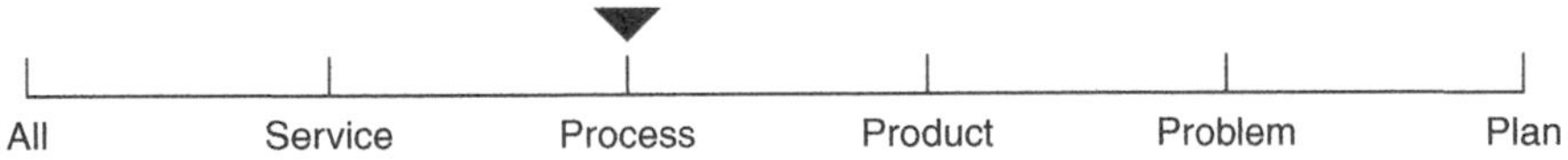

Improve the process of presenting the bill and collecting payment.

This time, the team members are restricted to making the current process for payment better. They do not have the freedom to devise a totally new method for the customer to pay.

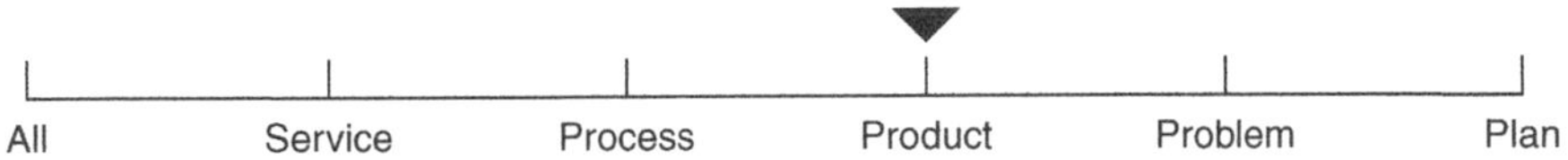

Improve the layout of the bill.

This goal is even more focused, and the team even more restricted, not just to a current process but to a current product: the piece of paper that functions as the bill. Although the team may completely change the format, its solution should still involve a piece of paper functioning as a bill. Note that no particular problem is defined in this statement.

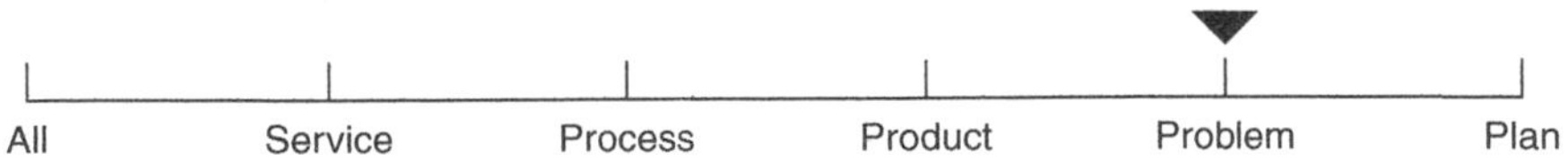

Make the bill format easier for customers to read.

This time a problem is implied: the current bill format is hard for customers to read. The team is not asked to make any other changes to the bill, such as improving its overall appearance.

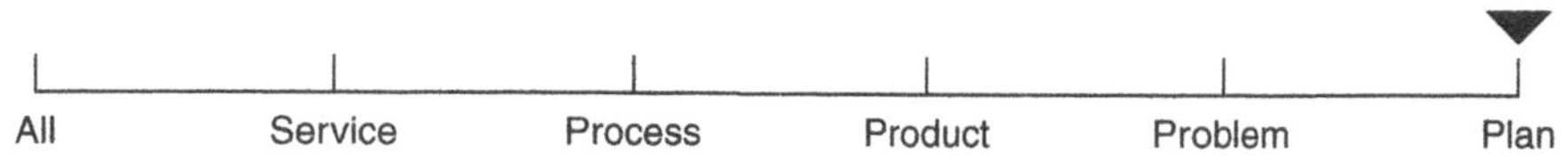

Train the staff on using the new bill.

This goal is the most focused of all. The team is not asked to create any changes or improvements, but merely to carry out a task identified by others.

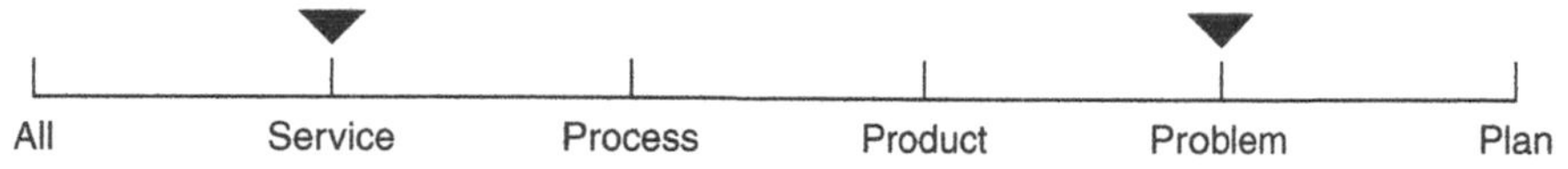

Improve customer satisfaction with private parties at a lower cost.

This is an unclear statement. Two goals—improve a service (private parties) and solve a problem (cost)—are mixed here. Is the focus on generally improving private parties, with cost being a limitation the team must work within? Or is the focus on reducing cost? The answer is not clear, so there should be discussion to clarify the intent, and the statement should be rewritten.

A project can have two separate goals at different points on the continuum, as long as they are separate and both are clear to everyone: "Improve customer satisfaction with private parties; as part of that improvement, reduce the net cost of private party service." This statement says that the basic goal is "improve a service," but there is a specific problem that everyone recognizes and wants the final plan to address.

The following example is from a group that was studying a process of preparing technical reports. This is the statement that showed the need for the continuum.

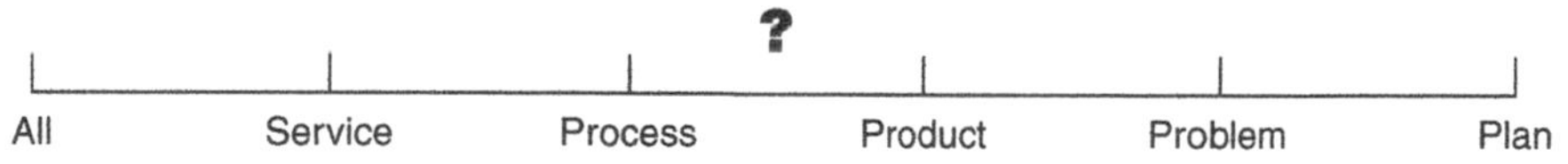

Improve report preparation through review of the steps in the process.

It seems to be a clear example of "improve the process." After several months, the team members had significantly improved the process and thought their work was done. Management said it was too early to disband, and the team did not understand why.

A facilitator led the team and the steering committee in a discussion about goals, using the continuum. Different expectations became visible to everyone. Although the charter provided by the steering committee seemed to clearly imply "improve the process," an early status report said, "The team was formed to recommend changes to the report-writing process that will reduce the time to issue a survey report." This sounds like "solve a problem." Despite significant improvements, the team had not reduced the time as much as management expected.

Early discussion of expectations, using the continuum, would have prevented the conflict and provided better focus for the effort. Also see the project charter checklist example on page 436 for another way the team could have discussed their assignment. See the Medrad story on page 58 for another continuum example.

Considerations

- If the continuum is new to participants, the facilitator of the meeting should start by explaining it, using an example such as the restaurant.
- If different people place a project in different places on the continuum, the issue is not "Who is right?" but rather "What is right?" The appropriate scope needs to be determined, then clarified, so everyone understands it the same way, and the project statement needs to be rewritten so it clearly conveys to everyone what is expected of the team.
- Where a goal statement belongs along the continuum depends not on the results of accomplishing the goal, but on the freedom or constraints the statement gives the team. For example, the goal, "Install new software XYZ on all PCs" may have the result of improving computer service. However, this statement does not allow the team to consider and choose how best to improve service; the team is asked only to implement a specific plan.
- Generally, more creativity and broader improvement result from goals farther left on the continuum. Always consider whether a goal's scope can be expanded to a point farther left on the continuum, unless the goal logically follows from work that already has been done. Think more broadly than the immediate problem at hand or the current way of doing things. Set goals that challenge you to develop creative alternatives that eliminate problems instead of just fixing them, that add value rather than just efficiency. This is especially critical between the two continuum points of improving a process and improving a service. Look just a little further than the process you currently operate to ask what service that process is intended to provide. Would you be willing to provide that service in a different way, if someone could think of one?

  Example: Instead of the restaurant asking a team to improve the payment process, the team could be asked to enable unobtrusive and thoughtful payment that does not disturb the relaxed, luxurious atmosphere. This subtle difference, focused on the restaurant's mission, frees the team to imagine creative ways to handle the financial side of the transaction.
- To facilitate discussion, try having each person describe which words in the mission statement caused the placement of his or her mark.
- When determining whether the scope is appropriate, consider such things as overall organizational goals, makeup of the team, budget constraints, and timing expectations.

- The team's and management's scopes must match, so that the team's efforts will support the organization's overall strategy and their solutions will be readily accepted and supported.

control chart

Also called: statistical process control
Variations:

variables charts: $\overline{X}$ and R chart (also called averages and range chart), $\overline{X}$ and s chart, chart of individuals (also called X chart, X-R chart, IX-MR chart, XmR chart, moving range chart), moving average–moving range chart (also called MA–MR chart), target charts (also called difference charts, deviation charts, and nominal charts), CUSUM (also called cumulative sum chart), EWMA (also called exponentially weighted moving average chart), multivariate chart (also called Hotelling T^2)

attributes charts: p chart (also called proportion chart), np chart, c chart (also called count chart), u chart.

charts for either kind of data: short run charts (also called stabilized charts or Z charts), group charts (also called multiple characteristic charts)

Description

The control chart is a graph used to study how a process changes over time. Data are plotted in time order. A control chart always has a central line for the average, an upper line for the upper control limit, and a lower line for the lower control limit. These lines are determined from historical data. By comparing current data to these lines, one can make conclusions about whether the process variation is consistent (in control) or is unpredictable (out of control, affected by special causes of variation).

There are many types of control charts. Each is designed for a specific kind of process or data. Control charts for variable data are used in pairs. The top chart monitors the average, or the centering of the distribution of data from the process. The bottom chart monitors the range, or the width of the distribution. If your data were shots in target practice, the average is where the shots are clustering, and the range is how tightly they are clustered. Control charts for attribute data are used singly.

When to Use

- When controlling ongoing processes by finding and correcting problems as they occur, or . . .
- When predicting the expected range of outcomes from a process, or . . .

- When determining whether or not a process is stable (in statistical control), or . . .
- When analyzing patterns of process variation from special causes (nonroutine events) or common causes (built into the process), or . . .
- When determining whether your quality improvement project should aim to prevent specific problems or to make fundamental changes to the process

Control Chart Decision Tree

Figure 5.20 is a decision tree for deciding which control chart should be used, depending on the type of data. The two broadest groupings are for variable data and attribute data.

Variable data are measured on a continuous scale. For example, time, weight, distance, or temperature can be measured in fractions or decimals. The only limit to the precision of the measurement is the measuring device. If you are using a measuring device that can only give you whole numbers, such as 78 degrees, or if you only need the precision of the nearest whole number, such as five days, you are still using variable data. The possibility of measuring to greater precision defines variable data.

Attribute data are counted and cannot have fractions or decimals. Attribute data arise when you are determining only the presence or absence of something: success or failure, accept or reject, correct or not correct. For example, a report can have four

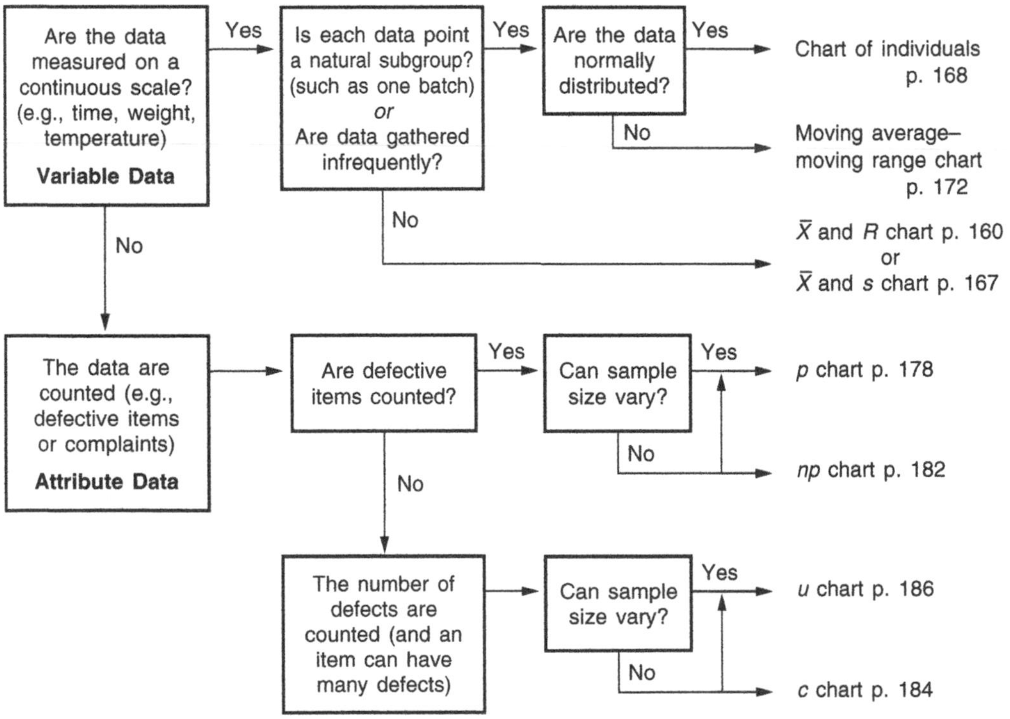

Figure 5.20 When to use the basic control charts.

errors or five errors, but it cannot have 4½ errors. A longer discussion of attribute data can be found on page 177.

Basic Procedure

1. Choose the appropriate control chart for your data.
2. Determine the appropriate time period for collecting and plotting data.
3. Follow the procedure for that control chart on the following pages to collect data, construct your chart, and analyze the data.
4. Look for out-of-control signals on the control chart. When one is identified, mark it on the chart and investigate the cause. Document how you investigated, what you learned, the cause, and how it was corrected.
   - A *single point* outside the control limits. In Figure 5.21, point sixteen is above the UCL.
   - *Two out of three* successive points are on the same side of the centerline and farther than 2σ from it. In Figure 5.21, point four sends that signal.
   - *Four out of five* successive points are on the same side of the centerline and farther than 1σ from it. In Figure 5.21, point eleven sends that signal.
   - A run of *eight in a row* are on the same side of the centerline. Or *ten out of eleven, twelve out of fourteen,* or *sixteen out of twenty.* In Figure 5.21, point twenty-one is eighth in a row above the centerline.
   - Obvious consistent or persistent patterns that suggest something unusual about your data and your process.

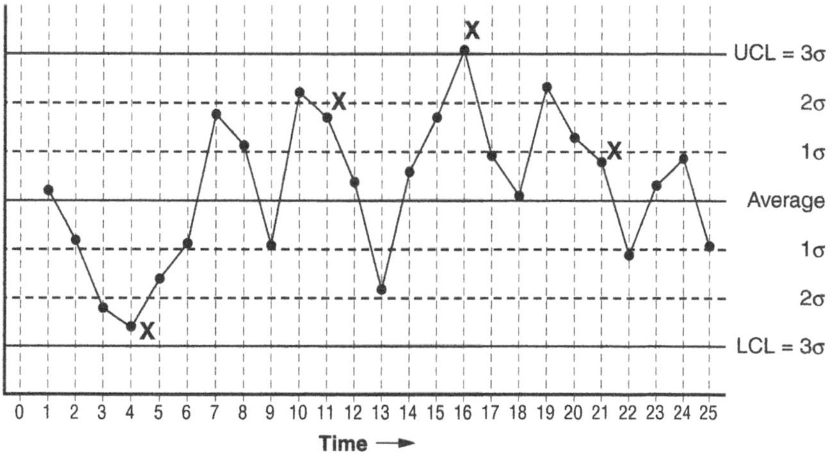

Figure 5.21 Out-of-control signals.

5. Continue to plot data as they are generated. As each new data point is plotted, check for new out-of-control signals.

6. When you start a new control chart, the process may be out of control. If so, the control limits calculated from the first 20 points are conditional limits. When you have at least 20 sequential points from a period when the process is operating in control, recalculate control limits.

Considerations

Out-of-Control Signals

- Some control charts have special requirements for their out-of-control signals. Check the analysis section of the particular control chart.
- The distribution of ranges is not symmetrical but skewed right. Therefore, one would expect more points below $\bar{R}$ than above it. To allow for this, the fourth out-of-control signal is often modified for range charts to require a run of 12 points in a row below $\bar{R}$.
- These signals are based on statistics. The points plotted from an in-control process are random but, taken together, match a predictable distribution, such as a normal curve. The out-of-control signals highlight patterns that are statistically unlikely to be random outputs of that distribution, such as too many points far from the centerline. In Figure 5.21, the last three out-of-control signals show more output above and far from the average than one would expect from a random process, an indication that the process average has shifted up.
- The signals do not indicate whether patterns are desirable or undesirable. A change in the process may be toward better performance or toward worse performance. It is just as important to do something to understand and maintain good performance as it is to eliminate bad performance.
- Additional out-of-control signals are often used. For example:
  - Fifteen points in a row within 1σ of the centerline
  - Six points in a row steadily increasing or decreasing
  - Fourteen points in a row alternating up and down
  - Eight points in a row all farther than 1σ from the centerline, on either side
  - A jump of 4σ
  - Additional rules for runs and trends

- However, using too many out-of-control signals can cause a false positive—a problem signaled when one doesn't really exist. (This is called a type I error. See *hypothesis testing* for more information.) Every out-of-control signal has a small probability of occurring from random chance. For example, an in-control process will generate a point outside the control limits 0.27 percent of the time. When many signals are used simultaneously, those small chances multiply. Using just four signals—single point out, two-out-of-three, four-out-of five, and eight-in-a-row—the chance of a false positive is one in twelve. Use no more than four or five signals, or you will be looking frequently for problems that don't exist. For most processes, just a few signals, plus thinking carefully about what the charts might be revealing about your process, will reveal many improvement opportunities.
- The out-of-control signals listed on the previous page and in the procedure are not in order of probability but rather in an order that aids remembering. For example, four out of five points on one side of the centerline is less probable than a single point outside the control limits. Thus, four out of five is a more sensitive signal than the single point outside. See Hoyer and Ellis in the Resources for a discussion of the probabilities and sensitivities of different signals.
- There is also such a thing as a type II error: an out-of-control process that does not show any of the out-of-control signals. For example, if the process average shifts slightly, the standard out-of-control signals will be slow to identify the change.
- Out-of-control signals should be marked on the chart. Put an X beside the point that completes the pattern, or circle the set of points that form the signal.

Autocorrelation

- The time period chosen for collecting and plotting data should be based on how fast variation occurs in the process. The process should have time to change between samples. If not, the data will be *autocorrelated.*
- For example, suppose liquid is continually flowing both into and out of a tank so that the contents are completely replaced every three hours. If you are monitoring concentration, you should not plot it more often than every three hours. Data taken more frequently will be autocorrelated. Temperature is a variable that is often autocorrelated because it usually does not change rapidly.
- A good test for autocorrelation is to use correlation analysis or a scatter diagram to graph the measurement taken at one time against the measurement taken at the previous time. If samples are being taken too frequently, the scatter diagram will show correlation. See *scatter diagram* or *correlation analysis* for more information.

Control Limits

- Control limits are *not* specification limits. In fact, specification limits should never be placed on a control chart. Specifications reflect customer requirements, whereas control limits reflect the historical performance of the process.

- Control limits should be recalculated only when the process has experienced a permanent change from a known cause and at least 20 plot points have been generated by the changed process. You should not recalculate limits after each 20 points or subgroups or after each page of the chart.
- Choose numerical scales that include the values for the control limits plus a little extra.

Other Considerations

- For controlling an ongoing process, control charts are most useful when plotted as soon as the data is generated and analyzed by the people who are working with the process. Collecting lots of data on a check sheet and plotting them by computer later will create charts that are prettier but of limited use.
- Computer software is available for generating control charts. Sometimes computerized control systems in manufacturing automatically generate control charts. These are useful to remove the labor required and to ensure that calculations are accurate. However, computers should never substitute for the most important aspect of control charting: people who understand the purpose and meaning of the charts and who watch them closely, using them as a tool to monitor and improve their processes.
- Blank control chart forms and worksheets are provided for each type of chart. Permission is granted to copy these charts for individual use.
- Control charts are easy and powerful to use, but they are often applied incorrectly. The more knowledgeably they are used, the more they can improve your processes. This summary contains only the basics of control charts, intended as a reference. The theory behind them and fine points about their use are beyond the scope of this book. Take a class, study some books, or obtain expert help to get the most value out of control charts in your processes.

variable control charts

$\bar{X}$ and R chart

Also called: averages and range chart

Description

The $\bar{X}$ (pronounced *X*-bar) and *R* chart is a pair of control charts used to study variable data. It is especially useful for data that does not form a normal distribution, although it can be used with normal data as well. Data are subgrouped, and averages and ranges for each subgroup are plotted on separate charts.

When to Use

- When you have variable data, and . . .
- When data are generated frequently, and . . .
- When you want to detect small process changes
- Especially in manufacturing, where a sample of four or five pieces may be used to represent production of several hundred or thousand pieces

Procedure

Construction

1. Determine the appropriate time period for collecting and plotting data. Determine the number of data points per subgroup (n). Collect at least $20n$ data points to start the chart. For example, with subgroup size of three, 60 data points will be needed to create 20 subgroups.
2. If the raw data do not form a normal distribution, check whether the averages of the subgroups form a normal distribution. (The normality test can be used.) If not, increase the subgroup size.
3. Calculate $\overline{\overline{X}}$ (X-double-bar), $\overline{R}$ (R-bar), and the control limits using the worksheet for the $\overline{X}$ and R chart or moving average–moving range chart (Figure 5.22) and the chart for $\overline{X}$ and R or moving average–moving range (Figure 5.23).
4. On the "Average" part of the chart, mark the numerical scale, plot the subgroup averages, and draw lines for the average of the averages and for the control limits for X. On the "Range" part of the chart, mark the numerical scale, plot the ranges, and draw lines for the average range and for the control limits for R.
5. Continue to follow steps 4, 5, and 6 of the basic procedure.

Analysis

1. Check the R chart for out-of-control signals. All the signals listed on page 157 can be used.
2. If the R chart is in control, check the $\overline{X}$ chart for out-of-control signals. All the signals listed on page 157 can be used.

Process: ______________________ Calculated by: ______________________

Data dates: ______________________ Calculation date: ______________________

Step 1. Calculate average $\bar{X}$ and range R (the difference between the highest and lowest values) for each subgroup. Record on chart.

Number of values in each subgroup = n = ______

Number of subgroups to be used = k = ______

Step 2. Look up control limit factors.

n	A_2	D_3	D_4
2	1.880	—	3.267
3	1.023	—	2.574
4	0.729	—	2.282
5	0.577	—	2.114
6	0.483	—	2.004
7	0.419	0.076	1.924

A_2 = ______

D_3 = ______

D_4 = ______

Step 3. Calculate averages ($\bar{\bar{X}}$ and $\bar{R}$).

Sum of the averages = $\Sigma\bar{X}$ = ______

Average of the averages = $\bar{\bar{X}}$ = $\Sigma\bar{X}$ ÷ k

= ______ ÷ ______ = ______

Sum of the ranges = ΣR = ______

Average of the ranges = $\bar{R}$ = ΣR ÷ k

= ______ ÷ ______ = ______

Step 4. Calculate control limits.

$3\hat{\sigma}$ estimate for $\bar{X}$ chart = $3\hat{\sigma}_{\bar{X}}$ = A_2 × $\bar{R}$

= ______ × ______ = ______

Upper control limit for $\bar{X}$ chart = $UCL_{\bar{X}}$ = $\bar{\bar{X}}$ + $3\hat{\sigma}_{\bar{X}}$

= ______ + ______ = ______

Lower control limit for $\bar{X}$ chart = $LCL_{\bar{X}}$ = $\bar{\bar{X}}$ − $3\hat{\sigma}_{\bar{X}}$

= ______ − ______ = ______

Upper control limit for R chart = UCL_R = D_4 × $\bar{R}$

= ______ × ______ = ______

Lower control limit for R chart = LCL_R = D_3 × $\bar{R}$

= ______ × ______ = ______

Figure 5.22 $\bar{X}$ and R chart or moving average–moving range chart worksheet.

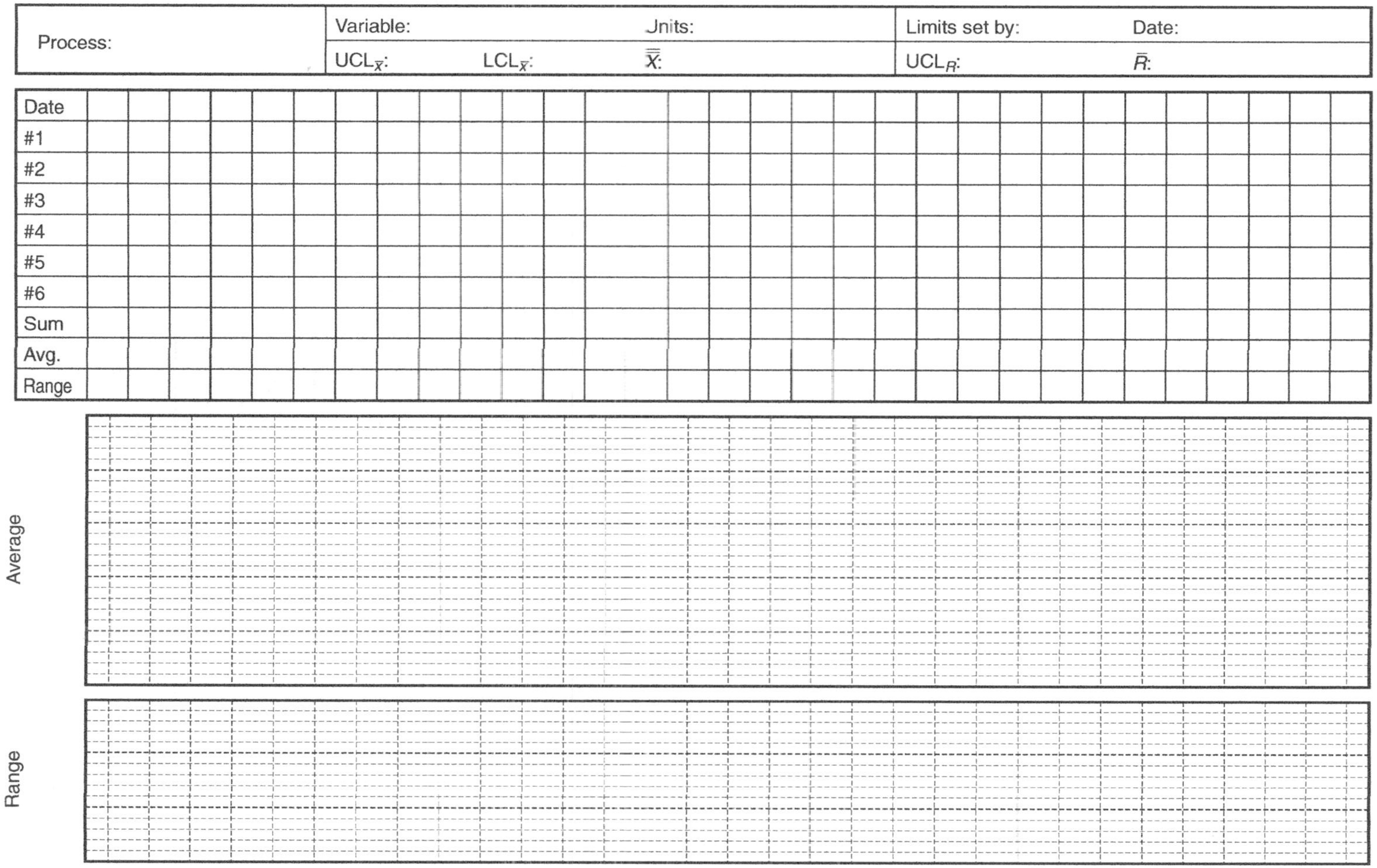
Process:
Variable:
Jnits:
Limits set by:
Date:
UCL$_{\bar{X}}$:
LCL$_{\bar{X}}$:
$\bar{\bar{X}}$:
UCL$_{\bar{R}}$:
$\bar{R}$:
Date
#1
#2
#3
#4
#5
#6
Sum
Avg.
Range
Average
Range

Figure 5.23 $\bar{X}$ and R chart or moving average–moving range chart.

Example

The ZZ-400 team collected a set of 40 values, arranged in time sequence, for product purity. The histogram in Figure 5.24 indicates that the data are slightly skewed, as data that are being pushed toward 100 percent or zero percent often are. The team will try subgrouping the data by twos. Figure 5.25 shows the subgrouped data.

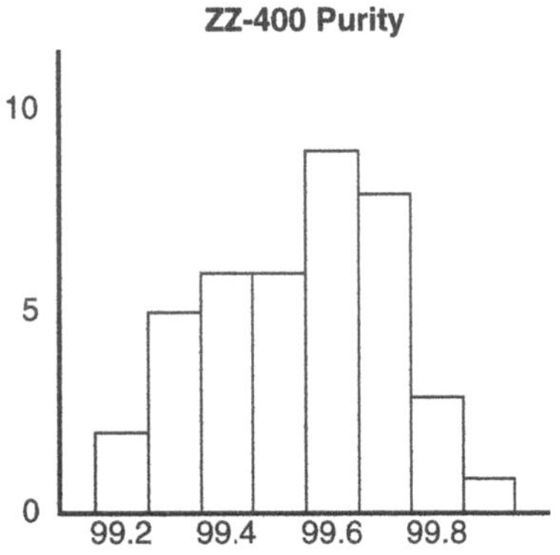

Figure 5.24 Histogram of $\bar{X}$ and R chart example.

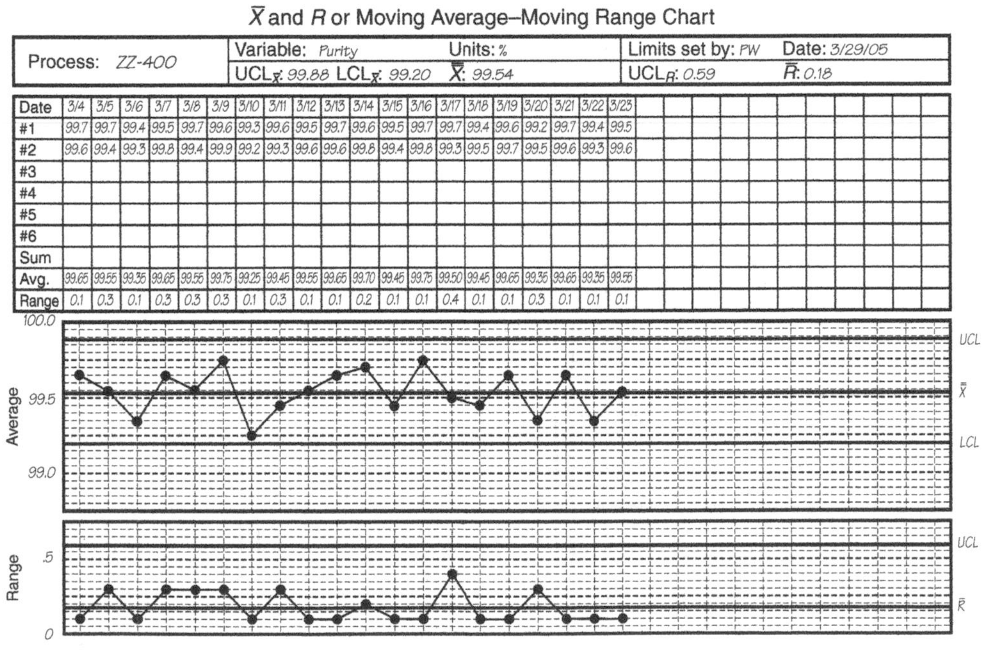

Date	3/4	3/5	3/6	3/7	3/8	3/9	3/10	3/11	3/12	3/13	3/14	3/15	3/16	3/17	3/18	3/19	3/20	3/21	3/22	3/23
#1	99.7	99.7	99.4	99.5	99.7	99.6	99.3	99.6	99.5	99.7	99.6	99.5	99.7	99.7	99.4	99.6	99.2	99.7	99.4	99.5
#2	99.6	99.4	99.3	99.8	99.4	99.9	99.2	99.3	99.6	99.6	99.8	99.4	99.8	99.3	99.5	99.7	99.5	99.6	99.3	99.6
#3																				
#4																				
#5																				
#6																				
Sum																				
Avg.	99.65	99.55	99.35	99.65	99.55	99.75	99.25	99.45	99.55	99.65	99.70	99.45	99.75	99.50	99.45	99.65	99.35	99.65	99.35	99.55
Range	0.1	0.3	0.1	0.3	0.3	0.3	0.1	0.3	0.1	0.1	0.2	0.1	0.1	0.4	0.1	0.1	0.3	0.1	0.1	0.1

Figure 5.25 $\bar{X}$ and R chart example.

The first subgroup contains values one and two: 99.7 and 99.6. Their average is 99.65 and their range is 0.1. The second subgroup contains values three and four: 99.7 and 99.4. Their average is 99.55 and their range is 0.3. The calculations are continued for all the data (see Figure 5.25). At this point, the team checks the distribution of subgroup averages with a normal probability plot. The distribution is approximately normal, so subgroup size of two is acceptable.

Now, using the worksheet, averages and control limits can be calculated. See the completed worksheet (Figure 5.26) for the calculations.

The subgroup averages and ranges are plotted, and the averages and control limit lines are drawn on the chart. Figure 5.25 shows the control chart. There are no out-of-control signals.

This example is part of the ZZ-400 improvement story on page 80.

Considerations

- This is the most commonly used type of control chart.
- The subgroup size and sampling method should be chosen to minimize the chance of variations occurring within the subgroup and to maximize the chance of variations occurring between the subgroups. However, each measurement should be independent of the others.
- The larger the subgroup size, the better the control chart will detect changes in the average.
- For most nonnormal distributions, a subgroup size of two or three will be adequate to have the averages form a normal distribution and, therefore, to have the chart calculations be accurate. If the data are highly nonnormal, a subgroup size of four or five might be needed.
- For subgroup sizes smaller than seven, LCL_R will be zero. Therefore, there will be no out-of- control signals below LCL_R.
- An out-of-control signal on the $\overline{X}$ chart indicates that the center of the process distribution has changed. An out-of-control signal on the *R* chart indicates that the width of the process distribution has changed, although the process averages (the $\overline{X}$ values) may show no unusual variation. Think of it like shooting at a target: *R* changes when your shots form a tighter or looser pattern around the bull's-eye. $\overline{X}$ changes when the pattern is no longer centered around the bull's-eye.
- When the chart is first drawn, if the *R* chart is out of control, the control limits calculated for the $\overline{X}$ chart will not be valid. Find and eliminate the source of variation in the *R* chart, then start over to establish new control limits.

Process: *ZZ-400 Purity* Calculated by: *PW*

Data dates: *3/4 – 3/23/05* Calculation date: *3/29/05*

Step 1. Calculate average $\overline{X}$ and range R (the difference between the highest and lowest values) for each subgroup. Record on chart.

Number of values in each subgroup = n = *2*

Number of subgroups to be used = k = *20*

Step 2. Look up control limit factors.

n	A_2	D_3	D_4
2	1.880	—	3.267
3	1.023	—	2.574
4	0.729	—	2.282
5	0.577	—	2.114
6	0.483	—	2.004
7	0.419	0.076	1.924

A_2 = *1.880*

D_3 = ______

D_4 = *3.267*

Step 3. Calculate averages ($\overline{\overline{X}}$ and $\overline{R}$).

Sum of the averages = $\Sigma\overline{X}$ = *1991*

Average of the averages = $\overline{\overline{X}}$ = $\Sigma\overline{X} \div k$ = *1991* ÷ *20* = *99.54*

Sum of the ranges = ΣR = *3.6*

Average of the ranges = $\overline{R}$ = $\Sigma R \div k$ = *3.6* ÷ *20* = *0.18*

Step 4. Calculate control limits.

$3\hat{\sigma}$ estimate for $\overline{X}$ chart = $3\hat{\sigma}_{\overline{X}}$ = $A_2 \times \overline{R}$ = *1.880* × *0.18* = *.34*

Upper control limit for $\overline{X}$ chart = $UCL_{\overline{X}}$ = $\overline{\overline{X}} + 3\hat{\sigma}_{\overline{X}}$ = *99.54* + *.34* = *99.88*

Lower control limit for $\overline{X}$ chart = $LCL_{\overline{X}}$ = $\overline{\overline{X}} - 3\hat{\sigma}_{\overline{X}}$ = *99.54* – *.34* = *99.20*

Upper control limit for R chart = UCL_R = $D_4 \times \overline{R}$ = *3.267* × *0.18* = *0.59*

Lower control limit for R chart = LCL_R = $D_3 \times \overline{R}$ = ______ × ______ = ______

Figure 5.26 $\overline{X}$ and R chart example worksheet.

$\overline{X}$ and s chart

Description

An $\overline{X}$ and s chart is very similar to an $\overline{X}$ and R chart, with one exception: the standard deviation of the sample, s, is used to estimate subgroup variation. Computer software is desirable for performing the calculations.

When to Use

- When you have variable data, and . . .
- When enough data is available that subgroup size is 10 or greater, or . . .
- When you must rapidly detect very small process changes

Procedure

Follow the procedure for constructing an $\overline{X}$ and R chart. However, instead of an R chart, you will construct an s chart. Substitute the following formulas:

s chart: $s = \sqrt{\dfrac{\Sigma(X_i - \overline{X})^2}{(n-1)}}$ calculated for each subgroup

$$\overline{s} = \Sigma s \div k$$

$$\text{UCL}_s = B_4\overline{s}$$

$$\text{LCL}_s = B_3\overline{s}$$

$\overline{X}$ chart:

$$\text{UCL}_{\overline{X}} = \overline{\overline{X}} + A_3\overline{s}$$

$$\text{LCL}_{\overline{X}} = \overline{\overline{X}} - A_3\overline{s}$$

Values of A_3, B_3, and B_4 can be found in Table A.2.

Example

The Joint Commission on Accreditation of Healthcare Organizations (JCAHO) is identifying standardized performance measures to support performance improvement within hospitals and to permit rigorous comparison between hospitals. In the focus area of pneumonia treatment, one of the measures is antibiotic timing: time from initial hospital arrival to first dose of antibiotic. While some studies have used attribute data, measuring the proportion of patients who received antibiotic within the recommended time, the Joint Commission chose to measure timing as variable data, both to emphasize continuous improvement and to be able to statistically analyze trends over time. In the United States, almost 1000 patients per month enter hospitals with pneumonia, so there is ample data for using an $\overline{X}$ and s chart.

$\overline{X}$ and s charts look just like $\overline{X}$ and R charts. Figure 5.33, page 182 shows an $\overline{X}$ and s chart used in a short-run situation.

Considerations

- To calculate control limits for the average chart, do not use an overall s calculated from all the data. This will inflate the control limits if the process is not in control. Instead, calculate s for each subgroup, then average those to calculate $\overline{s}$.

chart of individuals

Also called: X chart, X-R chart, IX–MR chart, XmR chart, moving range (MR) chart

Description

The chart of individuals is a pair of control charts used to study variable data that are not generated frequently enough for an $\overline{X}$ and R chart and that form a normal distribution. Data are not subgrouped. Individual data points are plotted on one chart, and the differences between successive points, the moving ranges, are plotted on the second.

When to Use

- When you have variable data, and . . .
- When the distribution of data from the process is normal, and . . .
- When you cannot use an $\overline{X}$ and R chart because frequent data are costly or not available, such as with destructive testing or a slowly changing process, or . . .
- When you cannot use an $\overline{X}$ and R chart because the measurement remains constant for a relatively long period before the process changes, such as in batch operations where conditions within each batch are constant

Procedure

Construction

1. Determine the appropriate time period for collecting and plotting data. Collect at least 20 data points, arranged in time order, from the process to be studied.
2. Determine if the distribution of the data is normal. See *histogram* for a discussion of normal distribution, and see *normality test* for ways to determine whether it is normal.

3. Calculate the average, called $\overline{X}$ (X-bar), and the control limits using the worksheet for chart of individuals (Figure 5.27) and the chart of individuals form (Figure 5.28).
4. On the "value" part of the chart, mark the numerical scale, plot the individual values, and draw lines for the average and the control limits for X. On the "moving range" part of the chart, mark the numerical scale, plot the moving ranges, and draw lines for the average moving range and the UCL_{MR}. (LCL_{MR} is zero.)
5. Continue to follow steps 4, 5, and 6 of the basic procedure.

Process: ____________________ Calculated by: ____________________

Data dates: ____________________ Calculation date: ____________________

Step 1. Calculate $\overline{X}$.

Number of values	=	k	=	_____				
Sum of the values	=	ΣX	=	_____				
Average	=	$\overline{X}$	=	ΣX	÷	k		
			=	_____	÷	_____	=	_____

Step 2. Calculate $\overline{MR}$.

Calculate the ranges (the difference between two consecutive data points) and record on the form. Ignore negative signs.

Number of moving ranges	=	$k - 1$	=	_____				
Sum of the moving ranges	=	ΣMR	=	_____				
Average moving range	=	$\overline{MR}$	=	ΣMR	÷	$(k - 1)$		
			=	_____	÷	_____	=	_____

Step 3. Calculate control limits.

Estimate 3 standard deviations	=	$3\hat{\sigma}_X$	=	2.66	×	$\overline{MR}$		
			=	2.66	×	_____	=	_____
Upper control limit for X chart	=	UCL_X	=	$\overline{X}$	+	$3\hat{\sigma}_X$		
			=	_____	+	_____	=	_____
Lower control limit for X chart	=	LCL_X	=	$\overline{X}$	−	$3\hat{\sigma}_X$		
			=	_____	−	_____	=	_____
Upper control limit for MR chart	=	UCL_{MR}	=	3.267	×	$\overline{MR}$		
			=	3.267	×	_____	=	_____

Figure 5.27 Chart of individuals worksheet.

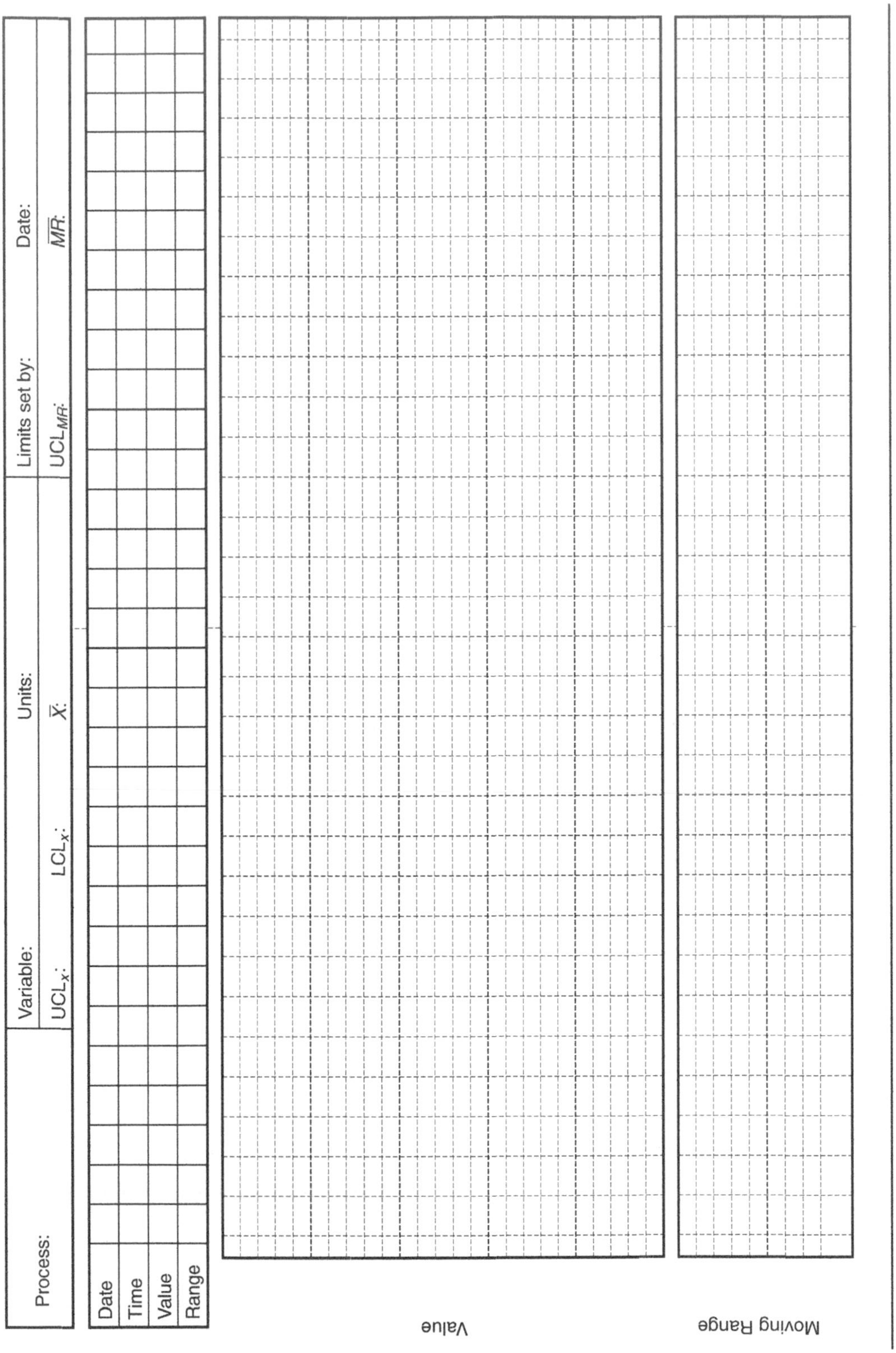

Figure 5.28 Chart of individuals.

Analysis

1. Check the "moving range" part of the chart for out-of-control signals. All the signals listed on page 157 can be used.
2. If the chart is in control, check the "value" part of the chart for out-of-control signals. All the signals listed on page 157 can be used.

Example

Figure 5.29 shows the "value" portion of the chart of individuals drawn by a team working to reduce the elapsed time between completing a technical survey and issuing the report. The team had more than two years of historical data. The initial control limits are calculated from data taken the first year. This chart shows the second half of the data.

There is an out-of-control signal around points 70 to 76: seven points in a row below the average. If the team had been keeping a control chart at the time, it should have investigated what was different then about the process. Notice the period between reports 95 and 117 when the process went wildly out of control. The team identified a special cause: new people in the group.

Immediately following that period, the team implemented its first improvements. Notice the drop in both the average and the UCL.

Also see the Medrad story on page 60 for another example of the use of a chart of individuals within the improvement process.

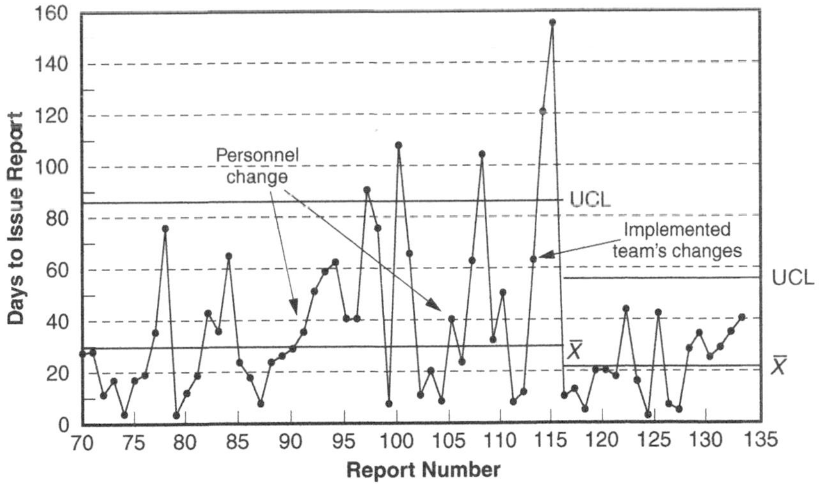

Figure 5.29 Chart of individuals example.

Considerations

- The chart of individuals does not pick up process changes as quickly as the $\bar{X}$ and R chart. If you have a choice between them, use the $\bar{X}$ and R chart.
- When data come from a nonnormal distribution (such as a skewed distribution), the control limits of a chart of individuals will not represent the process accurately. Depending on the situation, those limits will result in many more or many less out-of-control signals than are actually occurring. Use an $\bar{X}$ and R chart or moving average–moving range chart instead.
- Although the same word, range, is used in this chart and the $\bar{X}$ and R chart, they are calculated and interpreted differently. The chart of individuals does not separate variation in the process distribution's width from variation in the average, as the $\bar{X}$ and R chart does.

moving average–moving range chart

Also called: *MA–MR* chart

Description

The moving average–moving range (*MA–MR*) chart is a pair of control charts used to study variable data that are not generated frequently enough for an $\bar{X}$ and R chart, especially data that do not form a normal distribution. Data are subgrouped in a unique way: each successive subgroup drops the oldest measurement from the previous subgroup and adds one new one. This chart detects small process changes better than the chart of individuals.

When to Use

- When you have variable data, and . . .
- When you cannot use an $\bar{X}$ and R chart because frequent data are costly or not available, such as with destructive testing or a slowly changing process, or . . .
- When you cannot use an $\bar{X}$ and R chart because the measurement remains constant for a relatively long period before the process changes, such as in batch operations, and . . .
- When you cannot use a chart of individuals because the data are not normally distributed, or . . .
- When you want to detect small process changes

Procedure

Construction

1. Determine the appropriate time period for collecting and plotting data. Determine the number of data points per subgroup (n). Collect at least $20 + n - 1$ data points to start the chart. For example, with a subgroup size of three, you will need 22 data points.
2. Arrange data in subgroups and calculate subgroup averages and ranges, as follows. Subgroup averages and ranges are calculated differently for the *MA–MR* chart than for the $\overline{X}$ and R chart.
   a. Record the first n individual values in the first column on the moving average chart (Figure 5.23, page 163). These values form the first subgroup.
   b. Drop the oldest value from the subgroup and record the remaining values in the second column, starting at the top. Write the next value in the bottom position to complete the second subgroup.
   c. Continue the process for all the data. You should finish with $n - 1$ fewer subgroups than there are individual values.
   d. Calculate the average (*MA*) and range (*MR*) for each subgroup. See the example for a sample calculation.
3. If the raw data do not form a normal distribution, check whether the averages of the subgroups do. (See *normality test* for how to do this.) If not, increase the subgroup size.
4. Calculate $\overline{MA}$ (*MA*-bar), $\overline{MR}$ (*MR*-bar), and the control limits using the worksheet for the $\overline{X}$ and R chart or moving average–moving range chart (Figure 5.22, page 162).
5. On the "Average" part of the chart, mark the numerical scale, plot the subgroup averages, and draw lines for $\overline{MA}$ and for the averages' control limits. On the "Range" part of the chart, mark the numerical scale, plot the ranges, and draw lines for $\overline{MA}$ and the range control limits.
6. Continue to follow steps 4, 5, and 6 of the basic procedure.

Analysis

1. Check the moving range chart for out-of-control signals. "Points outside the control limits" is the only valid signal. Do *not* check for runs or trends. These signals are not valid because the same individual values are included in more than one subgroup.

2. If the moving range chart is in control, check the moving average chart for out-of-control signals. Again, the presence of points outside the control limits is the only valid signal.

Example

Individual values, in time order, are 3, 5, 6, 1, 2, 5. Table 5.2 shows columns of subgroups and the calculations for the averages and ranges. The control chart itself is not shown because it looks just like an $\overline{X}$ and R chart.

Considerations

- For subgroup sizes smaller than seven, LCL_R will be zero. Therefore, there will be no out-of-control signals below LCL_R.
- Just as on the $\overline{X}$ and R chart, an out-of-control signal on the *MA* chart indicates that the center of the process distribution has changed. An out-of-control signal on the *MR* chart indicates that the width of the process distribution has changed, although the process subgroup averages (the *MA* values) may show no unusual variation. Think of it like shooting at a target: *MR* changes when your shots form a tighter or looser pattern around the bull's-eye. *MA* changes when the pattern is no longer centered around the bull's-eye.
- When the chart is first drawn, if the *MR* chart is out of control, the control limits calculated for the *MA* chart will not be valid. Find and eliminate the source of variation in the *MR* chart and start over to establish new control limits.
- For situations where limited data are available because of short runs or frequently changing process targets, consider using target or short-run charts, described later in this section.

Table 5.2 *MA–MR* chart example calculation.

	Subgroup #			
	1	**2**	**3**	**4**
Value #1	3	5	6	1
Value #2	5	6	1	2
Value #3	6	1	2	5
Sum	14	12	9	8
Average (*MA*)	4.7	4.0	3.0	2.7
High	6	6	6	5
Low	3	1	6	1
Range (*MR*)	3	5	5	4

target chart

Also called: difference chart, deviation chart, or nominal chart

Description

A target chart is a variation that can be used with any of the basic variable charts. It allows the same characteristic from different parts or products to be plotted on one chart. The numbers plotted on the chart and used for calculating control limits are differences between the measured values and target values.

When to Use

- When you have variable data, and . . .
- When the same process generates different parts or products, each of which has a different target (set point, specification, and so on) for the characteristic you want to monitor, and . . .
- The variability (standard deviation) is the same for each different part or product

Procedure

1. Determine which basic control chart (chart of individuals, $\bar{X}$ and R, or $\bar{X}$ and s) is appropriate for the situation.
2. Determine the target value for each part or product. You may use historical data or specification limits. Common choices of targets are:
   a. The centerline from an existing control chart for the variable.
   b. The average from previous production data.
   c. The control chart centerline or production data average for a similar variable.
   d. The midpoint between upper and lower specification limits.
   e. A value that a process expert (operator, machinist, engineer) recommends.
   f. For one-sided specifications, a value at least three standard deviations away from the specification.
3. Follow the specific procedure for constructing the appropriate basic control chart, with the following changes. For each data point, calculate:

$$\bar{X}_t = \bar{X} - \text{Target} \qquad (\bar{X} \text{ and } R \text{ or } \bar{X} \text{ and } s \text{ chart})$$

$$\text{or } X_t = X - \text{Target} \qquad (\text{chart of individuals})$$

Plot $\overline{X}_t$ or X_t on the chart, and use it instead of $\overline{X}$ or X for all calculations. The vertical scale on the $\overline{X}_t$ or X_t chart will have zero at or close to the centerline, with values above and below zero showing the deviation from target. For an $\overline{X}$ and R or $\overline{X}$ and s chart, each subgroup must contain only data from one part or product. You cannot change parts or products within a subgroup. The range calculations will be the same as if you had used $\overline{X}$.

4. When the part or product changes, draw a vertical line on the chart and identify the part or product.

5. Analyze the charts for out-of-control signals. If the range chart is out of control, examine the culprit data separately. You may need to construct a separate control chart for the data with different variations. Alternatively, you may need to use a short-run control chart.

6. If the range chart is in control, analyze the $\overline{X}_t$ or X_t chart. In addition to the usual signals, examine the distance between the centerline and the zero line. This shows how far the process is from target. Notice whether particular parts or products are consistently off target in one direction or the other.

Example

Figure 5.30 shows a target chart of individuals for a paint-mixing process. Each point represents a different batch of paint. The variable being monitored is the solids content of the paint. Each of five types of paint, A through E, requires different solids content. By using

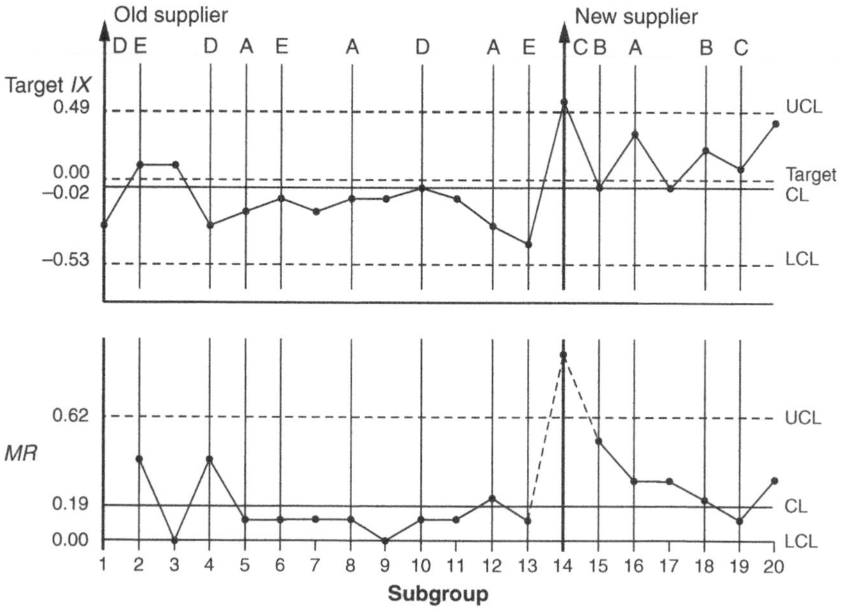

Figure 5.30 Target chart example.

a target chart and plotting the actual solids content minus that paint's target value, all data can be plotted on one chart. Vertical lines show where the paint type changed.

At subgroup 14, the *MR* chart was out of control. Out-of-control ranges cannot be used to calculate control limits for their accompanying *X* charts. Investigation showed that the out-of-control situation was caused by changing suppliers and should not recur. That point was removed and the remaining *MR* data was used to calculate control limits for the target *X* chart.

Considerations

- Target charts are useful for job shops, just-in-time, and short runs. They reduce the amount of paperwork, since separate charts aren't needed for each part or product, and let you start charting sooner, since you need only 20 data points from all production runs, not from each. Most important, they let you focus on changes in the process, regardless of part or product.
- By marking the chart every time the part or product changes, you can monitor trends or special causes affecting just one part as well as those affecting the entire process.
- Occasionally, circumstances require the process to be aimed at a point different from the midpoint between specifications. In these cases, the judgment of a process expert should guide the choice of targets.
- Target charts cannot be used for attribute data. Instead, use short-run charts.

attribute control charts

Two kinds of attribute data arise in control charting. In the first kind, each item is either entirely good or entirely bad. This is called the number or percentage of *defective* or *nonconforming items*. For example, a car coming off an assembly line is either accepted or rejected; paperwork is accurate or inaccurate. This kind of data is placed on *p* or *np* charts.

In the second kind of data, each item can have multiple defects or flaws. This is called the number or proportion of *defects* or *nonconformities*. For example, you may count how many bubbles are in a car's paint or how many errors are found in a report. This kind of data is placed on *c* or *u* charts.

The same situation can be analyzed for nonconforming items or for nonconformities. It depends upon which measurement you decide is most useful to study. You could construct a *p* chart for the number of inaccurate reports. It would not show how many errors there are. Or you could construct a *u* chart for the number of errors. It would not show how many reports are inaccurate.

Attribute data contain less information than variable data. For example, a part may be nonconforming because it is the wrong length. Attribute data tells us only that the part is nonconforming. Variable data tell us exactly how long the part is and therefore how far off it is. Charting lengths would give much more information about the process than charting the number of nonconforming items. Whenever possible, chart variable data instead of attribute data.

The terms "defect" and "defective" can have legal implications. Therefore, this book uses the more generic terms "nonconformity" and "nonconforming item."

p chart

Also called: proportion chart

Description

The *p* chart is an attribute control chart used to study the proportion (fraction or percentage) of nonconforming or defective items. Often, information about the types of nonconformities is collected on the same chart to help determine the causes of variation.

When to Use

- When counting nonconforming items (not total number of nonconformities) and . . .
- When the sample size varies

Procedure

Construction

1. Gather the data.
   a. Define a sample. It should be large enough to contain several nonconforming items.
   b. Determine the number of samples (k) to collect before calculating control limits. Collect data from at least 20 samples.
   c. For each sample, count the total number of items (n) and the number of nonconforming items (X). Calculate the proportion nonconforming, $p = X/n$.
   d. Record the values of X, n, and p on the attribute chart (Figure 5.31).
2. Mark the numerical scale and plot the values of p on the chart. Connect consecutive points.
3. Calculate the average proportion nonconforming $\bar{p}$ and the control limits, using the p chart worksheet (Figure 5.32). If sample sizes are within 20 percent of each other, calculate only one set of control limits, using an average sample

Process:	Data being counted:		Limits set by:	Date:
	UCL:	LCL:	Average:	

Date																				
Time of lot																				
Number nonconforming																				
Lot size																				
Proportion nonconforming																				
	1	2	3	4	5	6	7	8	9	10	11	12	13	14	15	16	17	18	19	20

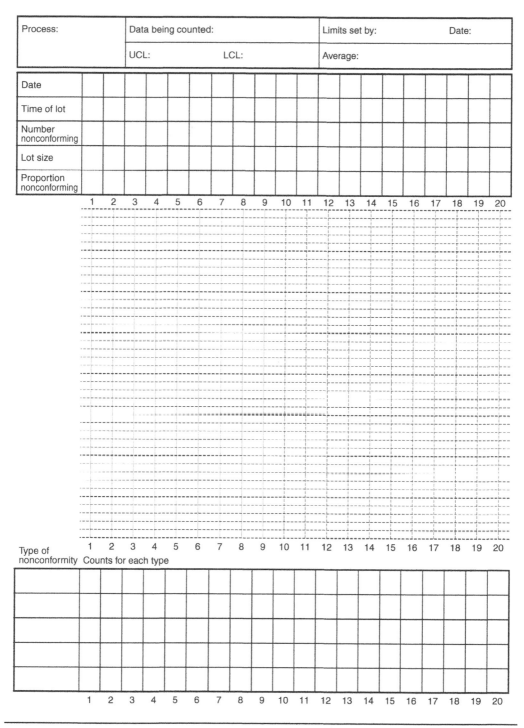

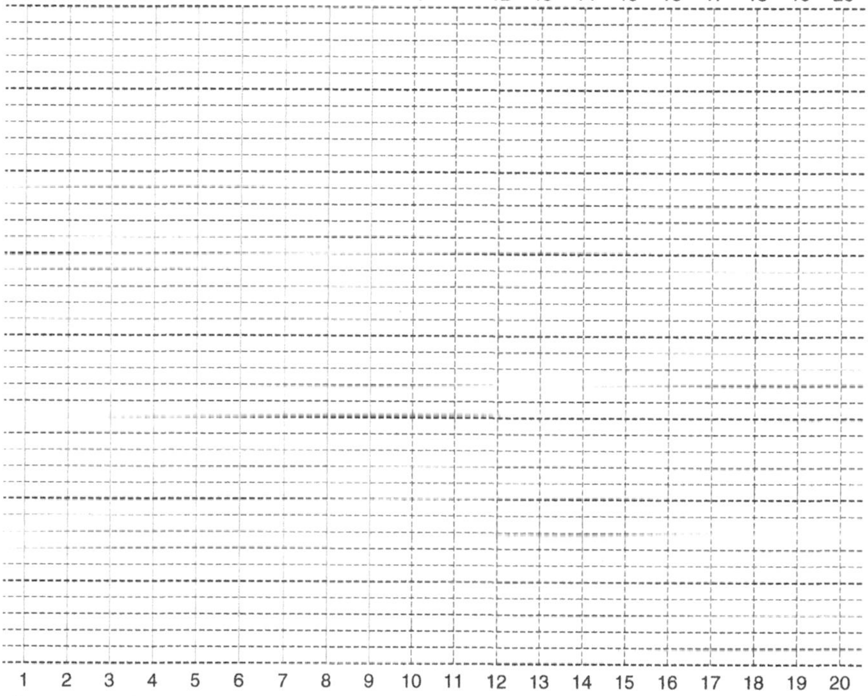

Type of nonconformity	Counts for each type																			
	1	2	3	4	5	6	7	8	9	10	11	12	13	14	15	16	17	18	19	20

Figure 5.31 Attribute chart.

p Chart Worksheet

Proportion nonconforming items; sample size varies

Process: ______________________ Calculated by: ______________________

Data dates: ______________________ Calculation date: ______________________

- The number of items, n_i, in each sample can be different for each sample.
- The number of nonconforming items in each sample is X_i.
- Plot $p_i = X_i \div n_i$.

Step 1. Calculate average.

Total number of nonconforming items $= \Sigma X_i =$ ______

Total number of items in all samples $= \Sigma n_i =$ ______

Average proportion nonconforming items $= \bar{p} = \Sigma X_i \div \Sigma n_i =$ ______ $\div$ ______ $=$ ______

Step 2. Calculate $3\hat{\sigma}_p$.

If sample size varies by less than 20%, you may use an average sample size $\bar{n}$.

$\bar{n} = \Sigma n_i \div k$ *k = number of samples*

$=$ ______ $\div$ ______ $=$ ______

Use only the left column of calculations, and use $\bar{n}$ instead of n_s. Omit step 4.

Smallest sample size n_s calculates outer $3\hat{\sigma}_{po}$

$n_s =$ ______

$1-\bar{p} = 1-$ ______ $=$ ______

$3\hat{\sigma}_{po} = 3\sqrt{\bar{p} \quad (1-\bar{p}) \div \quad n_s}$

$= 3\sqrt{\_\_\_\_ \times \_\_\_\_ \div \_\_\_\_}$

$=$ ______

Largest sample size n_l calculates inner $3\hat{\sigma}_{pi}$

$n_l =$ ______

$3\hat{\sigma}_{pi} = 3\sqrt{\bar{p} \quad (1-\bar{p}) \div \quad n_l}$

$= 3\sqrt{\_\_\_\_ \times \_\_\_\_ \div \_\_\_\_}$

$=$ ______

Step 3. Calculate control limits.

Outer Limits

Upper control limit UCL_{po}

$= \bar{p} + 3\hat{\sigma}_{po}$

$=$ ______ $+$ ______ $=$ ______

Lower control limit LCL_{po}

$= \bar{p} - 3\hat{\sigma}_{po}$

$=$ ______ $-$ ______ $=$ ______

Inner Limits

Upper control limit UCL_{pi}

$= \bar{p} + 3\hat{\sigma}_{pi}$

$=$ ______ $+$ ______ $=$ ______

Upper control limit UCL_{pi}

$= \bar{p} - 3\hat{\sigma}_{pi}$

$=$ ______ $-$ ______ $=$ ______

Step 4. If a point is between inner and outer limits, calculate and plot exact limits for its sample size n_i:

$UCL_p = \bar{p} + 3\sqrt{\bar{p} \quad (1-\bar{p}) \div \quad n_i}$

$=$ ______ $+ 3\sqrt{\_\_\_\_ \times \_\_\_\_ \div \_\_\_\_}$

$=$ ______ $+$ ______

$=$ ______

$LCL_p = \bar{p} - 3\sqrt{\bar{p} \quad (1-\bar{p}) \div \quad n_i}$

$=$ ______ $- 3\sqrt{\_\_\_\_ \times \_\_\_\_ \div \_\_\_\_}$

$=$ ______ $-$ ______

$=$ ______

Figure 5.32 *p* chart worksheet.

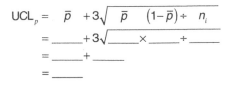

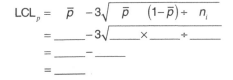

size $\bar{n}$. Use just the left column of the worksheet. If sample sizes vary by more than 20 percent, calculate two sets of control limits: inner and outer upper control limits and inner and outer lower control limits.

4. Draw lines for $\bar{p}$ and the control limits on the chart and label them.
5. Continue to follow steps 4, 5, and 6 of the basic procedure.

Analysis

Check the chart for out-of-control signals. All the signals listed on page 157 can be used, except for this change to the "single point outside" signal when using inner and outer limits:

- If a point falls between the inner and outer limits, calculate exact control limits for that point. Use the actual sample size for that point and the formulas on the *p* chart worksheet. The process is out of control if the point falls outside these limits.

Examples

Here are some data requiring a *p* chart:

- Number of inaccurate invoices per day. The number of invoices processed daily varies.
- Number of off-color batches per week. Batch processing time varies.
- Number of defective assemblies per shift. Production rate varies.

Figure 5.33 shows a *p* chart used to monitor invoices missing cost center numbers. This chart is from the Medrad Freight Processing Team story on page 60. The dramatic drop occurred when the team introduced a change to the process. The chart is out of control, but process changes were new. The control limits are conditional.

On Medrad's chart, control limits are shown for each point. This helps you see how changing sample size affects the limits. To calculate just inner and outer control limits (as in step 3) for the set of data after the process change, the largest sample size, 1498 for point 21, and the smallest sample size, 451 for point 13, would be used. The inner and outer limits are the limits calculated for those points. Exact lower limits need to be calculated only for points falling between 4.1 and 5.8 (points 11, 16, 21, 22, and 27) and exact upper limits only for points falling between 10.0 and 11.7 (point 24).

Considerations

- If the number of items per sample varies, you must use a *p* chart. If all the samples have the same number of items, you can use either the *np* chart or *p* chart.
- Each item is either conforming or nonconforming. Do not count multiple nonconformities per item. If you wish to count several nonconformities per item, use a *c* chart or *u* chart.

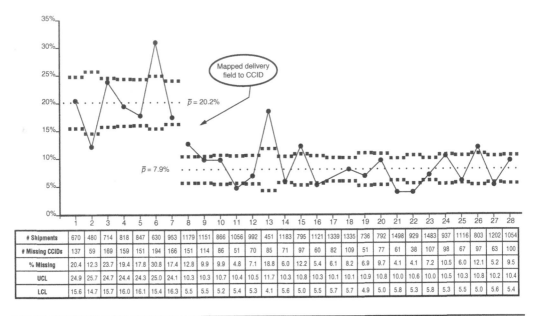

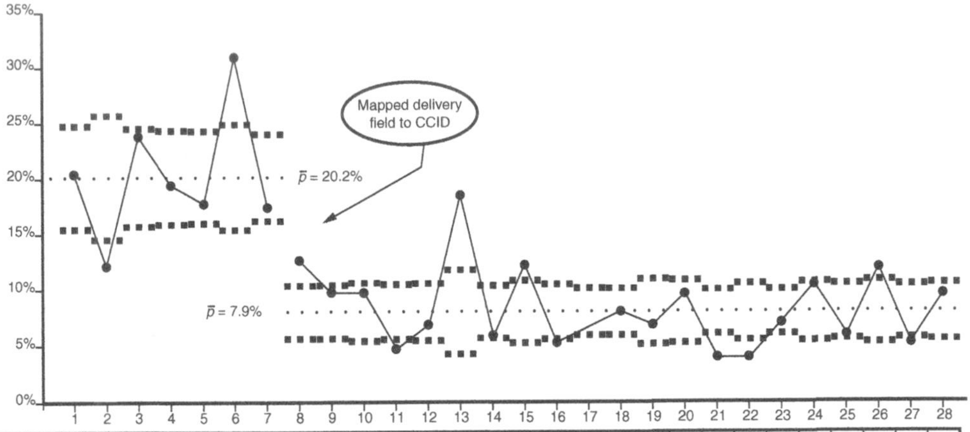

# Shipments	670	480	714	818	847	630	953	1179	1151	866	1056	992	451	1183	795	1121	1339	1335	736	792	1498	929	1483	937	1116	803	1202	1054
# Missing CCIDs	137	59	169	159	151	194	166	151	114	86	51	70	85	71	97	60	82	109	51	77	61	38	107	98	67	97	63	100
% Missing	20.4	12.3	23.7	19.4	17.8	30.8	17.4	12.8	9.9	9.9	4.8	7.1	18.8	6.0	12.2	5.4	6.1	8.2	6.9	9.7	4.1	4.1	7.2	10.5	6.0	12.1	5.2	9.5
UCL	24.9	25.7	24.7	24.4	24.3	25.0	24.1	10.3	10.3	10.7	10.4	10.5	11.7	10.3	10.8	10.3	10.1	10.1	10.9	10.8	10.0	10.6	10.0	10.5	10.3	10.8	10.2	10.4
LCL	15.6	14.7	15.7	16.0	16.1	15.4	16.3	5.5	5.5	5.2	5.4	5.3	4.1	5.6	5.0	5.5	5.7	5.7	4.9	5.0	5.8	5.3	5.8	5.3	5.5	5.0	5.6	5.4

Figure 5.33 *p* chart example.

- Theoretically, control limits should be calculated for each different sample size. Charts done this way show a control limit line that looks like a city skyline, changing around every data point, as illustrated in the example. Practically, using an average sample size works as long as sample sizes do not vary by more than 20 percent. When they vary more than that, using inner and outer control limits reduces the labor of calculation to just the occasions when points fall between those lines.
- If a sample is larger or smaller than the ones used to calculate inner and outer limits, recalculate those limits or calculate exact limits for that point.
- See also *short run charts* for another way to simplify control limits.

np chart

Description

The *np* chart is an attribute control chart used to study the number of nonconforming or defective items. Often, information about the types of nonconformities is collected on the same chart to help determine the causes of variation.

When to Use

- When counting nonconforming items (not total number of nonconformities), and . . .

- When the sample size does not vary

Procedure

Construction

1. Gather the data.
   a. Define a sample. It should be large enough to contain several nonconforming items, and its size must be constant. Determine the sample size, n.
   b. Determine the number of samples to collect before calculating control limits. Collect data from at least 20 samples.
   c. For each sample, count the number of nonconforming items, X.
   d. Record the values of X and n on the attribute chart (Figure 5.31, page 179). Ignore the row for proportion nonconforming.
2. Mark the numerical scale and plot the values of X on the chart. Connect consecutive points.
3. Calculate the average number of nonconforming items $n\bar{p}$ and the control limits, using the np chart worksheet (Figure 5.34).
4. Draw lines for $n\bar{p}$ and the control limits on the chart and label them.
5. Continue to follow steps 4, 5, and 6 of the basic procedure.

Analysis

Check the chart for out-of-control signals. All the signals listed on page 157 can be used.

Example

Here are some data requiring an *np* chart:

- Number of defective drums per lot. The lot size is always 60.
- Number of broken cookies in a 24-cookie package.
- Number of defective CDs in a random sample of 100 CDs.

Considerations

- For an *np* chart, all samples must have the same number of items. If the number of items in a sample varies, use the *p* chart.
- Each item is either good or bad. If you wish to count multiple nonconformities per item, use a *c* or *u* chart.

np Chart Worksheet

Number of nonconforming items; sample size is constant

Process: ____________________ Calculated by: ____________________

Data dates: ____________________ Calculation date: ____________________

- The number of items, n_i, in each sample must be the same.
- Plot X_i, the number of nonconforming items in a sample.

Step 1. Calculate average.

Sample size $= n =$ ______

Number of samples $= k =$ ______

Total number of nonconforming items $= \Sigma X_i =$ ______

Average number of nonconforming items $= n\bar{p} = \Sigma X_i \div k =$ ______ $\div$ ______ $=$ ______

Step 2. Calculate $3\hat{\sigma}_{np}$.

$$\bar{p} = n\bar{p} \div n$$
$$= \_\_\_\_\_\_ \div \_\_\_\_\_\_$$
$$= \_\_\_\_\_\_$$

$$3\hat{\sigma}_{np} = 3\sqrt{n\bar{p} \times (1 - \bar{p})}$$
$$= 3\sqrt{\_\_\_\_\_\_ \times (1 - \_\_\_\_\_\_)}$$
$$= 3\sqrt{\_\_\_\_\_\_ \times \_\_\_\_\_\_}$$
$$= \_\_\_\_\_\_$$

Step 3. Calculate control limits.

Upper control limit UCL_{np}

$$= n\bar{p} + 3\hat{\sigma}_{np}$$
$$= \_\_\_\_\_\_ + \_\_\_\_\_\_ = \_\_\_\_\_\_$$

Lower control limit UCL_{np}

$$= n\bar{p} - 3\hat{\sigma}_{np}$$
$$= \_\_\_\_\_\_ - \_\_\_\_\_\_ = \_\_\_\_\_\_$$

Figure 5.34 *np* chart worksheet.

c chart

Also called: count chart

Description

The *c* chart is an attribute control chart used to study the number of nonconformities or defects. Often, information about the types of nonconformities is collected on the same chart to help determine the causes of variation.

When to Use

- When counting nonconformities (not the number of nonconforming items), and . . .
- When the sample size does not vary, and . . .
- When the opportunity for nonconformities does not vary from one sample to the next

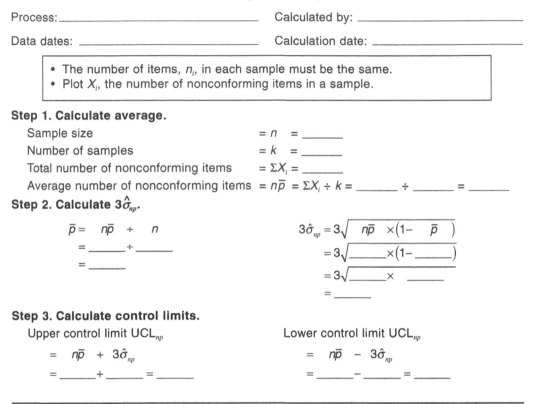

c Chart Worksheet

Number of nonconformities; sample size is constant

Process: ____________________ Calculated by: ____________________

Data dates: ____________________ Calculation date: ____________________

- The number of items in each sample must be the same.
- Plot c_i, the total number of nonconformities in a sample.

Step 1. Calculate average.

Number of samples $= k =$ ______

Total number of nonconformities $= \Sigma c_i =$ ______

Average number of nonconforming items $= \bar{c} = \Sigma c_i \div k =$ ______ $\div$ ______ $=$ ______

Step 2. Calculate $3\hat{\sigma}_c$.

$$3\hat{\sigma}_c = 3\sqrt{\bar{c}}$$
$$= 3\sqrt{\_\_\_\_\_\_}$$
$$= \_\_\_\_\_\_$$

Step 3. Calculate control limits.

Upper control limit UCL_c

$= \bar{c} + 3\hat{\sigma}_c$

$=$ ______ $+$ ______ $=$ ______

Lower control limit LCL_c

$= \bar{c} - 3\hat{\sigma}_c$

$=$ ______ $-$ ______ $=$ ______

Figure 5.35 *c* chart worksheet.

Procedure

Construction

1. Gather the data.
    a. Define a sample. It should be large enough that nonconformities have a chance to occur, and its size must be constant.
    b. Select the number of samples to be collected before calculating control limits. Collect data from at least 20 samples.
    c. For each sample, count the number of nonconformities (c).
    d. Record the values of c on the attribute chart (Figure 5.31, page 179). Ignore the rows for sample size and proportion nonconforming.
2. Mark the numerical scale and plot the values of c on the chart. Connect consecutive points.
3. Calculate the average number of nonconformities $\bar{c}$ and the control limits, using the c chart worksheet (Figure 5.35).

4. Draw lines for $\bar{c}$ and the control limits on the chart and label them.
5. Continue to follow steps 4, 5, and 6 of the basic procedure.
6. On the bottom section of the chart, tally different types of nonconformities.

Analysis

Check the chart for out-of-control signals. All the signals listed on page 157 can be used.

Examples

Here are some data requiring a *c* chart:

- Number of errors per 100 invoices
- Fabric flaws per 10 square feet
- Number of black specs in one gram of product

Considerations

- Each sample must have the same opportunity for nonconformities to occur. For example, number of errors per report would not be appropriate for a *c* chart if the reports vary in length. A more appropriate sample would be number of errors per 10 pages.
- For a *c* chart, more than one nonconformity can be counted per item. If you wish to count how many items are nonconforming, use a *p* or *np* chart.
- The counts of each type of nonconformity at the bottom of the worksheet can be used to construct a Pareto chart or do other analysis.

u chart

Description

The *u* chart is an attribute control chart used to study the proportion of nonconformities or defects. A proportion is used instead of a count when the opportunity for nonconformities changes from one sample to another, and so the fair way to express frequency of nonconformities is on a per-unit basis. Often, information about the types of nonconformities is collected on the same chart to help determine the causes of variation.

When to Use

- When counting nonconformities (not the number of nonconforming items), and . . .
- When the sample size varies, or . . .
- When the opportunity for nonconformities changes from one sample to the next

Procedure

Construction

1. Gather the data.
   a. Define a sample. It should be large enough so that nonconformities have a chance to occur.
   b. Select the number of samples to be collected before calculating control limits. Collect data from at least 20 samples.
   c. For each sample, determine the total sample size n, that is, the opportunity for nonconformities within the sample. Determine the number of nonconformities (c). Calculate the proportion of nonconformities per unit, $u = c / n$.
   d. Record the values of c, n, and u on the attribute chart (Figure 5.31, page 179).
2. Mark the numerical scale and plot the values of u on the chart. Connect consecutive points.
3. Calculate the average number of nonconformities $\bar{u}$ and the control limits, using the u chart worksheet (Figure 5.36). If sample sizes are within 20 percent of each other, calculate only one set of control limits, using an average group size n. Use just the left column of the worksheet. If sample sizes vary by more than 20 percent, calculate two sets of control limits: inner and outer upper control limits and inner and outer lower control limits.
4. Draw lines for $\bar{u}$ and the control limits on the chart and label them. If you have calculated inner and outer limits, draw all four control limits.
5. Continue to follow steps 4, 5, and 6 of the basic procedure.
6. On the bottom section of the chart, tally different types of nonconformities.

Analysis

Check the chart for out-of-control signals. All the signals listed on page 157 can be used, except for this change to the "single point outside" signal when using inner and outer limits:

u Chart Worksheet

Proportion nonconformities; sample size varies

Process: ______________________ Calculated by: ______________________

Data dates: ______________________ Calculation date: ______________________

- The sample size, n_i, for each sample can be different for each sample.
- The number of nonconformities in each sample is c_i.
- Plot $u_i = c_i \div n_i$.

Step 1. Calculate average.

Total number of nonconformities $= \Sigma c_i =$ ______

Total of all samples' sizes $= \Sigma n_i =$ ______

Average proportion nonconformities $= \bar{u} = \Sigma c_i \div \Sigma n_i =$ ______ $\div$ ______ $=$ ______

Step 2. Calculate $3\hat{\sigma}_u$.

If sample size varies by less than 20%, you may use an average sample size $\bar{n}$.

$\bar{n} = \Sigma n_i \div k$ $\qquad$ *k = number of samples*

$=$ ______ $\div$ ______ $=$ ______

Use only the left column of calculations, and use $\bar{n}$ instead of n_s. Omit step 4.

Smallest sample size n_s calculates outer $3\hat{\sigma}_{uo}$

$n_s =$ ______

$3\hat{\sigma}_{uo} = 3\sqrt{\bar{u} \div n_s}$

$= 3\sqrt{\_\_\_\_\_\_ \div \_\_\_\_\_\_}$

$=$ ______

Largest sample size n_l calculates inner $3\hat{\sigma}_{ui}$

$n_l =$ ______

$3\hat{\sigma}_{ui} = 3\sqrt{\bar{u} \div n_l}$

$= 3\sqrt{\_\_\_\_\_\_ \div \_\_\_\_\_\_}$

$=$ ______

Step 3. Calculate control limits.

Outer Limits

Upper control limit UCL_{uo}

$= \bar{u} + 3\hat{\sigma}_{uo}$

$=$ ______ $+$ ______ $=$ ______

Lower control limit LCL_{uo}

$= \bar{u} - 3\hat{\sigma}_{uo}$

$=$ ______ $-$ ______ $=$ ______

Inner Limits

Upper control limit UCL_{ui}

$= \bar{u} + 3\hat{\sigma}_{ui}$

$=$ ______ $+$ ______ $=$ ______

Lower control limit LCL_{ui}

$= \bar{u} - 3\hat{\sigma}_{ui}$

$=$ ______ $-$ ______ $=$ ______

Step 4. If a point is between inner and outer limits, calculate and plot exact limits for its sample size n_i:

$UCL_u = \bar{u} + 3\sqrt{\bar{u} \div n_i}$

$=$ ______ $+ 3\sqrt{\_\_\_\_\_\_ \div \_\_\_\_\_\_}$

$=$ ______ $+$ ______

$=$ ______

$LCL_u = \bar{u} - 3\sqrt{\bar{u} \div n_i}$

$=$ ______ $- 3\sqrt{\_\_\_\_\_\_ \div \_\_\_\_\_\_}$

$=$ ______ $-$ ______

$=$ ______

Figure 5.36 *u* chart worksheet.

- If a point falls between the inner and outer limits, calculate exact control limits for that point. Use the actual sample size for that point and the formulas on the *u* chart worksheet. The process is out of control if the point falls outside these limits.

Examples

Here are some data where the opportunity for nonconformities, *n*, changes, requiring a *u* chart:

- Defects per window pane. *n* = pane area (several sizes are manufactured)
- Bad solders per circuit board. *n* = number of solders on a board (boards vary in complexity)
- Changes to engineering drawings per week. *n* = number of drawings finished per week

Considerations

- For a *u* chart, more than one nonconformity can be counted per sample. If you wish to count the number of items nonconforming, use a *p* or *np* chart.
- Theoretically, control limits should be calculated for each different sample size. Charts done this way show a control limit line that looks like a city skyline, changing around every data point. Practically, using an average sample size works as long as sample sizes do not vary by more than 20 percent. When they vary more than that, using inner and outer control limits reduces the labor of calculation to just the occasions when points fall between those lines.
- If a sample is larger or smaller than the ones used to calculate the inner and outer limits, recalculate those limits or calculate exact limits for that point.
- See also short run charts for another way to simplify control limits for *u* charts.
- The counts of each type of nonconformity at the bottom of the worksheet can be used to construct a Pareto chart or do other analysis.

short run chart

Also called: stabilized chart, Z chart

Description

A short run chart is a variation that can be used with any of the basic charts. It allows a characteristic from different parts or products to be plotted on one chart, even when variability changes. The numbers plotted on the top chart are differences between the measured values and target values, divided by a target range or standard deviation. The numbers plotted on the bottom chart are range or standard deviation divided by a target range or standard deviation. This transforms the data to unitless ratios that can be compared.

When to Use

- When data are either variable or attribute, and . . .
- When the same process generates different parts or products, at least one of which has a different target (set point, specification, and so on) for the characteristic you want to monitor, and . . .
- The variability (standard deviation) or the unit of measure is different for at least one part or product

Procedure: Variable Data

1. Determine which basic control chart (chart of individuals, $\bar{X}$ and R, $\bar{X}$ and s) is appropriate for the situation.
2. Determine the target X or $\bar{X}$ for each part or product, using one of the methods described in step 2 of the procedure for target charts, page 175.
3. Determine a target $\bar{R}$ or $\bar{s}$ for each part or product. The best choice is $\bar{R}$ or $\bar{s}$ from an existing control chart for the variable. If this is not available, other ways of determining the target $\bar{R}$ or $\bar{s}$ are, in order of preference:

   a. Calculate from previous production data.

$$\text{Target } \bar{R} = s \times d_2 \div c_{4\text{m}}$$

$$\text{Target } \bar{s} = s \times c_{4\text{n}} \div c_{4\text{m}}$$

where d_2, c_{4n}, and c_{4m} are from Table A.2, using for n on the table:

for d_2 and c_{4n}: n = subgroup size of the short run chart

for c_{4m}: $n = m$ = number of measurements in the data used to calculate s

and where s is the sample standard deviation

$$s = \sqrt{\frac{\Sigma\left(X_i - \bar{X}\right)^2}{m-1}} \quad \text{calculated for each subgroup}$$

where $\bar{X}$ is calculated from the previous production data

b. Control chart or previous production data for a similar variable. Use the formulas in (a) if you use production data.

c. Calculation from process tolerance stated by a process expert (operator, machinist, engineer):

$$\text{Target } \bar{R} = d_2 \times \text{tolerance width} \div 6\ \text{Cpk}_{\text{GOAL}}$$

$$\text{Target } \bar{s} = c_4 \times \text{tolerance width} \div 6\ \text{Cpk}_{\text{GOAL}}$$

Here and in section d below, d_2 and c_4 are from Table A.2, using n = subgroup size of the short run chart and Cpk_{GOAL} is the goal for the process capability index.

d. Calculation from specification limits:

$$\text{Target } \bar{R} = d_2 \times \ (\text{USL} - \text{LSL}) \div 6\ \text{Cpk}_{\text{GOAL}}$$

$$\text{Target } \bar{s} = c_4 \times \ (\text{USL} - \text{LSL}) \div 6\ \text{Cpk}_{\text{GOAL}}$$

Or for one sided specifications:

$$\text{Target } \bar{R} = d_2 \times |\text{Specification limit} - \text{Target } X \text{ or } \bar{X}| \div 3\text{Cpk}_{\text{GOAL}}$$

$$\text{Target } \bar{s} = c_4 \times |\text{Specification limit} - \text{Target } X \text{ or } \bar{X}| \div 3\text{Cpk}_{\text{GOAL}}$$

4. Follow the specific procedure for constructing the appropriate basic control chart, with these changes. (For a chart of individuals, $\bar{X} = X$, since $n = 1$.)

$\bar{X}$ or X chart: Plot $(\bar{X} - \text{Target}) \div \text{Target } \bar{R}$

Centerline = 0

UCL = A_2

LCL = $-A_2$

R chart: Plot $R \div \text{Target } \bar{R}$

Centerline = 1

UCL = D_4

LCL = D_3

s chart: Plot $s \div \text{Target } \bar{s}$

Centerline = 1

UCL = B_4

LCL = B_3

Each subgroup must contain only data from one part or product. You cannot change parts or products within a subgroup.

5. When the part or product changes, draw a vertical line on the chart and identify the part or product.

6. Analyze the charts for out-of-control signals. When the chart is new, if either chart is out of control, the initial target values may have been inaccurate. Update the targets to reflect the actual process, or, if possible, adjust the process to meet the targets.

7. If the range chart is in control, analyze the $\overline{X}_t$ or X_t chart. In addition to the usual signals, notice whether particular parts or products are consistently off target in one direction or the other.

Example

Figure 5.37 shows a short run $\overline{X}$ and s chart. The torque of self-locking nuts is being measured for three different locking systems. Each locking system has a different target torque and a different target standard deviation, based on historical data. The production rate is high, so ten samples are taken per batch of nuts and an $\overline{X}$ and s chart is appropriate. Vertical lines show where production changed to a different locking system.

Standard deviation for types B and C is higher than the historical data. Type A's torque has decreased from historical data, while type B's has increased and type C's appears to be in control.

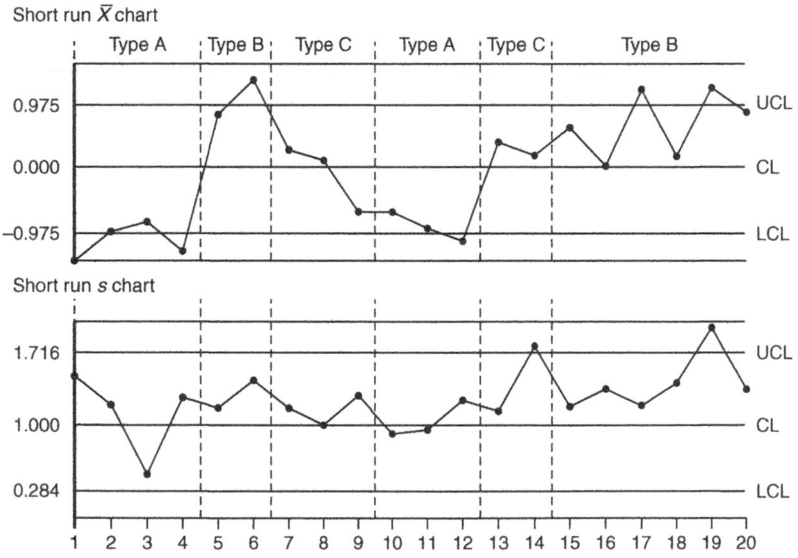

Figure 5.37 Short run $\overline{X}$ and s chart example.

Procedure: Attribute Data

1. Determine which attribute control chart (p, np, c, or u) is appropriate for the situation.
2. Using the worksheet for the appropriate chart, determine a target average ($\bar{p}$, $n\bar{p}$, $\bar{c}$, or $\bar{u}$) for each part or product. If historical data is not available, use one of the other methods described in step 2 of the procedure for target charts, page 175.
3. Each time data is collected from the process, use formulas from the attribute chart worksheet to calculate the appropriate attribute (p, np, c, or u) and σ. Then calculate:

$$Z = (\text{attribute} - \text{target average}) \div \sigma$$

Plot Z on the short run chart. UCL = 3, LCL = –3, and centerline = 0.

5. When the part or product changes, draw a vertical line on the chart and identify the part or product.
6. Analyze the charts for out-of-control signals. When the chart is new, if the chart is out of control, the initial target values may have been inaccurate. Update the targets to reflect the actual process, or, if possible, adjust the process to meet the targets. Notice whether particular parts or products are consistently off target in one direction or the other.

Considerations

- Short run charts are useful for job shops, just-in-time, and short runs. They reduce the amount of paperwork, since separate charts aren't needed for each part or product, and let you start charting sooner, since control limits are known before any data is collected. Most important, they let you focus on changes in the process, regardless of part or product.
- There are disadvantages to short-run charts. Compared to traditional charts, more calculations must be made every time a point is plotted. Also, because the charts are dimensionless and the numbers are different from process-generated numbers, it's harder to analyze the results and to catch errors.
- By marking the chart every time the part or product changes, you can monitor trends or special causes affecting just one part as well as those affecting the entire process.
- Short run p charts and u charts can be used to create consistent control limits, instead of the shortcuts described with the basic p chart and u chart procedures.
- Short run charts, target charts, and even the basic control charts can be used when a process is just starting, even before there are twenty data points. However, different constants must be used in the control limit formulas. See Pyzdek for more information.

group chart

Also called: multiple characteristic chart

Description

A group chart is a variation that can be used with any of the basic control charts. It allows the same type of measurement at different locations or from different process streams to be plotted on one chart. When group charts are combined with target or short run charts, even different types of measurements can be tracked on the same chart.

When to Use

- When you want to monitor on one chart similar measurements at multiple locations, such as diameters at different points along a part or temperatures at several thermocouples, or . . .
- When you want to monitor on one chart several types of measurements, such as all analytical results for a batch of chemicals or several different dimensions of a part

Procedure

1. Determine which basic control chart (chart of individuals, $\overline{X}$ and R, $\overline{X}$ and s, p, np, c, or u) is appropriate for the situation. Determine also whether a target chart or short run chart is required to handle different process aims or different standard deviations.
2. Collect data in groups. Each group contains all the measurements from one part, product, batch, or sampling time. If $\overline{X}$ and R or $\overline{X}$ and s is the appropriate basic chart, place the data in each group into subgroups, as usual, and calculate subgroup averages. Each subgroup must contain data from only one location or one type of measurement. If appropriate, use formulas from the target chart or short run chart procedures to adjust X, $\overline{X}$, or the attribute value.
3. In each group, choose the largest plot point value (MAX) and the smallest plot point value (MIN). Plot these values at the same x-axis position, one above the other. Label each point with its location or type of measurement. If a range chart is appropriate, similarly plot the largest and smallest ranges.
4. Follow the appropriate basic, target, or short run control chart procedure to calculate centerlines and control limits. Use all the measurements, not just the MAX and MIN, in the calculations. However, use control limits only if the data streams from which the measurements are taken are independent.

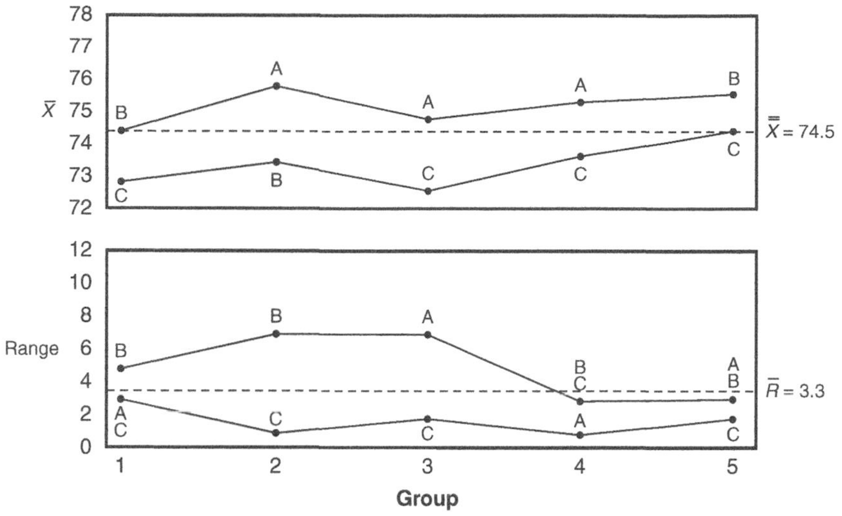

Figure 5.38 Group $\bar{X}$ and R chart example.

5. Analyze the charts for overall trends or patterns and any patterns within a group. If you are using control limits, analyze the charts for out-of-control signals.

Example

Figure 5.38 shows a group $\bar{X}$ and R chart. The diameter of a cylindrical part is measured at three different locations along the shaft to determine shaft uniformity. Four parts are measured for each subgroup. The average is calculated for each location, and then the highest and lowest averages are plotted on the $\bar{X}$ chart. Similarly, the highest and lowest ranges (not necessarily the ranges of the locations plotted on the $\bar{X}$ chart) are plotted on the R chart. Each point is coded A, B, or C to indicate which location it represents. Several points have two letters, indicating that two locations had the same value.

Control limits are not used because measurements from the three locations are not independent. A shaft that is smaller at location A is likely also to be smaller at locations B and C.

Location C's diameter is consistently smaller than A and B. Location C also consistently appears as the lowest value on the range chart, indicating its variation is less than that of A and B.

Considerations

- Group charts can be useful in situations where the sheer number of locations or process streams being tracked makes traditional SPC overwhelming. If a wall full of control charts is being ignored, or if charts are not being kept at all because there would be too many, group charts may be a useful compromise to ensure that significant process changes are identified.

- Data that fall between the MAX and MIN values are not seen on the charts. Therefore, many out-of-control situations signaled by patterns and trends are not visible. At any given time, only the two most deviant measurements show up on the charts. For this reason, groups should not contain more than about five locations, process streams, or types of measurements.
- Another disadvantage of group charts is the large amount of calculating required. Computerizing the calculations can help.
- Group short run charts can be used to follow a part or product throughout its processing.

other control charts

Many other control charts have been developed to fit particular situations. SPC software often generates a variety of control charts. Some of the more common ones are listed below. You can find more detail about these and other specialized charts in SPC references or courses.

- *CUSUM (cumulative sum).* This control chart is used with variable data to detect small drifts in the process average. It does not detect large shifts as well as the standard charts. Unlike standard charts, all previous measurements are included in the calculation for the latest plot point. The value plotted is the running sum of the differences between each subgroup's value and a target value. If the mean shifts, the CUSUM values will trend upward or downward. Sometimes a mask resembling a sideways V is used to test for out-of-control signals. With computerized CUSUM charts, a spreadsheet is used for the same purpose.
- *EWMA (exponentially weighted moving average,* also called *geometrically weighted moving average).* Like CUSUM, this control chart is used with variable data to detect or to compensate for small drifts in the process average. It does not detect larger shifts as well as the standard charts. Previous measurements are included in the calculation for the latest plot point. However, unlike CUSUM, newer values are given more weight than older ones. A weighting factor λ (lambda) is chosen to make the chart more or less sensitive to small shifts and noise in the data. λ is usually between 0.2 and 0.4. At $\lambda = 1$, the EWMA chart gives no weight to older values and becomes an $\overline{X}$ and R chart.
- *Multivariate* (also called *Hotelling T^2).* This control chart for variable data is used to monitor related characteristics. One characteristic might affect another or a third characteristic might affect both of them. All characteristics can be monitored together on one multivariate chart.

correlation analysis

Description

Correlation analysis helps quantify a linear relationship between two variables. The analysis generates a correlation coefficient *r* that tells whether the relationship is linear, how strong it is, and whether the correlation is positive or negative.

When to Use

- When you have paired numerical data, and . . .
- After drawing a scatter plot of the data that suggests a linear relationship, and . . .
- When you want a statistical measure of how well a line fits the data

Procedure

While correlation analysis can be done manually, computer software makes the calculations easier. Follow the instructions accompanying your software.

The analysis will calculate the correlation coefficient *r*, which lies between –1 and 1.

- If *r* is close to zero, the two variables are not linearly related.
- The closer *r* is to 1 or –1, the stronger the linear relationship of the two variables.
- Positive *r* means that *x* is small when *y* is small and *x* is large when *y* is large.
- Negative *r* means that *x* is small when *y* is large, and vice versa.

Example

Figures 5.39 through 5.42 show scatter plots of data with different relationships. Figure 5.39 shows an almost perfect positive linear relationship with $r = 0.98$. Figure 5.40 shows a weaker negative linear relationship with $r = -0.69$. The data points are scattered more widely than in Figure 5.39. In Figure 5.41, the two variables are not related, and $r = -0.15$. In Figure 5.42, correlation analysis calculated the same *r*, –0.15. However, in this case it is clear that the variables are related, although not linearly.

Considerations

- If $r = 0$, there is no linear relationship. However, there could be a perfect curved relationship, as in Figure 5.42. Always draw a scatter plot of the data first to see what kind of relationship to expect. Correlation analysis provides meaningful results only with linear relationships.

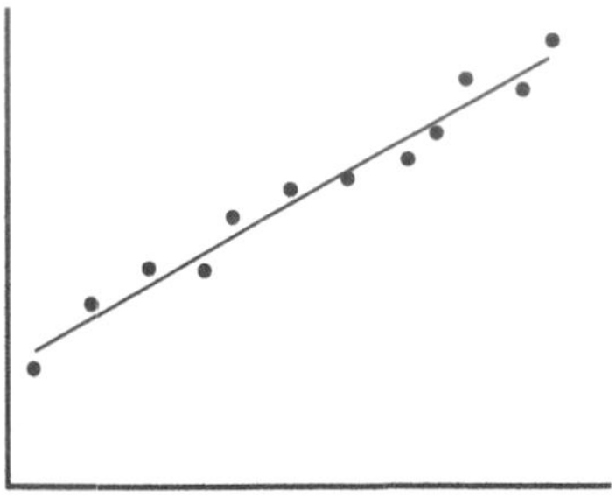

Figure 5.39 Strong positive linear correlation.

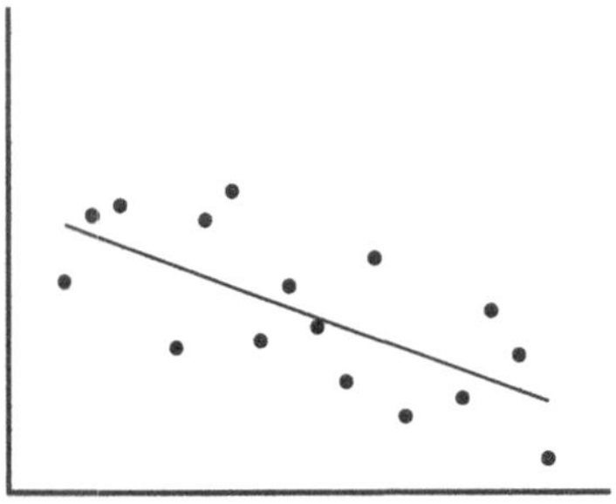

Figure 5.40 Weak negative linear correlation.

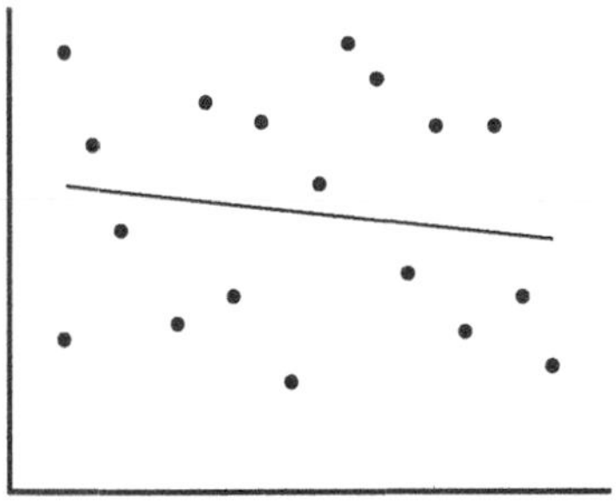

Figure 5.41 No correlation.

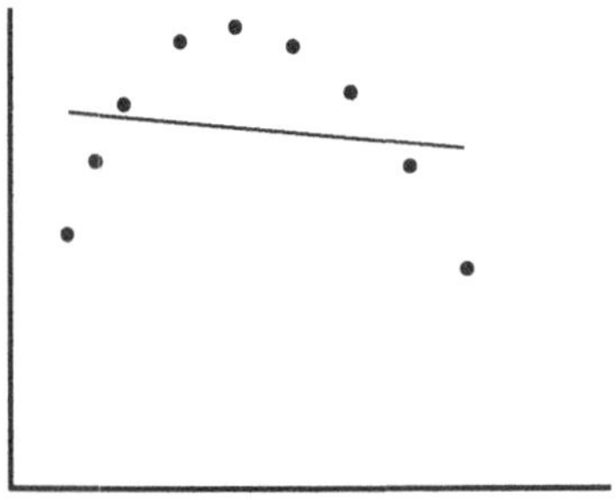

Figure 5.42 Nonlinear correlation.

- Correlation analysis quantifies the strength of the relationship between two variables, but does not calculate a regression line. If you want to find the best line through the data, see *regression analysis.*
- The coefficient of determination, r^2, calculated in regression analysis is the square of the correlation coefficient r.

cost-of-poor-quality analysis

Also called: cost-of-quality analysis, red and green circle exercise

Description

A cost-of-poor-quality analysis is a way of studying a process's flowchart to identify potential problems. *Cost of poor quality* means costs incurred because things are not done right the first time and every time. The analysis helps a team look critically at individual steps of a process to find opportunities for improvement.

When to Use

- When flowcharting a process, to be sure that cost-of-poor-quality activities are included, or . . .
- After flowcharting a process, to identify problems, potential causes, and areas in which to concentrate improvement efforts

Procedure

1. Obtain or draw a detailed flowchart of the process.
2. Identify all process steps (including recycle loops) that incur costs of quality: inspection, fix, and damage control. (See "Considerations" for definitions of these costs.) Draw a red circle (red for *Stop*) around those steps or recycles.
3. If few or no steps have red circles, ask, "What can go wrong? How do we tell if things go wrong? How does the process handle things going wrong?" If necessary, add missing steps to the flowchart that show how problems are handled.
4. For each red circle, ask, "What process step, done perfectly, would allow us to eliminate this red-circled step?" Draw a green circle (green for *Go*) around each step identified here.

5. The green circles show steps to examine for ways to prevent problems and to seek improvement in general. Green circles will contain the root causes of problems identified by the red circles.

6. (Optional) Determine actual cost data for all red-circled activities. Use this data to prioritize improvements.

Example

Figure 5.43 shows a cost-of-poor-quality analysis for the flowchart for filling an order. (This is the same example as the detailed flowchart, Figure 5.63, page 261.) Unfortunately, this book is printed in black and white, so you will have to imagine red and green.

Imagine the dotted circles are red: inspections, fixes (including rework), and damage control. They were drawn first. Note that on this flowchart, all the decision diamonds follow checking (inspection) steps. Also, the two recycle loops are fix steps.

There are no damage-control steps because the scope of this flowchart is not broad enough to show what happens when things go wrong with the external customer. What if a good customer is mistakenly refused credit? Or the customer is unhappy with the delivery date? Or the wrong product is shipped? Or the bill is incorrect? If those possibilities were shown, they would be circled in red.

At the next step in the analysis, preventions for each red (dotted) circle are sought in previous boxes on the flowchart. For "Inspect product" → "Is product good?" and the recycle loop flowing from the "No" answer, the prevention is "Make product." If product were made perfectly each time, the inspection would be unnecessary. Imagine that the solid circle around "Make product" is green.

For "Inspect materials" → "Are materials good?" and the recycle loop flowing from that decision diamond, the preventions are "Order materials" and "Vendor." Notice that although we don't control what the vendors do, their processes cause us to add an inspection step. That is why it is important to involve suppliers in your quality improvement efforts.

If the flowchart scope were large enough to include problems with the external customer, then green circles would be drawn around "Check credit," "Schedule shipment," "Ship product," and "Prepare bill." Doing these right prevents damage control later.

For three of the red (dotted) circles, no opportunities for prevention can be found on this flowchart. "Are materials in inventory?" depends on the material inventory process; "Is product in inventory?" depends on the product inventory process; "Is credit good?" depends on the credit process (and also on the customer). Names of those processes have been written in and circled with green. The team may decide that to improve the order-filling process, the material inventory process that links with it should be studied.

Notice that this analysis does not find all forms of waste in the process. Compare this analysis to the value-added analysis example, page 508, which uses the same order-filling process. There, the two wait steps are identified as waste.

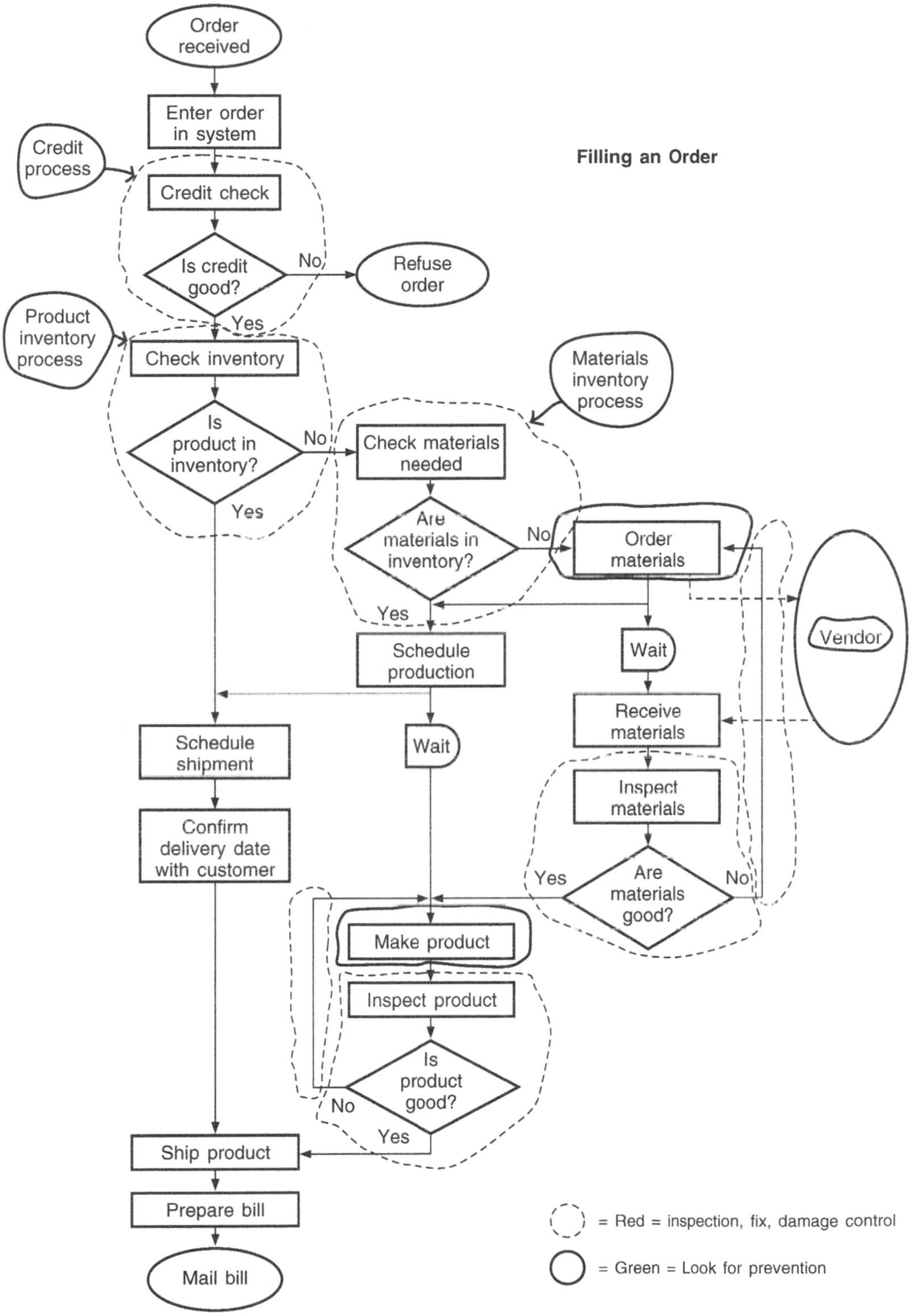

Figure 5.43 Cost-of-poor-quality analysis example.

Considerations

- Cost of poor quality has traditionally been called "cost of quality." More and more often, the phrase is changed to reflect the fact that it's *poor* quality that incurs costs, not quality.
- There are three types of activities that incur costs to be eliminated:

  Inspection: Testing, inspecting, or checking to determine whether a failure or error has occurred. For example, proofreading, countersigning a document.

  Fix: Correcting, reworking, or disposing of errors or poor output that are identified before the customer is affected. For example, redoing a report when wrong information is discovered in it.

  Damage control: Correcting, reworking, or disposing of errors or poor output that are identified after the customer is affected. Also, handling problems that arise because the customer has been affected, including explaining to the customer how we are solving the problem.
- There is a fourth type of cost of quality: *prevention.* It is preferable (cheaper and easier) to handle problems through prevention rather than waiting until inspection, fix, or damage control are needed. Quality improvement is based on preventing problems from occurring. Through a cost-of-poor-quality analysis, you can identify process steps that are inspection, fix, or damage control (red circles) in order to determine prevention measures (green circles) that can reduce or even eliminate those steps.
- The process steps circled in red may never actually disappear. We are all human, so we will never be perfect. However, thinking about the ideal, perfect world helps us to see where to focus improvement efforts. For example, we may always want to do some sort of inspection to check whether product is good. But to improve the process, we should work on getting better at making product, rather than getting better at checking.
- If a flowchart has few or no red circles, it may have been drawn with very little detail, as a macro flowchart. Cost-of-poor-quality analysis can be done only with flowcharts that include enough detail to show the problems and fixes.
- Often, opportunities for improvement will be found beyond the boundaries of your flowchart. A narrow scope may have been the best way to draw the part of the process in which you were interested. However, do not limit yourself to that scope when you identify opportunities for improvement.
- Cost data can be used to prioritize improvement opportunities. Be cautious, however, when you do this. As Dr. Deming said, "The true cost of poor quality is unknown and unknowable." The largest costs of fix and damage control, such as wasted capacity and lost customer goodwill, may be difficult or impossible to assess. Do not make a decision based on cost data if some costs are unknown,

and do not present such data to decision makers without educating them about unknown costs.

- As the example shows, not all forms of waste in the process will be found with this analysis. Costs of poor quality are waste, but other forms of waste abound, such as delays and unnecessary movement. This analysis does the best job of identifying costs of poor quality. Follow up with value-added analysis to find other forms of waste.

criteria filtering

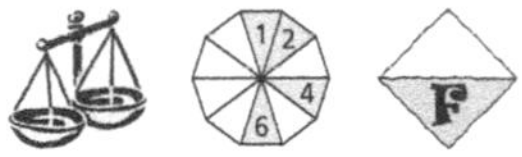

Variation: Battelle method

Description

Criteria filtering evaluates and prioritizes a list of options. The team establishes criteria and then evaluates each option against those criteria. This tool is less rigorous than some of the other evaluation tools, so it can be used quickly. Criteria filtering is often used as the third phase of list reduction.

When to Use

- After a list of options has been narrowed by the first two techniques of list reduction, and . . .
- When a list of options must be narrowed to a smaller list, or . . .
- When a list of options must be narrowed to a final choice, and the decision is not significant enough to spend the time required by a more rigorous tool

Procedure

Materials needed: flipchart, marking pen.

1. Brainstorm criteria for ranking the options. Record the criteria on a flipchart.
2. Discuss the criteria and decide on a final list.
3. Rank-order the criteria in terms of importance, most significant to least. The team should reach consensus on the rankings.
4. Create an L-shaped matrix of options and criteria. Evaluate each option against the first criterion by asking for yes or no votes to the question, "Does this idea satisfy this criterion?" Then evaluate all options against the second criterion, and

so on, until all the criteria have been addressed. If team members agree, simply write "yes" or "no." If there is disagreement, tally votes.

5. Tabulate the votes.
6. Discuss the results of this criteria ranking to help guide the team toward consensus. A numerical count of yes and no answers should not automatically generate the team's decision.
7. If too many choices still remain, go to one of these options:
   - Look for overlooked criteria.
   - Discuss the remaining choices to see if there is a clear favorite.
   - Use a more rigorous tool, such as the *decision matrix.*
   - Use *multivoting.*

Example

A team decided to conduct periodic surveys of their entire department, located in offices across the country. They needed to choose one survey method from those available. They listed the survey methods, then identified four criteria. Those criteria became the first four column headings of Figure 5.44, ranked in order of declining importance from left to right.

They evaluated each method first against their most important criterion, "Anonymous." Two methods did not meet that criterion. To save time, they immediately eliminated those two and did not evaluate them against the other criteria. Two of the remaining methods, face-to-face and telephone interviews, would only meet the anonymity criterion if the interviews and analysis were done by neutral individuals outside the department. From then on, the group evaluated those two methods with that approach in mind.

	Anonymous?	Easy for respondent?	Low cost to administer?	Low effort to administer?	Likely to be completed quickly?
Written questionnaire	Yes	Yes	Yes	Yes	No
Face-to-face interviews	Yes (non-dept. interviewers)	Yes	No	No	No
Telephone interviews	Yes (non-dept. interviewers)	Yes	Yes	No	Yes
Focus group	No	—	—	—	—
E-mail survey	No	—	—	—	—
Web-based survey	Yes	Yes	Yes	Yes	Yes

Figure 5.44 Criteria filtering example.

Their second most important criterion, "Easy for respondent," was met by all the methods. The third and fourth criteria differentiated the travel- and labor-intensive interviews from the written methods.

That still left two methods, written questionnaire and Web-based survey, which met all criteria. They could review the three criteria in a decision matrix, evaluating how well each option met their criteria. But first, they thought about the two remaining options and considered if an overlooked criterion differentiated the two methods. Someone suggested that a written questionnaire arriving in the mail was likely to be lost in a stack of papers, whereas a Web-based survey would be more likely to get an immediate response. The group realized they had missed a fourth criterion, "Likely to be completed quickly," which allowed them to select their preferred method.

See *list reduction* for another example of criteria filtering.

Battelle method

This variation consists of a series of filters, starting with easy, low-cost ones then moving to more difficult, higher-cost ones. This method reduces the number of options to which expensive analysis must be applied.

When to Use

- When there are many options to consider, and . . .
- When some of the criteria to be considered are costly (in time or resources) to evaluate

Procedure

1. Identify culling criteria that are easily answered yes–no questions. If there are many, group them into a series of filters.
2. Create an L-shaped matrix of options and criteria. Evaluate each option against each criterion in the first group. Eliminate any option receiving a "no" response to any of the criteria.
3. Repeat with each group of criteria, if there were more than one.
4. Identify rating criteria, which may be more costly to answer but are still yes–no questions. Group the criteria so that all in one group are equally costly to answer, in terms of time or resources.
5. Determine a cut-off number of "yes" answers, below which options will be eliminated.
6. Using an L-shaped matrix to record responses, evaluate each remaining option against each criterion, starting with the least costly group of criteria. Eliminate options rating below the cut-off.

7. Repeat with each group of criteria, if there were more than one.
8. Finally, identify scoring criteria, which may use numerical ratings, high–medium–low ratings, or other scoring methods. Again, group them so that all in one group are equally costly to answer. Assign weights to each of the criteria.
9. Determine a cut-off score below which options will be eliminated.
10. Once more, evaluate each remaining option against each criterion, starting with the least costly group of criteria. Multiply the rating by the criteria weight to determine each option's score for each criterion. Sum the totals for each option. Eliminate options scoring below the cut-off.

Considerations

- Criteria for project selection or problem solution often include these issues: time, cost, manpower required or available, knowledge required, training, safety, legality, authority, effectiveness, impact, enthusiasm (of the team and others).
- List reduction can be used in step 2 to quickly and methodically review the brainstormed list of criteria, eliminate duplicates and inappropriate ideas, and generate a final list of criteria.
- Phrase criteria as questions with yes or no answers, rather than qualities to be rated low, medium or high. That's what makes this tool quick. If you need to compare the degree to which options meet the criteria, use the decision matrix.
- Always phrase your criteria so that "yes" is desirable and "no" is undesirable. You don't want to eliminate an excellent option because you forgot that "no" on one question is good.
- If you discuss the most important criteria first, sometimes the answer—or a narrowed list of choices—will become obvious before you spend time on less important criteria.
- Sometimes this tool is used to separate easily accomplished activities, called "low-hanging fruit," from ones that are harder to accomplish. However, often the greatest benefit is obtained from the harder-to-accomplish projects. Therefore, ease of accomplishment—time and resources required—should be less important than effectiveness and impact criteria. One exception: When a new team is beginning work, a good way to learn is by tackling low-hanging fruit first.
- The Battelle method can efficiently reduce a list of options. Be careful to use initial screens that are appropriate for culling. The third part of this three-part process is actually a decision matrix, which is explained more thoroughly on page 219.

- Two tools—decision matrix and prioritization matrix—evaluate options more rigorously. One of them should be used instead of criteria filtering for final decisions about major issues, such as which project to select or which solution to implement. Also see the effective–achievable chart for a tool that separates ease-of-accomplishment criteria from effectiveness criteria and visually compares how well various options meet criteria. Paired comparisons can be used for a decision based on individual preferences rather than criteria.

critical-to-quality analysis

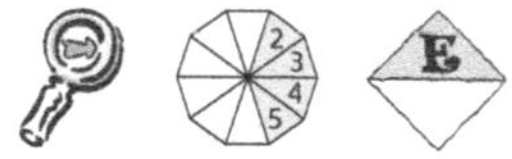

Description

A critical-to-quality (CTQ) analysis is a way of studying the flowchart of a process to identify quality features or characteristics most important to the customer and to find problems. The analysis studies inputs and outputs and identifies the steps that influence the quality of process outputs.

When to Use

- When identifying the quality features or characteristics most important to the customer, or . . .
- When drawing a detailed flowchart of a process, to be sure that steps critical to achieving quality are included, or . .
- After flowcharting a process, to identify problems and their potential causes, or . . .
- Whenever looking for opportunities for improvement in a process

Procedure

1. Obtain or draw a detailed flowchart of the process.
2. Study the output side of the flowchart and answer *who, what, when, where,* and *how.*
   - *Who* receives the output?
   - *What* is the output? (for example, information, printed form, authorization, service)

- *When* is it needed? (time and frequency)
- *Where* on the flowchart does it come from?
- *How* critical is it to the receiver? *How* critical is its timing?

Use customer input to answer the *when* and *how* questions. Write the answers to these questions on the flowchart, using whatever symbols your group finds meaningful. Common markings are a box or oval for *who,* an arrow for *where,* words on the arrow for *what,* diamonds or circles for *when* and *how.*

3. Look at the input side of your flowchart and answer *who, what, when, where,* and *how.*
   - *Who* provides the input?
   - *What* is the input? (for example, information, printed form, authorization, raw material)
   - *When* is it received? (time and frequency)
   - *Where* on the flowchart is it received? Used?
   - *How* critical is it? *How* critical is its timing?

   Write the answers to these questions on the flowchart, using your own symbols.
4. For each output, list your customer's needs. (For example: On time? Correct? Complete? Easy to use?) Again, use customer input to answer these questions. Evaluate whether the output meets those needs.
5. For each input, list your needs as a customer. Evaluate whether the input meets your needs.
6. Find critical-to-quality steps.
   - Steps where the quality of output can be affected: hurt or helped
   - Steps where an input determines what happens next in the process
   - Steps where you can measure whether inputs or outputs are meeting needs

   Mark or color these steps so they stand out.
7. Study these critical-to-quality steps to identify problems in your process.

Example

A team from the Parisian Experience restaurant was working on the problem of guests waiting too long. They had drawn a deployment flowchart (Figure 5.64, page 266) of the process for seating guests. They did a critical-to-quality analysis to look for problem areas. To simplify the analysis, they focused on subsections of the flowchart, starting with the maitre d's activities.

Four outputs were identified: acknowledgment of entering diners, information to the diners about table availability, information to the diners that their table is ready, and the service of seating the diners. Figure 5.45a shows the deployment chart section marked with symbols showing the team's answers to the *who, what, when, where,* and *how* questions.

When they looked at inputs, they discovered the process was less straightforward than they had assumed. The first input is information that a dining party has entered the restaurant. After debating "who," the team realized that it could be provided in a variety of ways, from diners announcing themselves to electronic signals. The second input, information that a table has become available, is used in two different places on the

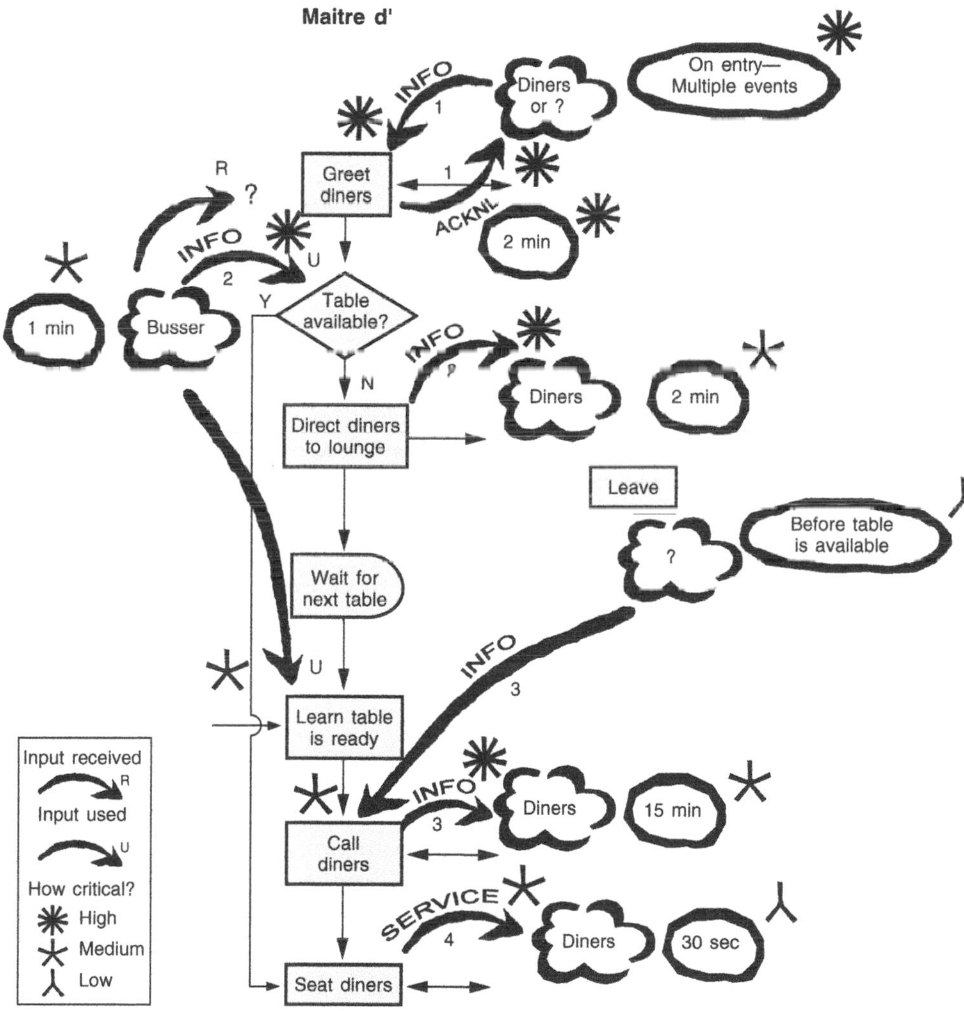

Figure 5.45a Critical-to-quality analysis example.

Output	Customer Needs	Met?
1	Within 2 minutes of entering restaurant Courteous Welcoming	✗ ✔ ✔
2	Within 2 minutes of entering restaurant Accurate time estimate Complete alternatives for wait Courteous Understanding	✗ ✗ ✔ ✔ ✔
3	Within 15 minutes of wait Easily heard Impossible to miss call	✗ ✔ ✗
4	Within 30 seconds of arrival from lounge Appropriate pace for individuals High style Courteous	✔ ✔ ✔ ✔
Input	**Our Needs**	**Met?**
1	Easy to hear or see entry, regardless of distractions or activity	✗
2	Within 1 minute after table ready Info captured whether or not maitre d' at station	✗ ✗
3	Know before table becomes available	✗

Figure 5.45b Criteria-to-quality analysis example.

flowchart, with the earlier one more critical, since diners are less committed and might leave. Also, the flowchart does not show where that information is received, because that part of the process is ill-defined.

Most of the inputs and outputs were identified by finding places on the deployment flowchart where flow moved between players. However, the third input, information that a dining party has left before their table became available, wasn't even on the chart. The team discovered it while discussing possible situations. The restaurant's current process doesn't have a way of getting that information to the maitre d'. Time could be wasted paging or searching for that party while other diners wait.

Figure 5.45b summarizes steps 4 and 5, the team's analysis of needs and their evaluation of whether those needs are currently met. Clearly, the maitre d' is not receiving needed input.

The team identified the following steps as critical-to-quality:

- *Greet diners.* Here the timing and tone of the output can create lasting impressions. An input directly affects this step.
- *Table available?* Here the input determines what happens next in the process and directly affects wait time, the key measurement of customer satisfaction.
- *Direct diners to lounge.* Here the way in which information is provided can keep or lose customers and make their waiting experience pleasant or frustrating.
- *Call diners.* Here preoccupied diners may not recognize their call, greatly lengthening their wait. Also, measurement of waiting time can be done here.

Considerations

- If you have trouble answering any of the questions, your flowchart may not be detailed enough. Do one of two things. You can include more detail on the flowchart, expanding each step into its substeps. Or, if the flowchart covers a broad scope, narrow your focus, choosing just one section of the process and expanding it.
- Customer input may be direct, such as interviews and focus groups. Or it may be indirect, from sources such as complaint data, warranty data, purchasing data, and customer observation.

critical-to-quality tree

Also called: CTQ tree, critical-to tree, CTX tree, critical-to flowdown

Description

The critical-to-quality tree helps translate the voice of the customer—customer needs and wants stated in their own words—into measurable product or process characteristics stated in the organization's terms and with performance levels or specifications that will ensure customer satisfaction.

When to Use

- After collecting voice of the customer (VOC) data, and . . .
- When analyzing customer requirements, and . . .
- Before determining which quality characteristics should be improved, especially when customer requirements are complex, broad, or vague

Procedure

1. List customer requirements for the product or service in their own words. Place each requirement in a box in the first tier of a tree diagram.
2. Address the first requirement. Ask questions to make the requirement more specific. Useful questions include:
   - What does this really mean to the customer?
   - What does this mean for each subsystem or step in our process?
   - How could we measure this?

   Don't get too specific too fast. Keep the answers only one step more detailed than the first tier. Write answers in a second tier of the tree diagram.
3. Do a "necessary and sufficient" check of the answers. Ask two questions:
   - "Is meeting each of these characteristics *necessary* in order for the customer to be satisfied that the initial requirement was met?" If the requirement can be achieved without meeting a characteristic, that characteristic should be removed.
   - "Would meeting all these characteristics be *sufficient* for the customer to be satisified that the initial requirement was met?" If the characteristics are not sufficient, decide what is missing and add it.
4. Repeat steps 2 and 3 for each answer in the second tier, creating a third tier. Continue until you have reached characteristics at a level of detail that are meaningful to the organization and can be measured.
5. Repeat steps 2 through 4 for each customer requirement identified in step 1. It is not necessary that each branch of the tree be the same length.
6. Check that all the characteristics at the end of each branch are measurable. Use operational definitions to clarify them. These are critical-to-quality (CTQ) characteristics.
7. Define targets for each measure.

Example

Ben-Hur's Pizza wishes to add home delivery to their services in order to expand their business. They have surveyed current and potential customers to determine what would make them order Ben-Hur pizza instead of a competitors' or instead of a different kind of food. Summarized VOC data told them that when customers order-in pizza, they want "hot pizza, now, with my choice of toppings and crusts, at a reasonable cost." To learn what this means in more detail, they constructed a CTQ tree. The branch for "now" is shown in Figure 5.46.

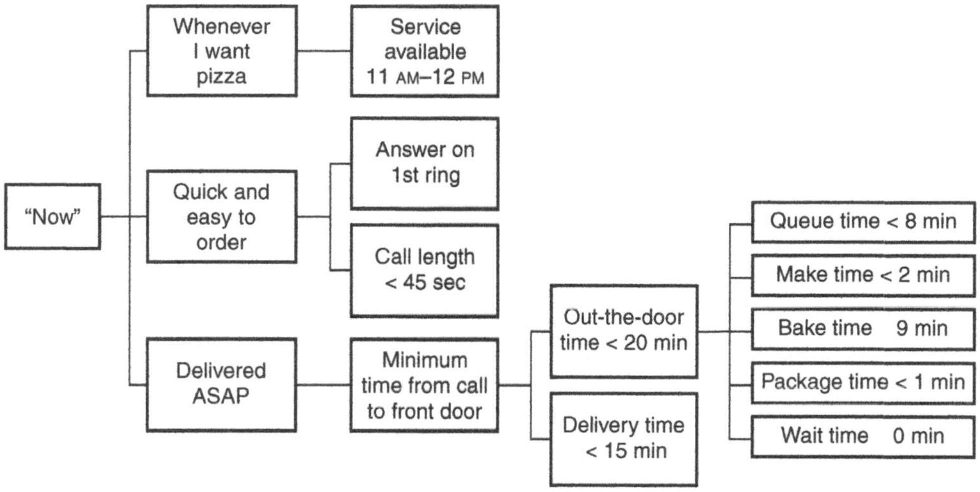

Figure 5.46 Critical-to-quality tree example.

"Now" could mean three things to the customer: "whenever I want pizza," "quick and easy to order," and "delivered ASAP." These characteristics relate to three broad aspects of pizza delivery service, from the customer's point of view: service availability, ordering the service, getting the service. Each of these chararacteristics is translated into statements that are measurable and more meaningful to Ben-Hur.

While the first two translate readily into actionable CTQs, "delivered ASAP" takes several more levels because "getting the service"—the customers' point of view—is composed of multiple process steps from Ben-Hur's point of view. Targets are assigned to each CTQ that will allow Ben-Hur to fine-tune its process to satisfy customers.

Later in its design process, Ben-Hur will assess how its current pizza-making process measures up against those targets and identify design elements of the new home-delivery service that will enable them to meet the targets. For example, they might design an order-taking checklist that allows an order to be taken with just a few checkmarks, and they might streamline the pizza-making process to eliminate wasted motion. Those great ideas—the *hows*—don't belong on the CTQ tree. When they came up during discussion of the CTQs, they were written on a separate flipchart and saved for later.

Considerations

- CTX is shorthand for "critical to something." Here, CTQ means "critical to quality." Sometimes characteristics affecting delivery, cost, or schedule are differentiated as CTD, CTC, or CTS. That's the reason for the name "critical-to tree," which could use any of these characteristics. (Be careful using acronyms, however. Sometimes CTC means "critical to customer" and CTS means "critical to satisfaction," both of which refer to the customer requirements at the beginning of the tree.)

- CTQs are specific characteristics of the process or product that are critical (essential) to the customers' perception of quality, that can be measured, and that the organization can control. CTQ target values are set so that when achieved, the customer's needs are met. In other words, CTQs are the customers' needs translated into the processes of your organization.
- VOC data often have been organized with tools such as the affinity diagram or its variation, thematic analysis. Look back at the affinity diagram when making the CTQ tree. Ideas within a broad grouping may become characteristics at lower levels of the tree.
- While VOC data usually must be grouped and summarized, retain the customers' words as much as possible before making the CTQ tree. For example, if customers say, "I want pizza now" do not summarize that as "Fast delivery." Such interpretations may be incomplete (as the example shows) or wrong, as well as a translation into the language of the organization's process, which should only happen later as you develop the tree.
- It's easy to get sidetracked into discussions of *how* to accomplish the CTQs. Stay with the *whats* during this procedure. When *hows* come up in discussion, write them down and get back to the *whats*.
- The CTQ tree may have only two levels or many more, depending on the complexity of the product or service. If the product is a computer, the tree will have many levels; if it is a bolt, it will have only a few.
- One way of determining when you have drilled down far enough is when it is absurd to go any further. In the pizza example, it would be absurd to ask, "What does it mean to bake for nine minutes?" That process step cannot be subdivided, and the characteristic is clear and measurable in the organization's terms, so it is a good CTQ.
- For a complex product or service, it may be useful to construct first a tree diagram that analyzes the product or service into its systems, subsystems, and basic components. Use as many or as few levels as necessary. Then, when making the CTQ tree, relate each of its levels to matching levels of the product or service tree. For the pizza example, the service was first divided, from the customer's point of view, into availability, ordering, and receiving. Later, the receiving process was analyzed from Ben-Hur's point of view into queue, making, baking, packaging, waiting, and delivery, each of which can have distinct CTQs.
- You need not drill down to the most fundamental level immediately. Working at a higher level can help you understand your system and be able to focus first on the most critical areas.
- Characteristics at intermediate levels of the tree are called *drivers.* These elements of the process can be controlled in order to meet the customers' needs or wants.

Often the drivers are called Xs and the characteristics they influence (on the previous level) are called Ys. Y = f(X) is a mathematical shortcut for saying "Y is a function of X," that is, Y depends upon X and adjusting X will change Y. At each level, there are Xs that affect the Ys of the previous level, which in turn are Xs affecting the Ys before them. If you follow this sequence from the ends of the branches to the beginning, you see that controlling the CTQs at the tips of the tree determines whether the customer requirements at the beginning are met. Eventually, process characteristics that affect CTQs will be identified. See *cause-and-effect matrix* for a related tool that does this.

- After completing the CTQ tree, if there are too many CTQ characteristics, decide which should be the focus of your improvement efforts. A decision-making tool such as decision or prioritization matrices can be used to narrow the list. Criteria would include which CTQs affect customer satisfaction the most and which are within the project scope. Alternatively, a cause-and-effect matrix can be used to link process steps to CTQ characteristics and identify which process steps affect the most important CTQ characteristics.
- When you have identified project CTQs, projects will be aimed at improving these factors, and your project metrics will be based on their measurements.
- The critical-to-quality concept is central to Six Sigma projects.
- A more structured way to identify CTQs is through quality function deployment. See *house of quality* and *quality function deployment* for more information.
- See *tree diagram* for more information on the generic tool. See *requirements table* and *requirements and measures tree* for additional tools for studying customer requirements and measures. Also, "Considerations" for the requirements-and-measures tree provides additional ideas about good measures.

cycle time chart

Variations: cost–cycle time chart, value/non-value-added cycle time chart

Description

A cycle time chart is a graph that visually shows how much time is spent at each step of a process. Often it also shows associated costs and/or whether the steps are value-adding or non-value-adding.

When to Use

- When studying a process to reduce its cycle time, and . . .
- When analyzing or communicating cycle time of "as-is" or "should-be" processes, or . . .
- After improving a process, to analyze or communicate the new cycle time

Procedure

1. Develop or obtain a detailed flowchart or deployment flowchart of the process. Number each step sequentially.
2. Determine how much time each step actually takes. Add the times for all steps to determine the total cycle time for the process.
3. Decide what information, in addition to cycle time, you wish to include on your graph. Some possibilities are: cost for each step, cumulative cost, and whether each step is value- or non-value-adding. Collect or assemble the information.
4. Draw a graph with the x-axis representing time. Determine the scale needed so that the full line equals the total cycle time. Using that scale, mark off distances to represent the cycle times of each step, starting with step 1 on the left. Label each space with the step number.
5. Draw bars for each step. Choose a method based on what additional information you wish to show.
   a. To show only cycle time: Draw bars for each step that are all the same height and as wide as the x-axis distance you marked off. For waits or delays, do not draw a bar.
   b. To show whether a step is value- or non-value-adding: Draw bars for value-adding steps above the line. Draw bars for non-value-adding steps below the line. Or, shade or color each bar to indicate real value-adding, organizational value-adding, and non-value-adding activities. Or use both methods. Again, for waits or delays, do not draw a bar. This method creates a *value/non-value-added cycle time chart.*
   c. To show each step's cost: Scale the y-axis so that the top of its range is slightly larger than the highest cost. For each step, draw a bar with the height representing its cost. You may shade or color each bar to indicate the value-added category. This method and the next one create *cost–cycle time charts.*
   d. To show cumulative cost: Scale the y-axis so that the top of its range is slightly larger than the total cost. For step one, draw a vertical line at the right side of its space, representing step one's cost. Draw a line between the origin and the top of that line. For step two, draw a vertical line at the right side of

its space, representing the sum of the costs for step one and step two. Draw a line across the top of that space, connecting the tops of the first vertical line and the second one. Continue for each step of the process. You may shade or color each bar to indicate the value-added category.

6. Write clarifying information on the chart: cycle time for each step, total cycle time, actual costs, or any other information you think is necessary to communicate clearly.

Example

A medical clinic improvement team is tackling office visit cycle time—the total elapsed time from the patient's arrival to departure. They flowcharted the process and did a value-added analysis, shown in Figure 5.47. (See *value-added analysis* on page 507 for more explanation of this step.) They collected data by having patients and/or staff record actual times during a visit. The average times for each process step are shown in the cycle time chart of Figure 5.48.

They chose to show real value-adding (RVA) steps unshaded above the line and organizational value-adding (OVA) steps unshaded below the line. Non-value-adding (NVA) steps would have been shaded below the line, but all NVA steps are waits, for which bars are not drawn. The chart shows that less than half of the cycle time is spent in activities that add value for the patient. Almost half is spent in totally non-value-added waiting.

Costs were allocated to each step by first calculating an hourly cost for staff and facilities then multiplying the time spent in that step by the hourly cost. Figure 5.49 shows a cumulative cost–cycle time chart. (They realize that the costs do not capture the costs of patient dissatisfaction, such as patients who switch to another clinic or those who arrive late for an appointment, disrupting the schedule, because they assume they will have to wait anyway.)

While the team already knew cycle time was a problem and that patients were dissatisfied, the charts helped them to grasp the extent of the problem and communicate it

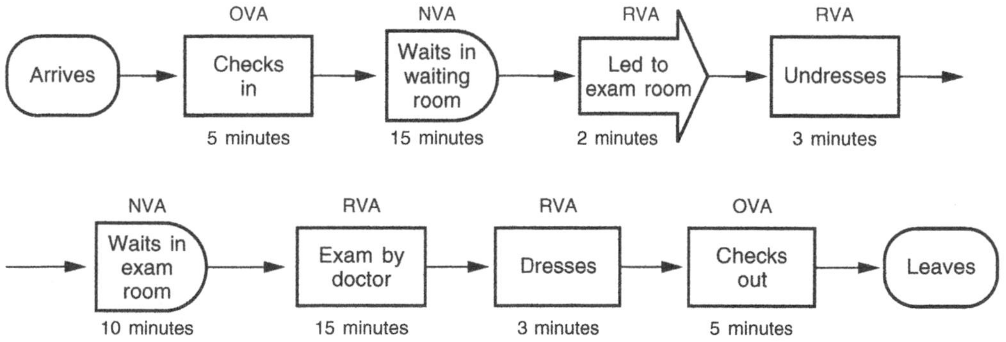

Figure 5.47 Value-added analysis for cycle time.

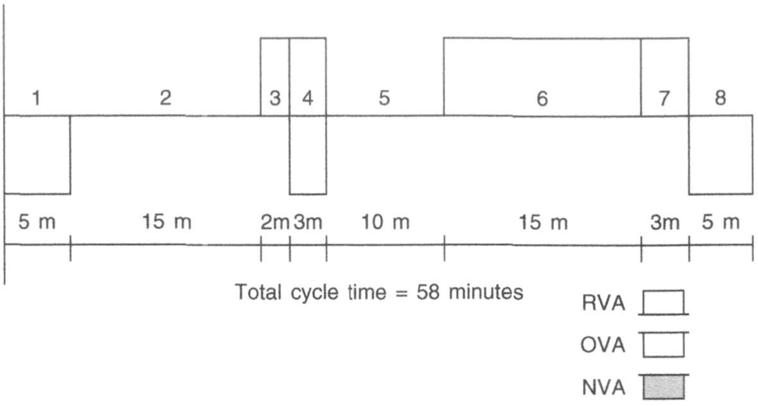

Figure 5.48 Cycle time chart example.

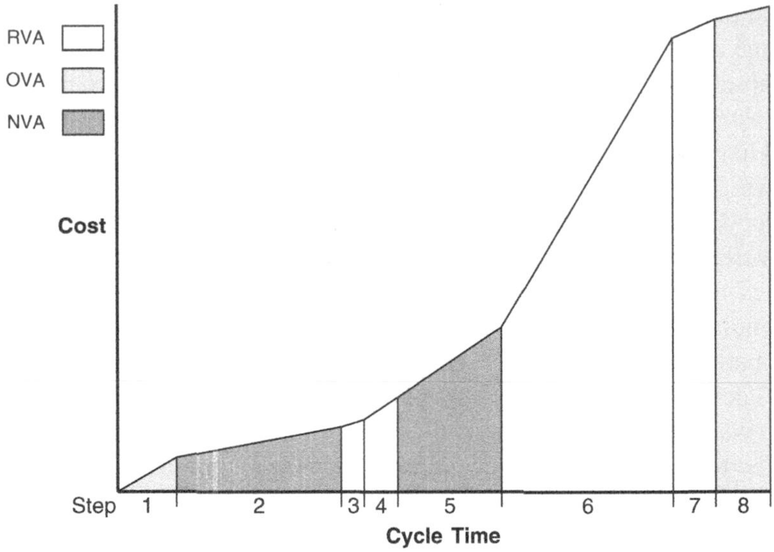

Figure 5.49 Cumulative cost–cycle time chart example.

to management and the rest of the staff. These charts will also be compared with similar charts drawn during and after process changes, to illustrate the extent of improvement.

Considerations

- Often the steps in a process vary from minutes to hours to days. It is not possible to show a step that takes only minutes in proportion to a step that takes days. In that case, show the longer steps proportional to one another and simply show the shorter ones as narrow as you can without their vanishing. You may also show a scale break along the x-axis for a step that is much longer than the rest.

- If a process branches into alternate or parallel activities, you will have to get creative. One option, for methods a and b above, is to branch the x-axis just as the process branches and show alternate or simultaneous steps one above the other. This approach will make it clear how long each sequence takes and where delays are caused by waiting for the other part of the process to be completed. Bar height can still represent cost, although you cannot use the y-axis to read costs directly. Another option with methods a or b is to identify the critical path and show only those activities. If you are using methods c or d, you must show all the steps in order to include all costs. In that case, put the steps in approximate sequence and don't worry about the fact that one parallel branch has to follow another.
- See *value-added analysis* for more information about classifying activities by their value.

decision matrix

Also called: Pugh matrix, decision grid, selection matrix or grid, problem matrix, problem selection matrix, opportunity analysis, solution matrix, criteria rating form, criteria-based matrix
See also: cause-and-effect matrix

Description

A decision matrix evaluates and prioritizes a list of options. The team first establishes a list of weighted criteria and then evaluates each option against those criteria. This is a variation of the L-shaped matrix.

When to Use

- When a list of options must be narrowed to one choice, and . . .
- When the decision must be made on the basis of several criteria, and . . .
- After the list of options has been reduced to a manageable number by list reduction

Typical situations are:

- When one improvement opportunity or problem must be selected to work on, or . . .
- When only one solution or problem-solving approach can be implemented, or . . .
- When only one new product can be developed

Procedure

1. Brainstorm the evaluation criteria appropriate to the situation. If possible, involve customers in this process.
2. Discuss and refine the list of criteria. Identify any criteria that must be included and any that must not be included. Reduce the list of criteria to those that the team believes are most important. Tools such as list reduction and multivoting may be useful here.
3. Assign a relative weight to each criterion, based on how important that criterion is to the situation. Do this by distributing 10 points among the criterion. The assignment can be done by discussion and consensus. Or each member can assign weights, then the numbers for each criterion are added for a composite team weighting.
4. Draw an L-shaped matrix. Write the criteria and their weights as labels along one edge and the list of options along the other edge. Usually, whichever group has fewer items occupies the columns.
5. Evaluate each choice against the criteria. There are three ways to do this:

   Method 1: Establish a rating scale for each criterion. Some options are:

1, 2, 3:	1 = slight extent, 2 = some extent, 3 = great extent
1, 2, 3:	1 = low, 2 = medium, 3 = high
1, 2, 3, 4, 5:	1 = little to 5 = great
1, 4, 9:	1 = low, 4 = moderate, 9 = high

   Make sure that your rating scales are consistent. Word your criteria and set the scales so that the high end of the scale (5 or 3) is always the rating that would tend to make you select that option: most impact on customers, greatest importance, least difficulty, greatest likelihood of success.

   Method 2: For each criterion, rank-order all options according to how well each meets the criterion. Number them with 1 being the option that is least desirable according to that criterion.

   Method 3, Pugh matrix: Establish a baseline, which may be one of the alternatives or the current product or service. For each criterion, rate each other alternative in comparison to the baseline, using scores of worse (–1), same (0), or better (+1). Finer rating scales can be used, such as 2, 1, 0, –1, –2 for a five-point scale or 3, 2, 1, 0, –1, –2, –3 for a seven-point scale. Again, be sure that positive numbers reflect desirable ratings.
6. Multiply each option's rating by the weight. Add the points for each option. The option with the highest score will not necessarily be the one to choose, but the relative scores can generate meaningful discussion and lead the team toward consensus.

Example

Figure 5.50 shows a decision matrix used by the customer service team at the Parisian Experience restaurant to decide which aspect of the overall problem of "long wait time" to tackle first. The problems they identified are customers waiting for the host, the waiter, the food, and the check.

The criteria they identified are "Customer pain" (how much does this negatively affect the customer?), "Ease to solve," "Effect on other systems," and "Speed to solve." Originally, the criteria "Ease to solve" was written as "Difficulty to solve," but that wording reversed the rating scale. With the current wording, a high rating on each criterion defines a state that would encourage selecting the problem: high customer pain, very easy to solve, high effect on other systems, and quick solution.

"Customer pain" has been weighted with 5 points, showing that the team considers it by far the most important criterion, compared to 1 or 2 points for the others.

The team chose a rating scale of high = 3, medium = 2, and low = 1. For example, let's look at the problem "Customers wait for food." The customer pain is medium (2),

Decision Matrix: Long Wait Time

Criteria → / ↓ Problems	Customer pain 5	Ease to solve 2	Effect on other systems 1	Speed to solve 2	
Customers wait for host	High—Nothing else for customer to do 3 × 5 = 15	Medium—Involves host and bussers 2 × 2 = 4	High—Gets customer off to bad start 3 × 1 = 3	High—Observations show adequate empty tables 3 × 2 = 6	28
Customers wait for waiter	Medium—Customers can eat breadsticks 2 × 5 = 10	Medium—Involves host and waiters 2 × 2 = 4	Medium—Customer still feels unattended 2 × 1 = 2	Low—Waiters involved in many activities 1 × 2 = 2	18
Customers wait for food	Medium—Ambiance is nice 2 × 5 = 10	Low—Involves waiters and kitchen 1 × 2 = 2	Medium—Might result in extra trips to kitchen for waiter 2 × 1 = 2	Low—Kitchen is design/space limited 1 × 2 = 2	16
Customers wait for check	Low—Customers can relax over coffee, mints 1 × 5 = 5	Medium—Involves waiters and host 2 × 2 = 4	Medium—Customers waiting for tables might notice 2 × 1 = 2	Low—Computerized ticket system is needed 1 × 2 = 2	13

Figure 5.50 Decision matrix example.

because the restaurant ambiance is nice. This problem would not be easy to solve (low ease = 1), as it involves both waiters and kitchen staff. The effect on other systems is medium (2), because waiters have to make several trips to the kitchen. The problem will take a while to solve (low speed = 1), as the kitchen is cramped and inflexible. (Notice that this criteria has forced a guess about the ultimate solution: kitchen redesign. This may or may not be a good guess.)

Each rating is multiplied by the weight for that criterion. For example, "Customer pain" (weight of 5) for "Customers wait for host" rates high (3) for a score of 15. The scores are added across the rows to obtain a total for each problem. "Customers wait for host" has the highest score at 28. Since the next highest score is 18, the host problem probably should be addressed first.

Also see the Medrad story on page 60 for an example of a decision matrix used to prioritize improvement opportunities.

Considerations

- A very long list of options can first be shortened with a tool such as multivoting or list reduction.
- Criteria that are often used fall under the general categories of effectiveness, feasibility, capability, cost, time required, support, or enthusiasm (of team and of others). Here are other commonly used criteria:

  For selecting a problem or an improvement opportunity: Within control of the team, financial payback, resources required (for example, money and people), customer pain caused by the problem, urgency of problem, team interest or buy-in, effect on other systems, management interest or support, difficulty of solving, time required to solve.

  For selecting a solution: Root causes addressed by this solution; extent of resolution of problem; cost to implement (for example, money and time); return on investment; availability of resources (people, time); ease of implementation; time until solution is fully implemented; cost to maintain (for example, money and time); ease of maintenance; support or opposition to the solution; enthusiasm by team members; team control of the solution; safety, health, or environmental factors; training factors; potential effects on other systems; potential effects on customers or suppliers; value to customer; potential problems during implementation; potential negative consequences.
- This matrix can be used to compare opinions. When possible, however, it is better to use it to summarize data that has been collected about the various criteria.

- Subteams may be formed to collect data on the various criteria.
- Several criteria for selecting a problem or improvement opportunity require guesses about the ultimate solution. For example: evaluating resources required, payback, difficulty to solve, and time required to solve. Therefore, your rating of the options will be only as good as your assumptions about the solutions.
- It's critical that the high end of the criteria scale (5 or 3) always is the end you would want to choose. Criteria such as cost, resource use, and difficulty can cause mix-ups: low cost is highly desirable! If your rating scale sometimes rates a desirable state as 5 and sometimes as 1, you will not get correct results. You can avoid this by rewording your criteria. Say "low cost" instead of "cost"; "ease" instead of "difficulty." Or, in the matrix column headings, write what generates low and high ratings. For example:

Importance	Cost	Difficulty
low = 1 high = 5	high = 1 low = 5	high = 1 low = 5

- When evaluating options by method 1, some people prefer to think about just one option, rating each criteria in turn across the whole matrix, then doing the next option, and so on. Others prefer to think about one criteria, working down the matrix for all options, then going on to the next criteria. Take your pick.
- If individuals on the team differ in their ratings, discuss them to learn from each other's views and arrive at a consensus. Do not average the ratings or vote for the most popular one.
- In some versions of this tool, the sum of the unweighted scores is also calculated and both totals are studied for guidance toward a decision.
- When this tool is used to choose a plan, solution, or new product, results can be used to improve options. An option that ranks high overall but has low scores on criteria A and B can be modified with ideas from options that score well on A and B. This combining and improving can be done for every option, and then the decision matrix used again to evaluate the new options.
- The prioritization matrices are similar to this tool but more rigorous. Also see the effective–achievable chart for a tool that separates ease-of-accomplishment criteria from effectiveness criteria and visually compares how well various options meet criteria. Paired comparisons can be used for a decision based on individual preferences rather than criteria. The cause-and-effect matrix is a variation of this tool used to understand the linkages between process steps and customer requirements.

decision tree

Also called: decision process flowchart, logic diagram, key

Description

A decision tree is a sequenced set of questions that lead to a correct decision or problem solution. It is a specialized tree diagram, but often it reads like a flow diagram. Typically, the tree is developed by people with expert knowledge of situations that are likely to occur repeatedly. Later, the tree is used by people without specialized knowledge to make decisions quickly without help.

When to Use

- When a situation requiring a decision or problem solution will arise repeatedly, and . . .
- When the thought process for making the decision is known and can be laid out as a series of questions
- Typical applications of decision trees include troubleshooting, emergency response, and documenting procedures that are complex, critical, or seldom used

Procedure

1. Define the kind of situation in which the decision tree will be used. Develop a statement of the decision to be made or problem to be solved. Write it on a sticky note and place it at the far left of the work surface.
2. Brainstorm questions that must be answered to reach the correct decision. For each question, note what the possible answers are. Usually, these will be yes–no or a small set of a choices. Write each question and its answers on a note and place it on the work surface. Let sequence guide you if that is helpful, but do not be too concerned about correct order yet.
3. Decide whether the questions must be asked in a particular sequence. If not, choose an efficient order. Sequence the questions by arranging the cards on the work surface. Show the link between an answer and the next question with an arrow.
4. Review the tree for missing questions or answers. Review the questions to be sure they will be clearly understood and correctly answered by others.

5. Test the tree. Create scenarios that reflect a range of different situations, and work through the tree for each one. Modify the tree if any problems are found.
6. Give people without expert knowledge the scenarios and ask them to use the tree to make decisions. If they do not reach the correct decision, identify the question(s) where the error occurred and modify the tree.

Example

Figure 5.20 on page 156, the control chart selection tree, and Figure 5.68 on page 276, the graph decision tree, are examples of decision trees.

Considerations

- Often certain questions must precede others.
- If there is no natural sequence for the questions, then choose an order that in most cases will lead to a decision as quickly as possible. Questions identifying common situations should come before less common ones.
- In a tree with a series of yes–no questions, try to arrange the tree so that the yeses and nos are positioned consistently. Otherwise users are likely to make errors if they don't notice that the position of the yeses and nos has switched.

design of experiments (DOE)

Also called: D of E, designed experiments

Description

Design of experiments is a method for carrying out carefully planned experiments on a process. By using a prescribed plan for the set of experiments and analyzing the data according to certain procedures, a great deal of information can be obtained from a minimum number of experiments. More than one variable can be studied at one time, so experimentation is less costly. In addition, interactions between variables can be identified. Usually, design of experiments involves a series of experiments that start by looking broadly at a great many variables and then focus on the few critical ones.

Described here is the "classical" approach. Other DOE methods, developed by Japanese engineer Dr. Genichi Taguchi, emphasize first minimizing variation and then achieving target values. Taguchi methods, sometimes called robust design, are applied

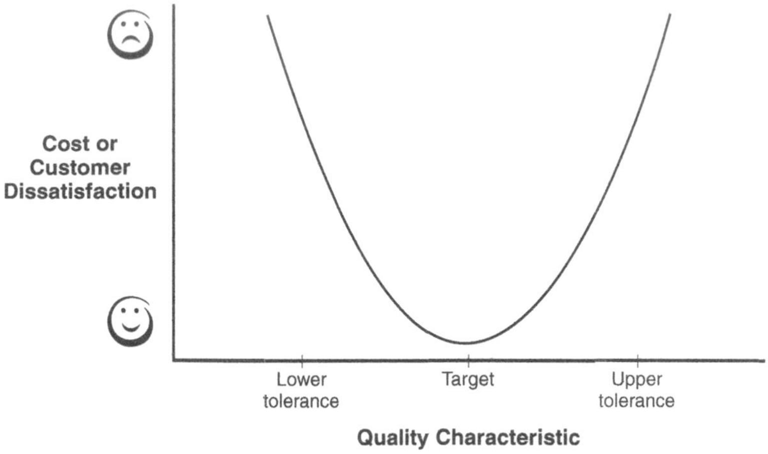

Figure 5.51 Taguchi's loss function.

to product and process design so that the process is robust, or insensitive to variation in uncontrollable variables, which Taguchi calls noise. His "loss function" (see Figure 5.51) teaches that the farther a quality characteristic is from its target because of variation, even if it is within the customer's tolerances, the greater the customer dissatisfaction and the cost to the producer.

A third set of methods, developed by American engineer Dorian Shainin and taught by Keki Bhote, addresses difficult chronic problems. This approach to DOE is part of a broader problem-solving methodology that includes other statistical tools such as multi-vari charts and components search. (One of these tools, paired comparisons, is different from the tool of that name in this book.) Mathematically simpler than the other two DOE approaches, the Shainin approach is used most often in manufacturing of component assemblies. A process of elimination identifies the *Red X,* the cause responsible for most of the process variation.

When to Use

- When you are studying a process whose output can be measured numerically, and . . .
- When you want to learn how key variables affect the output, and . . .
- When you want to know which variables are important and which are not, or . . .
- When you want to change the process average, or . . .
- When you want to reduce process variation, or . . .
- When you want to set process variables so the output will not be greatly affected by uncontrollable changes (you want to make the process robust)

Procedure

It is beyond the scope of this book to provide a procedure detailed enough that you could conduct and analyze designed experiments. This section will provide an overview to help you understand the ideas and methods involved. Before doing your own design of experiments, consult specialized books or secure training, and enlist the help of experts.

Usually, design of experiments is carried out in two broad phases:

Phase I: Screening Experiment. An experiment that studies many key variables. The purpose is to identify which ones significantly affect the output and which do not. When many variables are included, the results cannot provide good information about interactions of factors with each other.

Phase II: Optimization Study. One or more experiments that study just a few key variables. These experiments provide better information about interactions.

General Procedure*

1. Identify the process to be studied and the purpose of the study.
2. Identify the output measurement(s) that you want to improve. (This is called the *response.*)
3. Determine measurement precision and accuracy, using tools such as repeatability and reproducibility studies.
4. Identify potential key variables that you can control and that might affect the output. (These are called *factors.*) Use tools such as brainstorming, flowcharts, and fishbone diagrams. Identify each factor as A, B, C, and so on.
5. Choose the settings, or *levels,* for each factor. Usually two levels will be used for each. If the variable is quantitative, choose high and low levels. If the variable is qualitative, choose two different settings and arbitrarily call them high and low. Designate the high setting with + and the low one with –: A+, A–, B+, B–, and so on.
6. Determine and document the experimental design. This includes:
   - All the different combinations of levels (called runs or *treatments*), specifying which variables will be at which settings
   - How many times each treatment will be done (called *replication*)
   - The sequence of all the trials, preferably chosen using a method that ensures random order (called *randomization*)
7. Identify other variables that might interfere with the experiment. Plan how you will control or at least monitor them.

* This procedure is adapted from Larry B. Barrentine, *An Introduction to Design of Experiments: A Simplified Approach* (Milwaukee: ASQ Quality Press, 1999).

8. Run the experiment, following the design exactly.
9. Analyze the data and draw conclusions. Computer software or spreadsheets do the math for you. Graph the results and effects to better understand them. A Pareto chart of the effects also can help you compare effects and understand visually which are most important. See below for an overview of analysis.
10. If the conclusions suggest you make changes to improve the process, verify those results and then standardize the new process.
11. Determine what additional experiments should be run. Go back to step 5 to plan and carry them out.

Analysis

Here are the results that are returned:

Average response R. For each treatment, all responses for all replications are averaged. A common notation is to show the levels of the factors in parentheses. For example, the average response for the treatment with factors A and B both high is written (A+B+). The average response for all treatments with factor A high is (A+).

Main effects. The main effect for a factor shows whether changing that factor significantly affects the process response. An *effect* is the difference between two results. The main effect for a factor is the difference between the average response at the factor's high value and the average response at its low value. For example, $E(\text{A}) = \text{R}(\text{A}+) - \text{R}(\text{A}-)$.

Interaction effects. An interaction effect shows whether two factors together affect the response, although each factor alone might or might not affect it differently. An interaction effect is the difference between two effects. It's easier to write it in symbols than in words:

$$E(\text{AB}) = E(\text{AB}+) - E(\text{AB}-) = E(\text{BA}+) - E(\text{BA}-)$$

In words, it's the difference between the effect of A with B high and the effect of A with B low. (Which is the same as the effect of B with A high minus the effect of B with A low. Whew!)

Decision limits or P values. Hypothesis testing determines whether each effect is significant or whether it could have occurred through chance. Standard deviations of the test results are used to calculate either a set of decision limits or a P value. See *hypothesis testing* for more information. An effect is significant if it is beyond the decision limits—greater than the upper one or less than the lower one—or P is less than your chosen significance level α.

Example

Yummy Cookie Company has discovered that customers prefer their competitor's gingersnaps because they are crisper. They are running design of experiments on their gingersnap process with the objective of discovering how to maximize crispness. The response variable is crispness, which they can measure on a "crispometer."

They have already done screening studies on a wide range of variables and learned that baking temperature and shortening manufacturer are most significant. Baking temperature, a quantitative variable, will be set at levels of 350 and 400 degrees F. Shortening manufacturer, a qualitative variable, has two possibilities. They will arbitrarily assign the "high" value to Tall Shortening Company and the "low" value to Slick Oil Inc.

They will do a full factorial experiment, trying every combination of every level.

A+B+ A+B– A–B+ A–B–

Two factors at two levels each requires four treatments. Each will be replicated once, so the experiment will have eight trials in all. A random sequence is determined by writing each trial on a slip of paper and drawing slips. They document this experimental design.

Other process variables are fixed, and the experiment is run. The results are shown in Figure 5.52. The analysis is shown in Figure 5.53. The manufacturer's effect *E*(B) is

Shortening Manufacturers (B)		**Baking Temperature (A)** Low – 350°	**Baking Temperature (A)** High + 400°
	High + Tall	79 73	66 74
	Low – Slick	62 66	93 87

Figure 5.52 Design of experiments example results.

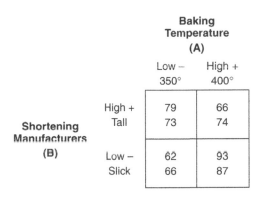

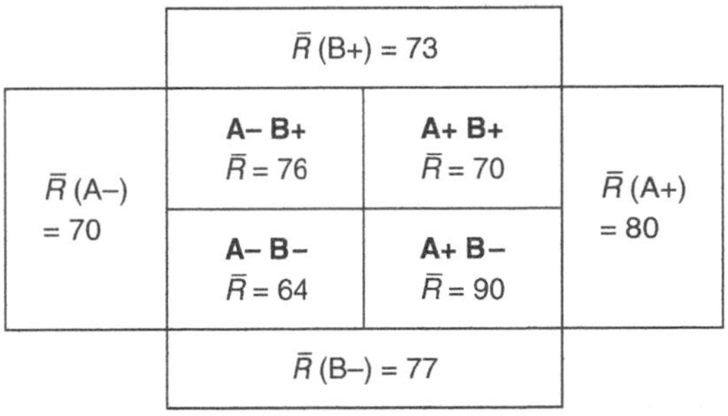

E (A) = 80 – 70 = 10

E (B) = 73 – 77 = –4

E (AB) = (70 – 76) – (90 – 64) = –16

Decision limits = ±8.6

Figure 5.53 Design of experiments example analysis.

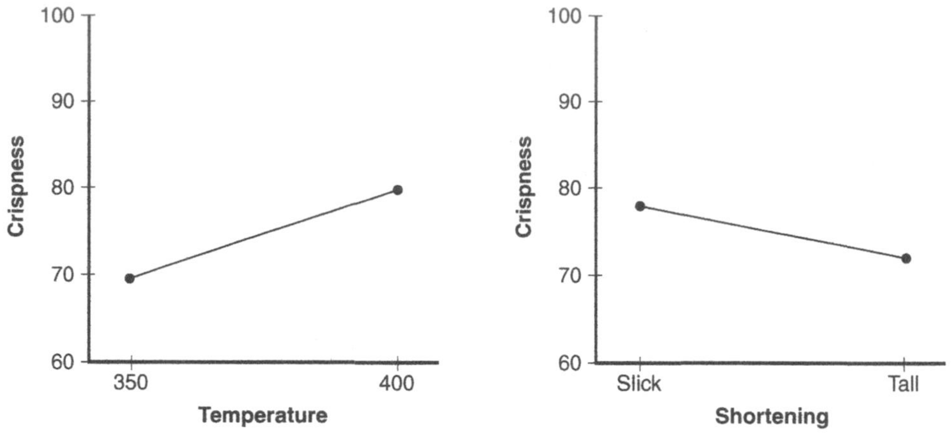

Figure 5.54 Design of experiments example: main effects.

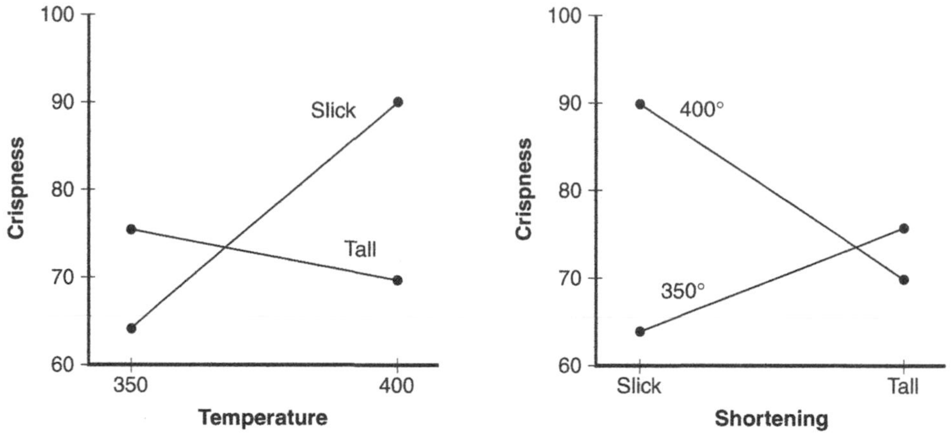

Figure 5.55 Design of experiments example: interaction effects.

not significant, since it is within the decision limits, but the temperature effect and the temperature–maufacturer interaction effect are beyond those limits and are significant.

Graphs of the temperature and manufacturer effects are shown in Figure 5.54 and the interaction effects in Figure 5.55. On average, crispness increases with temperature. However, for Tall's shortening, crispness *decreases* with increasing temperature. This is the temperature–manufacturer interaction effect. The crossed lines on the interaction graphs signal that interaction effects are at work.

The process was changed to bake the cookies at 400 degrees, using only shortening from Slick, since that combination of variables produced the crispest gingersnaps. Before making the change, the results were verified with additional runs.

Considerations

- Traditionally, experiments addressed one factor at a time. Not only did this lead to more trial runs, which meant higher cost, but this approach didn't discover interactions between factors. Design of experiments is therefore a more efficient and more powerful approach.
- Conducting designed experiments can be costly in terms of manpower, training, process time, and sometimes lost raw materials and product. However, the returns from an investment in this powerful technique can be substantial.
- Design the experiment carefully. Steps 1 through 7 of the procedure are all about planning, which is inexpensive. Step 8, actually running the experiment, is the costly part. Step 9, analysis, can be repeated or corrected. But after the experiment is run, the design cannot be changed.
- Also see *multi-vari chart* for a tool useful for preliminary analysis using historical or process data.

Screening Experiments and Optimization Studies

- *Fractional factorial* or *Plackett-Burman.* These are two different kinds of designs for screening experiments, carefully chosen to provide the most information for the smallest number of runs. For example, a Plackett-Burman screening design for 11 potential factors would require 12 runs. Not all combinations of all levels and all factors are tried, so not all interactions can be estimated. (See "confounding," below.) Statistical software packages and DOE books give treatment patterns for these designs.
- In addition to identifying which factors are critical, screening designs can tell you which factors don't really matter. Sometimes those factors can be set to more economical levels or can be left uncontrolled, leading to cost savings.
- Larger designs can provide lots of information efficiently. However, when experiments get too big, they become expensive and hard to manage.
- *Full factorial.* This is a design in which the treatments include all combinations of factors. This design is used for optimization studies. The number of treatments needed is the number of levels raised to the n power, where n is the number of factors. So, for an experiment with three factors and two levels each, a full factorial design would need $2^3 = 2 \times 2 \times 2 = 8$ treatments. An experiment with 11 potential factors would need 2^{11} or 2048 treatments. This is why screening experiments are done first to reduce the number of factors.

Planning a Designed Experiment

- Clear understanding of the experiment's objective is important. The type of analysis or the interpretation of the results may change with different objectives. In the cookie example above, if the experimenters did not care how crisp the gingersnap was as long as it did not vary much, they would decide to use shortening from Tall Shortening Company.

- *Replication* means doing a treatment more than once, repeating everything, including the complete setup. Replication reduces the effect of random variation on the results.
- *Randomization* means selecting a sequence by a means in which every possible sequence is equally likely. This is important to prevent unknown or uncontrolled variables from causing trends in the results. See *sampling* for more information about randomization. Sometimes it is impossible or too costly to randomize a factor. A procedure called *blocking* lets you do all the trials with one setting of the factor and then all the trials with the other setting. The statistical design needs to include the blocking.
- Different approaches are required when selecting levels for a screening experiment than for an optimization study. In screening experiments, levels are chosen to represent opposite extremes of possible operating conditions—without, of course, creating a dangerous or impossible situation. Wide levels make differences in the results more obvious. In optimization studies, levels are more tightly spaced near the level found to be best in the screening experiment.
- If levels of a factor are closely spaced, the difference between the resulting outputs (the *effect* of the factor) is smaller. As a result, more data is needed to prove that such differences are significant, and more replications will be needed to eliminate the influence of variation or experimental error.
- Using two levels produces only a linear model of the output, since two points define a line. You don't know anything about what happens in between. Using three levels is useful if you suspect the behavior between extremes is nonlinear.

Common Issues with Results

- *Interaction* means the effects of two factors acting together differs from what you would predict from the separate effects of either one. The response to one factor depends on the level of another factor. In the cookie example, the effect of increased temperature with Slick shortening is greater than expected, and the effect with Tall shortening is actually the reverse of what you would expect from looking at temperature alone. Discovering interactions is one of DOE's great benefits.
- An interaction effect can be graphed in two different ways, as shown in Figure 5.55. Try both to see which conveys key information better.
- *Confounding* is a problem that arises when fractional factorial or Plackett-Burman designs are used. These designs are much more efficient—12 runs instead of 2048 for 11 factors—but that efficiency comes with a price. Some effects cannot be distinguished. For example, the effect of the interaction of

factors A and B might actually be the interaction effect of factors C and D. AB and CD are said to be confounded. Techniques such as *reflection,* which doubles the number of trials, eliminate some of the confounding, but not all the confounded interaction effects can be eliminated. That is one reason that screening designs are usually followed by optimization studies using full factorial analysis. Screening designs will identify the key factors and possibly some major interactions. Optimization studies will clarify the interactions. (A full discussion of confounding is beyond the scope of this book.)

Analysis

- The math is straightforward arithmetic, but there's a lot of it. Data analysis is usually done with computers. Spreadsheets can be used, but statistical packages include design plans and make the analysis easier.
- However, software cannot substitute for a thorough understanding of the logic behind designed experiments in order to make correct conclusions and resulting process changes. Therefore, the hardest and most critical part of design of experiments is planning and carrying out the experiments.
- The data can be used to construct a mathematical model that will predict the response for any factor settings between the high and low levels.
- Variation analysis has not been included here. Consult a DOE book, training course, or expert for more information.

effective–achievable chart

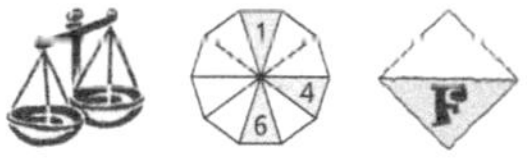

Also called: impact–effort chart, project desirability chart

Description

The effective–achievable chart is a two-dimensional chart used for prioritizing possible choices. It provides a visual comparison of multiple choices based on two dimensions: how effective each choice would be and how achievable it is.

When to Use

- When deciding among a list of options, such as goals to achieve, problems to investigate, methods to follow, or solutions to implement, and . . .
- When a quick comparison between options is desired, or . . .
- When communicating a decision to others in the organization

Procedure

1. For each choice, ask the following two questions:

   First, how *effective* will this choice be? That is,

   - How well will this method accomplish our purposes? or . . .
   - How well will this solution solve the problem? or . . .
   - How well will solving this problem contribute to overall improvement? or . . .
   - How well will achieving this goal support our overall mission?

   Second, how *achievable* is this choice? How easy will it be to accomplish? Consider what resources are required and the likelihood of success.

2. For each question, decide whether that choice rates "high" or "low." Finer distinctions are not necessary. Rate the choice in relation to the other choices, not on an absolute basis.
3. Write each choice in the box corresponding to its combination of effective and achievable ratings.
4. Choices should be chosen or prioritized in the following order:

   Highest—box A

   2—box B

   3—box C

   Lowest—box D

Example

Figure 5.56 shows an effective–achievable chart used by a team to decide how to gather employee opinions. The team's six options were placed on the chart, based on group consensus of effectiveness and achievability.

Face-to-face interviews of a 10 percent random sample were judged to be low in achievability because they require a lot of time and people, but they were judged to be highly effective at gathering opinions. Low achievability and high effectiveness places that option in box B. Focus groups over lunch are highly achievable—inexpensive and quickly accomplished—but they were not considered to be very effective at getting a good sample of employee opinions. High achievability and low effectiveness places that option in box C. The other options were placed on the chart similarly.

The group decided to start with telephone surveys, which was the only option both highly effective and highly achievable. If needed, group members would follow up with face-to-face interviews.

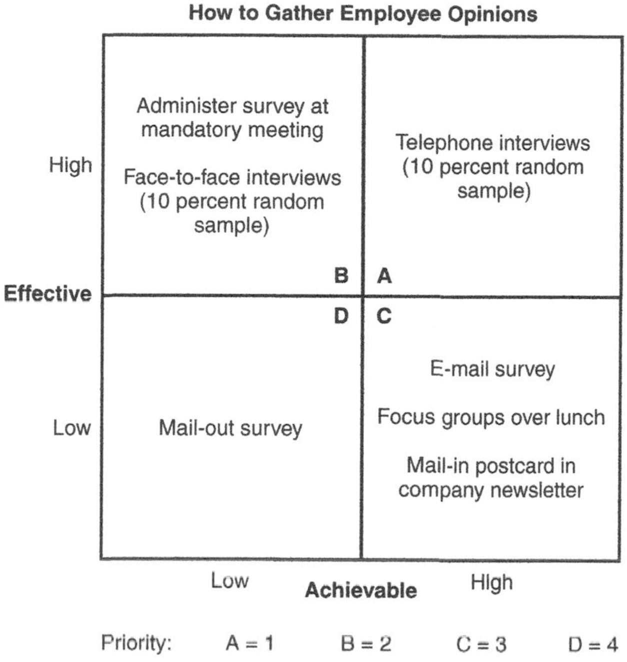

Figure 5.56 Effective–achievable chart example.

Considerations

- Ratings should be relative. Every choice is effective and achievable, or it would not be on your list. Compare choices to each other.
- Before using this tool, it can be useful to brainstorm evaluation criteria. What aspects of effectiveness need to be considered? Of achievability? You might use these aspects as headings for a decision matrix to evaluate each choice. Then use the effective–achievable chart to summarize your understanding.
- For project selection, sometimes this tool is called an impact–effort chart or project desirability chart, and the questions become: "How much impact will this project have?" and "How much effort will it require?"
- A similar chart called a control–impact chart asks the questions, "Is this action in our control?" and "How much impact would this action have?"
- This is not the best tool for making a decision involving many criteria or when some criteria are much more important than others. In those cases, use a decision or prioritization matrix.
- This is a variation of the generic *two-dimensional chart.* See its entry for more information.

failure modes and effects analysis (FMEA)

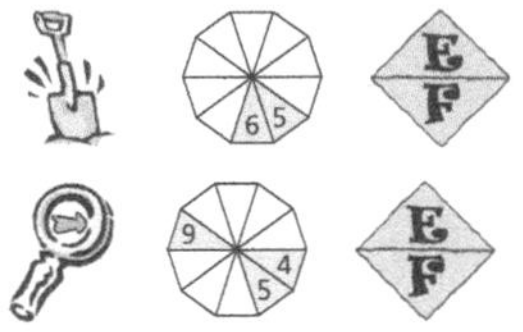

Also called: potential failure modes and effects analysis; failure modes, effects, and criticality analysis (FMECA)

Description

Failure modes and effects analysis (FMEA) is a step-by-step approach for identifying all possible failures in a design, a manufacturing or assembly process, or a final product or service. "Failure modes" means the ways, or modes, in which something might fail. Failures are any errors or defects, especially ones that affect the customer, and can be potential or actual. "Effects analysis" refers to studying the consequences, or effects, of those failures.

The failures are prioritized according to how serious their consequences are, how frequently they occur, and how easily they can be detected. The purpose of the FMEA is to take actions to eliminate or reduce failures, starting with the highest-priority ones. An FMEA also documents current knowledge and actions about the risks of failures, for use in continuous improvement.

FMEA is used during design to prevent failures. Later it is used for control, before and during ongoing operation of the process. Ideally, FMEA begins during the earliest conceptual stages of design and continues throughout the life of the product or service.

Begun in the 1940s by the United States military, FMEA was further developed by the aerospace and automotive industries. Several industries have formalized FMEA standards. What follows is an overview and reference. Before undertaking an FMEA process, learn more about standards and specific methods in your organization and industry through other references and training.

When to Use

- When a process, product, or service is being designed or redesigned, after QFD and before the design is finalized, or . . .
- When an existing process, product, or service is being applied in a new way, or . . .
- Before developing control plans for a new or modified process, or . . .
- When improvement goals are planned for an existing process, product, or service, or . . .

- When analyzing failures of an existing process, product, or service
- And periodically throughout the life of the process, product, or service

Procedure

This is a general procedure. Specific details may vary with standards of your organization or industry.

1. Assemble a cross-functional team of people with diverse knowledge about the process, product or service, and customer needs. Functions often included are: design, manufacturing, quality, testing, reliability, maintenance, purchasing (and suppliers), sales, marketing (and customers), customer service.
2. Identify the scope of the FMEA. Is it for concept, system, design, process, or service? What are the boundaries? How detailed should we be? Use flowcharts to identify the scope and to make sure every team member understands it in detail. (From here on, we'll use the word "scope" to mean the system, design, process, or service that is the subject of your FMEA.)
3. Fill in the identifying information at the top of your FMEA form. Figure 5.57 shows a typical format. The remaining steps ask for information that will go into the columns of the form.
4. Identify the *functions* of your scope. Ask, "What is the purpose of this system, design, process, or service? What do our customers expect it to do?" Name it with a verb followed by a noun. Usually you will break the scope into separate subsystems, items, parts, assemblies, or process steps and identify the function of each.
5. For each function, identify all the ways that it could fail to happen. These are *potential failure modes*. If necessary, go back and rewrite the function with more detail to be sure the failure modes show a loss of that function.
6. For each failure mode, identify all the consequences on the system, related systems, process, related processes, product, service, customer, or regulations. These are *potential effects of failure*. Ask, "What does the customer experience because of this failure? What happens when this failure occurs?"
7. Determine how serious each effect is. This is the *severity rating*, S. Severity is usually rated on a scale from 1 to 10, where 1 is insignificant and 10 is catastrophic. If a failure mode has more than one effect, write on the FMEA table only the highest severity rating for that failure mode.
8. For each failure mode, determine all the potential root causes. Use tools classified in this book as *cause analysis tools,* as well as the best knowledge and experience of the team. List all possible causes for each failure mode on the FMEA form.

Function	Potential Failure Mode	Potential Effects(s) of Failure	S	Potential Cause(s) of Failure	O	Current Process Controls	D	RPN	CRIT	Recommended Action(s)	Responsibility and Target Completion Date	Action Results					
												Action Taken	S	O	D	RPN	CRIT
Dispense amount of cash requested by customer	Does not dispense cash	Customer very dissatisfied Incorrect entry to demand deposit system Discrepancy in cash balancing	8	Out of cash	5	Internal low-cash alert	5	200	45								
				Machine jams	3	Internal jam alert	10	240	24								
				Power failure during transaction	2	None	10	160	16								
	Dispenses too much cash	Bank loses money Discrepancy in cash balancing	6	Bills stuck together	2	Loading procedure (riffle ends of stack)	7	84	12								
				Denominations in wrong trays	3	Two-person visual verification	4	72	18								
	Takes too long to dispense cash	Customer somewhat annoyed	3	Heavy computer network traffic	7	None	10	210	21								
				Power interruption during transaction	2	None	10	60	6								

Figure 5.57 FMEA example.

9. For each cause, determine the *occurrence rating*, O. This rating estimates the probability of failure occurring because of that cause during the lifetime of your scope. Occurrence is usually rated on a scale from 1 to 10, where 1 is extremely unlikely and 10 is inevitable. On the FMEA table, list the occurrence rating for each cause.
10. For each cause, identify *current process controls.* These are tests, procedures, or mechanisms that you now have in place to keep failures from reaching the customer. These controls might prevent the cause from happening, reduce the likelihood that it will happen, or detect failure after the cause has already happened but before the customer is affected.
11. For each control, determine the *detection rating,* D. This rating estimates how well the controls can detect either the cause or its failure mode after they have happened but before the customer is affected. Detection is usually rated on a scale from 1 to 10, where 1 means the control is absolutely certain to detect the problem and 10 means the control is certain not to detect the problem, or no control exists. On the FMEA table, list the detection rating for each cause.
12. (Optional for most industries) Is this failure mode associated with a *critical characteristic*? Critical characteristics are measurements or indicators that reflect safety or compliance with government regulations and need special controls. If so, a column labeled "Classification" receives a Y or ∇ to show that special controls may be needed. Usually, critical characteristics have a severity of 9 or 10 and occurrence and detection ratings above 3.
13. Calculate the *risk priority number,* RPN = S × O × D. Also calculate Criticality = S × O, by multiplying severity by occurrence. These numbers provide guidance for ranking potential failures in the order they should be addressed.
14. Identify *recommended actions.* These actions may be design or process changes to lower severity or occurrence. They may be additional controls to improve detection. Also note who is responsible for the actions and target completion dates.
15. As actions are completed, note results and the date on the FMEA form. Also, note new S, O, or D ratings and new RPN.

Example

A bank performed a process FMEA on their ATM system. Figure 5.57 shows part of it—the function "dispense cash" and a few of the failure modes for that function. The optional "Classification" column was not used. Only the headings are shown for the rightmost (action) columns.

Notice that RPN and criticality prioritize causes differently. According to the RPN, "machine jams" and "heavy computer network traffic" are the first and second highest risk. One high value for severity or occurrence times a detection rating of 10 generates a high RPN. Criticality does not include the detection rating, so it rates highest the only cause with medium to high values for both severity and occurrence: "out of cash." The team should use their experience and judgement to determine appropriate priorities for action.

Considerations

- It is not necessary to complete all of one step before moving on to the next. Begin with the information you have. Collect information that is available. Recycle and make your FMEA more complete as your team digs deeper or more information becomes available.
- Your organization or industry may have tables specifying the rating scales for severity, occurrence, and detection as well as the criteria for each number on the scale. Table 5.3 shows typical criteria. Sometimes a scale of 1 to 5 is used. Whatever scale and criteria you use, be sure to document them with your FMEA.

Table 5.3 Criteria for severity–occurrence–detection ratings.

Rating	Criteria		
	Severity	***Occurrence***	***Detection***
1	Not noticeable to customer.	Highly unlikely. < 1 in 1.5 million opportunities.	Almost certain to detect failure.
2	Some customers will notice. Very minor effect on product or system.	Extremely rare. 1 in 150,000 opportunities.	Excellent chance of detecting failure: 99.99%
3	Most customers notice. Minor effect on product or system.	Rare. 1 in 15,000 opportunities.	High chance of detecting failure: 99.9%
4	Customer slightly annoyed. Product or system slightly impaired.	Few. 1 out of 2000 opportunities.	Good chance of detecting failure: 95%
5	Customer annoyed. Noncritical aspects of product or system impaired.	Occasional. 1 out of 500 opportunities.	Fair chance of detecting failure: 80%
6	Customer experiences discomfort or inconvenience. Noncritical elements of product or system inoperable.	Often. 1 out of 100 opportunities.	Might detect failure: 50%
7	Customer very dissatisfied. Partial failure of critical elements of product or system. Other systems affected.	Frequent. 1 out of 20 opportunities.	Unlikely to detect failure: 20%
8	Customer highly dissatisfied. Product or system inoperable, but safe.	Repeated. 1 out of 10 opportunities.	Very unlikely to detect failure: 10%
9	Customer safety or regulatory compliance endangered, with warning.	Common: 1 out of 3 opportunites.	Highly unlikely to detect failure: 5%
10	Catastrophic. Customer safety or regulatory compliance endangered, without warning.	Almost certain. > 1 out of 2 opportunities.	Nearly certain not to detect failure, or no controls in place

- There are two basic types of FMEA: design and process. The design FMEA (DFMEA) looks at components, assemblies, parts, and other aspects of a product's design, with the objective of eliminating failures caused by poor design. Some references also describe a system FMEA, which is done very early in the design, at the conceptual stage, looking at the system broadly. Process FMEAs (PFMEA) address the steps of the process, with the objective of eliminating failures caused by the process and identifying process variables that should be controlled. Some references describe a service FMEA, which can be thought of as a process FMEA where the process is the delivery of a service. Both process and service FMEAs require studying the five Ms of a fishbone analysis: machines (equipment), manpower (people), materials, methods, and measurement.
- If you are doing a design FMEA, functions can be identified from customer expectations, safety requirements, government regulations, and your organization's requirements. QFD, which often precedes FMEA, is an excellent source of functions.
- If you are doing a process or service FMEA, functions are usually the steps of the process. Use block diagrams, process flow diagrams, and task analysis to help define them.
- Functions are actions. What does this thing *do?* If you don't have a verb in your function definition, then it isn't an action. Typical actions are open, filter, control, seal, move. Hint: If you can put "will not" or "cannot" in front of the word, then it's a verb. For example: will not open, cannot move.
- To expand your thinking to all potential failure modes, consider this checklist of ways that a function might be lost. Not all items on this checklist will apply in every situation, but going through the list may keep you from missing a failure mode. Examples are for a stovetop heating element and its function, "heats."

☐ No function at all	Does not get hot
☐ Partial function	Outer ring does not heat
☐ Complete function but only some of the time	Sometimes heats, sometimes does not
☐ Early or late function	Takes five minutes to get hot
☐ Gradual loss of function	Does not maintain temperature
☐ Too much function	Becomes superheated
☐ Something else unexpected happens along with function	Generates smoke

- What does the procedure mean in step 5: "If necessary, rewrite the function with more detail to be sure the failure modes show a loss of that function"? For the example of the stovetop heating element, the function statement must indicate how long it should take to achieve temperature. Otherwise, "takes five minutes to get hot" may or may not be a failure. Similarly, the temperature that the element is expected to achieve at each setting should be specified in order for "becomes superheated" to be a failure.

- If you brainstorm potential effects, your list may include extremely rare but catastrophic effects. For example, for the failure mode "light bulb burns out," an effect may be someone tripping, falling, and suffering severe injury—if the bulb burns out at night in a stairwell. One way to deal with these rare situations is to group all such effects together under a separate failure mode—"light bulb burns out under critical conditions"—and assign that mode its own severity, occurrence, and detection ratings. This is called *Procedure for Potential Consequences.*
- Because cause analysis can be difficult and time-consuming, begin with the highest-risk failures.
- Neither RPN nor criticality are foolproof methods for determining highest risk and priority for action. Severity, occurrence, and detection are not equally weighted; their scales are not linear; and the mathematics hides some surprises. For example, each of the three scales can have values from 1 to 10, so RPN can range from 1 to 1000—yet there are only 120 possible RPN values. Use RPN and criticality as guides, but also rely on your judgment. Effects with very high severity and any chance of occurrence must be addressed!
- Taking action transforms the FMEA from a paper chase into a valuable tool. By taking action, especially to reduce severity and occurrence ratings, you can prevent problems; reduce design, development, and start-up time and cost; improve customer satisfaction; and gain competitive advantage.
- The FMEA process relies on other tools to gather information or ideas. Some of the tools often used with FMEA are:
  - Tables, matrices, and checklists
  - Brainstorming, affinity diagrams, and other idea creation tools
  - Flowcharts and other process analysis tools
  - All the cause analysis tools, especially fault tree analysis, Pareto charts, why–why, and is–is not
  - Control charts, design of experiments, graphs, histograms, scatter diagrams, statistical analysis, and other data collection and analysis tools
- Software is available for preparing FMEA forms. These packages make it easier to revise and update the FMEA.
- Various industries, such as automotive and electromechanical, have developed detailed standards for FMEA. Ask for guidelines from your industry association or your industry's division of the American Society for Quality.

fault tree analysis

Also called: FTA

Description

Fault tree analysis uses a tree diagram to study a specific failure of a system, process, or product. The failure may have already happened, or it may be potential. Working backward from the failure, all the ways in which situations or events can combine to lead to the failure are identified, back to fundamental events or root causes. When probabilities of each cause are known, the probability of failure can be calculated. The primary purpose of fault tree analysis is to identify changes that can reduce or eliminate the chance of failure.

When to Use

- During design or redesign of a system, process, product, or service, to identify causes of a potential failure and look for ways to prevent that failure, or . . .
- After an accident, error, or other failure has occurred, to identify its causes and prevent future occurrences
- Especially when the system is complicated, with multiple interrelated causes of failure

Procedure

1. Identify the system or process that will be examined, including boundaries that will limit your analysis. Flow diagrams are useful here.
2. Identify the type of failure that will be analyzed. Define it as narrowly and specifically as possible. This is called the *top event.* Draw a rectangle at the top of the diagram and write a description of the failure in it.
3. Identify events that may be immediate causes of the top event. Write these events at the level below the event they cause.
4. For each event, ask "Is this a basic failure? Or can it be analyzed for its immediate causes?" If this event is a basic failure, draw a circle around it. If it can be analyzed for its own causes, draw a rectangle around it. If appropriate, use the other event symbols in Table 5.4.

Table 5.4 Event symbols.

Symbol	Event Name	Meaning
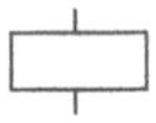	Event	Event that can be analyzed into more basic causes
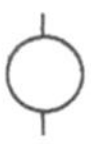	Basic event	Independent event that is not the result of other causes; root cause
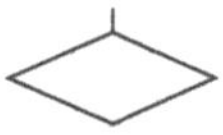	Undeveloped event	Event that will not be analyzed into causes because of insufficient information or unimportance
	Conditional event	Event that is the condition for an inhibit gate
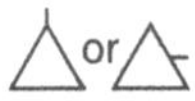	Transfer	Link to continuation of diagram in different place

Table 5.5 Gate symbols.

Symbol	Gate Name	Output Event Is Caused If:	To Reduce Risk Efficiently:
	AND	All input events occur at the same time.	Eliminate at least one of the input events, and failure is prevented.
	PAND Priority AND	All of the input events occur in sequence from left to right.	Eliminate at least one of the input events, or change the sequence in which they occur, and failure is prevented.
	OR	Any one of the input events occurs.	Eliminate as many input events as possible, beginning with the most likely one; each one reduces risk.
	XOR Exclusive OR	Only one of the input events occurs, not more than one.	Eliminate as many input events as possible, beginning with the most likely one; each one reduces risk. Or ensure that more than one event occurs, and failure is prevented.
	Inhibit	Input event occurs at the same time as conditional event.	Eliminate either input event or conditional event, and failure is prevented.
	Voting (m-out-of-n gate)	At least m (number) of the input events occur.	Eliminate input events, beginning with the most likely one; when there are only m – 1 input events left, failure is prevented.

5. Ask, "How are these events related to the one they cause?" Use the gate symbols in Table 5.5 to show the relationships. The lower-level events are the *input events.* The one they cause, above the gate, is the *output event.*

6. For each event that is not basic, repeat steps 3, 4, and 5. Continue until all branches of the tree end in a basic or undeveloped event.

7. (Optional) To determine the mathematical probability of failure, assign probabilities to each of the basic events. Use Boolean algebra to calculate the probability of each higher-level event and the top event. Discussion of the math is beyond the scope of this book. FTA software makes the calculations easier.

8. Analyze the tree to understand the relations between the causes and to find ways to prevent failures. Use the gate relationships to find the most efficient ways to reduce risk. Focus attention on the causes most likely to happen, using probabilities or your knowledge of the system.

Example

Figure 5.58 is a fault tree for a highway safety issue: single-vehicle accidents in which the vehicle overturns. Most of the gates are OR, but at the third level, events are linked by AND gates. For example, for a vehicle to overturn off the road, three events must occur: the vehicle must leave the road, no object must be in its path to stop it, and a sideways force must be imparted to the vehicle. "Vehicle leaves road" is followed by a transfer symbol, because that event has multiple causes, which are not explored on this section of the diagram. Similarly, the events on the bottom level are linked to continuations elsewhere.

The fault tree suggests several ways to reduce the risk of a vehicle overturning. At each AND gate, eliminating one of the input events will prevent the output event. Therefore, guardrails can prevent overturned vehicle accidents off the road by placing

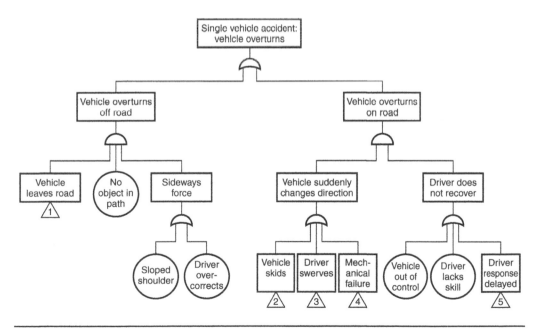

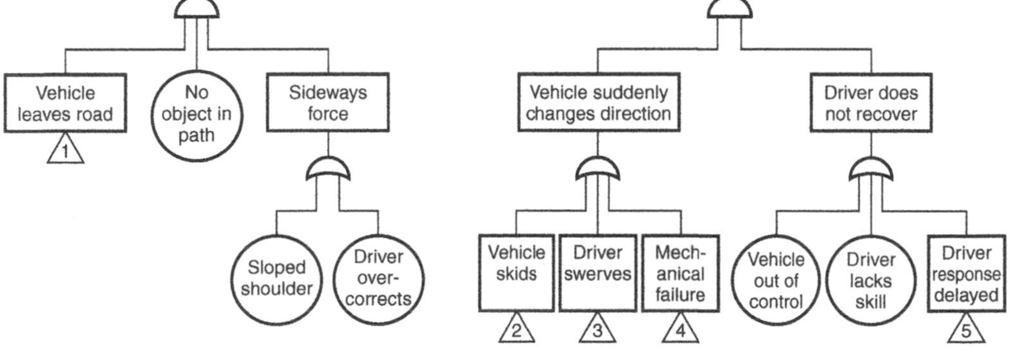

Figure 5.58 Fault tree analysis example.

an object in the vehicle's path. At OR gates, eliminating any of the input events reduces the risk of the failure. Level road shoulders, non-skid tires, and training drivers to recover from skids all can help reduce the risk of overturned vehicles.

Considerations

- Each level of the diagram should represent a tiny step away from the event above and toward the basic causes. Resist the urge to take large jumps, or you might miss important branches of the tree.
- Consider these possibilities when you are looking for immediate causes of a failure:
  - No input
  - Human error during installation or operation
  - Primary failure: age, fatigue, random failure
  - Secondary failure: operation under conditions outside design limits, stress from the environment (vibrations, earthquake, moisture)
  - Command fault: unintended operation after failure of a control
- The gates and events shown in the tables are the most commonly used. There are many others that can be used in complex situations.
- When an *inhibit* gate is used, the conditional event can be a fault or a situation. Only when the conditional event is present does the input event cause the output event. However, the conditional event is not a cause of the output event. With the AND gate, the two or more input events jointly cause the output event.
- A fault tree is one of the few tree diagrams that are usually drawn vertically, with the tree's trunk at the top and the branches below.
- If you are constructing a fault tree manually, consider using sticky notes for each event and for the gates. Different colors can be used for the different types of events and gates. This will let you change the tree easily while it is being developed. See *tree diagram* for more information on constructing trees.
- The earlier fault tree analysis is used, the more economical it is. Early in design, potential failures can be eliminated by changing drawings and specifications. Late in design or after the process is running, eliminating failures may require hardware or software changes or retraining of personnel. Waiting until a failure occurs is the most expensive option, because of the cost in dissatisfied customers, damaged equipment, or human life and limb.
- Computer software for generating fault trees makes the process easier. Use software if the tree will be complex or if you need to calculate probabilities.

- Another mathematical analysis defines the *minimal cut set*, which is the smallest combination of causes that results in the top event. Use software for that analysis also.
- Fault tree analysis shares features with the why–why diagram but is more structured. At each level of the fault tree, asking "Why?" leads to causes at the next lower level.
- Fault tree analysis is a deductive method. You start with a failure event and work your way backward to specific causes. In contrast, failure modes and effects analysis is an inductive method. You start with specific failures and work your way forward to consequences of those failures. As a result, FMEA is a broader process, used for analyzing entire systems or processes. FTA is a narrower process, applied to one failure at a time. It can be used within the FMEA process to understand the causes of failure modes.

fishbone diagram

Also called: cause-and-effect diagram, Ishikawa diagram
Variations: cause enumeration diagram, process fishbone, time-delay fishbone, CEDAC (cause-and-effect diagram with the addition of cards), desired-result fishbone, reverse fishbone diagram

Description

The fishbone diagram identifies many possible causes for an effect or problem. It can be used to structure a brainstorming session. It immediately sorts ideas into useful categories.

When to Use

- When identifying possible causes for a problem
- Especially when the team's thinking tends to fall into ruts

Procedure

Materials needed: flipchart or whiteboard, marking pens.

1. Agree on a problem statement (effect). Write it at the center right of the flipchart or whiteboard. Draw a box around it and draw a horizontal arrow running to it.
2. Brainstorm the major categories of causes of the problem. If there is difficulty here, use generic headings: methods, machines (equipment), people (manpower),

materials, measurement, environment. Write the categories of causes as branches from the main arrow.

3. Brainstorm all the possible causes of the problem. Ask "Why does this happen?" As each idea is given, the facilitator writes it as a branch from the appropriate category. Causes can be written in several places if they relate to several categories.
4. Ask again, "Why does this happen?" about each cause. Write subcauses branching off the causes. Continue to ask "Why?" and generate deeper levels of causes. Layers of branches indicate causal relationships.
5. When the group runs out of ideas, focus attention to places on the fishbone where ideas are few.

Example

Figure 5.59 is the fishbone diagram drawn by the ZZ-400 manufacturing team to try to understand the source of periodic iron contamination. This example is part of the ZZ-400 improvement story in Chapter 4. The team used the six generic headings to prompt ideas. Layers of branches show thorough thinking about the causes of the problem. For example, under the heading "Machines," the idea "materials of construction" shows four kinds of equipment and then several specific machine numbers. Some ideas appear in two different places. "Calibration" shows up under "Methods," as a factor in the analytical procedure, and also under "Measurement," as a cause of lab error. "Iron tools"

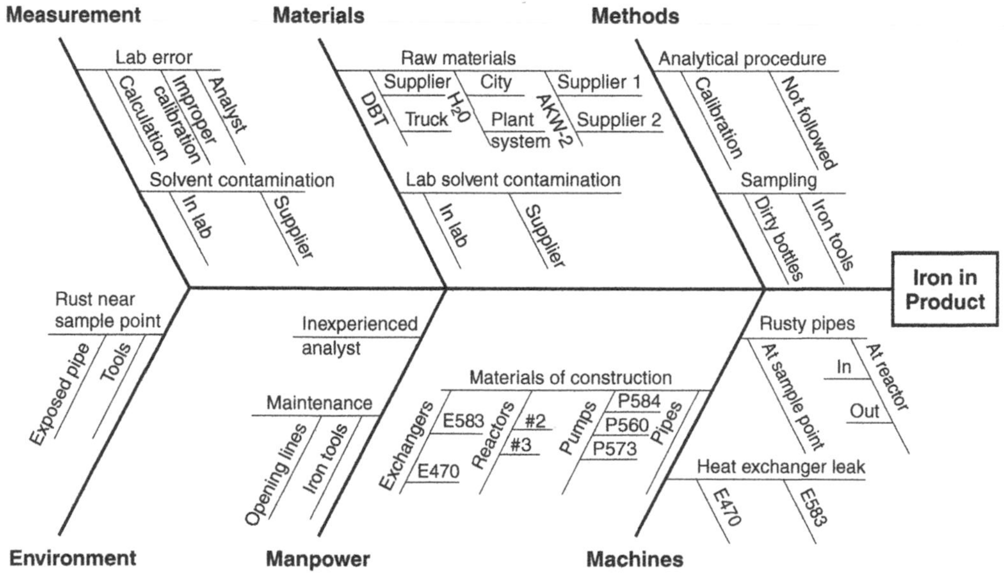

Figure 5.59 Fishbone diagram example.

can be considered a "Methods" problem when taking samples or a "Manpower" problem with maintenance personnel.

Also see the St. Luke's Hospital story on page 72 for another example of a fishbone diagram used within the improvement process. See the Medrad story on page 61 for an example of a nontraditional way to conduct a fishbone exercise.

cause enumeration diagram

Materials needed: sticky notes or cards, marking pens, large work surface (wall, table, floor), flipchart paper.

1. Agree on the problem statement.
2. Brainstorm all possible causes, using any brainstorming technique. (See *brainstorming* and *NGT.*) Record on sticky notes or cards. Continue until the group has run out of ideas.
3. Using an affinity diagram, group the causes and determine headings.
4. Using the headings as main causes, arrange the ideas on a fishbone drawn on flipchart paper.
5. Use the fishbone to explore for additional ideas, especially where there are few ideas on the fishbone.

process fishbone

Also called: production process classification diagram
Materials needed: flipchart paper, marking pens.

1. Identify the problem to be studied. Develop a flow diagram of the main steps of the process in which the problem occurs. There should be fewer than 10. Draw them as a series of boxes running horizontally across a flipchart page, with arrows connecting the boxes.
2. Draw a separate fishbone for each step of the process. Take each one in turn and brainstorm causes arising from each step. Also, consider the handoff from one step to another and the causes of quality problems that occur there.
3. Continue to brainstorm subcauses, as in the first procedure.

time-delay fishbone

Materials needed: flipchart paper, marking pens, tape, large wall space.

1. Begin a fishbone as in steps 1 and 2 of the main procedure.

2. Hang the diagram in a well-traveled area with pens nearby. Allow people to add to the diagram over a certain time frame, say a week or two.

   More people can participate this way. Even those who have not been trained in fishbones will catch on and join in.

Example

In the ZZ-400 unit, ideas were needed from everyone about the source of iron, but members of the four shifts were never all together. The team hung a flipchart page with the fishbone skeleton on the control room wall and asked people to contribute ideas whenever they could. The result was a detailed fishbone that everyone had thought about and discussed for two weeks. This example is part of the ZZ-400 improvement story in Chapter 4.

CEDAC (cause-and-effect diagram with the addition of cards)

CEDAC, developed by Dr. Ryuji Fukuda, is a disciplined problem-solving method wrapped around a fishbone diagram analysis. After an improvement target and metrics have been determined, causes are brainstormed, written on colored cards, sorted into categories, and placed on a fishbone diagram. Ideas for improvement are written on cards of a different color and also placed on the diagram. As in the time-delay fishbone, the diagram is located in a public area where a group can see it, think about it, and add to it over time. Every improvement idea is studied for its possible value to the organization. See Resources for additional information.

desired-result fishbone

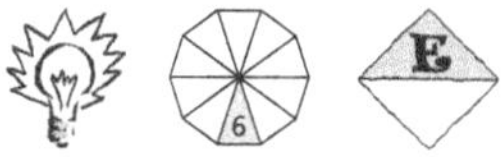

Instead of stating a problem, state a desired result. Then brainstorm ways to achieve the result. Or, state the desired result as a question, such as "How can we achieve our goal?"

reverse fishbone diagram

Also called: backward fishbone, action and effect diagram, AED, solution impact diagram

When to Use

- When choosing between alternative solutions, or . . .
- When evaluating the potential consequences of one or more solutions or actions

Procedure

Follow the basic fishbone diagram procedure, but instead of writing a problem statement, write a proposed solution or action. Brainstorm by asking, "What effect could this have?" Be sure to consider positive and negative effects, and distinguish between them by writing them in different colors or circling negative effects. If several actions are being evaluated, do a separate diagram for each one and compare effects.

Considerations

- The fishbone diagram broadens the team's thinking. We all tend to think in ruts at times. Determining major categories in advance, or using the generic ones, and then focusing on those categories where ideas are few will get a team out of its ruts.
- A traditional and widely used set of generic categories is the five Ms: machines, manpower, materials, methods, and measurement. Because they all begin with M, it is easy to remember them. However, many people consider the word *manpower* to be exclusionary. You may wish to use the five Ms to remember the categories, then substitute *people* for *manpower*. Other words that are sometimes substituted include *equipment, facilities,* or *technology* instead of *machines* and *policies* or *practices* instead of *methods*.
- Think broadly about the problem. Consider the environment, policies, external factors. Consider all causes, not just those within your control. It can be useful to understand them all.
- The hardest part of facilitating a fishbone session is deciding where to write an idea. Let the group tell the facilitator where to write it. This gets members thinking about the relationships between the causes and the effect. If members believe it belongs in different places, explore whether different ideas or different aspects of an idea are being considered.
- Have someone outside the team review the diagram for fresh ideas.
- Force field analysis can be used immediately after a fishbone diagram (cause-and-effect diagram) in a combination called CEFFA (cause-and-effect/force field analysis). Each cause identified on the fishbone becomes the subject of a force field analysis. Not only does this combination move a group quickly from problem to causes to potential solutions, but also overlooked causes can be uncovered during the force field analysis.
- If a problem occurs frequently, possible causes from a fishbone diagram can be made into a checklist for quickly pinpointing the source of the problem.

5W2H

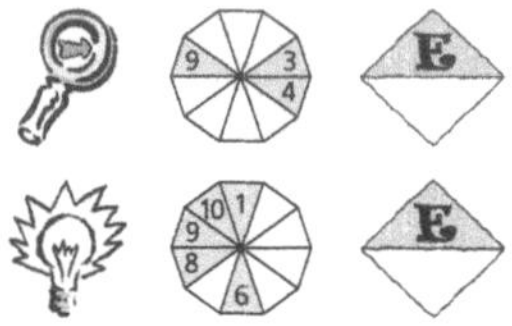

Also called: 5W2H method, 5W2H approach, five Ws two Hs

Description

5W2H is a method for asking questions about a process or problem. Its structure forces you to consider all aspects of the situation. The five Ws are *who, what, when, where,* and *why*. The two Hs are *how* and *how much* (or *many*).

When to Use

Most often, it is used:

- When analyzing a process for improvement opportunities, or . . .
- When a problem has been suspected or identified but must be better defined

But with modification, it can also be used in any of these situations:

- When planning a project or steps of a project (such as data collection or rolling out changes)
- When reviewing a project after completion
- When writing an article, report, or presentation

Procedure

1. Review the situation under study. Make sure everyone understands the subject of the 5W2H.
2. Develop appropriate questions about the situation for each of the question words, as in Table 5.6. The order of asking questions is not important.
3. Answer each question. If answers are not known, create a plan for finding them.
4. What you do next depends on your situation:
   - If you are planning a project, let your questions and answers help form your plan.
   - If you are analyzing a process for improvement opportunities, let your questions and answers lead you into additional questions about possible changes.

Table 5.6 5W2H questions.

5W2H	Typical Questions for Processes	Improvement Questions
Who?	Who does this? Who should be involved but is not? Who is involved but shouldn't be? Who has to approve?	Should someone else do it? Could fewer people do it? Could approvals be eliminated?
What?	What is done? What is essential?	Does every step have to be done? Are steps omitted?
When?	When is this activity started? When does it end? When is it repeated?	Can it be done at a different time? Can cycle time be shortened? Can it be done less frequently?
Where?	Where is this activity done?	Can it be done elsewhere?
Why?	Why do we do this?	Can it be eliminated? Can another group do it? Can it be outsourced?
How?	How is this done?	Is there a better way?
How Much?	How much does it cost?	How much less could it cost?

- If you are defining a problem, let your questions and answers lead you into cause analysis.
- If you are reviewing a completed project, let your questions and answers lead you into additional questions about modifying, expanding, or standardizing your changes.
- If you are preparing an article, report, or presentation, include answers to the questions in your text.

Example

Ben-Hur's Pizza plans to add home delivery to their services in order to expand their business. They use 5W2H to identify and begin to resolve some of the issues and details. First they brainstormed questions, shown in the center column of Figure 5.60. Once they had generated all the questions they could think of, they began to develop answers. Note that they didn't have all the answers. Those marked with * require more research or process analysis.

Considerations

- Table 5.6 shows typical questions. The third column shows additional questions you might ask in step 4 if you were seeking improvement opportunities. Don't let this list limit you. Your situation may suggest more or different questions.
- 5W2H is a kind of checklist. In this one, however, you have to come up with the questions. This checklist only provides the first word of each question to remind you of the kinds of questions you should ask.

5W2H	Questions	Answers
Who?	*Who will order-in pizza?*	*Working parents, school groups, offices, DINKs, shut-ins . . .*
	Who will take orders?	*Cashier, wait staff*
	Who will make the pizzas?	*Regular kitchen crew*
	Who will make deliveries?	*New hires: high school/college kids*
What?	*What pizzas should be on the menu?*	*Same selection as in-store*
	What should we call this service?	*Ben-Hur's Chariot Service*
When?	*When should we start this service?*	*May 15 (* do Gantt chart to confirm this date is feasible)*
	When (hours) should it be available?	*11 a.m. to midnight*
Where?	*Where should our delivery area be?*	*8 mile radius*
	Where will we store boxes?	*Cabinet by back door*
	Where will delivery vehicles park?	*Reserve spaces by back door*
Why?	*Why will people order-in pizza?*	*Parties, meetings, TV specials, too tired to go out or cook*
	Why should they order from us?	*Superior pizza, wide selection*
How?	*How will we take orders?*	*Checklist pad*
	How will we keep pizzas hot?	*Buy insulated sleeves*
	How will we hire delivery people?	*Ads in window; school placement offices*
	How will we handle super-high volume?	*Staff up; advertise pre-orders for known high-volume days*
	How will we advertise our service?	*Radio, TV, refrigerator magnets, introductory coupons*
How much? How many?	*How much should delivery cost?*	*Free*
	How much should the pizzas cost?	*(see separate price list)*
	How many delivery people do we need?	*Start with: afternoons 1, weekends and evenings 2*
	How much will delivery people earn?	*Minimum wage + tips*
	How much time from order to front door?	*35 minutes (*analyze process to see if feasible)*
	How many pizzas per day will be delivered?	**Study competition to determine*

Figure 5.60 5W2H example.

- The *is–is not matrix* is a similar tool used to study problems. Compare them to see which would be better for your situation.
- The "How much or how many" question can be aimed at cost or time (how much) or resources (how many). Answers to those questions will be an upper limit on the benefits possible from improvements.
- The 5W1H structure has been around a long time. Journalists have always been taught that a good newspaper story must include who, what, when, where, why, and how. This tool adds an extra H.

flowchart

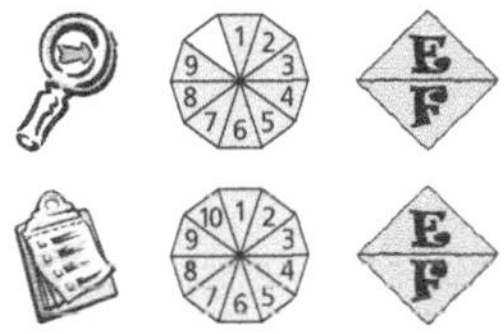

Also called: process flowchart, process flow diagram
Variations: macro flowchart, top-down flowchart, detailed flowchart (also called process map, micro map, service map, or symbolic flowchart), deployment flowchart (also called down-across or cross-functional flowchart), several-leveled flowchart
See also: arrow diagram, SIPOC diagram, and work-flow diagram

Description

A flowchart is a picture of the separate steps of a process in sequential order. Elements that may be included are: sequence of actions, materials or services entering or leaving the process (inputs and outputs), decisions that must be made, people who become involved, time involved at each step, and/or process measurements. The process described can be anything: a manufacturing process, an administrative or service process, a project plan. Usually listed as one of the seven QC tools, this is a generic tool that can be adapted for a wide variety of purposes.

When to Use

- To develop understanding of how a process is done, or . . .
- To study a process for improvement, or . . .
- To communicate to others how a process is done, or . . .
- When better communication is needed between people involved with the same process, or . . .

- To document a process, or . . .
- When planning a project

Basic Procedure

Materials needed: sticky notes or cards, a large piece of flipchart paper or newsprint, marking pens.

1. Define the process to be diagrammed. Write it at the top of the work surface. Discuss and decide on the boundaries of your process: Where or when does the process start? Where or when does it end? Discuss and decide on the level of detail to be included in the diagram.
2. Brainstorm the activities that take place. Write each on a card or sticky note. Sequence is not important at this point, although thinking in sequence may help people remember all the steps.
3. Arrange the activities in proper sequence.
4. When all activities are included and everyone agrees that the sequence is correct, draw arrows to show the flow of the process.
5. Review the flowchart with others involved in the process (workers, supervisors, suppliers, customers) to see if they agree that the process is as drawn.

Several variations follow. Still other flowcharts are listed as separate tools because they can be used for different situations. See: *arrow diagram, SIPOC,* and *work-flow diagram.*

Considerations

- Don't worry too much about drawing the flowchart the "right way." The right way is the way that helps those involved understand the process.
- Identify and involve in the flowcharting process all key people who are involved with the process. This includes those who do the work in the process: suppliers, customers, and supervisors. Involve them in the actual flowcharting sessions by interviewing them before the sessions and/or by showing them the developing flowchart between work sessions and obtaining their feedback.
- Do not assign a "technical expert" to draw the flowchart. People who actually perform the process should construct the flowchart.
- Computer software is available for drawing flowcharts. Software is useful for drawing a neat final diagram, but the method given here works better for the messy initial stages of creating the flowchart.
- The facilitator's role is to be sure all members participate, to ask the right questions to uncover all aspects of the process, and to help team members capture all their ideas in the language of the flowchart.

- Keep all parts of the flowchart visible to everyone all the time. That is why flipchart or newsprint should be used rather than transparencies or a whiteboard.
- Several sessions may be necessary. This allows members time to gather information or to reflect on the process. Even if the flowchart seems to be finished in one session, plan a review at a second session, to allow reflection time.

macro flowchart

Also called: high-level flowchart, high-level process map

Description

A macro flowchart shows only the major steps of the process.

When to Use

- When you need to understand or communicate a big-picture view of the major steps in a process, or . . .
- Before drawing a more detailed flowchart

Procedure

Follow the basic procedure above. At step 2, include only major activities.

Example

Figure 5.61 is a macro flowchart of an order-filling process. See the Medrad story on page 58 for another example of a macro flowchart.

Considerations

- If you have more than six or eight steps, consider whether you are including too much detail for a broad overview.

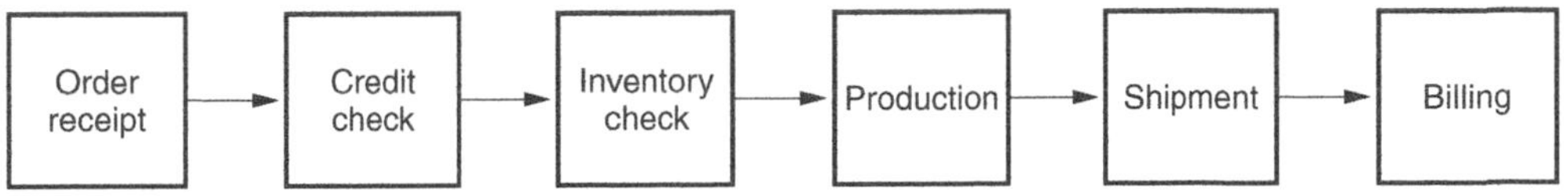

Figure 5.61 Macro flowchart example.

- Often the words on a macro flowchart are nouns, showing broad categories of actions or names of subprocesses, rather than verbs showing specific actions. Nouns are used in the example, such as "Credit check" and "Production." They could have been written as verbs: "Check credit" and "Make product." Thinking in verbs sometimes leads to excessive detail.
- Decisions, delays, and recycle loops are usually not shown on a macro flowchart. The only symbols you should need are rectangles and arrows. Inputs and outputs are sometimes shown.
- Macro flowcharts do not reveal costs of poor quality.

top-down flowchart

Description

A top-down flowchart shows the major steps of a process and the first layer of substeps.

When to Use

- When you need to understand or communicate the major steps of a process plus the key activities that constitute each step
- When you want to focus on the ideal process
- When you can't see the forest for the trees

Procedure

1. Define the process to be diagrammed, as in step 1 of the basic procedure.
2. Brainstorm the main steps in the process. There should be about six or eight. Write each on a card or note and place them in a line across the top of your paper.
3. Brainstorm the major substeps of the first main step. Again, there should be no more than six or eight. Write each on a card and place them in order vertically below the first main step.
4. Repeat for each main step.
5. When the group agrees that you have all the steps and substeps and that the sequence is correct, draw horizontal arrows leading from one main step to the next. Draw arrows pointing down from each main step to the first substep. You can draw arrows pointing down between substeps, but they are not necessary. Do not draw an arrow below the last substep of each vertical column.
6. Review the flowchart with others involved in the process (workers, supervisors, suppliers, customers) to see if they agree that the process is as drawn.

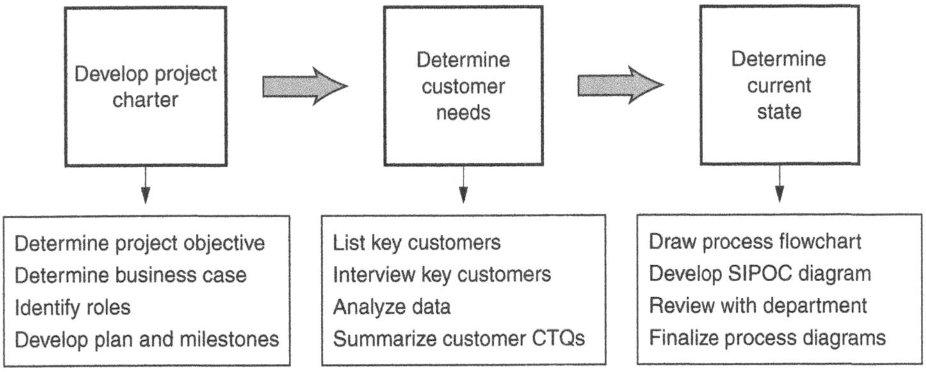

Figure 5.62 Top-down flowchart example.

Example

A group beginning a quality improvement project drew a top-down flowchart (Figure 5.62) to guide the first phase of the effort. The major steps are: develop project charter, determine customer needs, determine current state. Under each major step are four sub-steps that detail what must be done to accomplish each step. This example is part of the ZZ-400 improvement story in Chapter 4.

Considerations

- This is a fast method for flowcharting a process.
- This flowchart focuses on the essential components of a process or project. By comparing it to the way things are really done, the sources of waste and complexity can quickly become obvious.
- Use the top-down flowchart to get an overview of the entire process or project, then construct detailed flowcharts of particular steps as needed.
- The main steps, across the top of the chart, are the same as the main steps of a macro flowchart. The "Considerations" for a macro flowchart also apply here.

detailed flowchart

Also called: process map, micro map, symbolic flowchart

Description

The detailed flowchart shows major and minor activities, decisions, delays, recycle loops, inputs and outputs, and any other details necessary to capture exactly how a process is done.

When to Use

- When a team begins to study a process, as the first and most important step in understanding the process, or . . .
- When searching for improvement opportunities in a process (called *as-is*), or . . .
- When designing an improved process (called *should-be* or *to-be*), or . . .
- At any step in the improvement process, as a reference for how the process is done, or . . .
- When training people on a process, or . . .
- When documenting how the process is performed

Procedure

Construction

Follow steps 1 to 3 of the basic procedure. Then:

4. Have you included steps that take place to correct problems or when something goes wrong? Have you included decisions that must be made and the alternate actions that depend on the decisions? If not, add those.
5. List inputs and outputs at each step of the process. Write each on a note and place at the appropriate point in the process flow.

Conclude with steps 4 and 5 of the basic procedure.

Example

Figure 5.63 is a flowchart of the process for filling an order, from the time the order is received until the bill is mailed. An end point occurs early in the process if the potential customer's credit is not good. The flowchart also shows an output to and an input from the vendor in the middle of the process. The process is complicated enough that team members moved sticky notes around several times and added forgotten steps before they settled on the final flow.

Analysis reveals two delays ("Wait") and two recycle loops if raw materials or product aren't good. Cost-of-poor-quality analysis also will reveal several inspection steps. (See the cost-of-poor-quality analysis example and Figure 5.43, page 201, which shows the same process.) Compare this flowchart with two other flowcharts of the same process: Figure 5.61, a macro flowchart, and Figure 5.189, page 520, a work-flow diagram. Read all the improvement stories in Chapter 4 for other examples of detailed flowcharts.

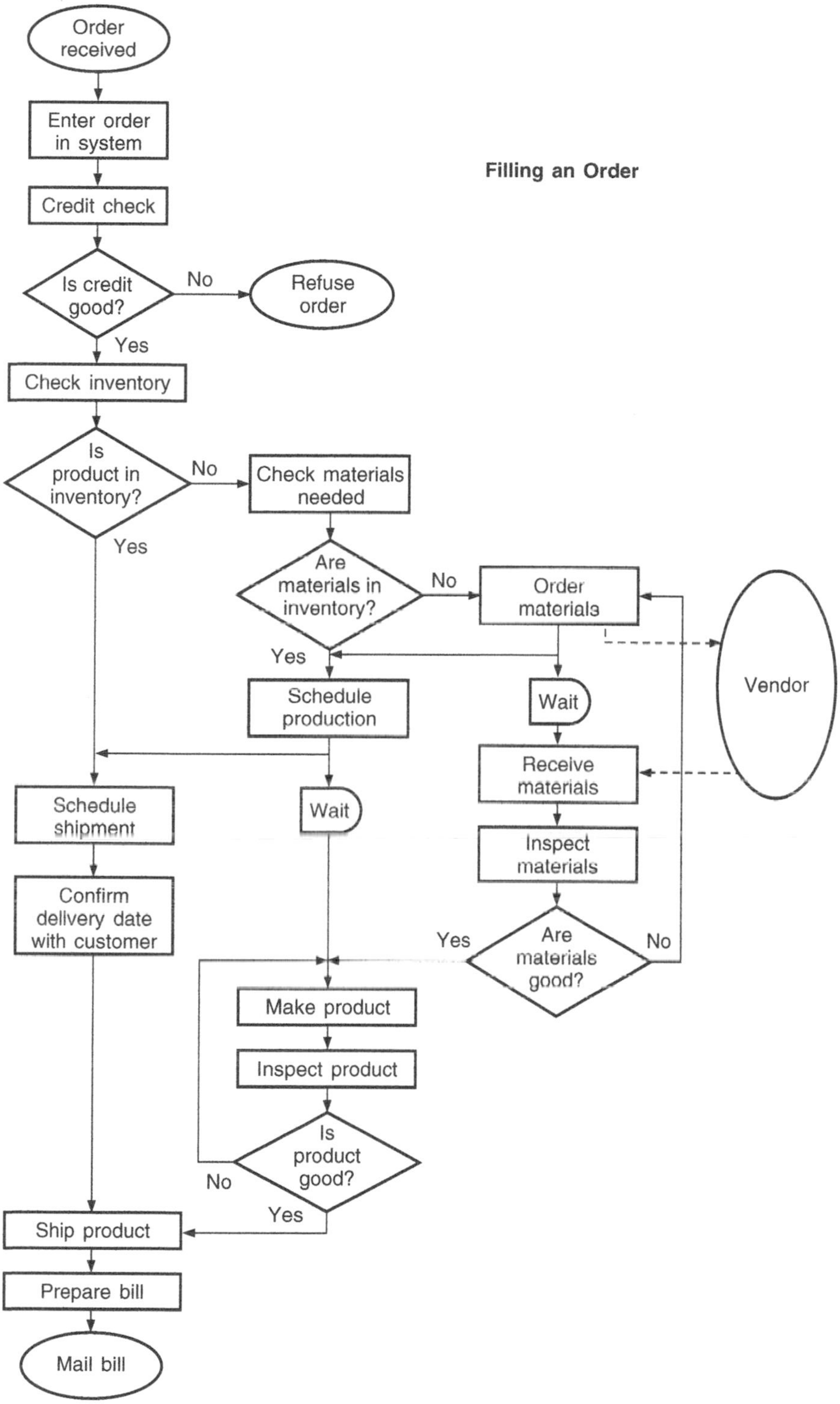

Figure 5.63 Detailed flowchart example.

Considerations

Construction

- The following symbols are commonly used in detailed flowcharts:

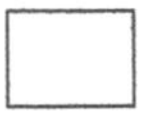 One step in the process; the step is written inside the box. Usually, only one arrow goes out of the box.

→ Direction of flow from one step or decision to another.

 Decision based on a question. The question is written in the diamond. More than one arrow goes out of the diamond, each one showing the direction the process takes for a given answer to the question. (Often the answers are *yes* and *no.*)

 Delay or wait

 Link to another page or another flowchart. The same symbol on the other page indicates that the flow continues there.

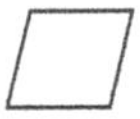 Input or output

 Document

 Alternate symbols for start and end points

- An *as-is* process map shows the process as it currently is, warts and all. This should include recycle loops, rework, delays, excessive handoffs, and any other non-value-added steps.
- To be sure you've included all as-is steps, ask questions like:
  - Is the flow the same for special times like month-end or year-end? What happens on a holiday? After normal business hours?
  - What would happen if a substitute or new employee operated the process?
  - What unexpected things might the customer do (especially in an interactive service process)?
  - Do some people do things differently than others?

  Use cost-of-poor-quality analysis, critical-to-quality analysis, or value-added analysis to be sure you've included all the warts in your current process.
- A *should-be* process map shows the ideal process. Non-value-added steps are eliminated, the flow is streamlined, and all other improvements are incorporated.

- Convert a macro or top-down flowchart into a detailed flowchart by identifying key steps or steps that cause problems, and flowchart each of these in more detail. It is often not necessary to create detailed flowcharts of an entire process.
- If a detailed flowchart of the entire process is needed, work on major steps separately. It is easier to focus on part of the process than to deal with the entire thing at once. Try using the approach of the several-leveled flowchart variation below.
- Detailed flowcharts can include additional information about each step. Time required (actual, estimated, and/or theoretical) can be added for cycle time analysis. Cost, method of measurement, and average measurement results are other information that can be included if they are relevant to the analysis or the way the flowchart will be used.
- Here are good questions to ask as a detailed flowchart is being developed.*
  - Where does the [service, material] come from?
  - How does the [service, material] get to the process?
  - Who makes this decision?
  - What happens if the decision is yes?
  - What happens if the decision is no?
  - Is there anything else that must be done at this point?
  - Where does the [product, service] of this process go?
  - What tests or inspections are done on the product at each part of the process?
  - What tests or inspections are done on the process?
  - What happens if the test or inspection is failed?

  Note that these are "what, where, how, who?" questions. It is not helpful to ask "why?" while flowcharting. Save that question for later.

Analysis

Some methods of analyzing detailed flowcharts are :

- Cost-of-poor-quality analysis (page 199), critical-to-quality analysis (page 207), and value-added analysis (page 507)
- Draw a flowchart of the ideal (should-be) process and compare it to the actual (as-is) process.
- Convert the flowchart to a SIPOC diagram (page 475): Identify inputs and outputs; define suppliers, customers, and their requirements.
- Convert to a deployment chart: Who does what when? (See variation following.)

* John T. Burr, "The Tools of Quality; Part I: Going with the Flow(chart)," *Quality Progress* 23, no. 6 (June 1990): 64–67. Reprinted with permission.

- Convert to a work-flow diagram (page 519) to identify physical inefficiencies.
- Classify each input as either controllable (C), noise (N), or standard methods (S). Controllable inputs can be changed to affect the outputs, for example, oven temperature in a pizza shop. For each controllable input, identify the target setting and its upper and lower limits. Noise are inputs that either cannot be controlled or are too expensive or difficult to control, for example, the humidity in the pizza shop. Standard methods are procedures for running the process, such as the pizza recipe.
- Analyze cycle time. Identify time requirements and label each step with actual time needed and theoretical time required. Identify holds or delays. Find the critical path. Collect data if necessary to determine true time requirements.
- Identify: large inventories, large lot sizes, long changeover times, long cycle time. (*Note:* These production-derived terms apply to office work, too.)
- Identify places in the process where those involved do not agree about correct steps or sequence.
- Mistake-proof the process. Identify common mistakes or problems at each step of process. What can go wrong? See *mistake-proofing*.

deployment flowchart

Also called: down-across, cross-functional flowchart

Description

A deployment flowchart is a detailed flowchart that also shows who (which person or group) performs each step.

When to Use

- When several different individuals or groups are involved in a process at different stages, or . . .
- When trying to understand or communicate responsibilities, or . . .
- When identifying supplier–customer relationships, internal or external, or . . .
- When studying how sequential or parallel steps affect cycle time, or . . .
- When allocating and tracking responsibilities on a project

Procedure

1. Begin with steps 1, 2, and 3 of the basic procedure. If a macro, top-down, or detailed flowchart has already been drawn, you have an excellent start.

2. On a flipchart page or newsprint, list all players (individuals or groups) involved in the process. List them across the top or down the left side, whichever is the narrower dimension of the paper. Draw lines between each, extending the full dimension of the paper. Optional: Include columns or rows beside the players for "Actual Time" and "Theoretical Time."
3. Starting at the beginning of the process, place the card or note with the first step of the process in the column (or row) of the player responsible for that step. Place the second step a little farther along, to indicate later time sequence, opposite that step's key player. Continue to place all steps opposite the person or group responsible. Place them as though along a timeline, with time moving away from the names. If two steps happen simultaneously or the sequence is unimportant, place the cards or notes at equal distances along the timeline.
4. Some process steps involve two players: "Joe telephones Sally." For these, make a second card to place opposite the second name. Write the action from the point of view of the second player: "Sally receives phone call."
5. Draw arrows between cards to show the flow of the process.
6. (Optional) In the column (or row) labeled "Actual Time," write opposite each step how long that step currently requires. In the column (or row) labeled "Theoretical Time," write opposite each step how long it should take in an ideal process. The theoretical time for a step that does not add value should be zero.
7. Review the flowchart with others involved in the process (workers, supervisors, suppliers, customers) to see if they agree that the process is as drawn.

Example

Figure 5.64 shows the deployment flowchart of the Parisian Experience restaurant's process for seating guests, drawn by the team trying to solve the problem of guests waiting to be seated. The key players are the maitre d', the diners, the busser, the waiter, and lounge staff. Notice that several steps involve two players and are written twice, from both points of view: "Order drinks" and "Take drink order"; "Call diners" and "Hear call"; "Seat diners" and "Take seats"; "Order meal" and "Take order."

Notice that the arrows follow the sequence of the process, not necessarily the actions of any player. For example, there is no arrow for the maitre d' linking "Call diners" and "Seat diners," because the maitre d' cannot seat them until they hear the call and leave the lounge. After the diners leave the lounge, the flow goes back to the maitre d' to seat them. Although this step is simultaneous with the diners' action of taking their seats, the diners cannot act until the maitre d' does.

The team can identify some opportunities for changes from this diagram. For example, being seated and ordering from the menu now are sequential steps. Perhaps the diners could be given their menus and have their orders taken while they are waiting for a table. The two places where the diner waits for a waiter are obviously opportunities for improvement.

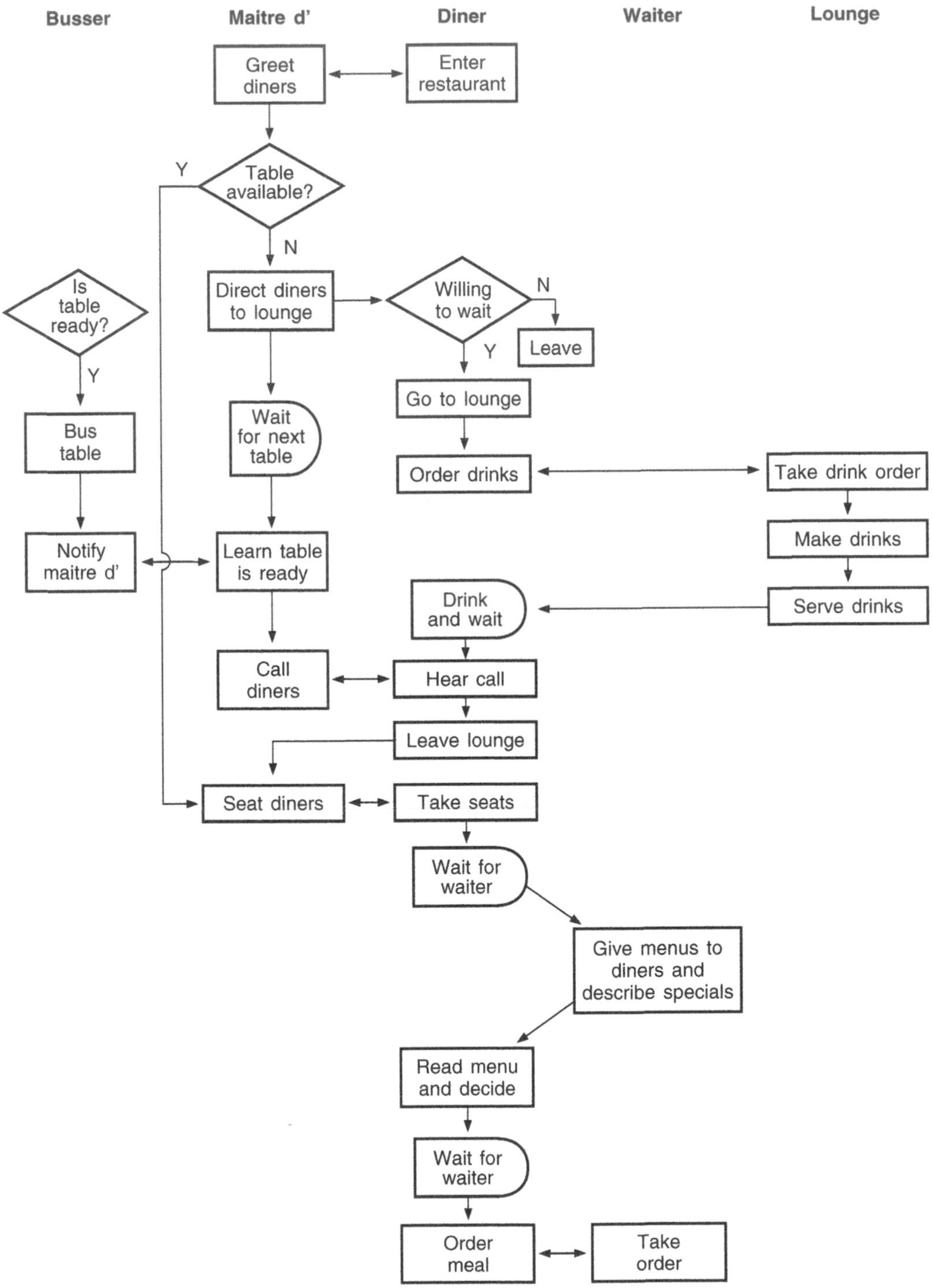

Figure 5.64 Deployment flowchart example.

Considerations

- You may wish to indicate primary and secondary or multiple responsibility by placing a step under two names with different symbols or colors.
- Optional step 6 (and the optional part of step 2) are for doing a cycle-time analysis.
- Often it is easier first to construct a detailed flowchart and then to convert it to a deployment chart—especially if this tool is new to the team.
- Look for sequential steps carried out by different players that could be done simultaneously.
- When the process flow moves between players, a customer–supplier relationship is involved.
- The "Considerations" for a detailed flowchart also apply here.

several-leveled flowchart

Description

The several-leveled flowchart combines macro and detailed flowcharts. This tool is actually multiple flowcharts of the same process. The first level is a macro flowchart, followed by detailed flowcharts each showing one increasing level of complexity. Each step is numbered, for example, 2.0. On a successive page, where more detail is shown, each step is given a number linking it to the original, such as 2.1, 2.2, and so on. Another level of detail for step 2.1 would be numbered 2.1.1, 2.1.2, 2.1.3, and so on.

When to Use

- When both high-level and detailed views of the same process need to be available, such as:
- When documenting a process, or . . .
- When training people on a process, or . . .
- When making a presentation about a process

Procedure

Use the procedures of the macro flowchart and the detailed flowchart, but combine them in this way:

1. Create a macro flowchart of the process. Number each step 1.0, 2.0, 3.0, and so on. This is level 0.
2. For each major step, create another flowchart showing the major substeps. These charts will be equivalent to the vertical sections of a top-down flowchart. Number each substep with the number of its major step, followed by a period and a sequential number: 1.1, 1.2, 1.3, and so on. These are level 1.

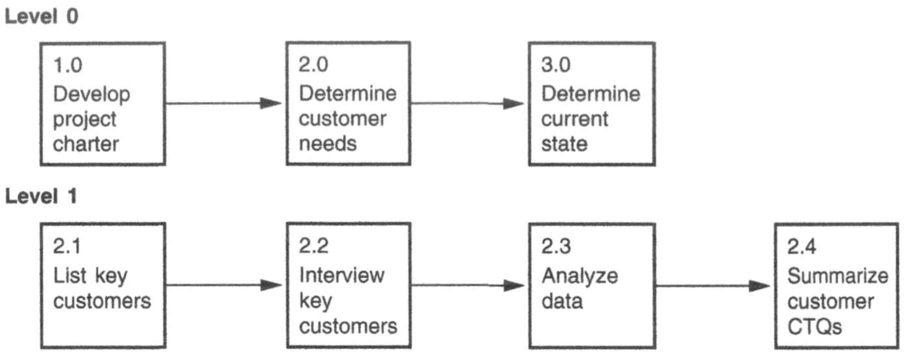

Figure 5.65 Several-leveled flowchart example.

3. Continue to develop lower-level charts showing increasing levels of detail. Number steps by adding a period and a sequential number to the number of the step being detailed. For example, a chart showing more detail of step 3.1.4 would be numbered 3.1.4.1, 3.1.4.2, 3.1.4.3, and so on.
4. As always, review the flowchart with others involved in the process (workers, supervisors, suppliers, customers) to see if they agree that the process is as drawn.

Example

For the group beginning a quality improvement project whose top-down flowchart is shown in Figure 5.62, page 259, two levels of a several-leveled flowchart would look like Figure 5.65.

force field analysis

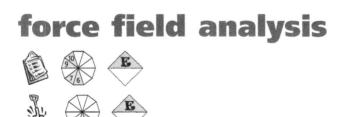

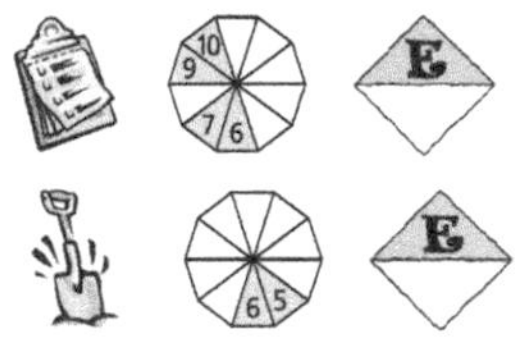

Description

Force field analysis clarifies opposing aspects of a desired change:

- Driving or positive forces that support an action or situation
- Restraining or negative forces that hinder it

When opposing forces are equal, no change can occur. For a change to happen, the driving forces must be stronger than the restraining forces. When all the forces have been considered, plans can be made that will encourage the desired change.

When to Use

- When the team is planning a change, such as implementation of a solution, or . . .
- When the team is identifying causes of a problem

Procedure

1. Write the desired change, or the problem, at the top of a flipchart or board. Draw a vertical line below it.
2. Brainstorm all the driving forces that support the change, or cause or enable the problem to occur. Write each one on the left side of the line. Decide how strong that force is, and draw between the words and the line right-pointing arrows whose size or length represent that strength.
3. Brainstorm all the restraining forces that prevent the action from happening, or prevent or hinder the problem from occurring. Write each one on the right side of the line. Again decide how strong each force is, and draw left-pointing arrows representing that strength.
4. For a desired change, discuss means to diminish or eliminate the restraining forces. For a problem, discuss means to diminish or eliminate the driving forces. Focus especially on the strongest forces.

Example

Figure 5.66 is a force field analysis created by someone who wanted to lose weight. Let's call him Sam. "Health threat," "negative self-image," and "clothes that don't fit" were among the forces making him want to lose weight. Unsympathetic friends and family, genetic traits, and years of bad eating habits were among the forces preventing it.

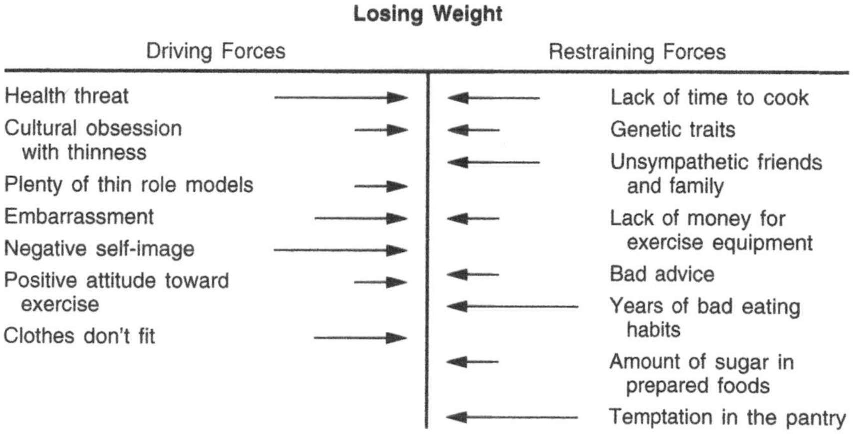

Figure 5.66 Force field analysis example.

For Sam, becoming more aware of health consequences, berating himself, or refusing to buy new clothes were not enough. Those actions only strengthened the driving forces. Instead, Sam explored ways to weaken or eliminate the restraining forces. For example, his friends and family could be coached on being supportive. Cooking lessons or cookbooks could teach how to make healthy food attractive and interesting, to counteract years of bad eating habits. Unfortunately, some restraining forces, such as Sam's genetic traits, cannot be changed. But if enough restraining forces are weakened, Sam can lose weight.

Considerations

- Think of the driving forces as pushing the centerline toward the right. Think of the restraining forces as pushing the centerline toward the left. When the two sets of forces are equal, the current state will be maintained, with the line staying in the center. To accomplish a change, the centerline must be moved by adjusting the forces.
- Strengthening the forces supporting a desired change often leads to a reaction that reinforces the opposing forces. First develop ways to reduce or eliminate the opposing forces, then ways to strengthen the driving forces.
- When you use force field analysis to study a problem, you must reverse your thinking. Driving forces are undesirable, causing the problem. Restraining forces are desirable, preventing the problem from occurring. Look for ways to reduce or eliminate the *driving* forces.
- Try to write a problem as too high or too low a level of something. This helps later to identify forces that raise or lower the level. For example, the problem "cumbersome update process" might be described as "too high a level of complexity in the update process."
- Instead of drawing arrows of different sizes, you can rate the forces' strengths on a scale from 1 to 5.
- If the group believes the same force needs to be written on both sides of the arrow, that force needs to be discussed more to understand which aspects drive the change and which aspects prevent it. Identify the driving and restraining elements and write them, not the force that was first named.
- Force field analysis can be used immediately after a fishbone diagram (cause-and-effect diagram) in a combination called CEFFA (cause-and-effect/force field analysis). Each cause identified on the fishbone becomes the subject of a force field analysis. Not only does this combination move a group quickly from problem to causes to potential solutions, but also overlooked causes can be uncovered during the force field analysis.

Gantt chart

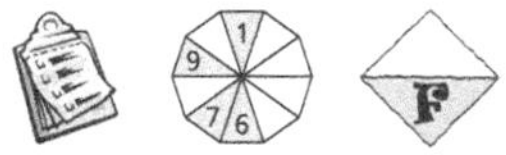

Also called: milestones chart, project bar chart, activity chart

Description

A Gantt chart is a bar chart that shows the tasks of a project, when each must take place, and how long each will take. As the project progresses, bars are shaded to show which tasks have been completed. People assigned to each task also can be shown.

When to Use

- When scheduling and monitoring tasks within a project, or . . .
- When communicating plans or status of a project, and . . .
- When the steps of the project or process, their sequence, and their duration are known, and . . .
- When it is not necessary to show which tasks depend upon completion of previous tasks

Basic Procedure

Construction

1. Identify the tasks that need to be done to complete the project. Also identify key milestones in the project. This may be done by brainstorming a list or by drawing a flowchart, storyboard, or arrow diagram for the project. Identify the time required for each task. Finally, identify the sequence. Which tasks must be finished before a following task can begin, and which can happen simultaneously? Which task must be completed before each milestone?
2. Draw a horizontal time axis along the top or bottom of a page. Mark it off in an appropriate scale for the length of the tasks (days or weeks).
3. Down the left side of the page, write each task and milestone of the project in order. For events that happen at a point in time (such as a presentation), draw a diamond under the time the event must happen. For activities that occur over a period of time (such as developing a plan or holding a series of interviews), draw a bar under the appropriate times on the timeline. Align the left end of the bar with the time the activity begins, and align the right end with the time the activity concludes. Draw just the outlines of the bars and diamonds; don't fill them in.
4. Check that every task of the project is on the chart.

Using the Chart

5. As events and activities take place, fill in the diamonds and bars to show completion. For tasks in progress, estimate how far along you are and fill in that much of the bar.

6. Place a vertical marker to show where you are on the timeline. If the chart is posted on the wall, an easy way to show the current time is with a heavy dark string and two thumbtacks.

Example

Figure 5.67 shows a Gantt chart used to plan a benchmarking study. Twelve weeks are indicated on the timeline. There are two milestone events, presentations of plans for the project and for the new process developed in the study. The rest of the tasks are activities that stretch over periods of time.

The chart shows the status at Tuesday of the sixth week. The team has finished seven tasks, through identifying key practices, measures, and documentation. This is the most hectic part of the project, with three time-consuming activities that must happen simultaneously. The team estimates it is one-fourth finished with identifying benchmark partners and scheduling visits; one-fourth of that bar is filled. Team members have

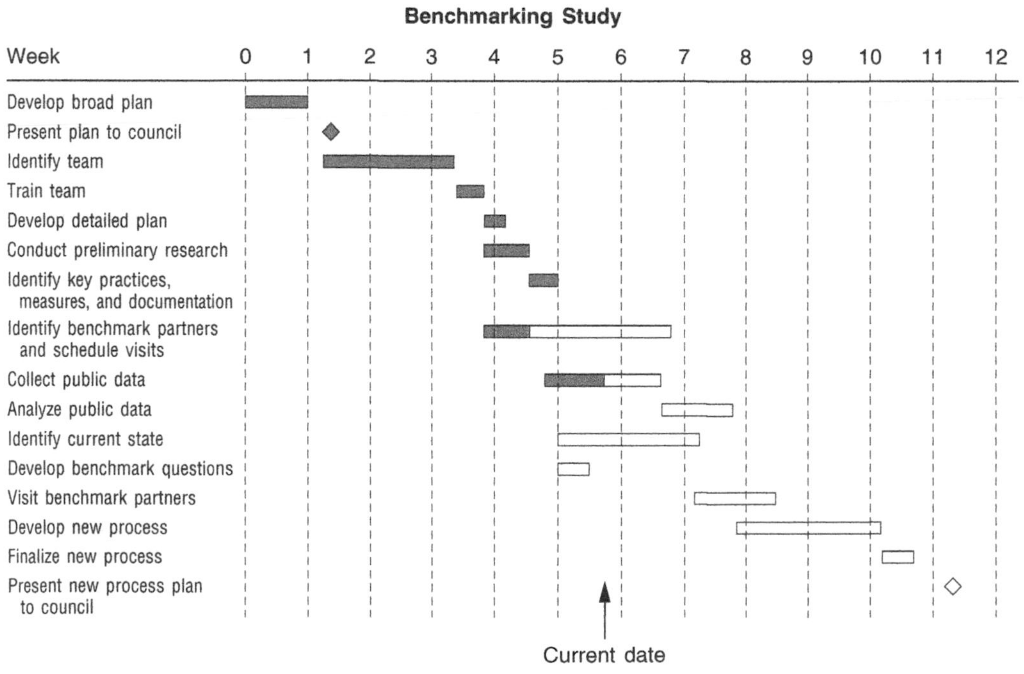

Figure 5.67 Gantt chart example.

not yet begun to identify the current state. They are behind schedule for those two tasks. They are halfway through collecting public data, which puts them slightly ahead of schedule for that task. Perhaps they need to reallocate their workforce to be able to cover those three activities simultaneously.

There is a fourth activity that could be happening now (develop benchmark questions), but it is not urgent yet. Eventually the team will have to allocate resources to cover it too, before visits can begin.

Compare this chart with the arrow diagram example (Figure 5.8, page 105). Both show the same project. Relationships and dependencies between tasks are clearer on the arrow diagram, but the Gantt chart makes it easier to visualize progress.

The St. Luke's Hospital story on page 74 has another example of a Gantt chart, used within the improvement process.

Variation

This version is similar to a deployment chart in that it shows responsibilities for tasks. This version can be done instead of the method above or in addition to it, as a second chart below the other one.

Complete steps 1 and 2 as described in the basic procedure.

3. Down the left side of the page, list the people or groups who will have direct responsibility for the project tasks.
4. Transfer each task of the project from the flowchart, storyboard, or arrow diagram. Represent it as a bar opposite the person or group responsible. Align the end tail of the bar with the date that activity begins, and align the right end with the date that activity concludes. Write the task above the bar.
5. Check that every task is on the activity chart as a bar. Review the chart with everyone named on it for their agreement.

Use the chart as described in the basic procedure.

Considerations

- The chart can be prepared in several layers, with major steps in a top layer and detail in subsequent layer(s). This method is analogous to a several-leveled flowchart, which could be used as a starting point for the Gantt chart.
- Sometimes Gantt charts are drawn with additional columns showing details such as the amount of time the task is expected to take, resources or skill level needed, or person responsible.
- The variation that shows tasks opposite people or groups responsible allows a quick assessment of who is not engaged in a project task or, conversely, who may be overloaded.

- Beware of identifying reviews or approvals as events unless they really will take place at a specific time, such as a meeting. Reviews and approvals often can take days or weeks.
- The process of constructing the Gantt chart forces group members to think clearly about what must be done to accomplish their goal. Keeping it updated as the project proceeds helps manage the project and head off schedule problems.
- Posting the chart in a visible place helps keep everyone informed and motivated.
- This chart fulfills some of the same functions as the arrow diagram but is easier to construct, can be understood at a glance, and can be used to monitor progress. The arrow diagram, however, analyzes the project's schedule more thoroughly, revealing bottlenecks and tasks that are dependent on other tasks.
- It can be useful to indicate the critical path on the chart with bold or colored outlines of the bars for the steps on the critical path. (See *arrow diagram* for a discussion of critical path.)
- Computer software can make constructing and updating the chart quick. Software can also easily construct a Gantt chart and arrow diagram from the same data, ensuring that the two are always consistent with each other as the project progresses and updates are made.
- This chart is sometimes called an "activity chart." However, that name has been used for the arrow diagram as well, potentially leading to confusion. Since several names exist for each of these diagrams, it seems simpler to avoid using the name "activity chart."

graph

Also called: plot or chart
Variations: dot chart, bar chart, pie chart, line graph, high-low graph
See also: box plot, control chart, histogram and other frequency distributions, multi-vari chart, Pareto chart, radar chart, run chart, scatter plot

Description

A graph, plot, or chart is a visual display of numerical data to achieve deeper or quicker understanding of the meaning of the numbers. Data alone, in lists or tables, can be overwhelming or meaningless. Arranging data in graphs helps you crack the code and reveal the information hidden in the data.

The data shown on graphs are in pairs, where each pair represents an observation or event. Graphs are drawn as rectangles (except for pie and radar charts) with one half of the data pair identified on the horizontal edge (called the x-axis) and the other half on the vertical edge (called the y-axis). Dots, lines, bars, or symbols show the location within the rectangle that represents the observation, the pair of data.

This is a generic tool. Many different kinds of graphs are available, depending on the kind of data and the purpose of the graph.

When to Use

- When analyzing data, especially to find patterns or trends in the numbers, or . . .
- When communicating data to others in a report or presentation

Graph Decision Tree

Figure 5.68 is a decision tree to help you decide what kind of graph would display your data most effectively. The correct graph depends on what kind of data you have and what the purpose of your graph will be.

Data can be *categorical* or *numerical*. There are two kinds of categorical data. *Nominal* data are names or labels of categories. *Ordinal* data have an order and are sometimes numbers, but the intervals between them are not necessarily uniform. Doing arithmetic with ordinal numbers isn't meaningful. Rankings and ratings are ordinal data. *Numerical* data can be integers or continuous (variable) numbers, including fractions or decimals.

If the pairs of data you wish to graph include categorical data, use the top part of the decision tree. Examples of such pairs:

- A list of problems (nominal data) and the number of times each problem occurred (numerical data): Pareto chart.
- Categories of customer service such as responsiveness and accuracy (nominal data) and performance ratings from 1 to 5 for each category (ordinal data): radar chart.
- Different zip codes (nominal data: even though they're named with numbers, the data are really locations) and the population in each (numerical data): bar or dot chart.
- Under 1, 1–5, 6–10, 11–20, over 20 years of experience (ordinal data) and the number of survey respondents in each (numerical data): bar or dot chart. Note that the year brackets are unequal.

When the data pairs are both numerical data, use the bottom part of the chart. For example:

- Data shows how long each of 200 customers waited for service. You want the graph to show how good or bad wait times are overall. The data pairs you will plot are minutes (numerical data) and how many customers waited that long (numerical data): histogram or polygon chart.

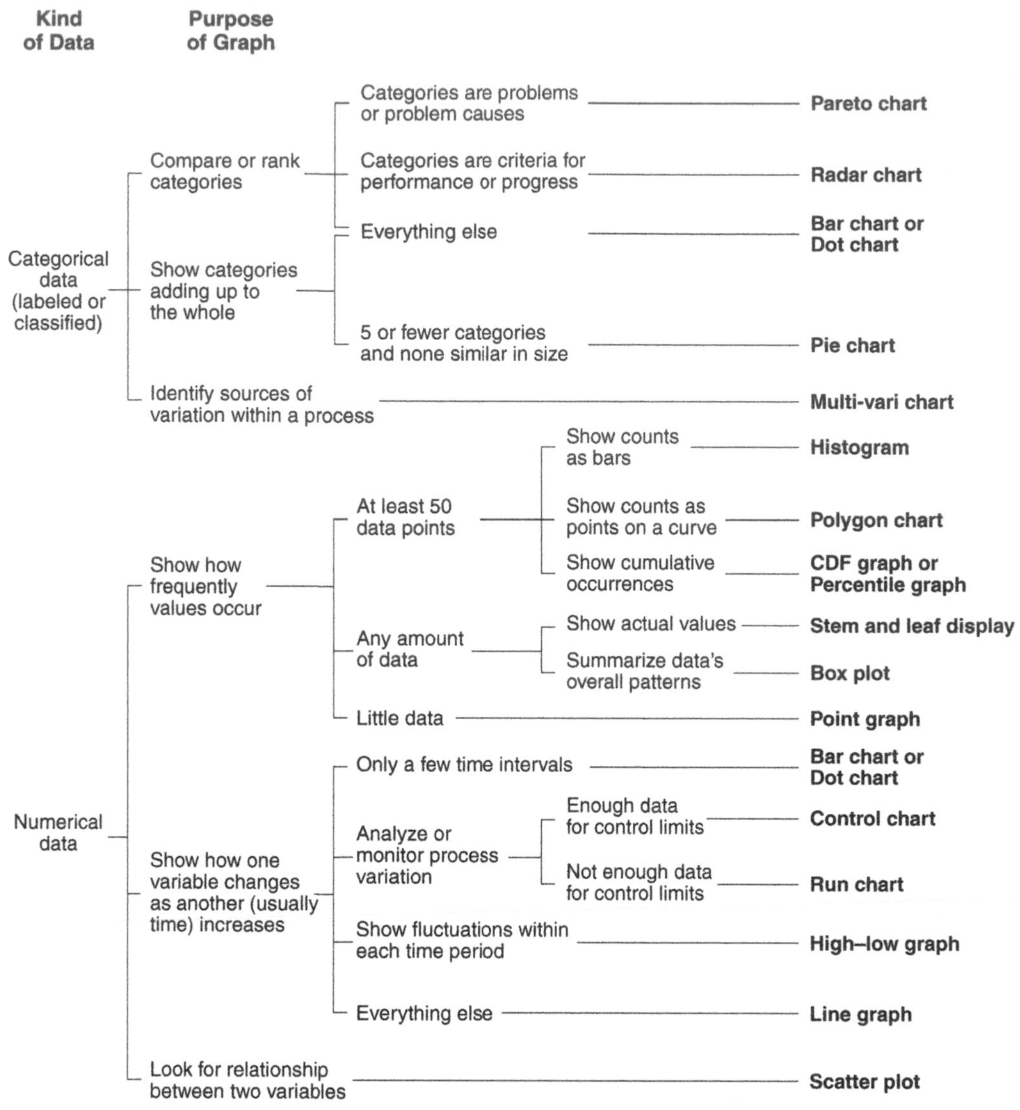

Figure 5.68 Graph decision tree.

- Same data. You want to see if wait times are changing over time. The data pairs you will plot are when each customer came in (numerical data) and how long that customer waited (numerical data): control chart.

Don't be confused when you have several data sets consisting of pairs of numerical data and categories labeling the different data sets. Despite the presence of categories, if it's numerical data you're graphing, use the lower part of the decision tree. The categories will become labels or symbols on your graph. For example:

- You wish to show the monthly values (numerical data) of the Dow Jones, NASDAQ, and S&P indexes (categorical data) over the last 20 years (numerical data). You will graph monthly values against months on a line graph. Each index forms a data set, and so will have its own labeled line.
- You wish to know whether class size (numerical data) affects test scores (numerical data). You have data grouped by school district (categorical data). You plot size against score on a scatter plot and use different symbols for the different districts.

Basic Procedure

1. Collect or assemble the data to be shown on your graph. Decide whether you have any categorical data or only numerical data. Decide what you want to study or to show with the graph.
2. Decide what kind of graph should be used. Through the rest of this basic procedure, also check the procedure and example sections of the chosen graph's entry for guidance.
3. Determine the range (lowest number to highest number) you need to show on the graph for each set of data. If two sets of data will be plotted on the y-axis, they must both have the same units of measure. Choose the scale for each axis to be as large as or slightly larger than the range.
4. Draw the axes (sometimes called scale lines) and tick marks to show the numerical scale or data labels. Write numbers or labels next to the tick marks and write the unit of measure for numerical data.
5. Determine appropriate symbols for the data. Plot the data. Add a legend for symbols.
6. If appropriate, draw reference lines for important values, such as the average, against which all the data must be compared. To note a significant number on the scale, such as the time a change was made, place a marker (an arrow and a note) along the scale line.
7. Complete the graph with title and date. Add notes, if necessary.
8. Analyze the graph. What does it teach? What additional graphing, analysis, investigation, or data collection does it suggest?

Considerations

Basic Graph Design Principles

- The two most important principles of good graphs are:
  - Make it easy for the viewer to quickly see the data.
  - Eliminate anything unnecessary.

- Unnecessary elements may include grid lines, many tick marks, many numbers opposite tick marks, notes and labels inside the graph area, cross-hatching, bars instead of lines.
- Make every mark on the graph do enough work to earn the right to be there.

Symbols

- Make symbols dark and big—easy to see.
- If several symbols will fall on top of one another and be hard to distinguish, use symbols that look like Y, X, or ✻ (called sunflowers). The number of lines radiating from the center of the symbol indicates how many data values fall on that spot. Or, you can use more sophisticated graphs, such as logarithmic scales, or plot residuals to eliminate overlap.
- If you must place labels next to data symbols, place them so that the eye can easily distinguish the data alone. Place the labels outside a central area occupied by just the data symbols. If that is not possible, make the data symbols very prominent and the label lettering thin and unobtrusive.
- Connect points only when showing time sequence on a line graph, run chart, or control chart. An exception: lines are often used on a multi-valued dot chart to make patterns stand out.

Scales and Scale Lines

- It is conventional to show time along the bottom edge of a graph, proceeding left to right. For cause-and-effect graphs, show cause along the bottom edge and effect along the left edge.
- Choose a scale that is only as large as or slightly larger than the range of data. Try not to have any areas of the graph that aren't working to show your data. The scale does not have to include zero. *Exception:* With a bar graph, the scale must include zero or the bar lengths are meaningless. Notice how in Figure 5.69, where the scale starts with 20, comparing bar lengths is misleading.
- If two graphs will be compared, their scales must be the same. If this is impossible because of very different absolute values (say, one graph with data around 10 and another with data around 1000) then at least be sure that the lengths that represent one unit are the same.
- Use a scale break when a large region of the graph will have no data in it. Scale breaks must extend across the entire graph, as in Figure 5.70, not just across the scale line. Do not connect data symbols across a scale break. Do not use scale breaks with bar graphs.

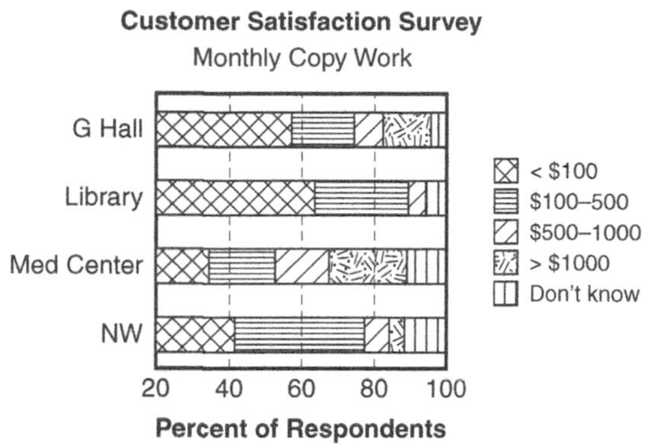

Figure 5.69 Stacked bar graph.

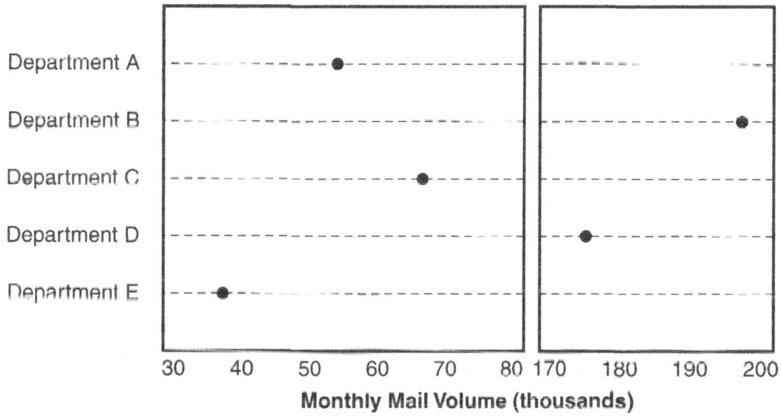

Figure 5.70 Scale break.

- Consider using two different scales on opposite sides of the graph when the data can be expressed in two different ways. For example, actual values on one scale and percent change on the other. Or grams and pounds. Or logarithmic values and original data values.
- Use a logarithmic scale when the purpose of the graph to show percent change or ratio comparisons.

Color, Pattern, and Special Effects

- Use colors or fill patterns sparingly, being aware that bold colors or patterns can distort the impression of the data. The color scale should be proportional to the numeric scale. For example, you might use dense to light cross-hatching to represent high to low values. Avoid highly contrasting colors or patterns with close values. The darker or bolder one will look larger than it really is and the lighter one smaller. See *pie chart* for an illustration of this. Also notice how the fill patterns in Figure 5.69 play tricks with the eye.

- Where colors or fill patterns are essential to designate multiple data sets (such as in a group bar chart), be sure that the colors or patterns are different enough to be distinguished.
- Avoid 3-D effects. They confuse the eye and are distracting.

Reproduction

- Graphs copied on a machine often become harder to read, especially if they are reduced. A good test: Copy it at two-thirds reduction, then copy the copy at two-thirds reduction. If it survives that test, it will be readable under most conditions.
- Don't use color to differentiate areas, bars, symbols, or lines if the graph might be copied on a black-and-white copier.

bar chart

Description

The bar chart is the most common way to show categorical data. Values are indicated by the length of a bar—a long, narrow rectangle—or sometimes of a line. The Pareto chart is a specialized bar chart.

When to Use

- When you have categorical data

Procedure

Follow the basic procedure on page 277, but replace steps 3 through 5 with the following:

3. Determine the highest number you need to show on the graph. You must start the scale with zero. If two sets of data will be plotted, they both must have the same units of measure. Choose the scale for the numerical axis to be as large as or slightly larger than the range.
4. Write labels for the categories either down the left side or across the bottom. Mark the numerical scale, starting with 0, on whichever edge does not have the labels.
5. Draw a bar or line for each value. Fill each bar with a light gray shading or fill pattern.

Examples

A university services department conducted a customer satisfaction survey. Figure 5.71 is a horizontal bar chart of the results from the library copy center. Notice that the bars start at zero, although none of them are shorter than five. For other examples of bar charts, see *Pareto chart* and the chart used in the St. Luke's Hospital improvement story on page 78.

Considerations

- The numerical scale *must* start at zero, or the length of the line, the visual indicator of numerical value, will be meaningless. Similarly, there should be no break in the scale line. There is one exception to this: If one bar is so much longer than the others that including its value in the scale would make it hard to see differences among the other bars, it can be broken. Its value should be clearly written at the top of the bar.
- Avoid crosshatching, which is visually distracting. Even better, eliminate the edges of the bar and draw only the shading. Or, eliminate the bar and use a line to indicate length. The bar creates redundant parallel lines which confuse the eye.
- Bars can be horizontal or vertical. Time is normally shown with vertical bars. One advantage of horizontal bars is that there is more room to write labels one above the other on the left side. Choose the format that makes the graph clearest to your viewers.
- There must be spaces between bars, to show that the scale is not continuous. In histograms, there is no space between bars.
- Most of the considerations for dot charts also apply to bar charts.

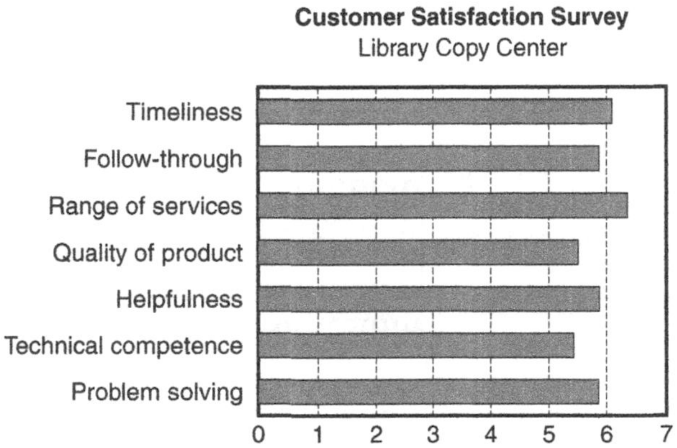

Figure 5.71 Bar chart example.

Variation: Pareto chart

The Pareto chart is a bar chart used to show relative importance of problems or problem causes. The measurement may be frequency of occurrence or cost expressed in money or time. Because this is such an important quality tool, it has its own entry on page 376.

Variations: grouped or clustered bar chart, stacked bar chart

These charts are different ways to show multi-way or cross-classified data. These are data that are classified one way into a set of categories and classified another way into a different set of categories. You want to show both groupings on one graph.

In the *grouped* or *clustered bar chart,* labels for one set of categories are shown along an axis. Beside each label are multiple bars, one for each category of the other classification. Usually, the grouped bars are coded with colors or fill patterns.

In the *stacked* or *divided bar chart,* one bar is divided into several sections along its length to show values of components. *This kind of chart is not recommended.* The human eye and brain cannot compare lengths that are not aligned.

Examples

Figure 5.72 is a grouped bar chart showing survey results for four different copy center locations. All results from the survey are shown together. Fill patterns were chosen so that adjacent bars are different from each other. Notice, however, how distracting the fill patterns are. Shades of gray or colors would have been better.

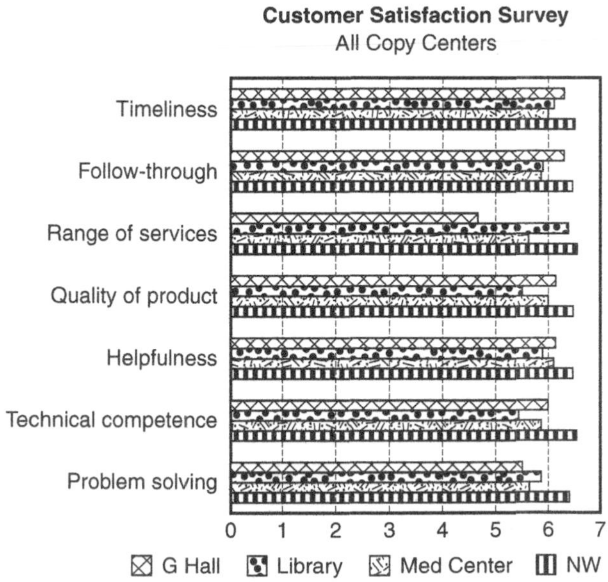

Figure 5.72 Grouped bar chart example.

This grouping, with criteria down the left edge and bars for each location, makes it easy to compare locations' ratings on each criterion. The chart could have been drawn with locations along the left edge and bars for each criterion. That arrangement would have made studying each location's results easier. Choose a grouping that makes it easy to do whichever analysis is needed.

Figure 5.69, page 279, showed a stacked bar chart reporting demographic data from the copy center survey. Which location had more customers spending between $100 and $500, G Hall or the Med Center? It's impossible to tell, which is why this kind of chart is not recommended.

dot chart

Also called: point chart

Description

The dot chart is an excellent graph for displaying categorical data. Values are indicated by positions of dots opposite a scale and sometimes also by the length of a line to the dot.

When to Use

- When you have categorical data

Procedure

Follow the basic procedure as on page 277 except:

4. At step 4, write the labels along the left edge of the graph. Place tick marks and their numbers along the top and/or bottom edges.
5. Place a large dot for each data point opposite its label. Draw faint dotted lines between the labels and the dots. If the numerical scale does not begin at zero, continue the line to the right side of the graph.

Example

In Figure 5.73, a dot chart shows the same data as the bar chart of Figure 5.71. Because a dot chart's scale need not start at zero, less space is wasted with this graph and the area of interest—between 5 and 7—can be enlarged. This chart has the values sequenced from smallest to largest, which makes it easy to see at a glance which criteria rated best and worst. The bar chart could have been sequenced like this too. However, if several charts will be compared, as in this example, one sequence for the categories should be used consistently.

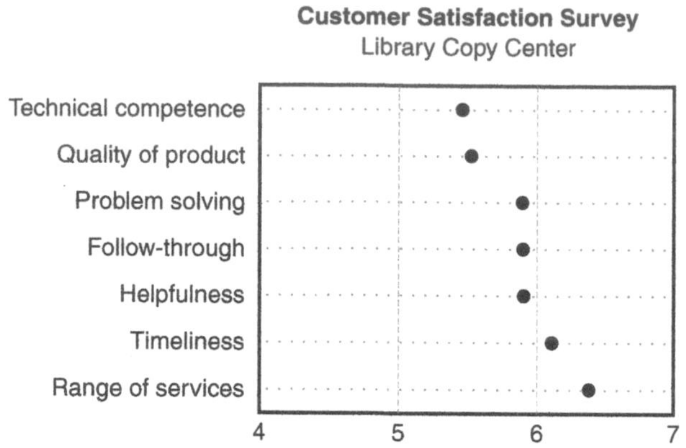

Figure 5.73 Dot chart example.

Considerations

- For a single chart, it is most useful to the viewer to arrange the categories so the values increase or decrease.
- Think about whether larger or smaller will be considered good, and translate the data so that a larger number is good. For example, don't graph "percent decrease in complaints" where a long line means a larger decrease and therefore fewer complaints. (Confused? The viewer of your graph will be too!) Instead, graph the actual number of complaints, where a series of shortening lines shows complaints decreasing.
- Dot charts, along with bar charts, are highly effective ways to present data because the viewer is comparing distances from a constant baseline. This is the easiest visual task—much easier than comparing distances from a moving baseline (stacked bar charts), angles (pie charts), or areas (stacked line graphs).
- If the numerical scale begins at zero, the length of the line as well as the position of the dot will signal to the viewer the numerical value. If the numerical scale does not begin at zero, the length of the dotted line is meaningless. To prevent the viewer from comparing meaningless lengths, continue the line across the graph.
- While dot and bar charts are usually used for categorical data, they can also be used for time data when there are five or fewer time intervals, such a graph showing results from four quarters of the year.

Variations: two-way dot chart; grouped dot chart; multi-valued dot chart

These charts are different ways to show multi-way or cross-classified data. These are data that are classified one way into a set of categories and classified another way into a different set of categories. You want to show both groupings on one graph.

In the two-way dot chart, each category from the second classification is drawn as a separate dot chart. The charts are placed side by side using one set of categories on the far left showing the first classification. Each separate chart is titled at the top to show the categories of the second analysis.

In the grouped dot chart, the left edge of the graph shows categories for both ways of slicing the data. All the categories for the second classification are grouped together repeatedly under each category of the first classification.

In the multi-valued dot chart, all the values for the second classification are placed on the same line opposite the categories for the first classification. Different symbols are keyed to the categories of the second classification. This is a common way of displaying satisfaction survey results, where areas to be rated are usually placed on the vertical edge and the Likert rating scale is placed along the horizontal edge. Different segments of the survey population are shown with different symbols. For this use, the chart is sometimes called "multiple rating matrix." See *survey* for more information.

Examples

Figure 5.74 shows a two-way dot chart analyzing the same customer satisfaction data as in the bar charts: seven criteria and four locations. The first analysis is the seven different criteria; those labels are at the far left. The data for each location is on a separate dot chart. This arrangement makes it easy to study each location separately.

Figure 5.75 shows a grouped dot chart of the same survey data. This time, all four locations for timeliness are grouped together, then all four locations for follow-through, and so on. This method of grouping allows easy location-to-location comparison for each criterion. Notice that the NW location ranks highest on all criteria, and G Hall ranks second-highest on all except range of services and problem solving.

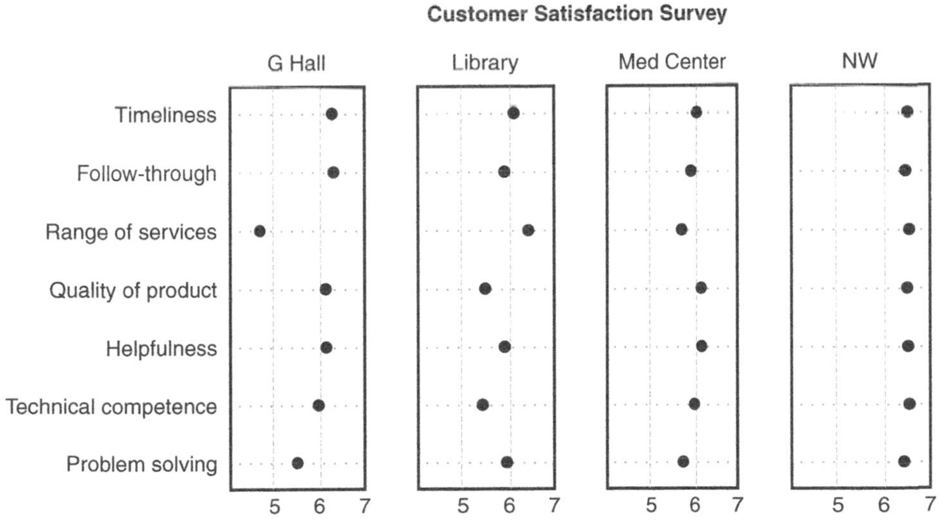

Figure 5.74 Two-way dot chart example.

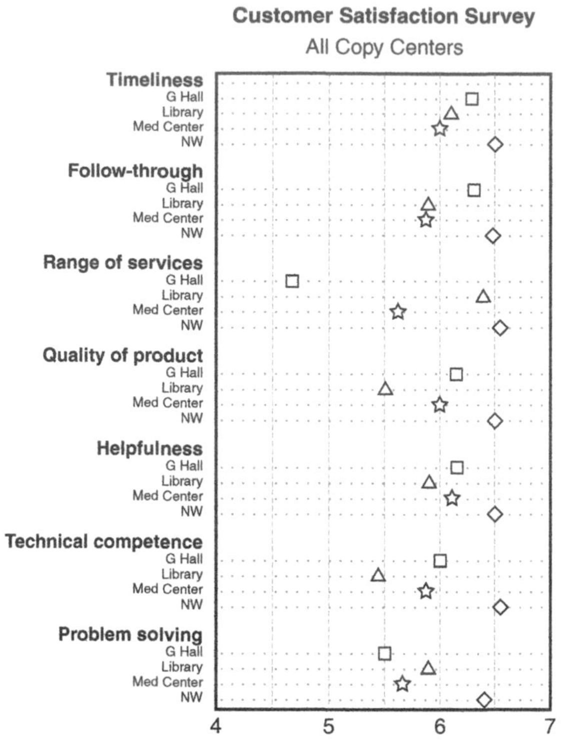

Figure 5.75 Grouped dot chart example.

The data could have been grouped the other way: All seven criteria for first location, then all seven criteria for the second location, and so on. That grouping—similar to the grouping of Figure 5.74—would have allowed easy criterion-to-criterion comparison for each location, but would have made it harder to see the pattern we noticed between locations. So it is useful to group your data different ways to see what you can learn.

Figure 5.76 shows a multi-valued dot chart of the same data. While the graph is more compact, it requires a legend, and patterns are harder to discern. Connecting dots on a graph usually represents a time sequence, but it is common with this type of chart to make patterns stand out.

radar chart

Also called: web chart, spider chart

The radar chart looks like a spiderweb, with spokes and connecting lines. It is used to track or report performance or progress on multiple criteria. Because its purpose is so specialized, it is described in its own entry on page 437.

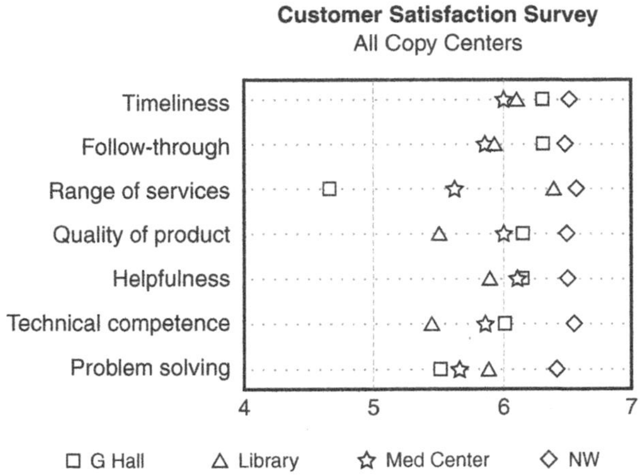

Figure 5.76 Multi-valued dot chart example.

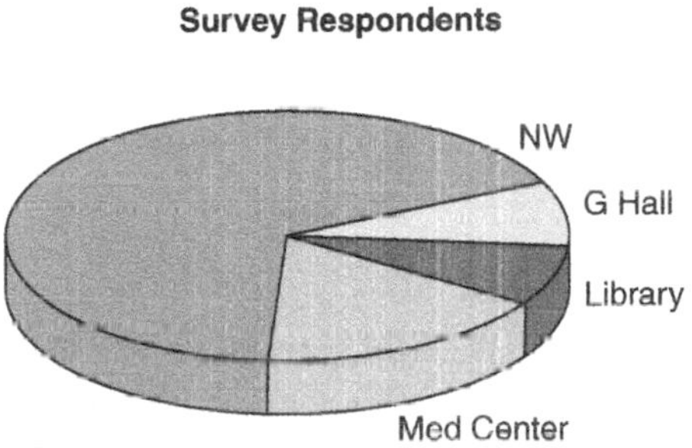

Figure 5.77 Poor pie chart.

pie chart

This kind of chart should be used with caution. This is a common way to show categorical data. A circle is divided into wedges to show relative proportions of categories that add up to a total. However, the human eye and brain cannot compare angles very well. In Figure 5.77, which location had more survey respondents, G Hall or the library? Instead of the pie chart, use a dot graph or a bar chart, which do better jobs of comparing values. In this pie chart, the problem is worsened, first because the similar wedges are positioned so that the 3-D perspective distorts their sizes, and also because strongly contrasting colors emphasize the library wedge.

The pie chart does one (and only one) thing well: it emphasizes that the parts sum to 100 percent. If you need to use a pie chart for this reason, be sure to write the values or percentages on or beside each wedge, as in Figure 5.78. But think hard about whether

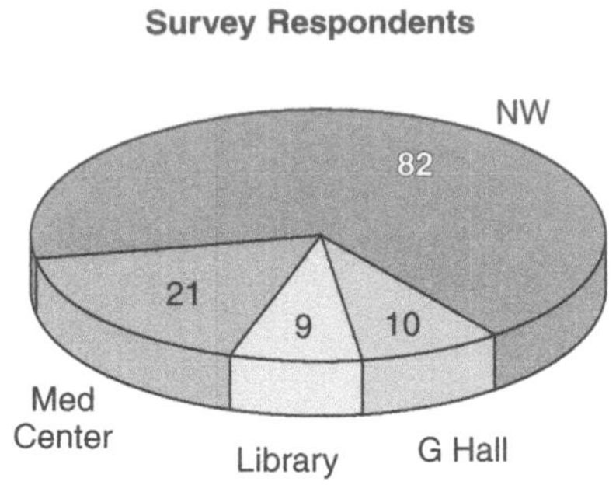

Figure 5.78 Acceptable pie chart.

a different chart would communicate your data better if the size difference between any two wedges is not obvious or if you have more than five wedges. Also, choose colors or fill patterns that are proportional to the wedge sizes, with the largest wedges being darkest. Don't use dark and light to differentiate closely sized wedges; the darker one will appear larger than it is, like the library in the pie in Figure 5.77.

frequency distributions

Frequency distributions count the number of times each value occurs in a set of data. For example, a frequency distribution of average temperature data would show how many times the average temperature was 50 degrees, how many times it was 51 degrees, and so on. Many different graphs are available to show frequency data.

A *histogram* graphs each value in a data set (on the x-axis) against how often the value occurs (on the y-axis). A histogram is the most commonly used graph to show frequency distributions. It looks very much like a bar chart, but there are important differences between them.

Because the histogram is such an important quality tool, it has its own entry on page 292. Other graphs that show frequency distributions are described in the same section.

The *polygon chart* (page 299) takes the shape of a histogram but links the frequency values with a line instead of showing bars.

The *stem-and-leaf display* (page 300) is a variation of the histogram where the actual value is used as the symbol for the data point.

The *point graph* (page 301) is useful with small data sets. It graphs each data point as a small circle along a vertical line.

The *cumulative polygon graph, percentile graph,* and *CDF (cumulative distribution function) graph* (page 302–4) add up frequencies of values as they increase, showing how many or what percentage or fraction of measurements are less than or equal to each value.

box plot

The *box plot* (page 121) summarizes significant features of a distribution but does not show all the data. It is especially useful for comparing two or more distributions or when there aren't enough data for a histogram.

line graph

Description

A line graph is the simplest kind of graph for showing how one variable, measured on the vertical y-axis, changes as another variable, on the horizontal x-axis, increases. The data points are connected with a line. The x-axis variable is usually time or its equivalent and is called the independent variable. The y-axis variable is called the dependent variable, because its value depends on the value of the independent variable.

When to Use

- When the pairs of data are numerical, and . . .
- When you want to show how one variable changes with another, continuous variable, usually time, and . . .
- Only when each independent variable is paired with only one dependent variable

Procedure

Follow the basic procedure on page 277, adding this at step 5:

> Connect points with lines. If several categories of dependent variables are being tracked against the same independent variable, label the lines or use different line types for each. Add a legend for line types.

Example

Figure 5.79 shows a line graph of SAT average scores over a ten-year period. There are two categories of scores, math and verbal. The growing difference between math and verbal averages is instantly obvious. Also see the line graph used in the St. Luke's Hospital improvement story on page 77.

Considerations

- If the dependent variable may have several values for each value of the independent variable, use a scatter plot.

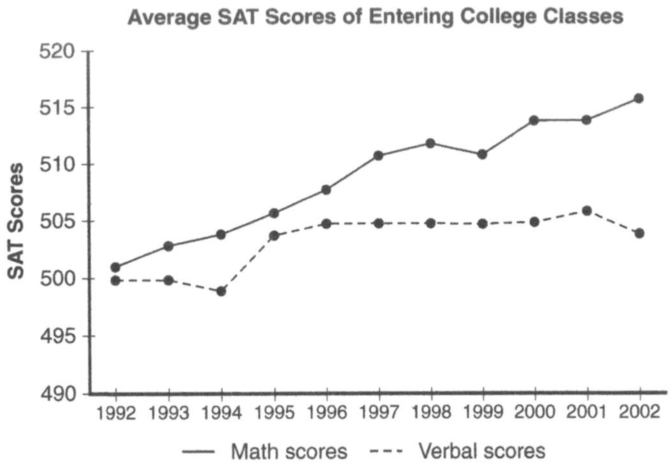

Figure 5.79 Line graph example.

- If your independent variable is an event that happens regularly and sequentially in time, you can consider it equivalent to time. Examples are: samples, batches, shifts.
- Use straight lines to connect points. Make sure the data points are big enough to show up through the line, unless individual values are not important.
- With multiple lines, be sure the viewer can tell the lines apart, especially where they cross. If this is a problem, place two separate graphs side-by-side. When you do this, there is always a tradeoff between the eye being able to distinguish the data sets and having to jump back and forth between the two graphs to make comparisons.
- Sometimes two lines are graphed with the intent of comparing their values. If the slopes of the lines are changing, the eye will not be able to make accurate comparisons. Instead, plot the differences between the two data sets.

Variation: cumulative line graph

Also called: area chart, stacked line graph, stratum chart

This kind of chart is not recommended. Cumulative line graphs are similar to stacked bar charts in that they show how various components add up to a total. The value of each component is measured from the top of the next-lower component to the bottom of the next-higher one.

The problem with these graphs is the same as the problem with stacked bar graphs: The eye and brain cannot compare distances that are not aligned. Also, since the purpose of all line graphs is to look for trends over time, trends within middle components are masked or exaggerated by the shifting baseline.

These graphs only work if the components are fairly regular, with no strong trends. A better alternative is to show each component as a separate line whose value is measured from the zero point of the vertical scale. A line for the total can be shown if needed.

control chart, run chart, multi-vari chart

Control charts and run charts are line graphs used to analyze process variation over time. A control chart is used when there is enough data to calculate control limits. A run chart is used when there is not. Because these are such important quality tools, they have their own entries on pages 155 and 463. A multi-vari chart is a unique chart used to study the sources of process variation, especially when the sources represent categorical data, such as different machines. It has its own entry on page 356.

high–low graph

Description

High–low graphs are used for data that fluctuate within each time period. Stock prices, weather data, and energy usage are typical applications.

Procedure

Follow the basic procedure on page 277, except for this change to step 5:

> The high and low values during each time period are plotted one above the other, and a vertical line is drawn connecting them. Do not connect points from different time periods.

Example

Figure 5.80 is a high–low graph of daily temperatures during one month.

scatter diagram

Also called: scatter plot, X–Y graph

A scatter diagram is used to look for relationships between variables. One variable may cause a second to change, or a third factor may affect both of them. Scatter diagrams can reveal such relationships or can verify that the variables are independent. Because this is such an important quality tool, it has its own entry on page 471.

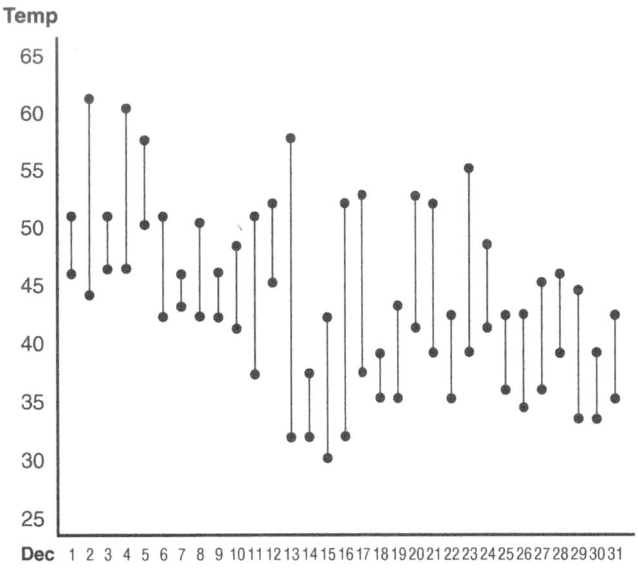

Figure 5.80 High–low graph example.

histogram and other frequency distributions

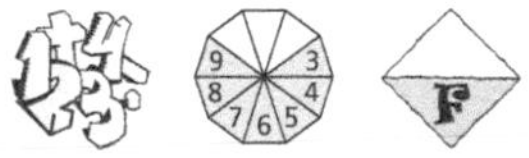

Includes polygon chart, stem-and-leaf display, point graph, percentile graph, CDF graph, cumulative polygon graph.

Description

A frequency distribution shows how often each different value in a set of data occurs. A histogram is the most commonly used graph to show frequency distributions. It looks very much like a bar chart, but there are important differences between them.

Other graphs for frequency distributions are included in this section. The *polygon chart* takes the shape of a histogram but connects the frequency values with a line instead of showing bars. The *stem-and-leaf display* preserves individual values by using them as symbols for the data point. The *point graph* shows each data point as a small circle along a vertical line. The *percentile graph* and *cumulative polygon graph* show how many measurements (or what percentage of measurements) are less than or equal to each value.

When to Use

- When the data are numerical, and . . .

- When you want to see the shape of the data's distribution, especially:
- When determining whether the output of a process is distributed approximately normally, or . . .
- When analyzing whether a process can meet the customer's requirements, or . . .
- When analyzing what the output from a supplier's process looks like, or . . .
- When seeing whether a process change has occurred from one time period to another, or . . .
- When determining whether the outputs of two or more processes are different, or . . .
- When you wish to communicate the distribution of data quickly and easily to others

The graph decision tree (Figure 5.68 on page 276) can help decide which graph is most appropriate for your data and purpose.

Procedure

Construction

1. Collect at least 50 consecutive data points from a process. If you don't have that much data, use the point graph variation.
2. Use the histogram worksheet (Figure 5.81) to set up the histogram. It will help you determine the number of bars, the range of numbers that go into each bar, and the labels for the bar edges. After calculating W in step 2, use your judgment to adjust it to a convenient number. For example, you might decide to round 0.9 to an even 1.0. The value for W must not have more decimal places than the numbers you will be graphing.
3. Draw x- and y-axes on graph paper. Mark and label the y-axis for counting data values. Mark and label the x-axis with the L values from the worksheet. The spaces between these numbers will be the bars of the histogram. Do not allow for spaces between bars.
4. For each data point, mark off one count above the appropriate bar with an X or by shading that portion of the bar. For numbers that fall directly on the edge of a bar, mark the bar to the right.

Analysis

1. Before drawing any conclusions from your histogram, satisfy yourself that the process was operating normally during the time period being studied. If any unusual events affected the process during the time period of the histogram, your analysis of the histogram shape probably cannot be generalized to all time periods.
2. Analyze the meaning of your histogram's shape. See "Considerations" for some typical shapes and their meanings.

Process: ______________________ Calculated by: ______________________

Data dates: ______________________ Date: ______________________

Step 1. Number of bars

Find how many bars there should be for the amount of data you have. This is a ballpark estimate. At the end, you may have one more or less.

Number of data points	Number of bars *(B)*	
50	7	
	8	
	9	
100	10	*B* = ______
	11	
150	12	
	13	
200	14	

Step 2. Width of bars

Total range of the data = *R* = largest value – smallest value

= ______ – ______ = ______

Width of each bar = *W* = *R* ÷ *B*

= ______ ÷ ______ = ______

Adjust for convenience. *W* must not have more decimal places than the data.

W = ______

Step 3. Find the edges of the bars

Choose a convenient number, *L*1, to be the lower edge of the first bar. This number can be lower than any of the data values. The lower edge of the second bar will be *W* more than *L*1. Keep adding W to find the lower edge of each bar.

*L*1	*L*2	*L*3	*L*4	*L*5	*L*6	*L*7	*L*8	*L*9	*L*10	*L*11	*L*12	*L*13	*L*14
___	___	___	___	___	___	___	___	___	___	___	___	___	___

Figure 5.81 Histogram worksheet.

Example

The Bulldogs bowling team wants to improve its standing in the league. Team members decided to study their scores for the past month. The 55 bowling scores are

103	107	111	115	115	118	119	121	122	124	124
125	126	127	127	129	134	135	137	138	139	141
142	144	145	146	147	148	148	149	150	151	152
153	153	154	155	155	155	156	157	159	160	161
163	163	165	165	167	170	172	176	177	183	198

Using the table on the histogram worksheet, they estimate B, the number of bars, to be 7. The highest score was 198 and the lowest was 103, so the range of values is

$$R = \text{largest} - \text{smallest}$$

$$R = 198 - 103 = 95$$

Then the width of each bar is

$$W = R \div B$$

$$W = 95 \div 7 = 13.6$$

The bowling scores have no decimal places, so the bar width must have no decimal places either. They round 13.6 up to 14. Because 14 is an awkward number to work with, they decide to adjust W to 15. Choosing 100 to be the lower edge of the first bar, the lower edges of the other bars are

$$100 + 15 = 115$$

$$115 + 15 = 130\text{, and so on}$$

Figure 5.82 shows the histogram they drew. They seem to have a double-peaked, or bimodal, distribution: a group of players who score in the low 100s and another more talented group that scores in the mid-100s. To improve the team's standing, members can try to improve everyone's score, which would shift the entire histogram to the right. Or they could focus their efforts on improving the poorer players, which would narrow the distribution, making the team as a whole more consistent.

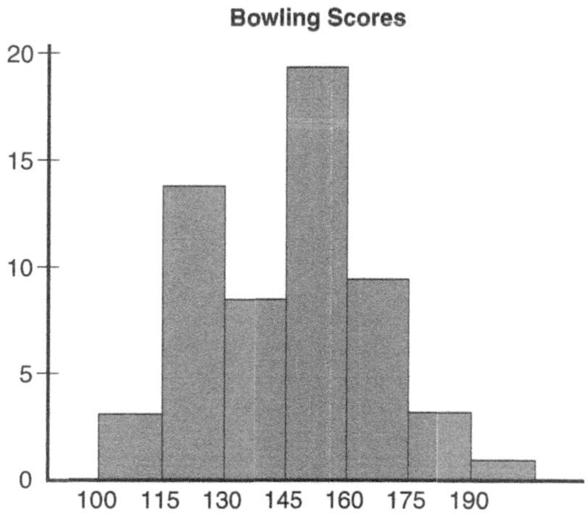

Figure 5.82 Histogram example.

Considerations

- Here are typical histogram shapes and what they mean.

  Normal. A common pattern is the bell-shaped curve known as the *normal distribution* (Figure 5.83). In a normal distribution, points are as likely to occur on one side of the average as on the other. Be aware, however, that other distributions look similar to the normal distribution. Statistical calculations such as the normal probability plot or goodness-of-fit tests must be used to prove a normal distribution. However, if the histogram has a different shape, it proves that the distribution is not normal.

  Don't let the name "normal" confuse you. The outputs of many processes—perhaps even a majority of them—do not form normal distributions, but that does not mean anything is wrong with those processes. For example, many processes have a natural limit on one side and will produce skewed distributions. This is normal—meaning typical—for those processes, even if the distribution isn't named normal!

  Skewed. The skewed distribution (Figure 5.84) is unsymmetrical because a natural limit prevents outcomes on one side. The distribution's peak is off center toward the limit and a tail stretches away from it. For example, a distribution of analyses of a very pure product would be skewed, because the product cannot be more than 100 percent pure. Other examples of natural limits are holes that cannot be smaller than the diameter of the drill bit or call-handling times that cannot be less than zero. These distributions are called right- or left-skewed according to the direction of the tail. Figure 5.84 is right-skewed.

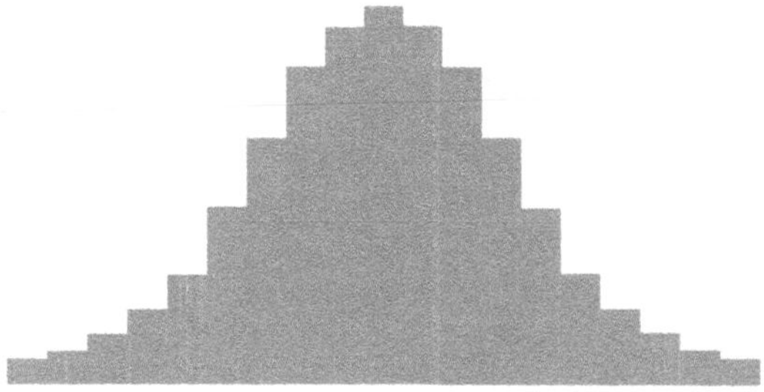

Figure 5.83 Normal distribution.

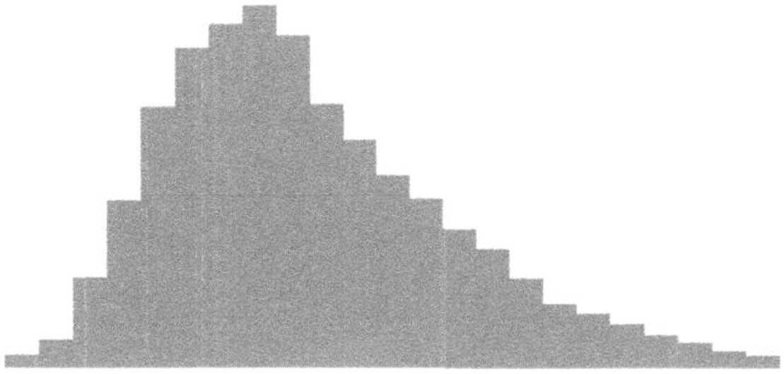

Figure 5.84 Skewed distribution.

Double-peaked or bimodal. The bimodal distribution (Figure 5.85) looks like the back of a two-humped camel. The outcomes of two processes with different distributions are combined in one set of data. For example, a distribution of production data from a two-shift operation might be bimodal, if each shift produces a different distribution of results. Stratification often reveals this problem.

Plateau. The plateau (Figure 5.86) might be called a *multimodal* distribution. Several processes with normal distributions are combined. Because there are many peaks close together, the top of the distribution resembles a plateau.

Edge peak. The edge peak distribution (Figure 5.87) looks like the normal distribution except that it has a large peak at one tail. Usually this is caused by faulty

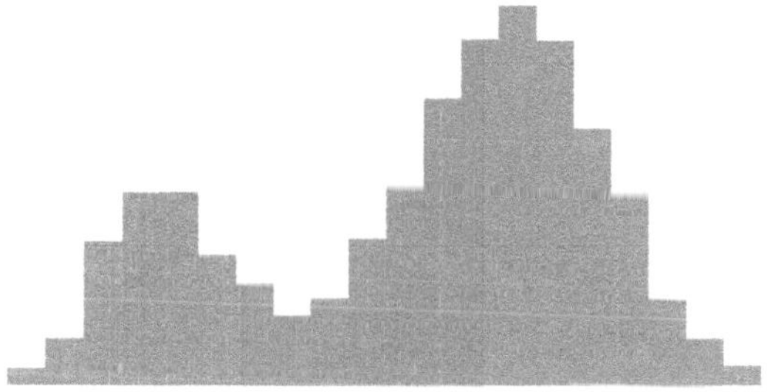

Figure 5.85 Bimodal (double-peaked) distribution.

Figure 5.86 Plateau distribution.

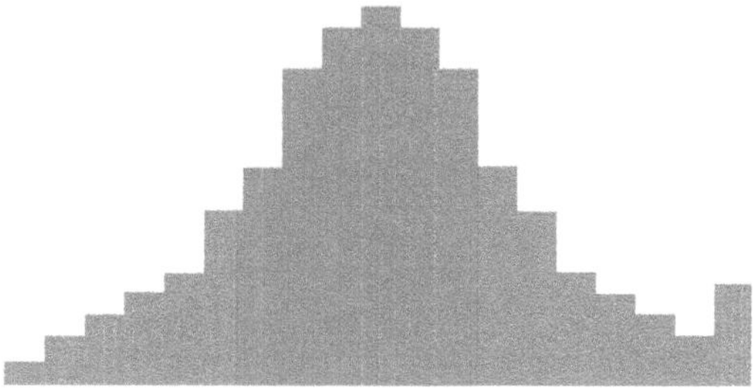

Figure 5.87 Edge peak distribution.

construction of the histogram, with data lumped together into a group labeled "greater than. . . ."

Comb. In a comb distribution (Figure 5.88), the bars are alternately tall and short. This distribution often results from rounded-off data and/or an incorrectly constructed histogram. For example, temperature data rounded off to the nearest 0.2 degree would show a comb shape if the bar width for the histogram were 0.1 degree.

Truncated or heart-cut. The truncated distribution (Figure 5.89) looks like a normal distribution with the tails cut off. The supplier might be producing a normal distribution of material and then relying on inspection to separate what is within specification limits from what is out of spec. The resulting shipments to the customer from inside the specifications are the heart cut.

Dog food. The dog food distribution (Figure 5.90) is missing something—results near the average. If a customer receives this kind of distribution, someone else is

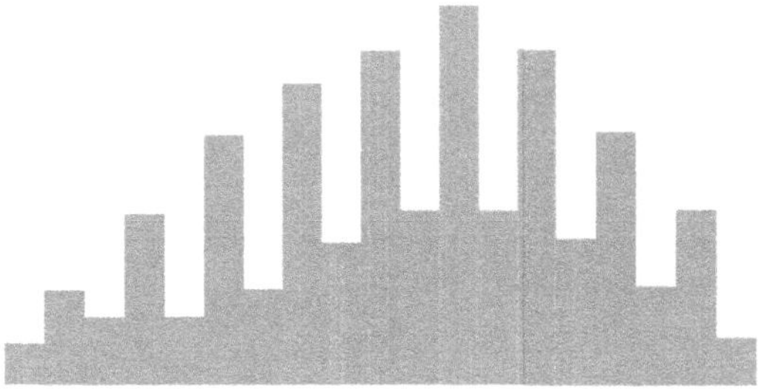

Figure 5.88 Comb distribution.

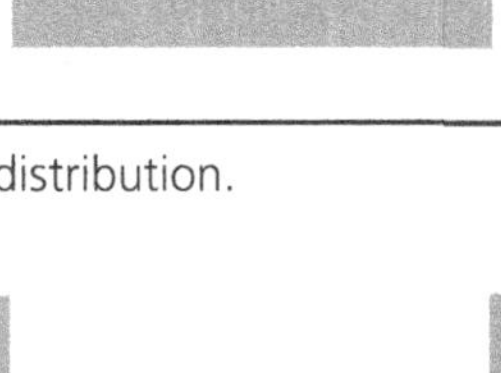

Figure 5.89 Truncated or heart-cut distribution.

Figure 5.90 Dog food distribution.

receiving a heart cut. Even though what the customer receives is within specifications, the product falls into two clusters: one near the upper specification limit and one near the lower specification limit. This variation often causes problems in the customer's process.

- A histogram is the appropriate graph to use when the data are numerical. If the data are categorical (nominal or ordinal) use a bar chart. The bars in a bar chart have spaces between them. The bars in a histogram touch to indicate that the numerical scale is continuous.
- Take care before acting based on your histogram if the data is old. The process may have changed since your data were collected.
- If there are few data points, interpret the histogram cautiously. Any conclusions drawn from a histogram with less than 50 observations should be seriously questioned.
- Any interpretation of a histogram shape is only a theory that must be verified by direct observation of the process.
- A histogram cannot be used to definitely conclude that a distribution is normal. There are other distributions that are similar in appearance. See *normal probability plot* for more information.
- If a process is stable, the histogram can predict future performance. If a process is not stable, the histogram merely summarizes past performance. If you discover that an unusual event occurred during the time period of your histogram, then any analysis of the histogram may apply only to that time period.
- Another tool, the box plot, is used to summarize the most important characteristics of a batch of data. It can be used instead of a histogram, especially when there are too few data points for a histogram. See *box plot*.
- See *graph* for more information about constructing clear and useful graphs.

polygon chart

Description

The polygon chart is very similar to a histogram. Instead of bars representing the count for each value, a point marks the count, and the points are connected with lines. The result is a polygon outlining the distribution's shape. Sometimes you will see a polygon chart called a histogram, especially when there is so much data that the line becomes smooth.

Procedure

Follow the histogram procedure, except change step 4:

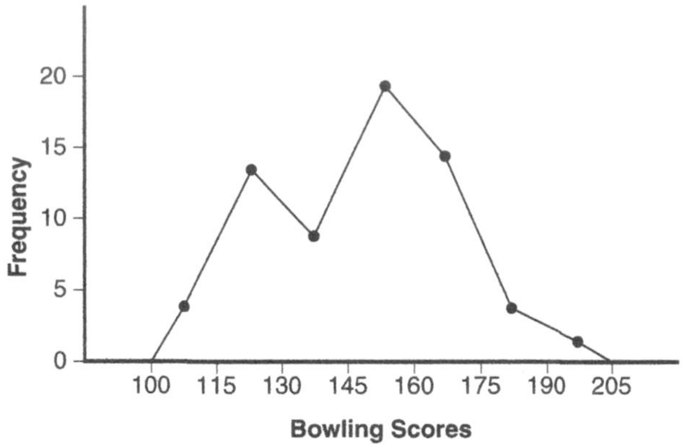

Figure 5.91 Polygon chart example.

4. Above the midpoint of each interval on the x-axis, place a point opposite the y-axis mark that represents the appropriate count. Draw straight lines between adjacent points. Draw straight lines from the outermost points to the x-axis.

Example

Figure 5.91 is a polygon chart of the Bulldog's bowling scores.

stem-and-leaf display

Description

The stem-and-leaf display is a type of histogram that shows individual data values. The last significant digit of each value becomes the symbol marking that value on the graph.

Procedure

1. Decide which digits in the data are changing. Out of this group, choose the two or three digits on the left that are most significant. Of these, the digit on the right will be the *leaf,* and the one or two on the left will be the *stem.*
2. Draw a vertical line on the page. To the left of the line, write the stems in order from smallest to largest in a column.
3. To the right of the line, write opposite each stem the leaf of each data value that has that stem. Do not use any digits to the right of the leaf.
4. Write a legend on the graph indicating how to read the numbers.

Bowling Scores	
10	37
11	15589
12	124456779
13	45789
14	124567889
15	012334555679
16	0133557
17	0267
18	3
19	8

Leaf = 1
10 | 3 = 103

Figure 5.92 Stem-and-leaf display example.

Example

Figure 5.92 shows the Bulldogs' bowling scores in a stem-and-leaf display. Only the two right digits are changing, but the Bulldogs would *like* the left digit to become a two (and it could become a zero), so they decide to include it for the future. They decide all digits are significant. The rightmost digits (ones) are the leaves: 3, 7, and so on. The other two digits (tens and hundreds) are the stems: 10, 11, 12, and so forth. The first row 10 | 37 indicates the scores 103 and 107. The display shows the same double-peaked distribution we saw with the histogram, even though the intervals are not the same.

point graph

Description

A point graph is useful for showing frequency distributions when there are few data points. Each piece of information is graphed along a vertical line as a small circle. This graph becomes ineffective when circles start overlapping so much that the viewer cannot tell how many circles lie together. A histogram should be used in that case.

Procedure

1. Mark a line with tick marks to show the numerical scale for the data. The line can be horizontal or vertical. Write the unit of measure.
2. In a line parallel to the scale line, draw a small circle to represent each data point. If two points have the same value, separate them slightly so both are visible.
3. If you are comparing more than one data set, draw the circles for each data set on separate, parallel lines.

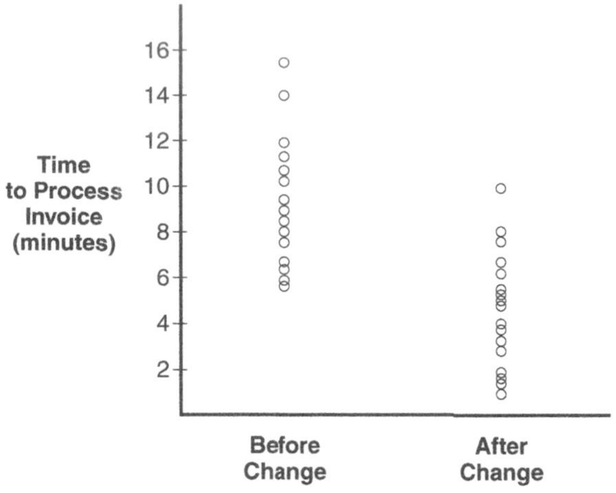

Figure 5.93 Point graph example.

Example

Figure 5.93 shows two point graphs comparing two distributions: the time required to process invoices before making changes and the time required after the process has been streamlined. The improvement is clear.

percentile graph

Also called: quantile graph

Description

The percentile graph shows what percentage of observations are less than or equal to each value. As you move from left to right along the x-axis, the percentages accumulate. The value plotted at the far right side of the graph is 100 percent. Sometimes, the percentile graph is drawn with percentile on the x-axis and values on the y-axis.

When to Use

- When you are more interested in what percentage of the data lies at or below each value than the individual counts for each value

Procedure

1. List all the values in the data set in sequence from smallest to largest. Give them sequence numbers from one, the smallest, to the largest. If there are n values altogether, the percentile for each number is

$$\text{percentile} = \frac{\text{sequence number} - 0.5}{n} \times 100$$

When there are multiple occurrences of the same value, use the highest sequence number to calculate the percentile for that value.

2. Draw the axes and tick marks for the scales. The y-axis range will be the range of values in the data set. The x-axis scale will range from 0 to 100 percent.

3. Mark a dot in the graph opposite the value on the y-axis and the percentile on the x-axis.

Example

To make the calculation clear, this example uses only ten numbers. The method is more valuable with a larger data set. The heights of ten children are (in inches) 62, 57, 55, 56, 57, 53, 58, 60, 57, 55. These are listed in sequence from smallest to largest, with the sequence numbers and the percentile.

Sequence Number	Value	Percentile
1	53	0.5 ÷ 10 × 100 = 5
2	55	
3	55	2.5 ÷ 10 × 100 = 25
4	56	3.5 ÷ 10 × 100 = 35
5	57	
6	57	
7	57	6.5 ÷ 10 × 100 = 65
8	58	7.5 ÷ 10 × 100 = 75
9	60	8.5 ÷ 10 × 100 = 85
10	62	9.5 ÷ 10 × 100 = 95

The resulting percentile graph is Figure 5.94. With it, you can answer questions such as: "Three-quarters of the heights are below what value?" The 75th percentile is

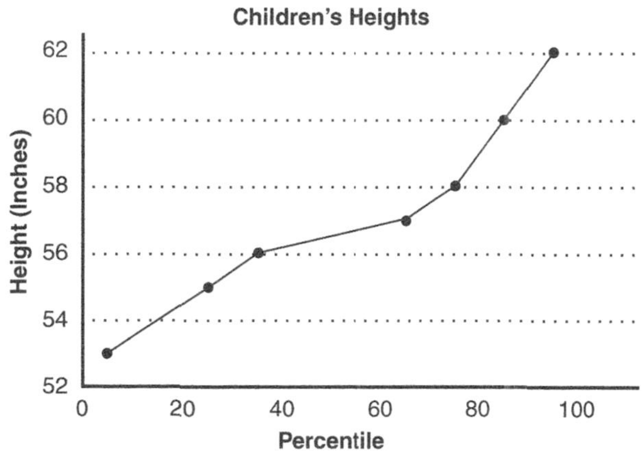

Figure 5.94 Percentile graph example.

58, so three-quarters of the heights are below 58. Notice that the points cluster in the middle, spacing apart at each end. This shows that just a few children are much shorter or taller than the rest. Of course, with only ten points, you could easily figure that out by looking at the numbers. But with a large data set, the percentile graph is useful for showing where the data clump together.

cumulative distribution function graph

Also called: CDF graph

A CDF graph shows the proportion of observations that are less than or equal to each value. Omit the multiplication by 100 in the percentile graph calculation to obtain fractions. Unlike the percentile graph, values are always shown on the x-axis and proportions on the y-axis. Figure 5.95 is a CDF graph of the children's height data used in the percentile graph example. Compare the two graphs.

cumulative polygon graph

Also called: ogive graph, cumulative frequency graph

A cumulative polygon graph shows cumulative frequencies instead of percentiles or proportions. Sometimes a cumulative polygon graph is superimposed on a histogram or polygon curve to show both types of information simultaneously. Figure 5.96 is a cumulative polygon graph of the Bulldogs' bowling scores.

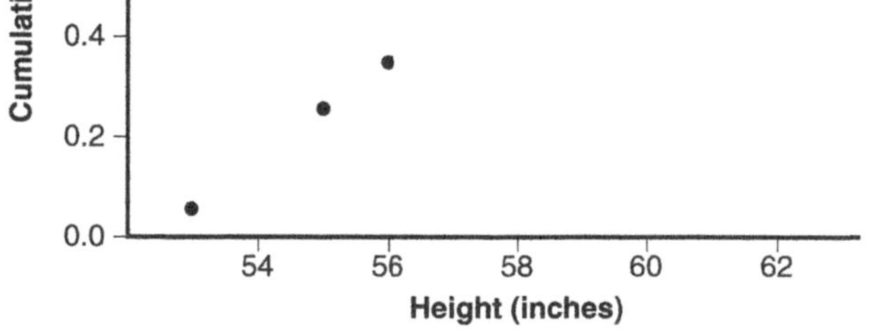

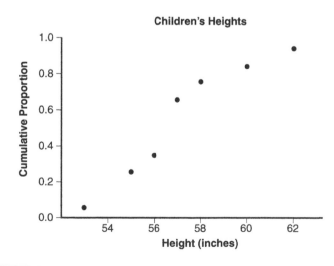

Figure 5.95 CDF graph example.

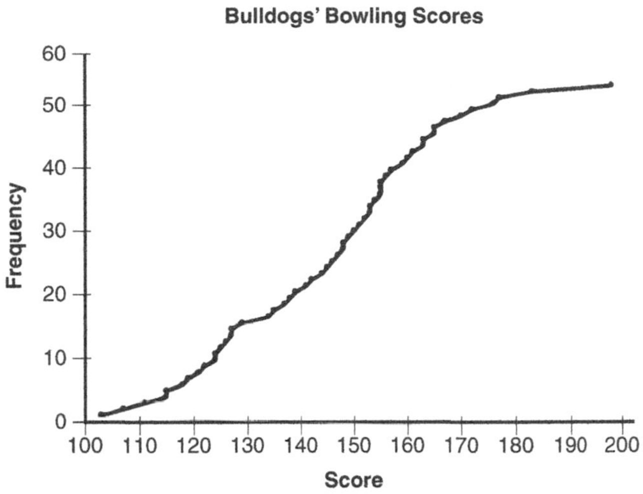

Figure 5.96 Cumulative polygon graph example.

house of quality

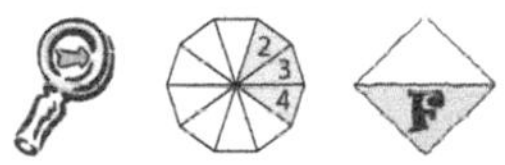

Also called: HOQ, QFD matrix, requirement matrix

Description

The house of quality is the first diagram used in quality function deployment (QFD) to plan design or improvement of a product or service. It combines several matrices and charts into one massive diagram. The structure of the "house" is shown in Figure 5.97.

Building the house of quality starts with analyzing information about what the customer wants, in the rows of the matrix. Added to this is information about quality attributes of the product or the service, in the columns. Relationships among these sets of information, in the central part of the matrix and at the edges, guide decisions about the new design.

As a result, customer needs are translated into technical specifications for the product or service, for a design that maximizes customer satisfaction.

When to Use

- When analyzing customer requirements, and . . .
- When customer requirements must be translated into the organization's language, especially . . .

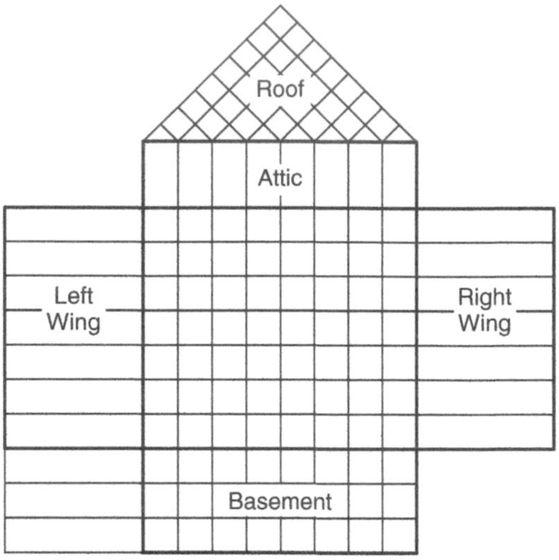

Figure 5.97 House of quality structure.

- When conflicting requirements will require trade-offs, or . . .
- When you are beginning design of a new product or service

Procedure

1. Assemble a cross-functional team. The team should include people knowledgeable about the customer and others knowledgeable about the product or service. In each of the following steps, the team must first obtain the information needed to build the house. This information may come from direct contacts with customers, from departmental studies (for example, marketing, engineering), or from the team members' job knowledge.

Customer Requirements—"Whats"

2. Write customer requirements in the left wing of the house as row labels. As much as possible, the customers' own wording should be used.
3. Add a column between the row labels and the center of the house for importance. 1 to 10 is a commonly used numerical scale, where 1 is unimportant and 10 is extremely important. Assign a number to each customer requirement, based on information you have obtained from customers themselves.
4. In the right wing of the house, record information about customers' perception of existing comparable products or services, your own and competitors'. Usually a scale of 1 to 5 is used. Use different symbols for your product and for each of the competitors' products.

5. (Optional) You may choose to add additional information about customer requirements. Information sometimes included: customer complaints, sales points, target value for competitive position, scale-up factor, absolute weight. See "Considerations" for more information.

Product or Service Characteristics—"Hows"

6. In the attic of the house, write product or service characteristics as column headers. Choose characteristics that directly affect the customer requirements and are measurable. Word them in the technical terms used within your organization.
7. Add another row in the attic with symbols indicating whether the characteristic needs to increase or decrease to better meet customer requirements. Common symbols are + and – or ↑ and ↓.
8. (Optional) You may choose to add more information about product or service characteristics. Information sometimes included: cost, cost of servicing complaints, technical difficulty.

Relationships

9. In the center of the house, use the matrix between customer requirements (rows) and product or service characteristics (columns) to identify relationships between them. Use symbols to indicate whether a relationship between the requirement and the characteristic is positive or negative and how strong it is.
10. On the roof of the house, use the matrix to identify correlations between product or service characteristics. Use symbols indicating whether a relationship is positive or negative and how strong it is.

Objective Measurements

11. In the basement of the house, create a row identifying the measurement unit for each of the product or service characteristics.
12. Also in the basement of the house, record data on the performance of existing comparable products or services, your own and competitors'. Or you can use a relative scale of 1 to 5 and different symbols for each organization, as the customers' assessments were done.

Analyze and Set Targets

13. (Optional) Determine weights for each of the product or service characteristics. Assign a numerical scale to the relationship symbols in the center of the house. Commonly used scales for weak, strong and, very strong relationships are 1, 3, 5, or 1, 3, 9. Starting with the first column, multiply each relationship number by the importance (or absolute weight, if used) of that customer requirement. Add the results down the entire column. The sum is the weight for that

characteristic. In the next row, rank-order the product or service characteristics, starting with 1 for the characteristic with the highest weight.

14. Determine targets for the measurements of each product or service characteristic. Synthesize all the information in the house of quality to decide on appropriate targets for the new design. Record the targets in a row in the basement of the house.

Example

A catalog clothing retailer used quality function deployment to plan improvements to their telephone customer service. The Great Calls Team, composed of customer service reps, trainers, product specialists, call center managers, and IS (information system) specialists, constructed the house of quality shown in Figure 5.98.

This company excels at providing product information to the customer, through ongoing training of telephone reps on new products plus a corps of product specialists to back them up. They have two leading competitors. Competitor 1 has a reputation for friendly telephone reps that customers enjoy talking to. These reps are also empowered

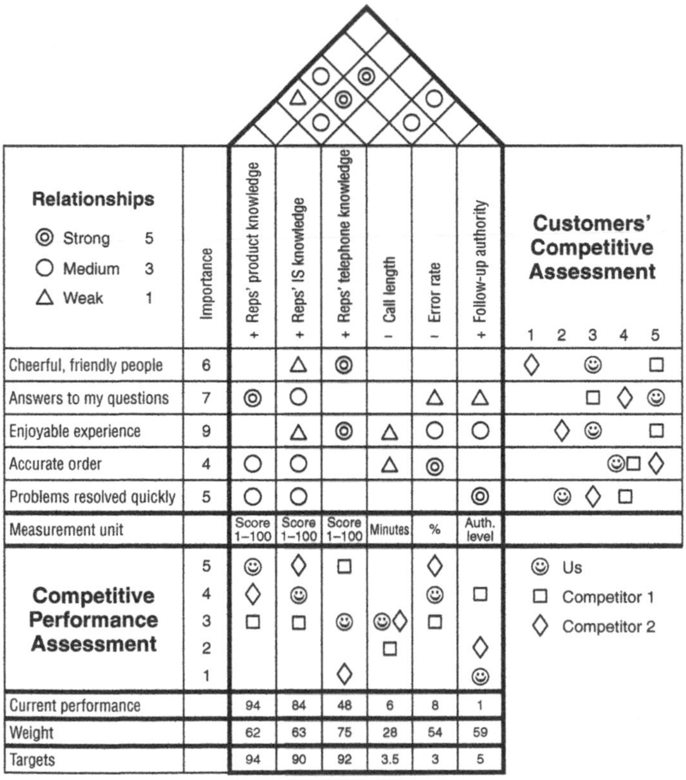

Figure 5.98 House of quality example.

to make decisions about resolving customers' problems, such as authorizing no-cost returns and replacements. Competitor 2 is known for running a very tight ship, with strict control on costs such as cost per call and error rate.

Through a series of customer interviews, the QFD team extracted the qualities customers experience during a telephone order that would make them choose one retailer over another. (They were careful to avoid characteristics of the products themselves, which was the subject of a different study.) These requirements went in the rows of the matrix. Customers' importance ratings and assessments of each retailer's performance were added to the matrix.

After considerable brainstorming, discussion, and list reduction, they identified service characteristics to place in the columns. "Telephone presence" was a score derived from monitoring calls to assess a list of items such as using the customer's name and saying key phrases. The relationships between IS knowledge and the requirements of "cheerful, friendly people" and "enjoyable experience" were hotly debated, with most arguing that they were totally unrelated. The customer service reps on the team convinced the others that it's hard to be cheerful and personable while you're struggling with the computer system. The managers realized that two of their favorite measurements, call length and error rate, were negatively related to each other.

Estimates of each competitor's performance on service characteristics were obtained through contracted market research studies. The results matched customers' perceptions. Predictably, competitor 2, with its heavy emphasis on managing costs, excelled at reducing error rate and rated very low on telephone presence. However, they also had limited follow-up authority within their autocratic organization, which hurt their call length measurements.

The Great Calls Team decided they would obtain the biggest competitive advantage from improving their customer service reps' telephone presence. Along with that, they planned to increase the reps' follow-up authority. Error rate and call length were targeted for moderate improvements. Product and IS knowledge were already strong points, so no new emphasis was planned for product knowledge and only moderate improvement in IS knowledge, through simplifying the order screens. With these changes, they hoped to position themselves as the catalog sales company providing the most exceptional service customers had ever received.

Considerations

- QFD and house of quality are complicated processes that require experience in order to gain the benefits. The information here is only a reference. Before using the house of quality, enlist experienced assistance or attend specialized training.
- When assembling a team, consider including these functions: engineering, R&D, production or manufacturing, sales, marketing, distribution, customer service, purchasing. For services, include the functions that actually perform the service as well as functions that support them. The product or service you are designing and your organization's structure will dictate which functions should be included.

- In the QFD process, the "Hows" of one matrix are frequently cascaded to another matrix, where they become the "Whats" and new "Hows" are developed.
- The house of quality is just one step in the QFD process. In fact, QFD sometimes can be done without the HOQ. The founder of QFD, Dr. Akao, said, "The house of quality alone does not make QFD." See Chapter 2 for more information about the entire QFD process.

Customer Requirements

- Customer requirements are sometimes called customer attributes, customer needs, demanded quality, voice of the customer, or just VOC.
- To identify customer requirements, voice of the customer tools are useful. An affinity diagram can be used to group similar requirements. A tree diagram is often used to delve into details. Sometimes the house of quality shows the levels of the tree diagram to the left of the requirements themselves. Be sure that all characteristics are from the same level of the tree.
- Consider requirements of customers who are not end users of your product. Requirements of parties such as regulatory agencies and retailers may be important. See *requirements matrix* for a tool to help identify customers.
- As much as possible, capture the exact language the customer uses to describe his or her requirements. The house of quality will translate that language into one more understandable to your organization for product and service characteristics. If customer requirements are translated before they reach the diagram, you introduce an opportunity for errors.
- Relative importance is usually based on what customers report as important to them. However, if it is possible to assess their actions, sometimes that more accurately indicates what they truly consider important.
- The customers' perception of existing comparable products or services is sometimes called "customer competitive assessment" or "competitive benchmarking."
- Study the patterns in the right wing of the house to identify advantages you have over your competition that you want to retain, weaknesses you want to eliminate, and opportunities to stand out from the competition.
- "Sales point" or "selling point" is sometimes used to indicate how much sales will improve if that customer requirement is satisfied. A requirement that will delight the customer can be given a score of 1.5 or 2 to show its greater competitive value. See the Kano model on page 17 for a discussion of why not all customer requirements are created equal. Using sales points does not reflect the nonlinear relationship between performance and customer satisfaction on requirements that are "delighters." Other mathematical methods, beyond the scope of this book, have been developed to handle nonlinearity.

- A competitive target value is sometimes used to show how the team would like its improved product or service to rank in customers' perceptions. The same 1 to 5 scale is used. The scale-up factor is the target value divided by the current customer rating of your product. For example, suppose on "easy to open," the customer rates your product 2. Your team decides the target value, how customers would rate the future product, should be 3. The scale-up factor is 1.5 (3 ÷ 2). The larger the scale-up factor, the harder the improvement will be to achieve but the more impressive it will be to the customer.
- If sales points or targets and scale-up factors are used, an absolute weight should be calculated. Simply multiply the importance rating by the scale-up factor and selling point.

Product and Service Characteristics

- These are often called engineering characteristics. That term is not being used here so that the house of quality's use in service applications will be clear. Another term sometimes used is quality attributes.
- Affinity and tree diagrams are useful here too. Use them to group characteristics and to develop detail. Sometimes the house of quality shows the levels of the tree diagram above the characteristics. Be sure that all characteristics are from the same level (usually third) of the tree. Other useful tools for developing these characteristics are brainstorming and list reduction techniques.
- Scores for technical difficulty or cost, sometimes placed in the basement of the house, reflect how hard it will be to change that characteristic.

Relationships

- The center of the house is often called the relationship matrix.
- A variety of different symbols and scales are used for relationships. Choose what makes sense and feels comfortable to your team. Common symbols are:

● Strong positive	✔ Strong positive	● Strong	◎ Strong positive
○ Positive	✓ Positive	○ Medium	○ Weak positive
x Negative	✗ Negative	△ Weak or possible	x Negative
xx Strong negative	✖ Strong negative		

- Sometimes different colors are used to differentiate relationships determined by experimentation or statistical calculations from ones determined by the teams' judgment or intuition.
- Odd patterns of relationship symbols in the center of the house can indicate possible problems.

Empty row—This customer requirement is not met by any product or service characteristics. A new characteristic must be identified.

Empty column—This characteristic is not necessary for any customer requirements. Did you miss a customer requirement? If not, eliminate the characteristic.

Row with no strong relationships—It will be difficult to meet customer requirements without at least one strong relationship to a characteristic. Rethink the relationships, or look for another characteristic.

Column with no strong relationships or entire matrix with few strong relationships—Characteristics should have a strong relationship to at least one requirement. Rethink the characteristic(s).

Identical rows or portions of rows—The customer requirements may not be from the same level of detail. Use a tree diagram to analyze the requirements.

Row or column with many relationships—The requirement or characteristic might be a cost, safety, or reliability issue. Remove the item from the matrix and save to use in an analysis devoted solely to that issue. Or, there maybe be a problem with levels, as in the previous item.

Relationships form diagonal line—Characteristics are simply requirements reworded, or vice versa. Reconsider both. Make sure the requirements are the voice of the customer, and the characteristics are the voice of the engineer or service designer.

- The roof of the house is frequently called the correlation matrix.
- Often the relationships in the roof of the house are simplified to positive (+) and negative (–).
- The roof matrix shows similar and conflicting characteristics. It also shows side effects and unintended consequences of characteristics. As a result, this part of the house reveals trade-offs between various design options that have to be considered.
- If trade-offs between characteristics are not thought through, the results could be unfilled customer requirements, poor quality, changes in your process and/or the product or service very late in the development process, and increased cost. Resolving the trade-offs during QFD leads to shortened development time.

Objective Measurements

- Gathering data about performance of your and competitive products or services is sometimes called technical competitive assessment or technical benchmarking. The tool called benchmarking usually goes beyond gathering this data to understanding how differences arise.

- The performance data in the basement should match the customer's performance perceptions in the right wing. If not, recheck your measurements. If they still don't match, consider whether customers are being influenced by preconceived ideas.

Analyze and Set Targets

- The calculated weights for characteristics reflect which ones are strongly linked to the most important customer requirements.
- The targets are the whole point of the house of quality. While many features of the house are optional, targets are not.
- When you set targets, you have the choice of leaving the current performance value unchanged, improving it a small amount, or improving it a lot.
- Do not use a range for targets. If you choose a range, the easiest end of the range is likely to become the target, which may not be ambitious enough. Choose a precise number.
- There is no fixed procedure for setting targets. Your situation is unique. Your team must study the insights the house of quality gives you and decide what goals make sense in your marketplace and for your organization. Ideally, you will choose to focus on things that are most important to customers and that your competition now does better than you.

Other Issues

- The size of the house of quality can quickly become overwhelming as the number of requirements and characteristics increases. If you have 20 requirements and 30 characteristics, the house will have more than 1000 relationships to determine. This problem has been the focus of many studies. If your house of quality is large, first eliminate any duplicate requirements. Then, consider eliminating trivial requirements, such as ones mentioned by only a few customers or ones with the lowest importance. Another approach is to group customer requirements and do a separate HOQ for each group. However, this might miss some correlations and trade-offs. Other solutions, mathematical methods beyond the scope of this book, can be found in the Resources and by searching the current quality literature.
- Another problem is multiple groups of customers with different, even contradictory, requirements. One solution is to design different products or services, using different houses of quality for each. If that is not possible, you could identify the most important customer group and satisfy only their requirements. This might lose current or potential customers. A third approach uses averages. First, assign each customer group a relative importance. Then calculate the overall importance rating for each requirement by multiplying the relative importance of each customer group by the importance rating that group gave to that requirement and adding the results for each group. No one's requirements are perfectly satisfied, but no one is completely ignored either.

- Another issue that has been studied is translating customers' verbal ratings (very important, somewhat important) into a numerical scale. Fuzzy arithmetic (yes, that's what mathematicians really call it) is used to avoid imposing too much precision on data that is imprecise and therefore reaching incorrect conclusions. The mathematical details, again beyond the scope of this book, can be found in the Resources and by searching the current quality literature.

hypothesis testing

Variations: *t*-test, *z*-test, F-test, chi-square test, ANOVA (analysis of variation)

Description

Hypothesis testing is an approach for analyzing data. It answers the question, "What is the probability that these results could have occurred through random chance?" The other side of that question is, "Is the effect we think we see in this data real?" Hypothesis testing is used when the question is about a large population but data are available only for a sample of the population. This approach is used to answer a wide variety of questions important in quality improvement, such as "Did the change we made to the process create a meaningful difference in the output?" or "Is customer satisfaction higher at location A than at our other locations?"

The most often used tests are: *z*-test, *t*-test, *F*-test, chi-square (χ^2) test, and analysis of variance tests, usually called ANOVA. These and other hypothesis tests are based on the fact that means, variances, proportions, and other statistics form frequency distributions with known patterns. The most well known distribution is the normal distribution, which is the basis for the *z*-test. The *t*-, F-, and chi-square tests are based on the *t*, F, and chi-square distributions.

When to Use

- When drawing conclusions about the average, proportion, variance, or some other characteristic of one or more sets of data, and . . .
- When the conclusion is based on samples taken from a larger population

For example:

- When deciding whether the mean or variance from a process has changed, or . . .

- When deciding whether the means or variances of several data sets are different, or . . .
- When deciding whether two proportions from two different data sets are different, or . . .
- When deciding whether the true proportion, mean, or variance is equal to (or greater or less than) a certain number

Procedure

The procedure for hypothesis testing has three parts: understanding the problem to be solved and planning the test (steps 1, 2, and 3 below), number-crunching that is usually done by computer (steps 4 and 5), and applying the numerical results to the problem (step 6). While computers can handle the math, understanding the concepts underlying hypothesis testing is crucial to the first and third parts.

If hypothesis testing is new to you, start by reading the definitions of terms in "Considerations." These definitions explain the concepts of hypothesis testing. Then come back and read the procedure.

Hypothesis testing cannot be covered in detail in this book. This procedure is an overview and a quick reference. For more information, consult a statistics reference or a statistician.

1. Determine the decision or conclusion that must be made about your data. Identify the appropriate test for the decision. The decision about which test to use depends on the purpose of the test and also upon the kind of data. Use Tables 5.7 and 5.8, which summarize frequently used hypothesis tests, or consult a statistician for help.
2. Phrase the null hypothesis and the alternative hypothesis. Determine also if the test is two-tailed, left-tailed, or right-tailed.
3. Choose the significance level, α.
4. Calculate the test statistic. Computer software is usually used.
5. Determine the P-value for the test statistic. Use a statistical table or a computer program for that statistic's distribution. For a z-test, you can use Table A.1, the area under the normal curve.
6. Compare the P-value to α for a right- or left-tailed test or to $\alpha/2$ for a two-tailed test. If the P-value is less, reject the null hypothesis and conclude that the alternate hypothesis is probably true. Otherwise, do not reject the null hypothesis, and conclude that there is not enough evidence to support the alternate hypothesis.

Table 5.7 Hypothesis tests for means, variances, and proportions.

Compare mean to a given value μ_o. H_0: $\mu = \mu_o$		
• When σ is known	$z = \dfrac{\bar{x} - \mu_o}{\sigma / \sqrt{n}}$	μ = true mean of population μ_o = hypothesized value of mean $\bar{x}$ = mean of sample σ = true standard deviation of population s = standard deviation of sample n = sample size df = degrees of freedom
• When σ is unknown *and* • When sample size $n \geq 30$	$z = \dfrac{\bar{x} - \mu_o}{s / \sqrt{n}}$	
• When σ is unknown *and* • When sample size $n < 30$ *and* • When data distribution is normal	$t = \dfrac{\bar{x} - \mu_o}{s / \sqrt{n}}$ $\text{df} = n - 1$	
Compare two means to each other. H_0: $\mu_1 = \mu_2$		
• When σ_1 and σ_2 are known *and* • When samples are independent	$z = \dfrac{\bar{x}_1 - \bar{x}_2}{\sqrt{(\sigma_1^2 / n_1) + (\sigma_2^2 / n_2)}}$	
• When σ_1 and σ_2 are unknown *and* • When sample sizes n_1 and $n_2 \geq 30$ *and* • When samples are independent	$z = \dfrac{\bar{x}_1 - \bar{x}_2}{\sqrt{(s_1^2 / n_1) + (s_2^2 / n_2)}}$	
• When σs are unknown *and* • When sample sizes n_1 or $n_2 < 30$ *and* • When data distribution is normal *and* • When samples are independent	*t*-Test: Formulas are complicated and depend on whether $\sigma_1 = \sigma_2$. Use computer software.	
• When samples are paired *and* • When sample sizes n_1 and $n_2 \geq 30$	$z = \dfrac{\bar{d}}{s_d / \sqrt{n}}$	$d = \bar{x}_1 - \bar{x}_2$ $\bar{d}$ = mean of d values s_d = standard deviation of d values n = number of sample pairs
• When samples are paired *and* • When sample size $n < 30$	$t = \dfrac{\bar{d}}{s_d / \sqrt{n}}$ $\text{df} = n - 1$	
Compare three or more means to one another. H_0: $\mu_1 + \mu_2 + \mu_3$ = etc.		
• When samples are independent *and* • When data distribution is normal *and* • When $\sigma_1 = \sigma_2 = \sigma_3$ = etc.	ANOVA: Consult statistician or statistics reference.	
Compare proportion to given value π_o. H_0: $\pi = \pi_o$		
• When samples are independent *and* • When expected frequencies $n\pi_o$ and $(1 - \pi_o)n \geq 5$	$z = \dfrac{p - \pi_o}{\sqrt{\pi_o(1 - \pi_o)/n}}$	π = true proportion in population π_o = hypothesized value of proportion p = proportion in sample n = sample size $\bar{p}$ = weighted average of p_1 and p_2
Compare two proportions to each other. H_0: $\pi_1 = \pi_2$		
• When samples are independent *and* • When n_1 and $n_2 \geq 30$	$z = \dfrac{p_1 - p_2}{\sqrt{\bar{p}(1 - \bar{p})((1/n_1) + (1/n_2))}}$ $\bar{p} = \dfrac{n_1 p_1 + n_2 p_2}{n_1 + n_2}$	
Compare standard deviation to given value σ_o. H_0: $\sigma = \sigma_o$		σ = true standard deviation of population σ_o = hypothesized standard deviation of population s = standard deviation of sample n = sample size df = degrees of freedom
• When data distribution is normal	$\chi^2 = \dfrac{(n-1)s^2}{\sigma_o^2}$ $\text{df} = n - 1$	
Compare two standard deviations to each other. H_0: $\sigma_1 = \sigma_2$		
• When data distribution is normal	$F = s_1^2 / s_2^2$ df for numerator $= n_1 - 1$ df for denominator $= n_2 - 1$	
For other purposes or when conditions are not met, consult a statistician or statistical reference.		

Table 5.8 Chi-square hypothesis tests for categorical data.

Chi-Square Tests

- When data are counts (frequencies) within categories, and
- When all $O \geq 1$, at least 80% of $O \geq 5$, and $N \geq 20$.

These tests are always right-tailed.

Compare several groups' frequency distributions (proportions) across several categories.

H_0: The groups do not differ in the distributions (proportions) of categories.

Determine whether two variables are independent.

H_0: Variable 1 and variable 2 are independent.

$\chi^2 = \sum \frac{(O-E)^2}{E}$ summed over all cells $df = (r-1)(c-1)$	First, construct contingency table with groups or variable 1 in r rows and with categories or variable 2 in c columns. Then for each cell: O = actual (observed) value of cell C = sum of all values in its column R = sum of all values in its row N = total of all observed values E = expected value of cell = $\frac{C \times R}{N}$

Compare two proportions to each other. H_0: $p_1 = p_2$

This is a specialized case of the test above, with $r = 2$ and $c = 2$.

The results are equivalent to a z-test for two proportions.

Goodness of fit: Compare an observed frequency distribution to an expected or theoretical distribution.

H_0: Observed distribution is the same as expected distribution.

$\chi^2 = \sum_k \frac{(O_i - E_i)^2}{E_i}$ $df = k - 1$	$O_i = x_{2i}$ = observed value in distribution 2 $E_i = n_2 p_{1i}$ = expected value in distribution 2 if its proportions were the same as distribution 1 k = number of categories

Alternate Procedure

Follow steps 1, 2, 3, and 4 of the previous procedure. Then:

5. Using a statistical table or a computer program, determine the critical value(s) of the test statistic and the rejection region as follows. The z-test is used as an example. For t, F, or chi-square test, replace the z with t, F, or χ^2.

Type of Test	Critical Value(s)	Rejection Region
left-tailed test	$-z_\alpha$	$z < -z_\alpha$
right-tailed test	z_α	$z > z_\alpha$
two-tailed test	$-z_{\alpha/2}$ and $z_{\alpha/2}$	$z < -z_{\alpha/2}$ or $z > z_{\alpha/2}$

6. Compare the test statistic to the rejection region. If the test statistic lies in the rejection region, reject the null hypothesis and conclude that the alternate hypothesis is probably true. Otherwise, do not reject the null hypothesis, and conclude that there is not enough evidence to support the alternate hypothesis.

Example: *t*-Test

A grocery store receives 50-pound boxes of apples from a supplier. They pay a fixed price per box. The supplier assures them that the mean weight of each box is indeed 50 pounds. The produce team randomly samples 10 boxes and weighs them. The weights are:

50.1	49.6	50.3	49.9	49.5	49.7	50.0	49.6	49.7	50.2

Are they being cheated?

Phrased statistically, the produce team's question is, "Is the mean weight of apple boxes we receive less than 50 pounds?" The null hypothesis is, "The mean of the boxes' weights equals 50 pounds" and the alternate hypothesis is, "The mean of the boxes' weights is less than 50 pounds." They decide to use a significance level of 5%.

On Table 5.7, they look at tests for comparing a mean to a given value. The true σ is unknown, the sample size is less than 30, and they assume the distribution of box weights is normal. Therefore, they will use a *t*-test. Because the alternate hypothesis includes "less than," they need a left-tailed test.

They enter their data in an online calculator and get the following results:

Sample mean = 49.86 Standard deviation = .28 $t = -1.583$ P = 0.07

Since P is larger than 0.05, they cannot reject the null hypothesis. They do not have evidence that they are being cheated. Figure 5.99 shows the t distribution, the test statistic $t = -1.583$, and the area under the curve beyond that value, P = 0.07.

Using the alternate procedure, they could have determined from a t table that for $\alpha = 0.05$ and df = 9, the critical value of $t_\alpha = -1.833$. With a left-tailed test, the rejection region is any z value less than –1.833. The test statistic, –1.583, does not lie in the rejection region, so they do not reject the null hypothesis.

Figure 5.100 shows the t distribution, the critical value, the rejection region, and the area under the curve equal to $\alpha = 0.05$. Comparing the two graphs shows how the two procedures are different ways of looking at the same situation. For this left-tailed situation, whenever the test statistic t is greater than the critical value t_α, the area under the curve to the left of t—the P value—will be larger than α, the area under the curve to the left of t_α.

Example: Chi-Square Test

A catalog clothing retailer wanted to know whether proposed changes to its product line would be similarly received in different parts of the country. They had randomly selected 750 customers, described the proposed new products, and asked them to estimate their likelihood of buying. They categorized the data by geographical groups and set up a contingency table with five rows and four columns, seen in Figure 5.17 for *contingency table* on page 149.

From Table 5.8, the appropriate test is chi-square, to compare several groups' distributions across several categories. This test is always right-tailed. The null hypothesis can be stated, "The five regional groups of customers do not differ in the distributions

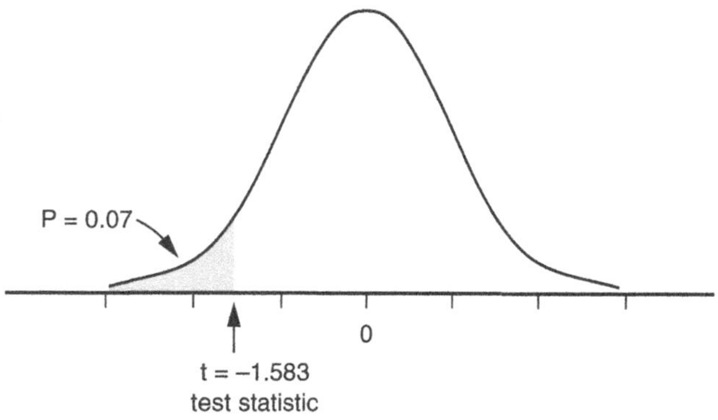

Figure 5.99 *t*-test example, first procedure.

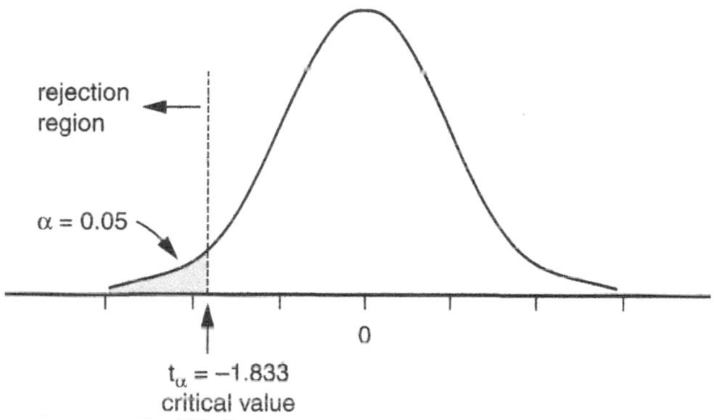

Figure 5.100 *t*-test example, alternate procedure

of their buying likelihoods." The alternate hypothesis is, "There is a difference in the five groups' buying likelihood distributions."

They decide on a significance level of 5% and calculate the degrees of freedom df = 12. Most chi-square tables are designed for the alternate procedure, in which you look up either α or $1 - \alpha$ and read the critical value. For $\alpha = .05$ and df = 12, the critical value of χ^2 is 21.026. If the test statistic is larger than that, they will reject the null hypothesis.

Using a spreadsheet, they calculate *E* for each cell. *E* represents the expected number of responses if the null hypothesis were true, that is, if the buying likelihood distribution for each region were the same as the overall buying likelihood distribution. Next they calculate $(O - E)^2 \div E$ for each cell and sum up all those values to obtain their test statistic $\chi^2 = 22.53$. This is larger than the critical value of 21.026, so they reject the null hypothesis. The distribution of buying likelihood varies regionally.

This test is equivalent to a test for whether two variables are independent. These results show that geographical region and buying likelihood, the two variables, are not independent. Knowing a customer's region, one can predict whether that customer is more likely to buy the new product line.

Another Example: Chi-Square Test

The same retailer planned to change the format and style of their catalog and wanted to know if the new format would be effective at increasing orders. As a test, they sent 200,000 copies of the spring catalog in the new format to randomly selected customers. The other 1,800,000 catalogs were the traditional version. See the contingency table example and Figure 5.18 on page 149 for the 2 × 2 contingency table used to organize their data.

The chi-square test will compare two proportions to each other. The null hypothesis is, "The proportion of customers buying from the test catalog is the same as the proportion buying from the standard catalog."

They use a significance level of 5%. For comparing proportions, there is one degree of freedom. The critical value of χ^2 for $\alpha = .05$, with 1 degree of freedom, is 3.841. The test statistic χ^2 is 278. Therefore, they reject the null hypothesis and conclude that the proportion of customers buying from the catalog with the new format is significantly different than those buying from the old catalog.

Considerations

- Like most subjects, statistics has specialized language that is a shortcut for expressing frequently used concepts. Here are definitions of terms used in this procedure:

  Test. One of the statistical tests, such as *z, t, F*, or chi-square. Knowing which test should be used is the trickiest part of the procedure. The choice depends upon both the kind of data and what kind of decision needs to be made about the data.

  Hypothesis. A statement expressed as fact, which will be proved or disproved by the test.

  Null hypothesis, H_o. The hypothesis that you want to test, that the data resulted from chance. It's called "null" because often (but not always) the null hypothesis says there is no difference between two sets of data or between a parameter calculated from the data and a given value.

  Alternate hypothesis, H_a. The hypothesis that must be true if the null hypothesis is false. Usually, the alternate hypothesis implies that the data resulted from a real effect, not chance.

  Statistic. A general term for any calculated number that summarizes some aspect of the sample data. Averages, means, variances, and proportions are all statistics.

  Test statistic. The calculated number used to test the null hypothesis. For each kind of test, there is a formula for the appropriate test statistic. It is chosen so that if the null hypothesis is true (the data resulted from chance), the statistic comes from a well known distribution, such as the normal distribution for the z-test.

Two-tailed, right-tailed, left-tailed. Describes whether the test is concerned with both sides of the frequency distribution (two-tailed) or just one side. If the alternate hypothesis includes ≠ (not equal to) in its expression, a two-tailed test is needed. If the alternate hypothesis includes < (less than), a left-tailed test is needed. If the alternate hypothesis includes > (greater than), a right-tailed test is needed. Chi-square tests are always right-tailed.

P value. The probability that the test statistic could have occurred by chance within the test's known distribution. The P value equals the area under the curve beyond the test statistic's value. (See Figure 5.99, page 319.) The smaller the P value, the more sure we can be that the effect is real and not merely chance. Because the various test distributions are well known, these probabilities can be found in tables or computer programs.

Significance level, α. A number reflecting how sure we must be that the effect is not due to chance before we will decide it is real. Usually, significance levels of 1%, 5%, or 10% ($\alpha = 0.01$, 0.05, or 0.10) are chosen. For example, if for a one-tailed test $\alpha = 0.05$, then we will decide the effect is real only if the result could be obtained by random chance less than 5% of the time, that is, $P < 0.05$.

Critical value. The value of the test statistic when its probability is exactly α. The area under the curve in the tail beyond the critical value(s) equals α. (See Figure 5.100.) For a two-tailed test, there are two critical values, one at each tail, where the area beyond each equals $\alpha/2$. The critical value, written as $\pm z_\alpha$ or $\pm z_{\alpha/2}$, is determined from a table or computer program.

Rejection region. The region of the frequency distribution where, if the test statistic falls there, the null hypothesis will be rejected. For a left-tailed test, these values are in the tail of the distribution less than the critical value. For a right-tailed test, these values are in the tail of the distribution greater than the critical value. For a two-tailed test, the region includes both tails.

Confidence level. $1 - \alpha$.

Confidence interval. A range of values that has a high probability of containing the test statistic if chance is causing the results. The rejection region is the area outside the confidence interval. Significance level, confidence level, and confidence interval are related in this way: If $\alpha = 0.05$ and the confidence level is 95%, then values lying within a 95% confidence interval are accepted as possibly occurring solely by chance. The null hypothesis cannot be rejected. The definition of confidence intervals is tricky. 95% confidence interval does not mean that 95% of all values in the distribution lie in that range. Instead, it means that if you have a value that is part of the distribution, 95% of all intervals one might construct based on sample data will contain that value.

- In hypothesis testing, you are looking at the curve of the test distribution and your calculated value of the test statistic, which lies somewhere along the horizontal axis of the graph. If the area under the curve beyond your test statistic, P, is small enough (less than the significance level α) then the test statistic probably did not come from that distribution.
- Because the curves are frequency distributions, the area under any portion of the curve is a measure of the number of occurrences that would have the values marked on the horizontal axis. That is why the area under the curve can be used to find α and P values.
- Because hypothesis testing is dealing with samples and probabilities, there are chances of drawing incorrect conclusions. *Type I error* occurs when the null hypothesis is true but is rejected. (See Figure 5.101.) We know the probability of a type I error: it is the significance level α. In the second example, there was a 5% probability that the difference in distributions among regions really was due to chance. *Type II error* occurs when the null hypothesis is false but is not rejected. If the grocery store really was being cheated on the weight of apples in the boxes, a type II error would have occurred. Calculating the probability of a type II error, β, is more complicated and beyond the scope of this book. Unfortunately, as α gets smaller, β gets larger. However, for a given α, increasing the sample size will make β smaller.
- Because of the possibility of type II error, if the null hypothesis is not rejected, you cannot conclude that the alternate hypothesis is false. You can only conclude that the data do not provide enough evidence to support the alternate hypothesis.

Reality \ Conclusion	H_O is true	H_O is not true
H_O is true	**Correct conclusion**	**Type I error** probability = α
H_O is not true	**Type II error** probability = β	**Correct conclusion**

Figure 5.101 Type I and type II errors.

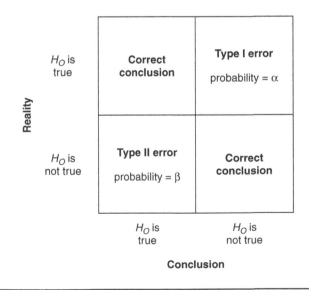

- Many Web sites have calculators that can compute test statistics and test distribution values. You input the data and the calculator computes the test statistic, probabilities, and/or critical intervals. However, it is crucial to know which test to use and how to interpret the results.
- "Paired" samples means two sample sets contain matched pairs of related observations, for example, measurements of the same sample before and after treatment or measurements of the same sample by two different instruments. The hypothesis being tested is usually that the means of the two sets of samples are equal, or in other words, the mean of the differences between the pairs is zero. This test is often called paired comparisons. It is one of the tools used in the Shainin design of experiments methodology. (See *design of experiments* for more information.) This is not the same tool as *paired comparison* described on page 372.
- See *contingency table* for a tool used to organize data before a chi-square test.
- Other hypothesis tests are available for other conditions. For example, when the data's distribution is not normal, a variety of nonparametric tests are available. It is beyond the scope of this book to summarize all hypothesis tests. If conditions on the tables do not match your data, consult a statistician for help with the appropriate test.

importance–performance analysis

Also called: customer window

Description

The importance–performance analysis studies customers' perceptions about both the importance and the performance of products or services. It can help organize discussions with customers about their needs and perceptions. It also compares customers' perceptions to those of the organization, looking for differences that might indicate incorrect priorities.

When to Use

- When surveying customer satisfaction, and . . .

- When setting priorities for product or service changes based on customer perceptions

Such as:

- When identifying products, services, or their features whose improvement would most increase customer satisfaction, or . . .
- When identifying products, services, or their features which should have changed emphasis: increased, reduced, or eliminated, or . . .
- When comparing customer satisfaction between customer segments or between your organization and a competitor

Procedure

Data Collection

1. Identify the products, services, or features that you want evaluated. Usually, you will study either the selection of products or services offered or the features of one particular product or service. From here on, the products, services, or features under study will be called "output."
2. Develop a survey. Set it up so that each question asks for an output to be rated twice on a numerical scale of 1 to 5 or 1 to 7. One rating is how important the output is to the customer (importance) and the other rating is the quality of that output (performance). Typical scales are:

Importance	Performance
5 = Critical	5 = Greatly exceeds expectations
4 = Important, but not critical	4 = Exceeds expectations
3 = Of some value	3 = Meets expectations; adequate
2 = Nice to have, but not necessary	2 = Needs some improvement
1 = Not needed	1 = Consistently lacking

3. Identify customer groups you want to survey, and select key customers or a representative sampling of customers. Administer the survey.
4. Identify key participants in the process of providing the outputs. (We'll call them "suppliers.") Administer the same survey to them. Their importance rating should reflect how important *they* think the output is *to the customer.* Their performance rating should reflect their opinion of the output's quality.
5. Calculate an average for each answer for the customers and the suppliers, and record the ratings on a table like Table 5.9. Alternatively, you may choose to record ranges of responses (highest to lowest) or, if the number of respondents is small, individual responses.

Table 5.9 Importance–performance analysis rating table.

Service	Importance by Supplier	Importance by Customer	Performance by Supplier	Performance by Customer

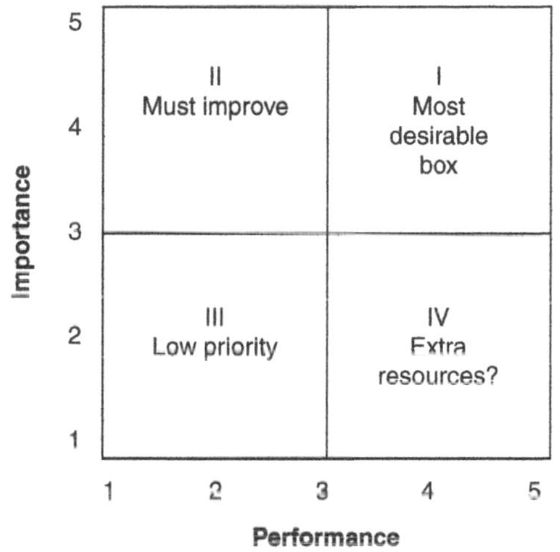

Figure 5.102 Customers' importance–performance comparison.

Analysis

1. *Importance–performance (I–P) chart.* For each output, plot the customers' importance and performance perceptions on a two-dimensional chart like Figure 5.102. Find the customers' importance rating on the left side and the customers' performance rating across the bottom. Then find the box where the importance row intersects with the performance column.

   - Box I is the most desirable. Outputs in that box are important to customers and, according to the customers, are being done well.
   - Box II contains the most pressing needs for improvement. Customers believe that important outputs are not being done well.
   - Improvement of outputs in box III may be possible, but the priority is lower than in box II. Customers believe performance could be better, but the outputs are not important.

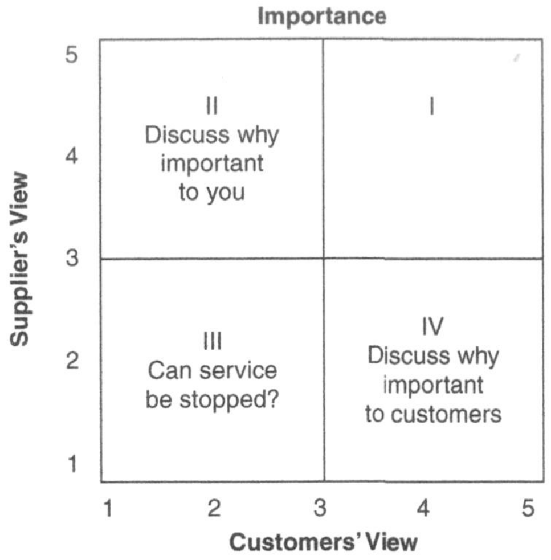

Figure 5.103 Supplier–customer comparison: importance.

- Box IV may indicate areas where resources can be shifted to other, more important outputs. Performance exceeds the customers' expectations for unimportant products or services.

2. *Importance comparison chart.* Plot a point for each output on an importance chart like Figure 5.103. The suppliers' view of the output's importance is on the left side, and the customers' view of its importance is across the bottom.

   - In boxes II and IV, the customers' and suppliers' views do not match. These could be fruitful areas for discussion with the customer.
   - In box III, both the customer and supplier rate importance 1 or 2. Talk with customers about eliminating the output or replacing it with something of more value.
   - Box I represents outputs that all agree are very important.

3. *Performance comparison chart.* Plot a point for each output on a performance chart like Figure 5.104. This time the views of performance are being compared between supplier and customers.

   - If both perceive a particular product or service to be a strength (box I), find out why. Maybe you will learn something that can be transferred to other services.
   - Again, areas of disagreement (boxes II and IV) are fruitful areas for discussion. Listen to the customers' perceptions to understand how they assess your performance.
   - If both supplier and customers think performance is poor (box III), be sure to talk about why.

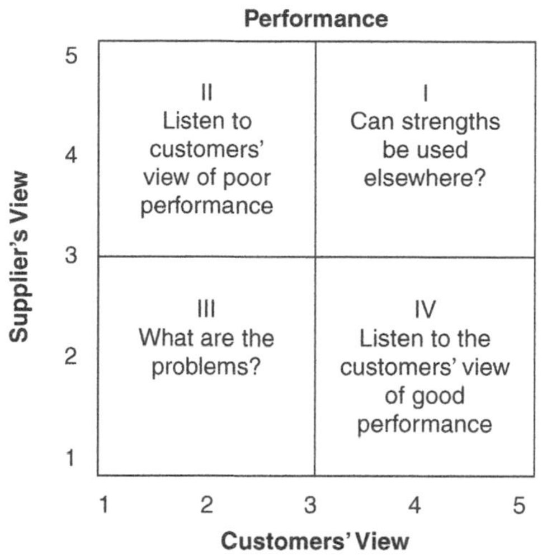

Figure 5.104 Supplier–customer comparison: performance.

Table 5.10 Importance–performance analysis example: rating table.

Service		Importance by Restaurant	Importance by Customers	Performance by Restaurant	Performance by Customers
Breakfast	(B)	4.1	4.2	4.9	3.9
Lunch	(L)	4.7	4.9	4.4	4.1
Dinner	(D)	4.9	4.9	4.6	4.4
Lounge/bar	(Bar)	2.2	3.2	3.1	4.8
Carryout	(C)	1.8	1.5	2.8	4.6
Private parties	(P)	3.9	4.7	4.1	2.5

Example

A restaurant identified six products and services that they wanted to study to improve customer satisfaction. An importance–performance questionnaire was given both to restaurant customers and to management and employees. The average ratings are shown in Table 5.10.

I–P Chart

For each service, a code placed on Figure 5.105 shows the customer's perception of importance and performance.

- Breakfast (B) was rated 4.2 in importance and 3.9 in performance, so it goes in box I. Lunch, dinner, and bar also fall in box I, which is most desirable.

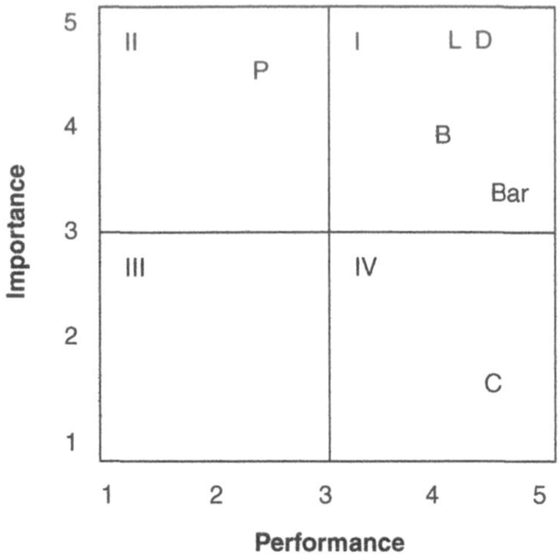

Figure 5.105 Importance–performance analysis example: customers' comparison.

- Private parties was rated high (4.7) in importance but low (2.5) in performance. It falls in box II. The restaurant should take action to understand the gap and to improve performance.
- Carryout (C) was rated low in importance (1.5) although it was high in performance (4.6), so it falls in box IV. The restaurant might want to divert resources from this area to improve private party service.

Importance Comparison Chart

A comparison of customers' and restaurant's importance ratings are shown in Figure 5.106.

- Both the customers and the restaurant consider carryout (C) relatively unimportant. The restaurant could redesign this service to provide greater customer value or eliminate it and assign those resources to more valued services.

Performance Comparison Chart

A comparison of customers' and restaurant's performance ratings are shown in Figure 5.107.

- Both the customers and the restaurant rate performance high for breakfast (B), lunch (L), and dinner (D). By discussing these services the restaurant may learn how to improve private parties (P), where the restaurant rates its performance higher than the customers do.
- The customers' high evaluation of the bar is interesting, especially compared to the medium rating by the restaurant. Maybe the bar employees are missing some well-deserved positive feedback. The restaurant may be able to learn something about its customers' expectations by discussing bar and carryout service with

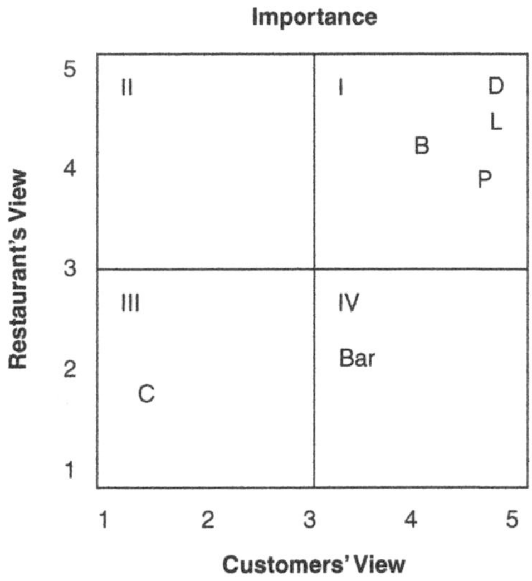

Figure 5.106 Importance–performance analysis example: importance comparison.

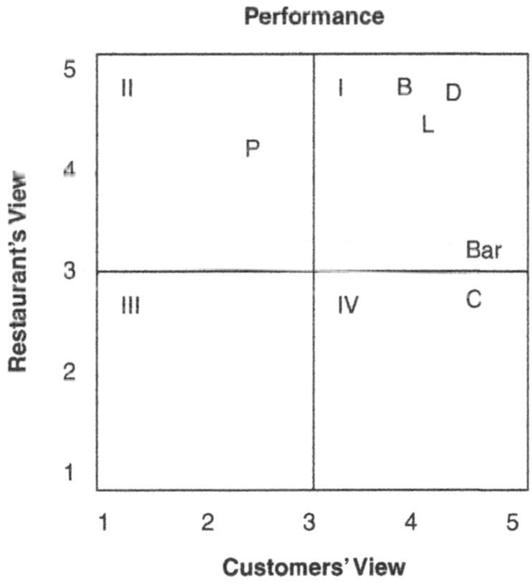

Figure 5.107 Importance–performance analysis example: performance comparison.

them. Customers think both are very well done while the restaurant thinks performance is just adequate.

- Overall, private parties (P) look like a great opportunity for improvement. Both customers and restaurant rate importance high, but in the eyes of the customers, the restaurant overrates its own performance.

Considerations

- The comparisons are made to learn from rather than to argue with the customer. The customer's perception is always "right." Our job is to learn why the customer feels that way and then to apply that knowledge toward improvement. Listen to the voice of the customer.
- Notice that there is no I–P chart for the suppliers' ratings. This is not an error. The suppliers' perceptions are important only in comparison to the customers'.
- Useful tools for identifying customers, products and services in step 1 include brainstorming, NGT, multivoting, requirements table, requirements-and-measures tree, and SIPOC.
- Numerical scales should be 1 to 5 or 1 to 7. Larger ranges make it harder for the respondent to differentiate between ratings; smaller ranges do not provide enough differentiation. Using an odd maximum number is desirable so there is a middle value that means "meets expectations."
- It might be useful to show data as ranges rather than points for averages, especially if the raw data shows great variation in responses to single questions. See *box plot* for ideas about showing ranges of data. Or, with small surveys, you could plot each respondent's answers individually.
- See *PGCV index* for a way to convert the importance and performance data into numerical summaries that can be statistically analyzed.
- See *surveys* for general information about planning, carrying out, and analyzing a customer survey.
- A perceptual map is a more general tool to study customer perceptions of any characteristics. See *two-dimensional charts* for more information.

is–is not matrix

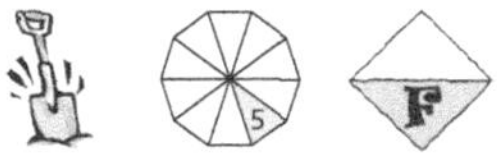

Description

The is–is not matrix guides the search for causes of a problem. By isolating who, what, when, where, and how about an event, it narrows investigation to factors that have an impact and eliminates factors that do not have an impact. By comparing what the problem *is* with what the problem *is not*, we can see what is distinctive about this problem, which leads to possible causes.

When to Use

- When looking for causes of a problem or situation

Procedure

1. Describe the event so that everyone clearly understands the problem. Describe it as a deviation from the way things should be. Write the problem statement in the upper left corner of the is–is not matrix (Figure 5.108).
2. Using the "Is" column of the matrix, describe what did or does occur.
   a. Determine *what* objects are affected and *what* exactly occurs. Be as specific as possible.
   b. Determine *where* the event occurs. This can be geographical (Houston), a physical location (the widget department or on machine 1), or on an object (at the bottom of the page or on the back of the widget). Or, it can be a combination (at the left side of each widget coming off line 2).

Problem Statement:	**Is** Describe what does occur	**Is Not** Describe what does not occur, though it reasonably might	**Distinctions** What stands out as odd?
What objects are affected? What occurs?			
Where does the problem occur? • Geographical • Physical • On an object			
When does the problem occur? When first? When since? How long? What patterns? Before, during, after other events?			
Extent of problem How many problems? How many objects or situations have problems? How serious is the problem?			
Who is involved? (Do not use this question to blame.) To whom, by whom, near whom does this occur?			

Figure 5.108 Is–is not matrix.

c. Determine *when* the event occurs. When did it happen first? When since? What patterns of occurrence have you noticed? Date, time, day, and season can all be important to the solution. Also, when the event occurs in relation to other events (before, after, during) can be significant.

d. Determine *how many or how much*—the *extent* of the problem. How many objects or occurrences had problems? How many problems? How serious are they?

e. Determine *who* is involved in the event. To whom, by whom, near whom does it occur? However, this analysis should never be used to assign blame—only to determine cause.

3. Use the "Is not" column of the matrix to identify circumstances that could occur but do not. Again use the what, where, when, how many or much (extent), and who questions.

4. Study the "Is" and "Is not" columns to identify what is different or unusual about the situations where the problem is compared to where it is not. What stands out as being odd? What changes have occurred? Write your observations in the column headed "Distinctions."

5. For each distinction, ask, "Does this distinction relate to a change we know about?" and "How could this distinction (or change) have caused our problem?" Write down all possible causes, including both what caused the problem and how it did so.

6. Test all possible causes by asking, "If this is the cause, does it explain every item in the "Is" and "Is not" columns?" The most likely cause must explain every aspect of the problem.

7. If possible, plan an experiment to verify the cause(s) you have identified. Depending on the cause, either try to duplicate the problem by "turning on" the cause, or try to stop the problem by reversing a change that caused it.

Example

In the ZZ-400 manufacturing unit, a subteam was trying to understand the source of iron in its product. (This example is part of the ZZ-400 improvement story in Chapter 4.) The problem, which occurred periodically, caused product purity to drop below specification. The team had learned from a stratified scatter diagram that the problem only occurred in reactors 2 and 3, not reactor 1. An is–is not matrix (Figure 5.109) brought together everything the team learned about the situation from a variety of other analysis tools, historical data, and its members' expert knowledge.

One of the mechanics noticed that the calendar tracking the problem looked just like the calendar for pump preventive maintenance (PM). So the team marked "Often

Problem Statement: Iron contamination in ZZ-400	**Is** Describe what does occur	**Is Not** Describe what does not occur, though it reasonably might	**Distinctions** What stands out as odd?
What objects are affected? What occurs?	Purity drops Iron increases	Other metal impurities Increase in moisture Change in color	Only iron increases
Where does the problem occur? • Geographical • Physical • On an object	Reactors 2 and 3	Reactor 1	Not reactor 1
When does the problem occur? When first? When since? How long? What patterns? Before, during, after other events?	Lasts about one shift About once a month Often during same week as pump PM When P-584C runs	Continuous Any particular shift Any other pump than P-584C	Just P-584C
Extent of problem How many problems? How many objects or situations have problems? How serious is the problem?	Purity drops from 99% range to 98% range Puts us out of spec Iron up to 0.65 (3 × normal)	Purity below 98.2% Pluggage problem Other specification problems	
Who is involved? (Do not use this question to blame.) To whom, by whom, near whom does this occur?	Mechanics doing pump PM		Mechanics usually around

Figure 5.109 Is–is not matrix example.

during same week as pump PM" and "Mechanic doing pump PM" on the matrix. Team members started investigating what happened during pump PM and what was different for reactor 1 pump PM. They discovered that the spare pumps were switched on whenever the main pump was being serviced. P-584C was used only for reactors 2 and 3; a different spare was used for reactor 1. Sure enough, the team could match incidents of high iron with times that P-584C had been run.

P-584C turned out to be made of carbon steel instead of the stainless steel necessary for that fluid. Apparently, the pump had been obtained from stores inventory during an emergency many years ago and was never replaced. When it was removed and a stainless steel pump installed, the incidents of high iron disappeared.

Considerations

- The problem statement must be concise and specific so that everyone can focus on the same issue. Be sure you are focusing on just one problem, rather than lumping together several things that happen in the same process. Also be sure you have described the problem at the level where it cannot be explained further. Instead of "copier won't work," say "paper jams in copier."

- The greater the level of detail in describing "Is" and "Is not," the greater the likelihood of determining the cause.
- Examples of specific descriptions of "Is": "The report begins normally then stops in loop three." "A grinding sound came from the engine, then it slowed to 20 mph."
- The "Is not" column requires thinking about what *could be* occurring but *is not.* For example: "Is not occurring at the Pittsburgh location." "Is not happening on weekends or at night." "Is not happening while daily reports print." "Is not continuous." "Is not happening with B team."
- If you cannot find a cause, either get more information for the "Is" and "Is not" columns, or work through the matrix again more carefully. Consider involving more people that have knowledge of the problem.

list reduction

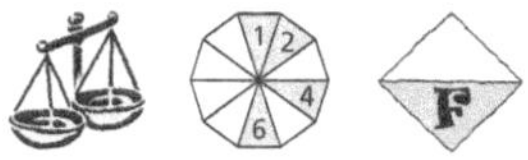

Description

List reduction is a set of techniques that are used to reduce a brainstormed list of options (such as problems, solutions, measures) to a manageable number.

When to Use

- After brainstorming or some other expansion tool has been used to generate a long list of options, and . . .
- When the list must be narrowed down, and . . .
- When the list of options may have duplicate or irrelevant ideas, and . . .
- When the group members together should think through the reasons for eliminating choices to reach consensus

Procedure

Materials needed: flipchart or whiteboard, marking pen

Wide Filter

1. Post the entire list of brainstormed ideas so that everyone can see all items.
2. For each item, ask the question, "Should this item continue to be considered?" Get a vote of yeses and nos. A simple majority of yes responses keeps the item on the list. If an item does not get a majority of yes votes, mark it with brackets.

3. After all items have been evaluated by the wide filter, ask the team members, "Does anyone want to put any of the bracketed items back on the list?" Any items that are mentioned by even one team member are left on the list. Remaining bracketed items are crossed off.

Combining Ideas

4. Label the first idea on the list number 1. Look at the second idea. Ask, "Does anyone think this is a different idea from number 1?" If one person thinks the second idea is different, label the second idea number 2. If all agree that the two items really are the same, eliminate one or develop new wording to combine the two ideas.
5. Similarly, compare item number 1 to all other items on the list, one at a time.
6. Take item number 2 and compare it to each item below it on the list. Continue to work down the list until all the ideas have been compared pairwise. Every idea should be either numbered or eliminated.

Criteria Filtering

Conclude with criteria filtering. For the procedure, see this tool's separate entry.

Example

A committee was formed by a local school board to identify opportunities for school–community partnerships. Initially, they had brainstormed answers to the question, "If there were no constraints, how would the 'ideal' partnership between school, community, and parents look in the future?" The list of 39 ideas had to be reduced to a manageable number of real possibilities for the next three years.

As the committee went through their list with a wide filter, they realized that many of the items were results they hoped to see. Those were bracketed, but labeled so the ideas could be saved as possible measurements of success. They also bracketed "pool" and "library," which were goals too big to start with.

When the facilitator asked, "Does anyone want to put any of the bracketed items back on the list?" members expressed concern that two "results" had no related initiative listed. So they put back reworded ideas, italicized. Someone else believed a community library of some sort might be achievable. The other bracketed ideas were crossed off.

Next they worked on combining ideas. Numbers 1, 2, and 3 seemed unique. Number 4, "volunteer coordinator," was combined with "volunteers-in-schools program." Many ideas on the list were about communication. By doing comparisons one at a time, according to the procedure, the group reduced six communication items to three.

Several items on the list were features of other items. For example, "lots of communication to community" and "wide open pipeline of information flow" could be part of any communication plan, Web site, or newsletter. That principle helped them combine other ideas. But when they discussed “transportation,” although it was a feature of several other ideas, several people thought the issue was so critical and difficult that it needed to be treated separately. It was left on the list.

After methodically combining ideas, the group’s flipchart looked like Figure 5.110. Thirty-eight ideas had been reduced to 18. That was still too many items to tackle, so the group moved into criteria filtering. They brainstormed and ranked a list of criteria for choosing three-year projects, shown in Figure 5.111, with 1 most important.

They made an L-shaped matrix (Figure 5.112) with criteria as the column headings and their remaining 18 options as row labels. Grants might be available for certain programs, so they wrote “Grant?” under the cost criterion. The library proponent proposed several feasible options, but cost was unknown without research. When they looked at

[~~eliminate drop-out problem~~] result
1 alternative education opportunities
[~~high college entry ratio~~] result
[~~more career opportunities here~~] result
[~~every business participating~~] result
2 *recruit business participation*
3 “service learning” projects for students
~~volunteer coordinator~~
[~~lots of volunteers in schools~~] result
4 *volunteers-in-schools program—with coordinator*
5 annual community surveys
~~newsletters~~
~~lots of communication to community~~
~~wide open pipeline of information flow~~
~~e-mails~~
6 *communication plan*
7 school–community Web site
8 community center equipped and functional
[~~lots of people using community center~~] result
9 educational opportunities for parents *and community with child care provided*
[~~lots of parents using these opportunities~~] result
[~~reduced drug problems~~] result
[~~reduced alcohol problems~~] result
[~~reduced child abuse problems~~] result
[~~reduced teen sex problems~~] result
[~~kids reading on grade level~~] result
10 [community library]
11 active preschool program
~~data collection for accountability~~
[~~children enter school with pre-literacy skills~~] result
12 community recreation district
13 after-school kids’ programs
14 transportation
15 mentoring program
16 tutoring program
~~child care for parents attending training~~
~~communitywide learning opportunities~~
17 community computer lab
~~place for kids to hang out safely~~
[~~swimming pool~~]
18 community newsletter publicizing needs, programs, honoring heroes, etc.—*frequent publication*

Figure 5.110 List reduction example.

Cost within our budget or grant available – 1
Community will support enthusiastically – 4
Builds community participation with schools – 2
Can have noticeable impact on educational results – 3

Figure 5.111 List reduction example: criteria.

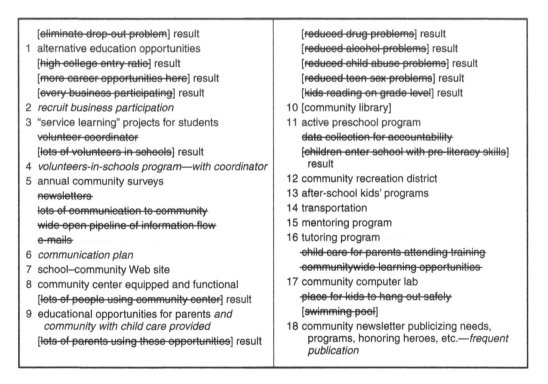

	Cost Feasible?	Builds Participation?	Education Impact?	Community Enthusiasm?
1. Alternative education opportunities	Yes	No	Yes	No
2. Recruit business participation	Yes	Yes	No	No
3. "Service learning" projects *	Yes	Yes	Yes	No
4. Volunteers-in-schools program *	Yes	Yes	Yes	No
5. Annual community surveys	Yes	No	No	No
6. Communication plan	Yes	Yes	No	No
7. School–community Web site *	Yes	Yes	If kids do it!	No
8. Community center equipped and functional	No	Yes	No	Yes
9. Educational opportunities for parents and community *	Grant?	Yes	Yes	Yes
10. Community library *	???	Yes	Yes	Yes
11. Preschool program	No	Yes	Yes	No
12. Community recreation district	No	Yes	No	No
13. After-school kids' programs *	Grant?	Yes	Yes	Yes
14. Transportation *	Grant?	Yes	When enables other programs	Yes
15. Mentoring program *	Yes	Yes	Yes	No
16. Tutoring program *	Yes	Yes	Yes	No
17. Community computer lab	No	Yes	No	No
18. Community newsletter *	Yes	Yes	If kids do it!	No

Figure 5.112 List reduction example: criteria filtering.

the "enthusiasm" criterion, they realized that most of the items would be a hard sell in their community, which was why their project was needed in the first place.

Each option was evaluated against all four criteria. The ten starred items had each received three or four yeses and became the working list of possible projects. These items were assigned to group members for further research and analysis.

Read the ZZ-400 story in Chapter 4 for another example of list reduction.

Considerations

- Any of the three sections can be used separately. However, because combining ideas and criteria filtering are more complex, they are easiest to use after the list has been reduced at least once by a wide filter.

- During wide filter and combining ideas, the opinion of one person is enough to keep an item on the list. It is valuable to listen to that person's reasons for keeping the item alive.
- In the third part of list reduction, use criteria filtering only to generate a reduced list or to select one item if the decision is not critical. Paired comparisons could also be used if the decision should be based on individual preferences rather than on criteria. To make final decisions about major issues, such as which project to select or which solution to implement, use a decision matrix or prioritization matrix, which evaluate options more rigorously.
- Another tool to narrow long lists is multivoting. The effective–achievable matrix is a tool that separates ease-of-accomplishment criteria from effectiveness criteria and visually compares how well various options meet criteria.

matrix diagram

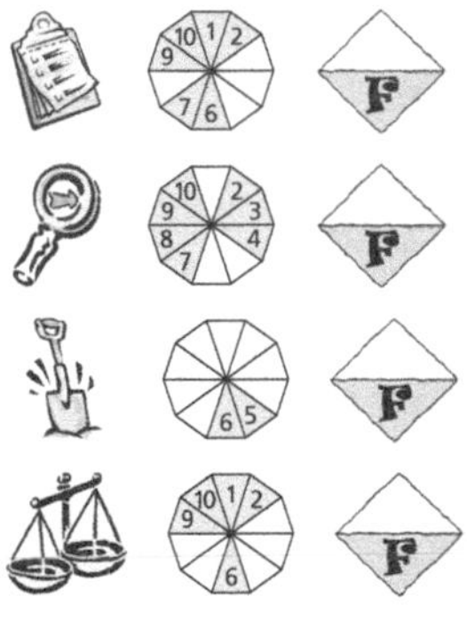

Also called: matrix, matrix chart

Description

The matrix diagram shows the relationship between two, three, or four groups of information. It also can give information about the relationship, such as its strength, the roles played by various individuals, or measurements. Six differently shaped matrices are possible: L-, T-, Y-, X-, C-, and roof-shaped, depending how many groups must be compared. This is a generic tool that can be adapted for a wide variety of purposes.

When to Use

- When trying to understand how groups of items relate to one another, or . . .
- When communicating to others how groups of items relate to one another

Common Uses of Matrices

- When distributing responsibilities for tasks among a group of people (sometimes called a responsibility or accountability matrix)
- When linking customer requirements to elements of a process (sometimes called a critical-to-quality or CTQ matrix)
- When sorting out which problems are affecting which products or which pieces of equipment
- When looking for cause-and-effect relationships
- When looking for reinforcement or conflicts between two plans that will be executed together

When to Use Each Shape

Table 5.11 summarizes when each type of matrix is used.

- An L-shaped matrix relates two groups of items to each other (or one group to itself).
- A T-shaped matrix relates three groups of items: groups B and C are each related to A. Groups B and C are not related to each other.
- A Y-shaped matrix relates three groups of items. Each group is related to the other two in a circular fashion.
- A C-shaped matrix relates three groups of items all together simultaneously, in 3-D.
- An X-shaped matrix relates four groups of items. Each group is related to two others in a circular fashion.
- A roof-shaped matrix relates one group of items to itself. It is usually used along with an L- or T-shaped matrix.

Table 5.11 When to use differently-shaped matrices.

L-shaped	2 groups	A ↔ B (or A ↔ A)
T-shaped	3 groups	B ↔ A ↔ C but not B ↔ C
Y-shaped	3 groups	A ↔ B ↔ C ↔ A
C-shaped	3 groups	All three simultaneously (3–D)
X-shaped	4 groups	A ↔ B ↔ C ↔ D ↔ A but not A ↔ C or B ↔ D
Roof-shaped	1 group	A ↔ A when also A ↔ B in L or T

Basic Procedure

1. Decide what groups of items must be compared.
2. Choose the appropriate format for the matrix.
3. Draw the lines forming the grid of the matrix.
4. List the items in each group as row labels and column headings.
5. Decide what information you want to show with the symbols on the matrix. See "Considerations" for commonly used symbols. Create a legend describing the symbols.
6. Compare groups, item by item. For each comparison, mark the appropriate symbol in the box at the intersection of the paired items' column and row.
7. Analyze the matrix for patterns. You may wish to repeat the procedure with a different format or a different set of symbols to learn more about the relationships.

Example: L-Shaped Matrix

Figure 5.113 is an L-shaped matrix summarizing customers' requirements. The team placed numbers in the boxes to show numerical specifications and used check marks to show choice of packaging. In this and the following examples, the axes have been shaded to emphasize the letter that gives the matrix its name. The L-shaped matrix actually forms an upside-down L. This is the most basic and most common matrix format. This example is part of the ZZ-400 improvement story on page 80. Also see the St. Luke's Hospital story on page 70 and the Medrad story on page 58 for other examples of L-shaped matrices.

Customer Requirements

	Customer *D*	Customer *M*	Customer *R*	Customer *T*
Purity %	> 99.2	> 99.2	> 99.4	> 99.0
Trace metals (ppm)	< 5	—	< 10	< 25
Water (ppm)	< 10	< 5	< 10	—
Viscosity (cp)	20–35	20–30	10–50	15–35
Color	< 10	< 10	< 15	< 10
Drum		✔		
Truck	✔			✔
Railcar			✔	

Figure 5.113 L-shaped matrix example.

Example: T-Shaped Matrix

Figure 5.114 is an example of a T-shaped matrix relating four product models (group A) to their manufacturing locations (group B) and to their customers (group C).

The matrix can be examined in different ways to focus on different information. For example, concentrating on model A, it is produced in large volume at the Texas plant and in small volume at the Alabama plant. Time Inc. is the major customer for model A, while Arlo Co. buys a small amount. If we choose to focus on the customer rows, we learn that only one customer, Arlo, buys all four models. Zig buys just one. Time makes large purchases of A and D, while Lyle is a relatively minor customer.

The ZZ-400 story has on page 88 an example of a T-shaped matrix used within the improvement process.

Example: Y-Shaped Matrix

Figure 5.115 is a Y-shaped matrix showing the relationships between customer requirements, internal process metrics, and the departments involved. Symbols show the strength of the relationships: primary relationships, such as the manufacturing department's responsibility for production capacity; secondary relationships, such as the link between product availability and inventory levels; minor relationships, such as the distribution department's responsibility for order lead time; and no relationship, such as between the purchasing department and on-time delivery.

The matrix tells an interesting story about on-time delivery. The distribution department is assigned primary responsibility for that customer requirement. The two metrics most strongly related to on-time delivery are inventory levels and order lead time. Of the two, distribution has only a weak relationship with order lead time and none with inventory levels. Perhaps the responsibility for on-time delivery needs to be reconsidered. Based on the matrix, where would you put responsibility for on-time delivery?

Products—Customers—Manufacturing Locations

Texas plant	●		○	○
Mississippi plant		●		○
Alabama plant	○			●
Arkansas plant		○	●	
● Large volume ○ Small volume	Model A	Model B	Model C	Model D
Zig Corp.		●		
Arlo Co.	○	○	○	●
Lyle Co.			○	○
Time Inc.	●			●

Figure 5.114 T-shaped matrix example.

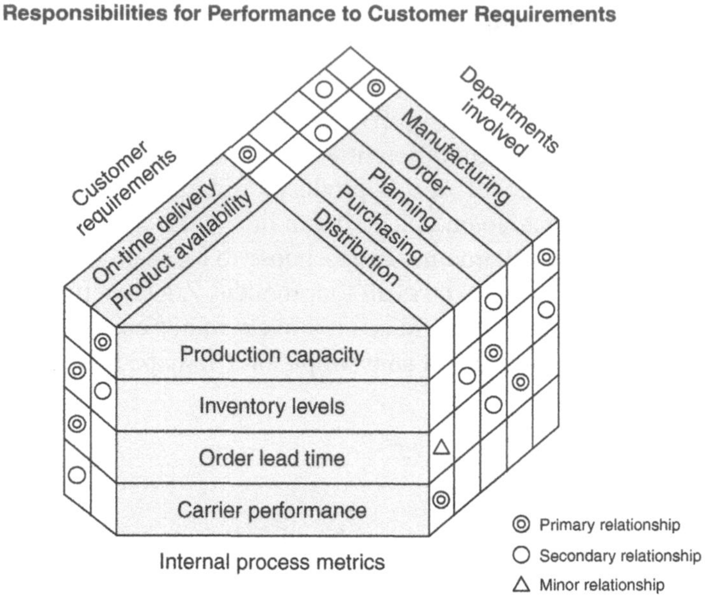

Figure 5.115 Y-shaped matrix example.

Example: C-Shaped Matrix

Think of C meaning *cube*. Because this matrix is three-dimensional, it is difficult to draw and infrequently used. If it is important to compare three groups simultaneously, consider using a three-dimensional model or computer software that can provide a clear visual image. Figure 5.116 shows one point on a C-shaped matrix relating products, customers, and manufacturing locations. Zig Company's model B is made at the Mississippi plant.

Example: X-Shaped Matrix

Figure 5.117 extends the T-shaped matrix example into an X-shaped matrix by including the relationships of freight lines with the manufacturing sites they serve and the customers who use them. Each axis of the matrix is related to the two adjacent ones, but not to the one across. Thus, the product models are related to the plant sites and to the customers, but not to the freight lines.

A lot of information can be contained in an X-shaped matrix. In this one, we can observe that Red Lines and Zip Inc., which seem to be minor carriers based on volume, are the only carriers that serve Lyle Co. Lyle doesn't buy much, but it and Arlo are the only customers for model C. Model D is made at three locations, while the other models are made at two. What other observations can you make?

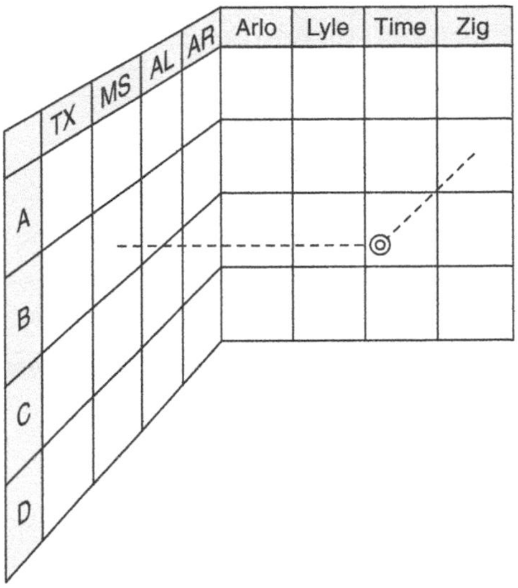

Figure 5.116 C-shaped matrix example.

Manufacturing Sites—Products—Customers—Freight Lines

○		●	○	Texas plant	●		○	○
	○	●	●	Mississippi plant		●		○
		●	●	Alabama plant	○			●
○	○		○	Arkansas plant		○	●	
Red Lines	Zip Inc.	World-wide	Trans South		Model A	Model B	Model C	Model D
		●	○	Zig Corp.		●		
			●	Arlo Co.	○	○	○	●
○	○			Lyle Co.			○	○
	○	●		Time Inc.	●			●

● Large volume
○ Small volume

Figure 5.117 X-shaped matrix example.

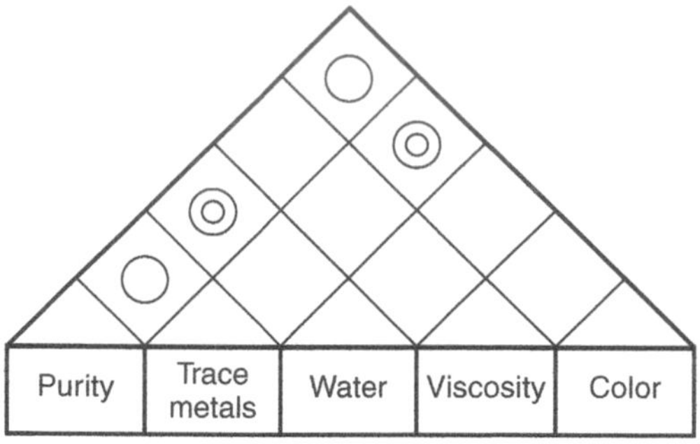

Figure 5.118 Roof-shaped matrix example.

Example: Roof-Shaped Matrix

The roof-shaped matrix is used with an L- or T-shaped matrix to show one group of items relating to itself. It is most commonly used with a house of quality, where it forms the "roof" of the "house." In Figure 5.118, the customer requirements are related to one another. For example, a strong relationship links color and trace metals, while viscosity is unrelated to any of the other requirements.

Considerations

- Sometimes relationship is indicated with just an X or a blank, to indicate yes or no: involved or not, applicable or not, relationship exists or not. But the matrix is more useful with symbols that give more information about the relationship between the items. Two kinds of information are most common.
  - How strong the relationship is between the two items
  - If the items are a person and an activity, the role the person plays in that activity
- Here are some frequently used sets of symbols.

◎ Strong relationship ○ Moderate relationship △ Weak or potential relationship No relationship	+ Positive relationship ○ Neutral relationship – Negative relationship
S Supplier C Customer D Doer O Owner	↑ Item on left influences item at top ← Item at top influences item on left The arrows usually are placed next to another symbol indicating the strength of the relationship.

Create your own symbols or letters that are instantly recognizable to you or your audience.

- The L-shaped matrix is the most common, but it might not be the most useful for your application. Think about the groups of data you have and their possible relationships, review the other formats, then consider whether a different matrix might give more insight or show your point more clearly.
- Many tools in this book are matrices that are designed for a specific purpose. For example, the decision matrix is designed to relate potential choices to a list of criteria. For these specialized matrices, steps 1, 2, and sometimes 5 of this procedure are already done for you.
- Some tools that have "matrix" in their names are really two-dimensional charts. See *two-dimensional matrix* for more information about this generic tool.

meeting evaluation

Description

The meeting evaluation helps a team work better together by identifying areas where the team could be more effective or efficient.

When to Use

- When a group is meeting together regularly to work on a project, and . . .
- At the end of each meeting, or . . .
- At a midpoint of each meeting, or . . .
- Every several meetings

Procedure

1. Decide on a set of evaluation questions. This can be done by the team, the facilitator, or by a department or organization for use by all teams. Reproduce blank forms.
2. As a team, decide whether you want to evaluate each meeting (recommended for a new team) or do periodic evaluations.
3. Decide whether you want to do the evaluations at the end of the meeting or earlier in the agenda.

4. Every attendee fills out the form.
5. Briefly discuss the answers. Focus on areas that are rated lower and on significant differences between attendees' ratings. Decide on appropriate changes or actions.
6. The facilitator should collect and summarize the responses.

Example

Figure 5.119 shows a typical meeting evaluation form.

Considerations

Evaluation Methods

- Evaluations are usually done at the end of meetings. However, people are tired and eager to leave by then, so taking a break mid-meeting and evaluating how you are doing sometimes works better.
- New teams or teams that are having trouble making progress should evaluate every meeting. Teams that have been in place for a while may choose to do evaluations only every few meetings.

Team:		Date:			
	Disagree				Agree
1. We managed time well.	1	2	3	4	5
2. We communicated clearly and respectfully.	1	2	3	4	5
3. Everyone participated.	1	2	3	4	5
4. We stayed on track.	1	2	3	4	5
5. The right people were at this meeting.	1	2	3	4	5
6. We used appropriate tools and methods.	1	2	3	4	5
7. We accomplished this meeting's goals.	1	2	3	4	5
8. This meeting was worth my time.	1	2	3	4	5
This meeting would have been better if:					

Figure 5.119 Meeting evaluation example.

- Some teams find it useful to focus on trouble spots. Choose an aspect of team performance that the team would like to improve, such as staying on topic during discussions. For several meetings, evaluate only that aspect. When performance improves, choose another aspect to address.
- Alternatively, different evaluation forms could be used at different meetings to probe details of communication, time management, teamwork, project management, using quality principles, and so on.

Evaluation Form

- There are as many evaluation formats and sets of questions as there are teams. Decide what the group is comfortable with and finds useful.
- Questions should address two different areas of performance: team effectiveness and team efficiency. Team effectiveness is about *what* you're doing, such as identifying significant issues, applying quality improvement principles, and accomplishing goals. Team efficiency is about *how* you're doing that, such as communicating well, spending time on the right matters, sharing work, and using appropriate tools.
- Use mainly questions that can be answered on a numerical scale so the evaluation can be quickly filled out by circling numbers. Scales of 1 to 5 or 1 to 6 work best. Yes–no responses are less desirable because they don't allow for shades of gray.
- Write each question so that the same end of the scale is always preferable. For example, do not combine "The meeting started on time" with "We spent too much time off topic." Instead, write "We stayed on topic" so that for both questions, "agree" is the preferred response.
- Another way to write items is with paired descriptors and a numerical scale between them. For example:

Stayed on topic	5	4	3	2	1	Went off topic
Managed time well	5	4	3	2	1	Managed time poorly

- Following are typical subjects included on evaluations. Use the ones that your team finds most helpful. Different items can be written for each subject. For example, "good communication" can mean a variety of things such as "We listened well to each other's opinions" or "We were open in expressing ourselves."

  Efficiency

  Good time management

  Good meeting pace

Having clear purpose(s)

Having and following agenda

Full participation of all attendees

Staying on topic

Good communication between attendees

Courtesy and respect between attendees

Using appropriate quality tools

Team members working between meetings

Staying on schedule with project plan

Effectiveness

Working on the right things

Accomplishing assigned work

Focusing on work processes and system issues

Looking for root causes

Using data

Involving the right people

- Keep your evaluation form short. One way to do this is to use one broad question for a subject rather than several detailed ones. Then, if the team rates itself low on that subject, you can use discussion or later evaluations to probe for the source of the problem. For instance, the example starts with, "We managed time well." If the team rates time management low, an evaluation with more detailed items could be used to study the problem, such as: "We started on time." "We ended on time." "Our discussions stayed on topic." "We spent the majority of our time on the most important issues."
- A few open-ended questions can be useful. Examples:

  What made this meeting more useful?

  What suggestions do you have for the facilitator or meeting leader?
- See *survey* for more information about writing questions and rating scales.
- The wide subject of team behavior and performance is critical to quality improvement teams but is beyond the scope of this book. Be sure your team gets training and guidance in how teams form and interact.

Mind Map®

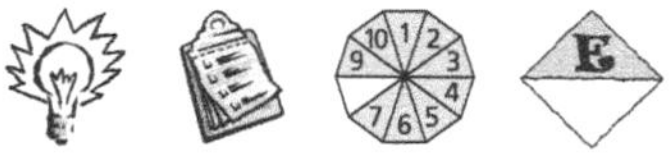

Also called: mind flow, brain web

Description

A Mind Map* is a graphic representation of ideas. A main subject is shown in the center of the page. Main topics radiate from it like spokes of a wheel, and related ideas branch from those. The graphic uses images and symbols, lines and arrows, various text fonts and sizes, color, and humor. Think of it as a multidimensional outline, instead of a traditional linear one. Or a brainstorm with form, instead of a page of random ideas.

The viewer can see key ideas and links between ideas at a glance, then look closer at specific sections for details. Also, ideas are expressed in a way that matches the way the brain works, making understanding and remembering easier. This tool is useful whenever clear thinking, learning, or memory are required.

When to Use

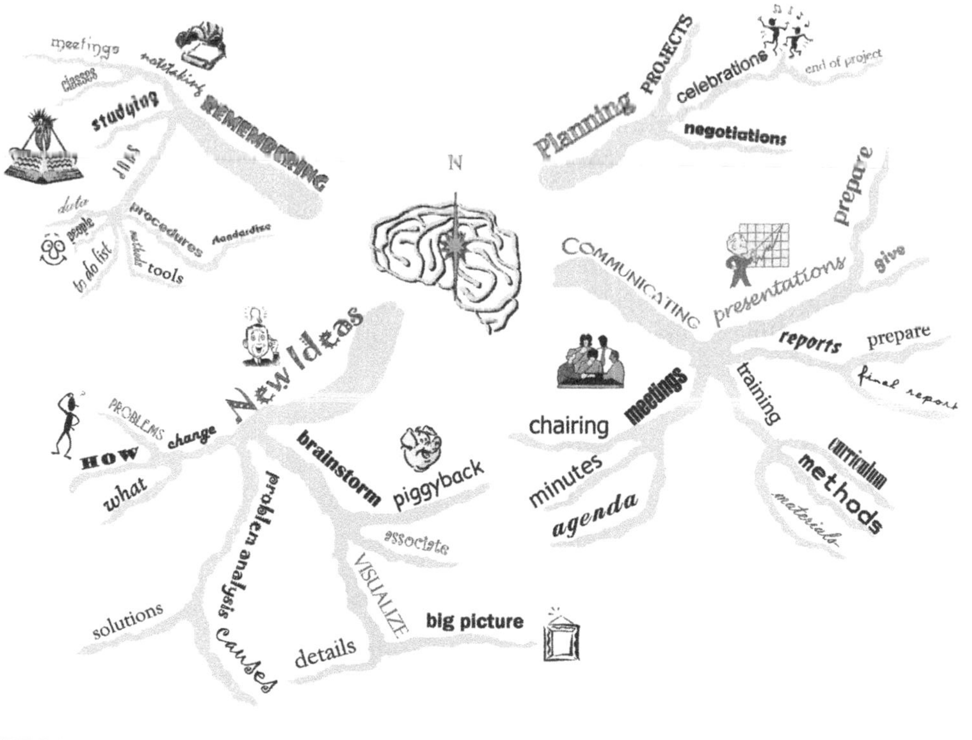

Figure 5.120 When to use a Mind Map.

* Mind Map is a registered trademark of the Buzan Organisation used with enthusiastic permission.

Procedure

1. Assemble a selection of pens or markers of various colors and thicknesses. Turn a sheet of flipchart paper to landscape orientation or use a dry-erase board.
2. Clarify the subject of the Mind Map. Draw an unframed image of the subject in the center of the work surface. Use three colors.
3. Add thicker and organic (wavy) branches radiating from the central image. Label each branch with a main topic closely associated with the central image. Print them in upper or lower case. Add images if you can.
4. Continue to draw branches radiating from the ends of the larger branches to show other associated ideas and details. Arrange idea branches in the work space to achieve clarity, association, and emphasis.
5. Use freely: images, symbols, colors, thicknesses, writing styles, lines, and arrows.
6. When the Mind Map is complete, make a second pass to reorganize, edit, and enhance the map.
7. If the Mind Map was created on a dry-erase board, capture it on paper.

Example

Figure 5.120, showing when to use this tool, and Figure 5.121, showing how to use it, are examples of Mind Maps used for communicating information. The original drawings were drawn in various colors to create emphasis and show relationships. For example, each of the four main branches of the "When to Use" Mind Map were different colors, although the green of Planning was repeated on the "prepare" branches of "Communicating." The color could not be reproduced in this book.

Notice that the key words showing broad categories of ideas are immediately grasped, and viewers automatically know where to look for additional details that they want. A linear outline of the same information would not be so easily accessible, nor would the relationships between ideas be so clear.

Considerations

- Put each key word and/or image on its own line. Make the line only as long as the word.
- Connect the lines as they radiate out from the central image. Make the lines thicker at the center, thinner as they approach the edges. Make the lines wavy, organic, like true branches, not artificially straight.
- Use colors freely. Write words in color, or outline, underline, or circle them. Devise your own color code, such as red for action, blue for ideal state.
- Create emphasis on more important ideas using color, pen thickness, different fonts, images, or symbols.

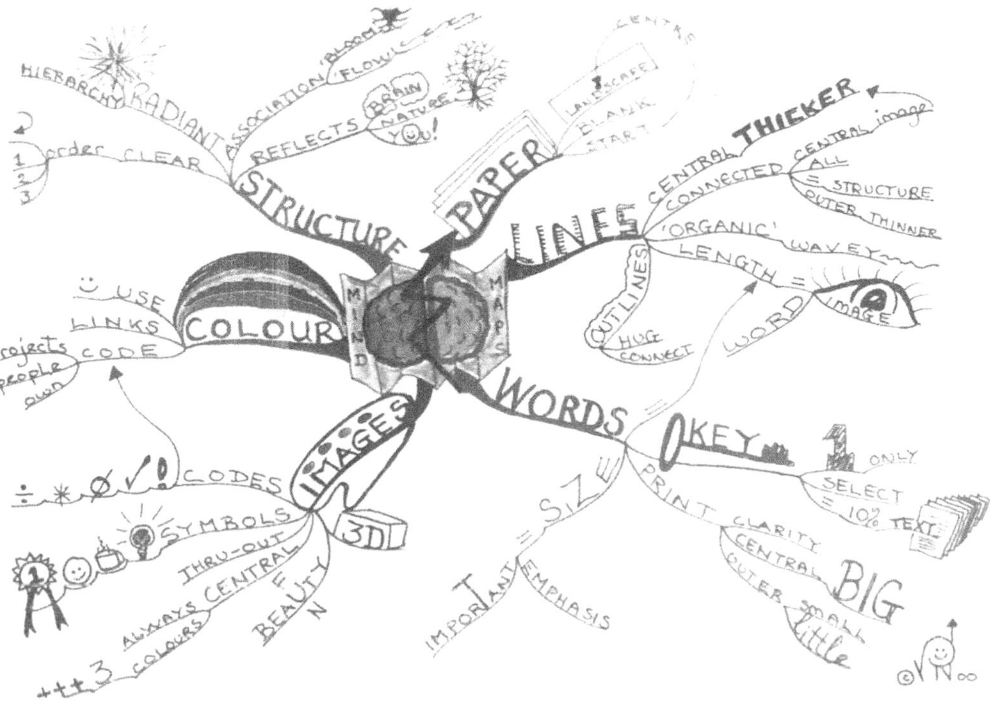

Figure 5.121 Mind Map example.
Source: Buzan Centres Ltd. Reproduced with permission.

- Show associations between ideas using lines, arrows, color codes, and similar fonts.
- The more of these suggestions you use, the better you will tap into the creative processes of your brain.
- Be creative! Find ways of creating Mind Maps that work best for you and the way your brain thinks.

mistake-proofing

Also called: poka-yoke, fail-safing

Description

Mistake-proofing, or its Japanese equivalent poka-yoke (pronounced to rhyme with "mocha okay"), is the use of any automatic device or method that either makes it impossible for an error to occur or makes the error immediately obvious once it has occurred.

When to Use

- When a process step has been identified where human error can cause mistakes or defects to occur, especially:
- In processes that rely on the worker's attention, skill, or experience, or . . .
- In a service process, where the customer can make an error which affects the output, or . . .
- At a hand-off step in a process, when output or (for service processes) the customer is transferred to another worker, or . . .
- When a minor error early in the process causes major problems later in the process, or . . .
- When the consequences of an error are expensive or dangerous

Procedure

1. Obtain or create a flowchart of the process. Review each step, thinking about where and when human errors are likely to occur.
2. For each potential error, work back through the process to find its source.
3. For each error, think of potential ways to make it impossible for the error to occur. Consider *elimination*—eliminating the step that causes the error; *replacement*—replacing the step with an error-proof one; or *facilitation*—making the correct action far easier than the error.
4. If you cannot make it impossible for the error to occur, think of ways to detect the error and minimize its effects. Consider inspection method, setting function, and regulatory function. (See "Considerations" for more information.)
5. Choose the best mistake-proofing method or device for each error. Test, then implement it.

Example

The Parisian Experience Restaurant wished to ensure high service quality through mistake-proofing. They reviewed the deployment chart of the seating process shown in Figure 5.122 and the critical-to-quality analysis shown in Figure 5.45 (page 209). They identified human errors on the part of restaurant staff or customers that could cause service problems.

The first potential error occurs when customers enter. The maitre d' might not notice that a customer was waiting if the maitre d' were escorting other customers to their table, checking on table status, or conferring with kitchen staff. The mistake-proofing device was an electronic sensor on the entrance door sending a signal to a vibrating

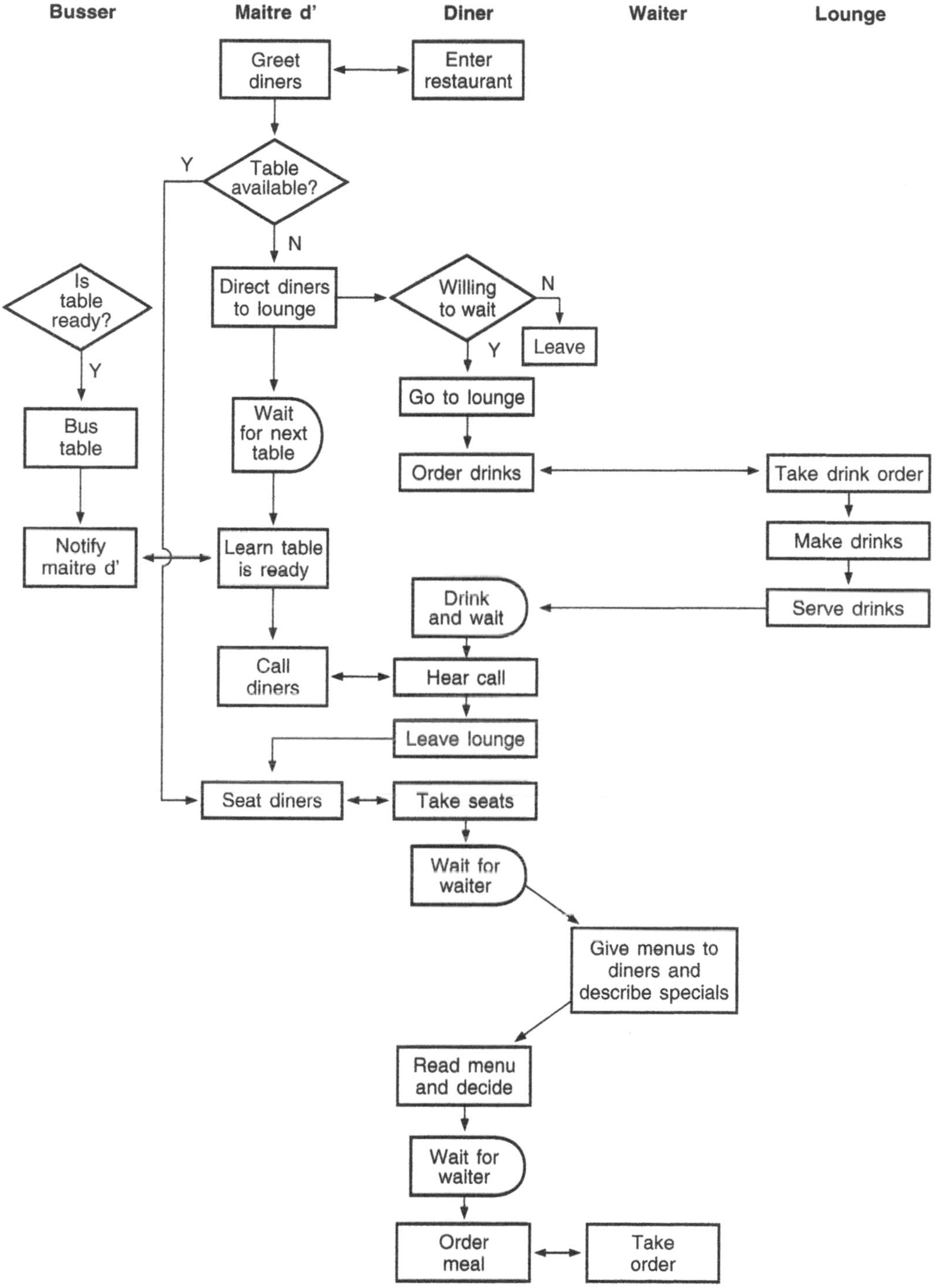

Figure 5.122 Restaurant's deployment chart.

pager on the maitre d's belt, to ensure that the maitre d' would always know when someone entered or left the restaurant. Other mistake-proofing methods replaced the process steps requiring the maitre d' to leave the front door to seat customers.

A possible error on the customers' part was identified at the step when diners are called from the lounge when their table is ready. They might miss the call if the lounge is noisy, if they are engrossed in conversation, or if they are hard-of-hearing. The mistake-proofing chosen by the team was to replace the step of the process in which the maitre d' called the customer's name over the loudspeaker. Instead, during the greeting step, the maitre d' notes a unique visual identifier of one or more members of the party. When the table is ready, the busser notifies the waiter, who comes to the maitre d' and learns how to identify the customers. The waiter finds the customers in the lounge, escorts them to their table, gives them menus, and takes additional drink orders. Not only does this mistake-proofing method eliminate a customer-caused problem, it improves the restaurant ambiance by eliminating the annoying loudspeaker, keeps the maitre d' at the front door to greet customers, creates a sense of exceptional service when the waiter magically knows the customers, and eliminates additional waiting time at the handoff between maitre d' and waiter.

This example shows mistake-proofing used in a service process, because service applications are less common and less obvious than those for production. Other examples are provided below to clarify the concepts discussed in "Considerations." Also see the Medrad story on page 62 and the St. Luke's Hospital story on page 70 for other examples of this tool.

Considerations

- The basic classes of mistake-proofing devices are limit switches, counters, guide pins, alarms, and checklists. See *checklist* for more information on that tool. For the others, consult a book on mistake-proofing. Hundreds of mistake-proofing devices have been developed for use on assembly lines and other manufacturing processes.
- Although developed in production industries, mistake-proofing is equally applicable to service processes. For example, mistake-proofing has been creatively applied in healthcare, where human error can be fatal.
- When reviewing the process for potential errors, look for handoffs, transfer of materials and information, or data entry. Look for steps where identity matching is critical or conditions must be just right. In service processes, look for errors in the task, in the treatment of customers, or in physical aspects of the service environment. Customers also can cause errors by failing to do things right on their end: bring appropriate materials, provide the right information, specify their needs, follow instructions, provide feedback, or follow up after the transaction.
- A key principle of mistake-proofing is minimizing the time between error and feedback. If an error is made at the beginning of the process but not detected until

the end, not only is the output defective, but also the error might have been repeated many times, or it might be too late to find the reason for the error. With mistake-proofing, the error is caught immediately and both its consequences and its cause can be corrected immediately. Shigeo Shingo said, "Errors will not turn into defects if feedback and action take place at the error stage," and, "Defects are errors passed downstream."

- Mistake-proofing methods and devices are categorized in three dimensions: inspection methods, setting functions, and regulatory functions.

Inspection Methods

- Three kinds of inspection are used to provide rapid feedback. They are classified based on who does the operation and where in the process the inspection takes place relative to the possible error.
- Successive inspection is done at the next step of the process by the next worker, as when the waiter checks the food on the tray before delivering it to the table.
- Self-inspection means a worker checks his own work immediately after doing it. An example is a surgical tray specific to the type of operation, with indentions for each instrument, which the surgeon checks to make sure all instruments were removed from the patient. Self-inspection is preferable to successive inspection, although both are after-the-fact.
- Source inspection checks before the process step takes place that conditions are correct. Often it is automatic and keeps the process from proceeding until conditions are right, such as an interlock that will not allow an oven door to be opened after self-cleaning until temperature drops to a safe level. Source inspection is preferable to the other kinds, because it keeps errors from occurring.
- Note that these types of 100% inspection, called "informative inspection," are very different from "judgment inspection," in which bad product is sorted from good at the end of the process. Judgment inspection is not an effective approach for managing or improving quality.

Setting Functions

- Setting functions are the methods by which a process parameter or product attribute is inspected for errors.
- The *contact* or *physical* method checks a physical characteristic such as diameter or temperature, often using a sensor. For example, a computer cable's pin pattern ensures it can only be plugged in the right way into the right location. The temperature interlock on self-cleaning ovens is another example.
- The *motion-step* or *sequencing* method checks the process sequence to make sure steps are done in order. Power saws will not operate unless a brake release button is pushed concurrently with the trigger.

- The *fixed-value* or *grouping and counting* method counts repetitions or parts or weighs to ensure completeness. The surgical tray discussed above is an example, as well as weighing a complex order before shipping to ensure all parts are enclosed.
- A fourth setting function is sometimes added: *information enhancement*. This makes sure information is available and perceivable when and where required. An example is giving car repair customers pagers so they can be notified when their car is ready.

Regulatory Functions

- Regulatory functions are signals that alert the workers that an error has occurred.
- *Warning* functions are bells, buzzers, lights and other sensory signals. Consider using color-coding, shapes, symbols, and distinctive sounds. Warning functions require a person to take action to prevent or remedy the error.
- *Control* functions prevent the process from proceeding until the error is corrected (if the error has already taken place) or conditions are correct (if the inspection was a source inspection and the error has not yet occurred). Control functions are more restrictive and more powerful than warning functions. Self-checkout stations in stores will not allow a second item to be scanned until the customer places the first item in the bag, which rests on a sensor. The stations also issue a warning and prevent further action if an item is placed in the bag without being scanned.

multi-vari chart

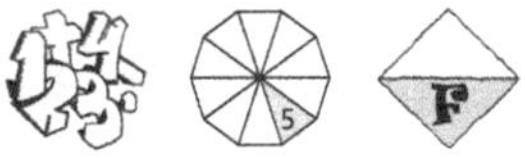

Description

The multi-vari chart shows which of several sources of variation are the greatest contributors to total variation and may reveal other patterns in the variation.

When to Use

- When studying sources of variation in a process, and . . .
- When you wish to identify the most important sources of variation, and . . .
- When the output characteristic is a variable measurement

Procedure

1. Identify the possible sources of variation you wish to study. Create a sampling tree showing the combinations of settings for sources A, B, and C.
2. Create a graph with the y-axis representing the output characteristic. Divide the x-axis into a section for each setting of the A source, the top level of the sampling tree. Divide each of those sections into subsections for each setting of the B source. Points plotted in a vertical line show the measurements at different settings of source C.
3. Calculate the mean of the values in the first line of points. Mark the value along the line with a symbol different from the points. Repeat for each B group in the first section of the chart.
4. Connect the means of the B groups within the first section of the chart.
5. Calculate the mean of all values in the first section. (This will be the average of the means calculated in step 3.) Mark the value with a third symbol at the centerline of the section.
6. Repeat steps 3, 4, and 5 for each section of the chart representing each setting of source A.
7. Connect the symbols for the overall means for each section.
8. Analyze the chart to identify the greatest source of variation and any patterns of variation.

Example

Figure 5.123 shows a sampling tree set up by Yummy Cookie Company as they try to understand why the crispness of their gingersnaps varies so much. Source A is the day of the week. Source B is different batches within the day. Source C is different cookies within the batch. They will sample three cookies per batch, two batches per day, for three days.

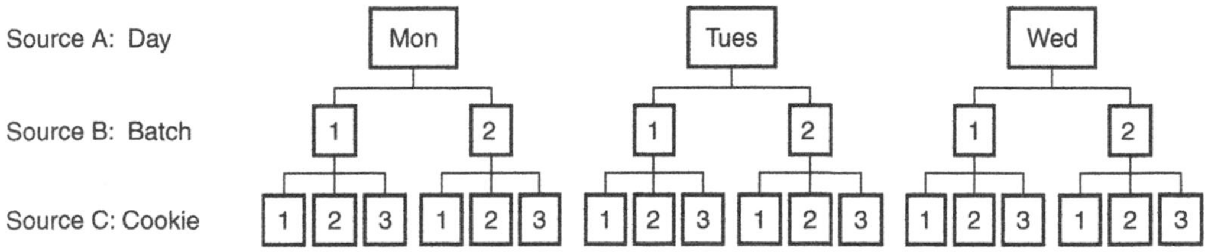

Figure 5.123 Multi-vari sampling tree example.

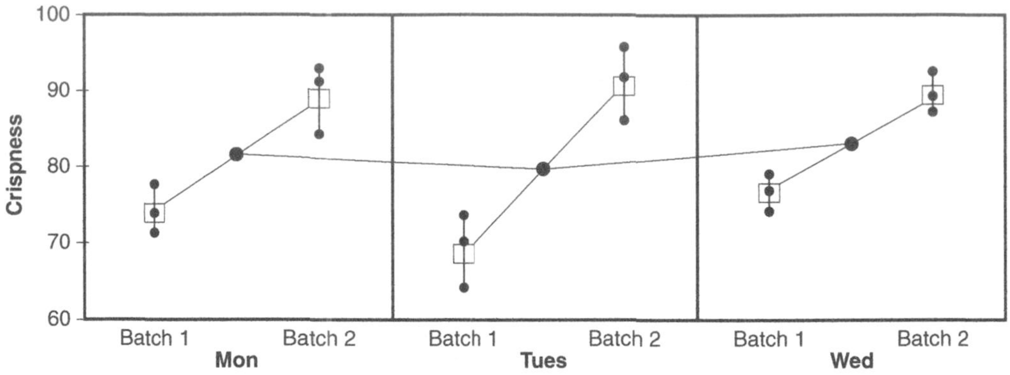

Figure 5.124 Multi-vari chart example.

The results are shown in Figure 5.124. Open squares show the average crispness of the three cookies in a batch. Large dots show the average crispness of all cookies sampled in a day.

Variation within a batch and from day to day are much smaller than between batches during a day. Interestingly, each day the second batch has higher values than the first. They will plan more studies to identify differences between batches and the reason for the consistent pattern.

Considerations

- The data used in this analysis can be historical data or data taken during operation of the process. Select representative data that spans about 80 percent of the output variable's typical range.
- The sources contributing to variation are treated as categorical data, even if they could be variable, such as location on a machine or time. Examples of categorical settings are high–low; short–long; on–off; lots 1, 2, and 3; day 1, 2, and 3; positions 1 through 5.
- At least three samples are recommended at the lowest level of the sampling tree.
- The ideal sampling plan is balanced, with the same number of B settings for each A setting, and the same number of C settings for each B. However, even if the sampling plan is not balanced, every possible combination must have at least one data point.
- A *nested* design means that one source of variation is found within another. Most typical are nested designs with three types of variation: *positional,* variation from one position to another within a unit; *cyclical,* variation from one unit to another; and *temporal,* variation from one time to another. The cookie example shows a nested design with variation within a batch, from batch to batch,

and from day to day. With nested designs, data should be kept in the order it is generated.

- A *crossed* design means that the three sources of variation can be manipulated independently. For example, if Yummy Cookie Company studied two different sweeteners, two different ovens, and two different baking times, they would use a crossed design. Interactions can be identified with crossed designs. (See *design of experiments* for more information about interactions.) With crossed designs, the data can be placed in different order to reveal different patterns.
- The variation sources can be either controllable or noise. In the example, all the sources are controllable. If Yummy decided to study ambient temperature as a possible source of variation, that would be noise: an uncontrollable environmental factor. Samples would be taken with room temperature at the high and low ends of its typical range.
- If source C data points are so close that they are hard to distinguish, place the vertical line at a slight angle to separate the points better.

multivoting

Also called: NGT voting, nominal prioritization
Variations: sticking dots, weighted voting, multiple picking-out method (MPM)

Description

Multivoting narrows a large list of possibilities to a smaller list of the top priorities or to a final selection. Multivoting is preferable to straight voting because it allows an item that is favored by all, but not the top choice of any, to rise to the top.

When to Use

- After brainstorming or some other expansion tool has been used to generate a long list of possibilities, and . . .
- When the list must be narrowed down, and . . .
- When the decision must be made by group judgment

Procedure

Materials needed: flipchart or whiteboard, marking pens, 5 to 10 slips of paper for each individual, pen or pencil for each individual.

1. Display the list of options. Combine duplicate items. Affinity diagrams can be useful to organize large numbers of ideas and eliminate duplication and overlap. List reduction may also be useful.
2. Number (or letter) all items.
3. Decide how many items must be on the final reduced list. Decide also how many choices each member will vote for. Usually, five choices are allowed. The longer the original list, the more votes will be allowed, up to 10.
4. Working individually, each member selects the five items (or whatever the number of choices allowed) he or she thinks most important. Then each member ranks the choices in order of priority, with the first choice ranking highest. For example, if each member has five votes, the top choice would be ranked five, the next choice four, and so on. Each choice is written on a separate paper, with the ranking underlined in the lower right corner.
5. Tally votes. Collect the papers, shuffle them, then record on a flipchart or whiteboard. The easiest way to record votes is for the scribe to write all the individual rankings next to each choice. For each item, the rankings are totaled next to the individual rankings.
6. If a decision is clear, stop here. Otherwise, continue with a brief discussion of the vote. The purpose of the discussion is to look at dramatic voting differences, such as an item that received both 5 and 1 ratings, and avoid errors from incorrect information or understandings about the item. The discussion should not become pressure on anyone to change their vote.
7. Repeat the voting process of steps 4 and 5. If greater decision-making accuracy is required, this voting may be done by weighting the relative importance of each choice on a scale of 1 to 10, with 10 being most important.

Example

A team had to develop a list of key customers to interview. First, members brainstormed a list of possible names. Since they wanted representation of customers in three different departments, they divided the list into three groups. Within each group, they used multivoting to identify four first-choice interviewees. This example shows the multivoting for one department.

Fifteen of the brainstormed names were in that department. Each team member was allowed five votes, giving five points to the top choice, four to the second choice, and so on down to one point for the fifth choice. The votes and tally are shown in Figure 5.125. (The names are fictitious, and any resemblance to real individuals is strictly coincidental.) Although several of the choices are emerging as agreed favorites, significant differences are indicated by the number of choices that have both high and low rankings. The team will discuss the options to ensure that everyone has the same information, then vote again.

Votes in rank order:		
Rhonda's votes: 4, 9, 12, 2, 8	Martha's votes: 10, 8, 15, 12, 11	
Terry's votes: 6, 10, 12, 9, 15	Al's votes: 8, 6, 11, 10, 4	
Pete's votes: 2, 9, 14, 4, 6		
1. Buddy Ellis	6. Albert Stevens 5 + 1 + 4 = 10	11. Mike Frost 1 + 3 = 4
2. Susan Legrand 2 + 5 = 7	7. Greg Burgess	12. Luke Dominguez 3 + 3 + 2 = 8
3. Barry Williams	8. Joan McPherson 1 + 4 + 5 = 10	13. Joe Modjeski
4. Lisa Galmon 5 + 2 + 1 = 8	9. Donald Jordan 4 + 2 + 4 = 10	14. Paul Moneaux 3
5. Steve Garland	10. Sam Hayes 4 + 5 + 2 = 11	15. Chad Rusch 1 + 3 = 4

Figure 5.125 Multivoting example.

sticking dots

Materials needed: flipchart paper, marking pens, colored round adhesive labels.

Follow the basic procedure, except individuals vote by placing round adhesive labels next to their choices, one dot per choice, and writing the ranking on the dot. This method can be used for voting by a large group that cannot be assembled at one time but who can all visit the flipchart with the list of options. Another use of this method is to differentiate the choices of different segments of the voting population by giving them different colored dots.

weighted voting

This variation normally results in fewer ties and allows the members to express the relative strength of their preferences better than 3–2–1 voting. The procedure is the same as multivoting except step 4:

4. Each member has a number of points equal to either the number of items or 1½ times the number of items. The members distribute these points over their choices. For example, one person may assign all points to one choice about which that person cares fervently, while another may distribute points equally among three top choices.

Example

This time the survey team gives each member 15 points (same as the number of names) to distribute. The results are shown in Figure 5.126.

Notice that Al's strong opinion about Joan is expressed by giving her almost half his votes. Terry and Pete are at the other extreme: they have equally divided their points among their choices. Rhonda and Martha show preferences among their choices.

This voting method separated the choices more than the other method did. The group should next discuss these results.

Rhonda's votes:	6 to #4, 5 to #9, 4 to #12
Terry's votes:	3 to #6, 3 to #10, 3 to #12, 3 to #9, 3 to #15
Pete's votes:	5 to #2, 5 to #9, 5 to #14
Martha's votes:	4 to #10, 4 to #8, 4 to #15, 3 to #12
Al's votes:	7 to #8, 3 to #6, 3 to #11, 2 to #10

1. Buddy Ellis	6. Albert Stevens 3 + 3 = 6	11. Mike Frost 3
2. Susan Legrand 5	7. Greg Burgess	12. Luke Dominguez 4 + 3 + 3 = 10
3. Barry Williams	8. Joan McPherson 4 + 7 = 11	13. Joe Modjeski
4. Lisa Galmon 6	9. Donald Jordan 5 + 3 + 5 = 13	14. Paul Moneaux 5
5. Steve Garland	10. Sam Hayes 3 + 4 + 2 = 9	15. Chad Rusch 3 + 4 = 7

Figure 5.126 Weighted voting example.

Notice that the top choices are in a different order than with multivoting. Then, 10 was top; this time, 10 is third. This is why it is always a good idea to discuss the top group of choices to ensure consensus.

multiple picking-out method (MPM)

In this method, one person with a strong opinion can keep an idea in consideration. It is useful when a huge number of items must be reduced quickly.

Materials needed: sticky notes, cards, slips of paper or just flipchart paper; marking pen; large work surface (wall, table, or floor); marking pen or set of colored adhesive dots for each individual; masking tape or correction fluid.

1. Write each option on a separate sticky note, card, or slip of paper. Alternatively, list the options on multiple sheets of flipchart paper with space next to each option. Spread the sheets out.
2. Decide how many items must be on the final list. This target must be an even multiple of the number of people in the group. For example, if 5 people are present, there could be 5, 10, 15, 20 (and so on) items on the final list.

Rounds of Multiple-Mark Choosing:

3. Every person gets a marker or adhesive dots. Each person marks a solid dot or places an adhesive dot beside any item he or she wants included in the final list. *Very important:* If someone else has already placed a dot during this round, don't put another one.
4. Remove or cross out any items with no dots added during this round. For example, during round one, items with no dots are eliminated. During round two, items with one dot are eliminated.

5. If any remaining items were marked twice (that is, have more dots than the round number), correct that with masking tape or correction fluid.
6. Count the number of items remaining. If it is less than 1.3 times the target number for the final list, go to step 7. Otherwise, return to step 3 for another round.

Rounds of Single-Mark Choosing:

7. Go around the room, with each person choosing one item to include in the final list. Continue until the target number of items has been chosen.
8. Review all the items that have been eliminated. If any team member thinks an eliminated item is needed, the team discusses it and by consensus can add it back to the list. Do this only for an agreed-upon number of items, typically one or two.

Considerations

- Multivoting does not guarantee consensus. Consensus is a decision that all individuals can live with and support. Discussion after multivoting will indicate whether the team has reached consensus. All individuals should have the opportunity to voice their opinions and arguments for or against an option. Consensus is reached when everyone is willing to actively support the decision, even though the option may not be that individual's preference.
- Multivoting incorporates several principles that have been shown to be important in arriving at accurate group decisions. First, independent voting takes place before discussion. The pattern is "judgment–talk–judgment." Second, voting by ranking rather than "majority rules" leads to more accurate decisions.
- Discussion after each round of multivoting can be useful in arriving at the best choice and in reaching consensus. Are the results surprising? Are there objections? Does the group want to discuss pros and cons of the top choices and vote again?
- It can be valuable to begin a multivoting session by brainstorming and discussing a list of criteria that are important to the current decision.
- To determine the most critical problem or the most significant root cause, multivoting should not be used in place of good data collection and analysis.
- To make final decisions about major issues, such as which project to select or which solution to implement, use the decision matrix or prioritization matrix, which evaluate options more rigorously. Paired comparisons can be used for a decision based on individual preferences rather than criteria.
- Another tool for narrowing long lists is list reduction. The effective–achievable matrix is a tool that separates ease-of-accomplishment criteria from effectiveness criteria and visually compares how well various options meet criteria.

nominal group technique (NGT)

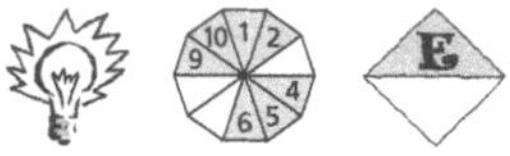

Description

Nominal group technique (NGT) is a structured method for group brainstorming that encourages contributions from everyone.

When to Use

- When some group members are much more vocal than others, or . . .
- When some group members think better in silence, or . . .
- When there is concern about some members not participating, or . . .
- When the group does not easily generate quantities of ideas, or . . .
- When all or some group members are new to the team, or . . .
- When the issue is controversial or there is heated conflict

Procedure

Materials needed: paper and pen or pencil for each individual, flipchart, marking pens, tape.

1. State the subject of the brainstorming. Clarify the statement as needed until everyone understands it.
2. Each team member silently thinks of and writes down as many ideas as possible, for a set period of time (5 to 10 minutes).
3. Each member in turn states aloud one idea. Facilitator records it on the flipchart.
   - No discussion is allowed, not even questions for clarification.
   - Ideas given do not need to be from the team member's written list. Indeed, as time goes on, many ideas will not be.
   - A member may "pass" his or her turn, and may then add an idea on a subsequent turn.

   Continue around the group until all members pass or for an agreed-upon length of time.

4. Discuss each idea in turn. Wording may be changed only when the idea's originator agrees. Ideas may be stricken from the list only by unanimous agreement. Discussion may clarify meaning, explain logic or analysis, raise and answer questions, or state agreement or disagreement.
5. Prioritize the ideas using multivoting or list reduction.

Considerations

- NGT usually concludes with a prioritization process. However, using the brainstorming process alone is valuable, so this book separates the two processes.
- Discussion should be equally balanced among all ideas. The facilitator should not allow discussion to turn into argument. The primary purpose of the discussion is clarification. It is not to resolve differences of opinion.
- Keep all ideas visible. When ideas overflow to additional flipchart pages, post previous pages around the room so all ideas are still visible to everyone.
- See *brainstorming* for other suggestions to use with this tool.

normal probability plot

Variations: probability plot, quantile–quantile (Q–Q) plot

Description

This graph is used to check whether a set of data takes the form of a normal distribution. It is a scatter diagram of the actual data versus a number that represents the normal distribution. If the data are from a normal distribution, the plot will form a straight line. In general, probability plots can be used to determine whether the data form any known distribution, such as binomial or Poisson.

When to Use

- When a tool or method you wish to use requires the data to be from a normal distribution, and . . .
- For best results, when you have 50 or more data points

For example:

- To determine whether a chart of individuals is appropriate for the data

- When choosing subgroup size for an $\overline{X}$ and R chart, to determine whether that subgroup size is large enough that the subgroup averages form a normal distribution
- Before calculating process capability index Cp or Cpk
- Before choosing a hypothesis test that is valid only for normal distributions

Procedure

Usually, you will simply enter your data into computer software, which will generate the plot. Calculation details are provided here so that you can understand what the program is doing and how to interpret it, and so you can do simple plots yourself.

1. Sequence the data from the lowest to the highest value. Number them from one to n.
2. Calculate the quantile number for each value. If i is the sequence number:

$$\text{Quantile number} = \frac{i - 0.5}{n}$$

3. Find the value of the normal distribution that matches each quantile number. On the table of area under the normal curve (Table A.1), locate the quantile number in the body of the table. Then read the z value from the left and top edges of the table.
4. Plot pairs of values as in a scatter diagram: each data value with its matching z value. Put the data value on the y-axis and the normal distribution's z on the x-axis. You will plot n pairs of numbers.
5. Draw a line that fits most of the data. If your data were a perfect normal distribution, the dots would form a straight line. Examine the shape of the dots compared to the line you have drawn to judge how well your data compare to the normal distribution. See "Considerations" for typical shapes. A correlation coefficient may be calculated to determine how well the line fits the plot.

Example

This example uses only 20 values to make the calculations easier to follow. Table 5.12 has the 20 values in numerical order, in the column labeled "process data." They have been numbered from 1 to 20.

The next step is calculating the quantile numbers. For the first value, 9, the calculation is

$$\text{Quantile number} = \frac{i - 0.5}{n} = \frac{1 - 0.5}{20} = \frac{0.5}{20} = 0.025$$

Table 5.12 Normal probability plot example calculations.

	Process data	Quantile number	*z*
1	9	.025	–1.960
2	11	.075	–1.440
3	11	.125	–1.150
4	13	.175	–0.935
5	14	.225	–0.755
6	15	.275	–0.595
7	16	.325	–0.455
8	17	.275	–0.320
9	19	.425	–0.190
10	21	.475	–0.065
11	23	.525	0.065
12	25	.575	0.190
13	26	.625	0.320
14	26	.675	0.455
15	28	.725	0.595
16	32	.775	0.755
17	36	.825	0.935
18	37	.875	1.150
19	43	.925	1.440
20	62	.975	1.960

Similarly, for the second value, the calculation is

$$\text{Quantile number} = \frac{1.5}{20} = 0.075$$

And you can see the pattern now: the third quantile number is 2.5 ÷ 20, the fourth is 3.5 ÷ 20, and so on to the last one, 19.5 ÷ 20.

Now look up the z values on the normal curve table. In this table, the first two digits for z are at the left edge of the row, and the last digit is at the top of the column. The first quantile number, 0.025, is in the row starting with –1.9, in the column headed by 0.06. So $z = -1.96$. Each quantile number is looked up in the same way.

Some of the quantile numbers are between two values on the table, so they need to be interpolated. For example, the fourth quantile number, 0.175, is about halfway between 0.1736 and 0.1762. The z for 0.1736 is –0.94, and the z for 0.1762 is –0.93. Halfway between those numbers is $z = -0.935$.

Now plot pairs of process data and z values. Figure 5.127 shows the result, with the best straight line drawn through the points. Notice that the points are above the line at

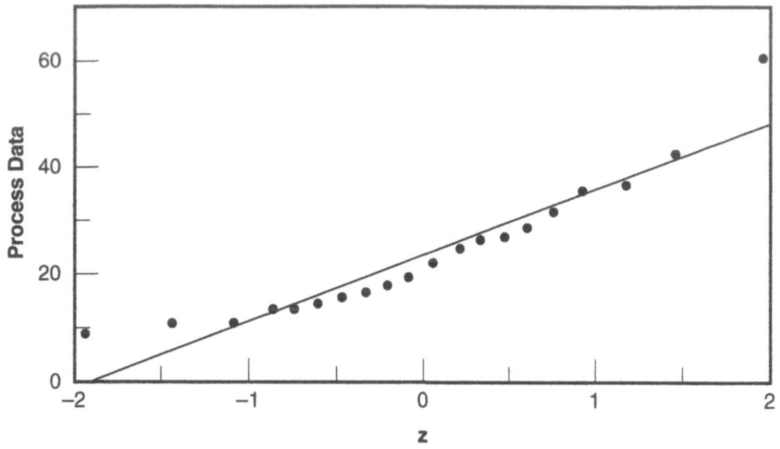

Figure 5.127 Normal probability plot example.

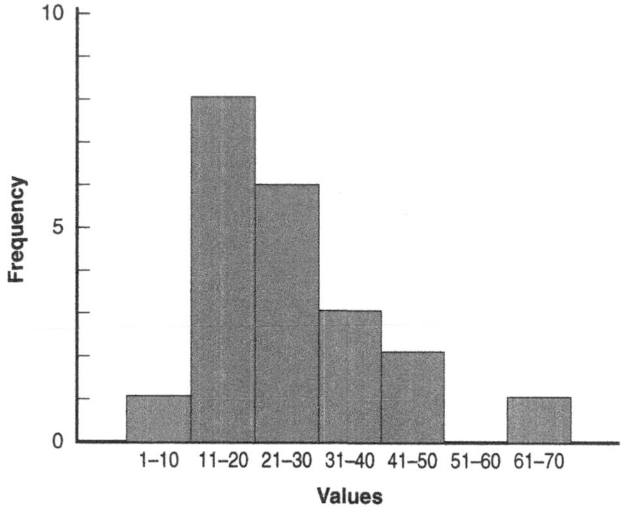

Figure 5.128 Histogram of normal probability plot data.

either end. That is typical of right-skewed data. Compare with the histogram of the data, Figure 5.128.

probability plot

This method can be used to check data against any known distribution. Instead of looking up the quantile numbers on the table of the normal distribution, look them up on a table for the distribution in which you are interested.

quantile-quantile plot

Similarly, any two sets of data can be compared to see if they come from the same distribution. Calculate quantile numbers for each distribution. Plot one on the x-axis and the other on the y–axis. Draw a 45-degree reference line. If the two sets of data come from the same distribution, points will lie close to the reference line.

Considerations

- Normal probability plots can be drawn in a variety of ways. In addition to the procedure given here, the normal distribution can be represented by probabilities or percentiles. The actual data may be standardized or placed on the x-axis.
- In this format, if the data does form a straight line, the mean of its normal distribution is the y-axis intercept, and the standard deviation is the slope of the line.
- Figure 5.129 shows how some common differences from the normal distribution look on a normal probability plot.

  Short tails. If tails are shorter than normal, the shape of the points will curve above the line at the left and below the line at the right—an S (for *short*) if you tip your head to the right. This means the data is clustered closer to the mean than a typical normal distribution.

  Long tails. If tails are longer than normal, the shape of the points will curve below the line at the left and above the line on the right—a backward S. This means there are more outlying data points than in a typical normal distribution. A bimodal distribution can also take this shape.

  Right-skewed. A right-skewed distribution has a short tail on the left and a long tail on the right. The shape of the probability plot forms an upward curve, or U, compared to the line. A distribution truncated on the left also will show this shape.

  Left-skewed. A left-skewed distribution has a long tail on the left and a short tail on the right. The shape of the points forms a downward curve compared to the line. A distribution truncated on the right also will show this shape.

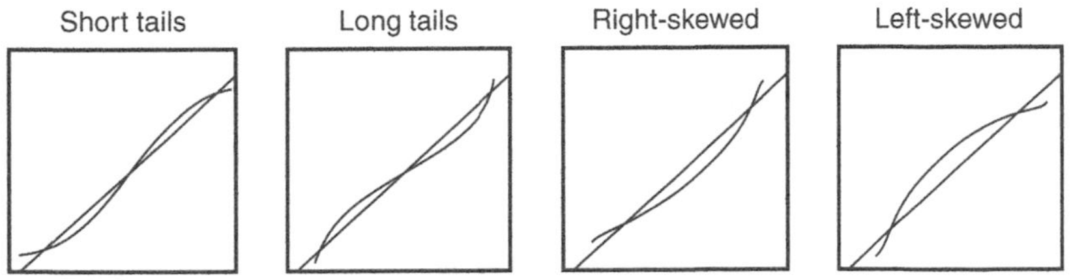

Figure 5.129 Shapes of normal probability plots.

- If a normal probability plot is drawn with the axes reversed, the distortions also will be reversed. For example, a left-skewed distribution will form a U-shaped curve.
- Remember that your process should be in control to make valid judgments about its shape.
- While drawing a histogram can be a first step toward understanding your data's distribution, histograms are not good for determining whether the data form a particular distribution. The eye is not good at judging curves, and other distributions form curves that are similar. Also, histograms of small data sets may not look normal when actually they are. Normal probability plots are better ways to assess the data.
- Another way to determine whether data fits a distribution is using goodness-of-fit tests, such as the Shapiro-Wilk W test, the Kolmogorov-Smirnov test, or the Lilliefors test. The details of such tests, done by most statistical software, are beyond the scope of this book. Consult a statistician for guidance in choosing the correct test and interpreting the results. See *hypothesis testing* to understand the general principles behind these tests and the results they generate.
- Your best choice is to have the statistical software generate a normal probability plot *and* perform a goodness-of-fit test. This combination will give you visual understanding of your data as well as statistical criteria for deciding whether it is or is not normal.

operational definition

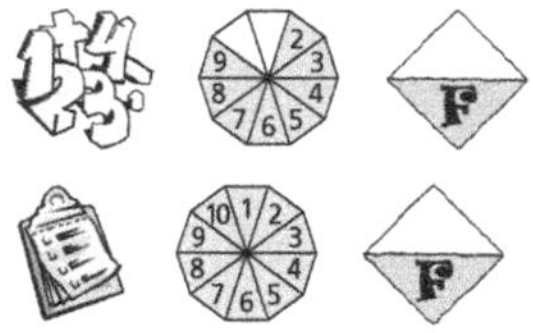

Description

Operational definitions are usable definitions of important terms and procedures in measurement or data collection. They remove ambiguity from words or operations that can be interpreted in different ways. The purpose of operational definitions is obtaining results that are consistent and meaningful.

When to Use

- When developing and using any form of data collection, or . . .
- When establishing metrics to monitor performance, or . . .
- When establishing customer requirements or product specifications

Procedure

1. Ask:
   - What characteristic will we measure?
   - How will we measure this characteristic?
2. Mentally or actually run through the data collection procedure. Note opportunities for differences in equipment, procedures, criteria, and the data collectors themselves. Identify all potential variation and decide how each will be handled. An operational definition must be defined for each potential source of variation.
3. Consider three elements for each operational definition:
   a. How will data be sampled?
   b. What procedure or test will be followed?
   c. What are the criteria to be satisfied?
4. Discuss the definitions until consensus is reached.
5. If possible, conduct a trial run, with all data collectors observing the same data. Compare the data collected and resolve any differences. Repeat the trial run until variation is minimized.

Example

An organization had a goal of increasing customer satisfaction by 25 percent. It needed operational definitions for measuring this goal. Discussion led to these definitions:

> We will survey a random sample of 10 percent of our customer base each quarter, using questionnaire B. The customer satisfaction rating will consist of the overall average of all sampled customers' scores and the standard deviation of the responses (formulas attached). Our goal is to increase the average by 25 percent while maintaining or reducing the standard deviation.

A team member pointed out that of the three elements in step 3, the test (3b) was defined by the survey and the criteria (3c) were defined by the stated method of calculation, but the method of sampling still was not defined. The team added the following operational definition for *random sample of 10 percent:*

> Say we have *n* customers. With a random-number generator we will generate a set of *n*/10 random numbers between one and *n*. Using the numbered customer database, we will choose the customers whose numbers correspond to those random numbers.

Also read the ZZ-400 story on page 79 for an example of using the concept of operational definition to think about a group's goals.

Considerations

- Operational definitions are easily overlooked because we all make assumptions without realizing it.
- Here's another way to understand the term "operational definition":

  Operational: Do we all agree what to do?

  Definition: Do we all agree what every word means?

paired comparison

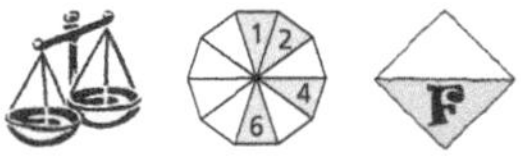

Variations: forced choice, numerical prioritization

Description

Paired comparison narrows a list of options to the most popular choice. It combines the individual preferences of all team members to arrive at a group decision. Normally, evaluation criteria are not explicitly considered when this tool is used. (Paired comparison is also the name of a statistical tool. See *hypothesis testing* for a description of this other "paired comparison," and see *design of experiments* for a description of the Shainin methodology, which uses this tool.)

When to Use

- When a list of options must be narrowed to one choice, and . . .
- After the list of options has been reduced to a manageable number by list reduction or some other wide screening tool, and . . .
- When individual preferences are more important than criteria or an objectively "correct" decision

Procedure

Materials needed: flipchart or whiteboard, marking pen.

1. List options as row labels down the left side of an L-shaped matrix. Number or letter each option sequentially. For column headings, write all pairs of comparisons. To do this methodically, pair each option with all succeeding options. If there are n options, there will be $n \times (n-1)/2$ pairs.
2. Start with the first column, option A versus option B. Each group member votes for one of these two options. Votes are tallied in the two rows opposite these two options.
3. Continue to vote one column at a time until all pairs have been voted upon.
4. Add across each row to obtain the total votes for each option. (Each column's sum should equal the number of people voting.) If the top vote-getters are close, check to be sure the comparison of those options matches the results of the total vote.

Example

A keynote speaker must be selected for a conference. The seven-member program committee whittled a long brainstormed list of names down to four possibilities. Suppose we could see inside their minds and know that the committee members ranked the four choices like this, in order from most to least preferred:

Member 1: C B A D

Member 2: B D C A

Member 3: B C A D

Member 4: A D C B

Member 5: B A D C

Member 6: C D B A

Member 7: C B D A

After they vote, the paired comparison matrix shown in Figure 5.130 is the result.

	A vs. B	A vs. C	A vs. D	B vs. C	B vs. D	C vs. D	Total
A. Dave Barry	|	||	||||	—	—	—	7
B. Sally Ride	~~||||~~ |	—	—	|||	~~||||~~	—	14
C. Steven Spielberg	—	~~||||~~	—	||||	—	||||	13
D. Ralph Nader	—	—	|||	—	||	|||	8

Figure 5.130 Paired comparison example.

Sally Ride has the most total votes, so she wins. Or does she? Ride and Spielberg were very close in total votes, and Spielberg was preferred in a direct comparison of the two. Strong preferences for Ride versus the other two options skewed the numbers. (Members' individual preferences are given above so you can verify for yourself that those opinions lead to this result.) The committee decided Spielberg was the preferred choice. Now, does he have an opening in his schedule?

forced choice

This tool also compares each option to every other option, but split opinions are not recorded, only the majority vote.

Procedure

1. List all options so they are visible to everyone.
2. Compare the first option to the second. Put a tally mark beside the better option. Compare the first to the third, the fourth, and so on, until the first option has been compared to all others.
3. Compare the second option to the third, the fourth, and so on.
4. Continue to compare each option to every one below it on the list, tallying which is preferred.
5. The option with the most tally marks is preferred.

Example

With the committee members privately ranking the choices as in the first example, the result for forced choice is shown in Figure 5.131.

numerical prioritization

This method allows everyone to make choices silently before individual results are combined.

Dave Barry	|
Sally Ride	||
Steven Spielberg	|||
Ralph Nader	

Figure 5.131 Forced choice example.

Procedure

Materials needed: flipchart, marking pen, paper and pen or pencil for each individual.

1. List all the options so they are visible to everyone. Code them with numbers or letters.
2. Each individual draws columns on a sheet of paper, one fewer columns than there are options. In the first column, choice A is paired with every other choice. In the second column, choice B is paired with every other choice following B, and so on, as shown in Figure 5.132.
3. Each individual starts with the first column, comparing the first option with every other option and circling his or her preference. In the second column, the second option is compared with all later options, and so on.
4. Each person counts how many times each number is circled and writes it next to that option's code.
5. When everyone has finished, combine the results. All individual counts for each option are added to obtain a group total. This can be done in two ways: Go around the room, have people call out their numbers, and tally them on a flipchart. Or turn in the papers and have one person tally results. The option with the largest total is the most popular. As in paired comparison, if two options are close, check that a direct comparison of those two matches the final result.

Example

The chart filled out by member 1 is shown in Figure 5.132. Member 1's numbers would be added with the other members' to arrive at the group's total.

Considerations

- This tool does not explicitly use decision criteria. However, criteria are likely to be in people's minds, so a discussion of appropriate criteria before voting may be useful. Sometimes this tool is adapted to evaluate criteria by doing multiple

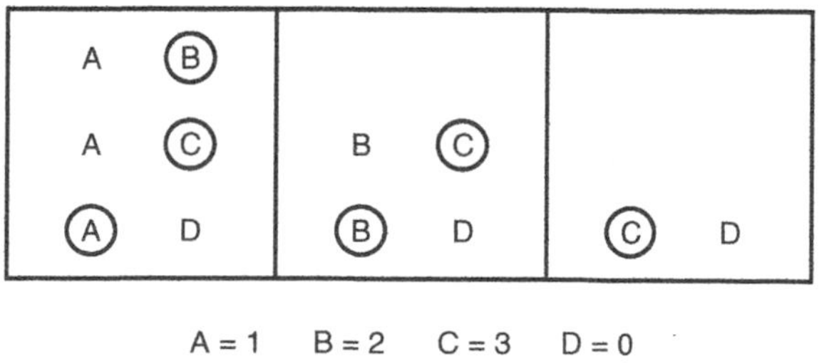

Figure 5.132 Numerical prioritization example.

rounds of voting, each time considering just one criterion. This is not recommended. If criteria are important, numerous, or not simple, choose a decision-making tool that explicitly evaluates and weights criteria. See *decision matrix, prioritization matrix, effective–achievable matrix,* or *criteria filtering.*

- The number of comparisons increases rapidly with the number of options. Six options require 15 comparisons; nine options require 36. Reduce the list first!
- Numerical prioritization is done silently and individually, so it is useful when group members want to reflect upon their choices or if some members tend to pressure others vocally.

Pareto chart

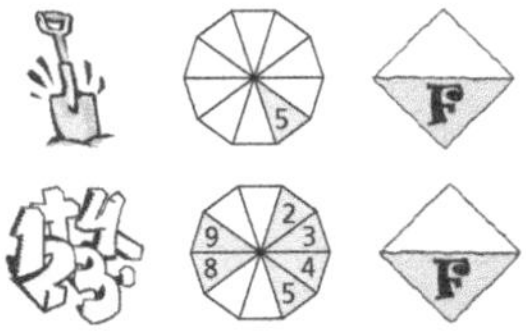

Also called: Pareto diagram, Pareto analysis
Variations: weighted Pareto chart, comparative Pareto charts

Description

A Pareto chart is a bar graph. The length of the bars represent frequency or cost (money or time), and they are arranged in order from longest on the left to shortest on the right. Therefore, the chart visually shows which situations are more significant.

When to Use

- When analyzing data about the frequency of problems or causes in a process, and . . .
- When there are many problems or causes and you want to focus on the most significant, or . . .
- When analyzing broad causes to their specific components, or . . .
- When communicating with others about your data

Procedure

1. Decide what categories you will use to group items.

2. Decide what measurement is appropriate. Common measurements are frequency, quantity, cost, or time.
3. Decide what period of time the chart will include.
4. Collect the data, recording the category each time. Or assemble data that already exists.
5. Subtotal the measurements for each category.
6. Determine the appropriate scale for the measurements you have collected. The maximum value will be the largest subtotal from step 5. If you will do optional steps 8 and 9, the maximum value will be the sum of all subtotals from step 5. Mark the scale on the left side of the chart.
7. Construct and label bars for each category. Place the tallest at far left, then the next tallest, and so on. If there are many categories with small measurements, they can be grouped as "other."

Steps 8 and 9 are optional but are useful for analysis and communication.

8. Calculate the percentage for each category: the subtotal for that category divided by the total for all categories. Draw a right vertical axis and label it with percentages. Be sure the two scales match. For example, the left measurement that corresponds to one-half should be exactly opposite 50 percent on the right scale.
9. Calculate and draw cumulative sums: Add the subtotals for the first and second categories, and place a dot above the second bar indicating that sum. To that sum add the subtotal for the third category, and place a dot above the third bar for that new sum. Continue the process for all the bars. Connect the dots, starting at the top of the first bar. The last dot should reach 100 percent on the right scale.

Example

Figures 5.133 and 5.134 are *nested Pareto charts.* Figure 5.133 shows how many customer complaints were received in each of five categories. Figure 5.134 takes the largest category, "documents," from Figure 5.133, breaks it down into six categories of document-related complaints, and shows cumulative values.

If all complaints cause equal distress to the customer, working on eliminating document-related complaints would have the most impact, and of those, working on quality certificates should be most fruitful.

Also see the Medrad story on page 56 for an example of Pareto analysis used to identify an improvement project.

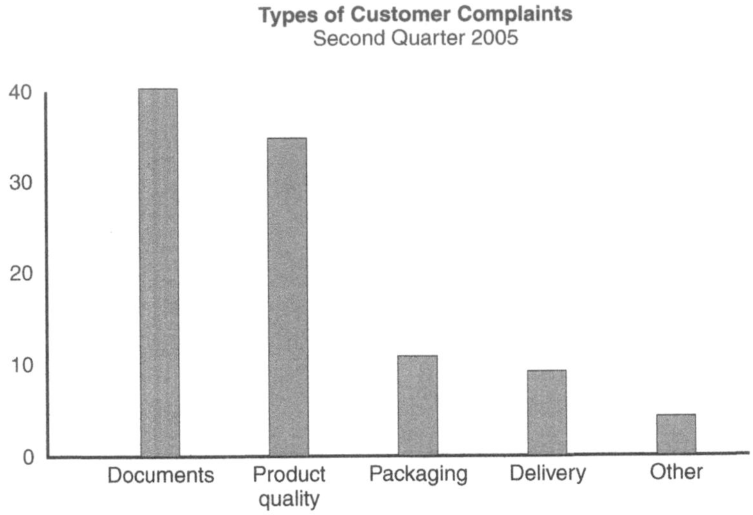

Figure 5.133 Pareto chart example.

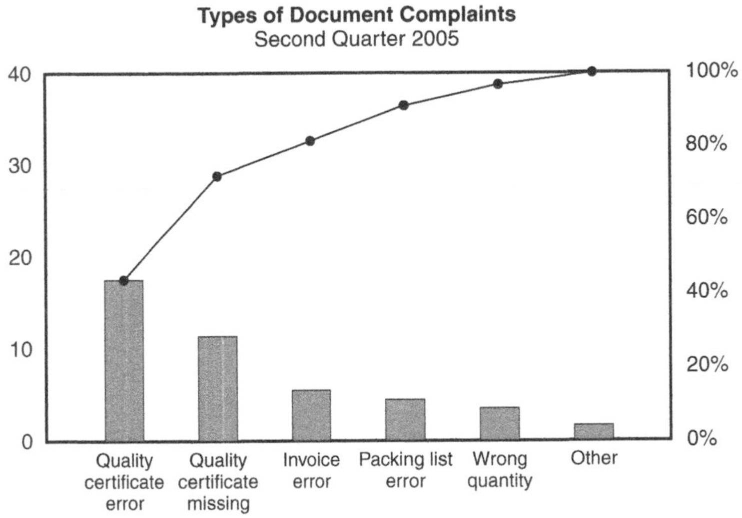

Figure 5.134 Pareto chart with cumulative line.

weighted Pareto chart

Description

In a weighted Pareto chart, each category is assigned a weight, which lengthens or shortens the bars. This reflects the relative importance or cost of each category.

When to Use

- When a Pareto analysis is appropriate, and . . .
- When the categories do not result in equal cost or pain to the organization, or . . .
- When there are more opportunities for one category to occur than another

Procedure

1. Decide what categories you will use to group items.
2. Decide what measurement is appropriate. Common measurements are frequency, quantity, cost, or time. Assign weights to each category to reflect their relative importance, cost, or opportunity for occurrence.
3. Decide what period of time the chart will include.
4. Collect the data, recording the category each time. Or assemble data that already exists.
5. Subtotal the measurements for each category. Multiply each subtotal by that category's weight.
6. Determine the appropriate scale for the measurements you have collected. The maximum on the scale will be the largest number you calculated in step 5. If you will do the optional steps, the maximum will be the sum of all the results from step 5. Mark the scale on the left side of the chart.
7. Construct and label bars for each category. Place the tallest at far left, then the next tallest, and so on. If there are many categories with small measurements, they can be grouped as "other."

Continue with optional steps 8 and 9 as in the basic procedure, using the calculated numbers from step 5 (subtotal × weight) to calculate percentages and cumulative sums.

Example

Document errors are a minor hassle for the customer, but product quality problems can cause real headaches. Weights have been estimated for customer complaints to reflect their importance.

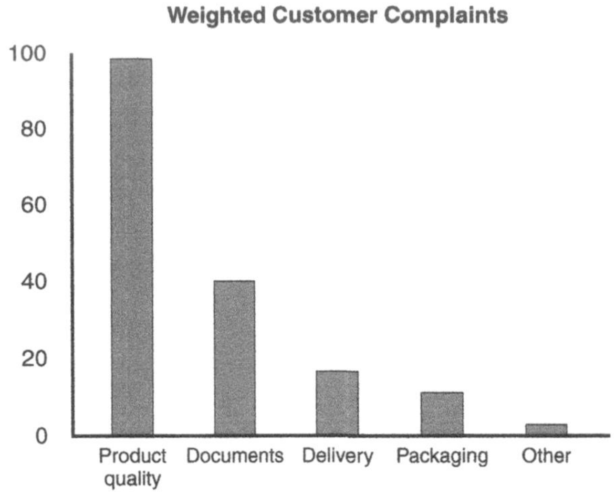

Figure 5.135 Weighted Pareto chart example.

Documents	1
Product quality	3
Packaging	1
Delivery	2
Other	0.5

Figure 5.135 shows a weighted Pareto chart using the same data as Figure 5.133. Product quality has become the first priority.

comparative Pareto charts

Description

Comparative Pareto charts place two or more Pareto charts side by side to compare two or more sets of data. The data differ in some fundamental way but can be analyzed using the same categories.

When to Use

- When a Pareto or weighted Pareto chart is appropriate, and . . .
- When you wish to compare two or more sets of data, and . . .
- When all sets of data will be analyzed using the same categories, and . . .
- When each set of data differs from the others on a significant variable

Such as:

- Comparing data from different processes
- Comparing data from different times
- Comparing data before and after process changes, to assess improvement

Procedure

Follow the basic procedure. In step 6, choose one chart to be the base, whose bars you will sequence in descending height from left to right. Usually choose the one with the largest vertical scale. For the other charts, keep the same sequence of categories on the horizontal scale and the same vertical scale. Comparison of charts will be easier.

Example

Figure 5.136 shows customer complaints before and after a project aimed at eliminating quality-related complaints.

Considerations

- The Pareto chart was developed by Dr. Joseph Juran. He named it after a 19th-century Italian economist named Vilfredo Pareto, whose work provided the first example of the unequal distribution Juran called the Pareto principle: 80 percent of an effect comes from 20 percent of the causes. While the percentages aren't always exactly 80/20, usually a "vital few" can be identified that will provide greater payback than the "useful many." (Juran first said "trivial many" but later renamed them to indicate that all problems are worth eradicating.)

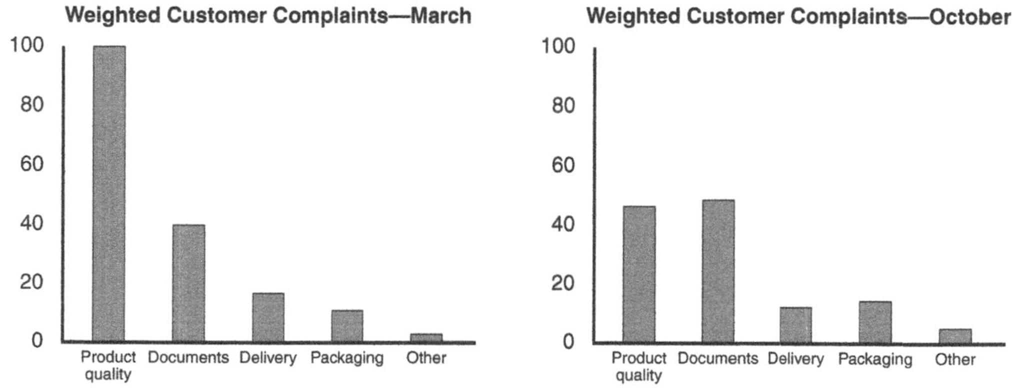

Figure 5.136 Comparative Pareto chart example.

- The best Pareto chart uses a measurement that reflects cost to the organization. If every category identified has equal cost or pain to the organization, then frequency is a good measurement. If not, it is more useful to measure dollars, time, or some other indicator of cost. If it is possible to measure cost or time directly, do so. If not, use a weighted Pareto chart and estimate weights for each category to translate frequency into cost.
- Weights need to be used if there is more opportunity for occurrence in one category than another. Suppose you are monitoring telephone response time between different departments. If customer support receives 800 calls a day, while billing receives only 80 calls a day, customer support has 10 times more opportunity to respond quickly (or slowly). The data from each group should be weighted to enable accurate comparisons. Dividing by the number of calls puts everything on an equal basis. The general rule then is: multiply by a weight of "1/area of opportunity."
- Check sheets are often used at step 4 of the procedure to collect data.
- Pareto charts are a wonderful communication tool. Tables of data draw yawns, but a Pareto chart instantly tells a story.
- Figure 5.137 is a blank worksheet that can be used to collect data or to draw a horizontal Pareto chart. Permission is granted to copy it for individual use.

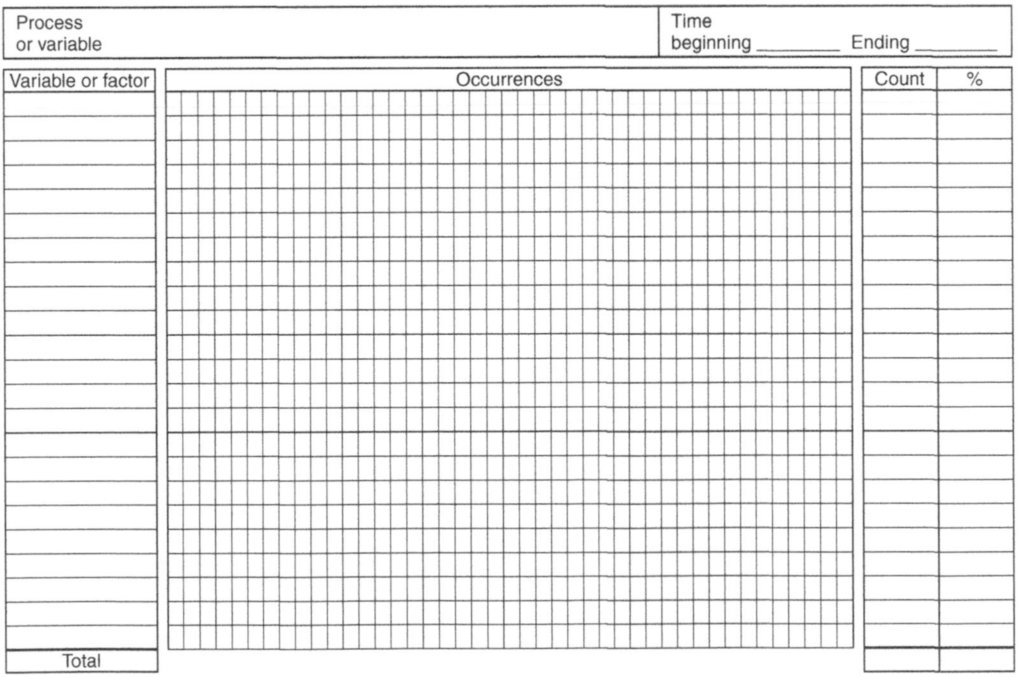

Figure 5.137 Pareto chart worksheet.

performance index

Also called: objectives matrix

Description

The performance index monitors performance when several characteristics contribute to overall quality. A family of several measurements is rolled into one overall number.

When to Use

- When monitoring step-by-step improvement over time toward a goal, and . . .
- When there are several customer requirements to consider as you measure overall performance, or . . .
- When monitoring large amounts of information for problems and trends

Procedure

Setting Up the Chart

1. Define the mission or objectives of the organization, system, or job to be measured.
2. Select indicators of key performance criteria based on the mission or objectives. Three to seven is a typical number of indicators. Write the performance indicators in the boxes across the top of the chart (Figure 5.138).
3. Establish current performance levels for each indicator. As a rule of thumb, about three months' performance should be reviewed to determine current performance. Enter current performance levels in the boxes on the chart opposite the number 3.
4. Establish goals. Choose a specific time period (say, a year) and determine realistic goals for that period. Enter these goals in the boxes opposite the number 10.
5. Establish mini-goals to correspond to the boxes between 3 and 10. Consider whether different resources and effort will be required to accomplish each mini-goal. If so, you may want to make the numerical distance between mini-goals smaller when more resources or effort are required. If there are no such considerations, you can set the mini-goals so the distance from each one to the next is equal. Write the mini-goals in the boxes opposite the scores of 4 to 9.
6. Establish minimum performance levels. These will be the lowest performance levels you can imagine occurring. Enter these levels in the boxes opposite the 0.

Date: ____________________

							Criteria
							Performance
							10
							9
							8
							7
							6
							5
							4
							3
							2
							1
							0
							Score
							Weight
							Value
						Index =	

Jan	Feb	Mar	Apr	May	Jun	Jul	Aug	Sep	Oct	Nov	Dec

Figure 5.138 Performance index.

Source: James Riggs and Glenn Felix, *Productivity by Objectives,* ©1983, pp. 225, 233. Adapted by permission of Prentice Hall, Englewood Cliffs, N.J.

7. Determine performance levels for scores of 1 and 2, taking into consideration slow periods and slumps. Write these in the appropriate boxes.
8. Assign weights to each indicator to show relative importance. The sum of the weights must equal 100. Enter weights in the boxes opposite "weight" at the bottom of the chart.
9. Make copies of this chart to use each time you calculate the index.
10. Monitor performance for several cycles of your process(es). If needed, adjust the current performance levels, the weights, the number of indicators, and so on.

Using the Chart

11. On a regular predetermined frequency (weekly, monthly, quarterly) collect data and record results on a blank chart. Enter the actual measure for each indicator opposite "performance."
12. Circle the actual performance for each indicator. If a mini-goal is not obtained, circle the next worse performance level.
13. For each indicator, determine the performance score (0–10) opposite the circled performance number. Write that number opposite "score" at the bottom of the chart.
14. Multiply each score by its weight factor to generate the weighted value. Write this number opposite "value."
15. Add all the "values." The sum is the performance index.
16. You may wish to plot performance indexes on a graph, with time on the x-axis and the index on the y-axis. When you begin, the index should be 300, because that is the score for current performance. The goal is 1000.

Example

Figure 5.139 is the performance index form for June for a manufacturing group. The italicized numbers were written on the photocopied form this month. Note that pounds made, with a performance score of 2, has deteriorated since the chart was established, although the other measures have remained steady or improved. Read the ZZ-400 improvement story on page 83 to see how this tool was used within their improvement process.

Considerations

- Do not change the numbers used to calculate the chart within a monitoring period, or comparisons will be meaningless.

Date: *June, 2005*

% Rework	Hours Downtime	Pounds Made	Safety & Environmental Flags	$ / Pound			Criteria
26	*28*	*42*	*9*	*278*			Performance
0	0	≥ 70	0	≤ 143			10
2	5	69	1	164			9
6	10	67	2	185			8
11	15	65	4	206			7
16	20	60	6	227			6
21	25	55	8	248			5
24	30	50	10	269			4
28	40	45	12	290			3
31	50	40	13	311			2
34	60	35	14	332			1
≥ 36	≥ 70	≤ 30	≥ 15	≥ 353			0
3	*4*	*2*	*4*	*3*			Score
25	15	15	20	20			Weight
75	*60*	*30*	*100*	*60*			Value

Index = *325*

Jan	Feb	Mar	Apr	May	Jun	Jul	Aug	Sep	Oct	Nov	Dec
			300	*315*	*325*						

Figure 5.139 Performance index example.

Source: James Riggs and Glenn Felix, *Productivity by Objectives,* ©1983, pp. 225, 233. Adapted by permission of Prentice Hall, Englewood Cliffs, N.J.

- You may wish to revise the chart annually or on some other long time frame. When you do, ask whether important current indicators are missing, whether some are no longer needed, whether priorities (weights) have changed. Also ask whether customers are still satisfied with the goals and whether you have become capable of better performance than the goals indicate.
- Best-of-class benchmarking can be valuable for setting challenging but realistic goals.
- Involve your customers when you determine indicators and set goals, minimums, and weights.
- Don't let an overall good score mask difficulties on a single indicator.

PGCV index

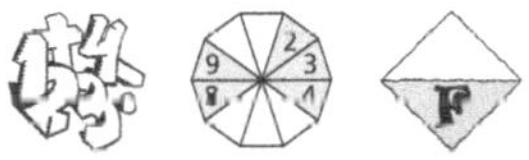

Also called: potential gain in customer value index

Description

The PGCV index numerically summarizes customers' perceptions about both the importance and the performance of products or services. It allows statistical analysis of customer satisfaction data.

When to Use

- When surveying customer satisfaction, and . . .
- When setting priorities for product or service changes based on customer perceptions, and . . .
- When a numerical summary is desirable for compactness, precise comparison, or further analysis

Such as:

- When identifying products, services, or their features whose improvement would most increase customer satisfaction, or . . .
- When identifying products, services, or their features that should have changed emphasis: increased, reduced, or eliminated, or . . .
- When comparing customer satisfaction between customer segments or between your organization and a competitor

Procedure

1. Identify the products, services, or features that you want evaluated. Usually you will study either the selection of products or services offered or the features of one particular product or service. (From here on, the products, services, or features under study will be called "output.")
2. Develop a survey. Set it up so that each question asks for an output to be rated twice on a numerical scale of 1 to 5 or 1 to 7. One rating is how important the output is to the customer (importance) and the other rating is the quality of that output (performance). Typical scales are:

Importance	Performance
5 = Critical	5 = Greatly exceeds expectations
4 = Important, but not critical	4 = Exceeds expectations
3 = Of some value	3 = Meets expectations; adequate
2 = Nice to have, but not necessary	2 = Needs some improvement
1 = Not needed	1 = Consistently lacking

3. Identify customer groups you want to survey, and select key customers or a representative sampling of customers. Administer the survey.
4. For each output, calculate "achieved customer value" (ACV), "ultimately desired customer value" (UDCV), and "potential gain in customer value" (PGCV).

$$\text{ACV} = \text{I} \times \text{P}$$

$$\text{UDCV} = \text{I} \times \text{P}_{\text{max}}$$

$$\text{PGCV} = \text{UDCV} - \text{ACV}$$

I = Importance rating

P = Performance rating

P_{max} = highest possible performance rating

5. Compare the PGCV values of all the outputs. Higher PGCV values represent outputs whose improvement would provide the greatest boost in customer satisfaction.

Example

A restaurant identified six products and services that they wanted to study to improve customer satisfaction. An importance–performance questionnaire was given to restaurant customers. (This is the same example as the related importance–performance analysis.) The average customer ratings and the PGCV calculations are shown in Figure 5.140.

Service	I	P	I × P	I × P_{max}	PGCV
Breakfast	4.2	3.9	16.4	21.0	4.6
Lunch	4.9	4.1	20.1	24.5	4.4
Dinner	4.9	4.4	21.6	24.5	2.9
Lounge/bar	3.2	4.8	15.4	16.0	0.6
Carryout	1.5	4.6	6.9	7.5	0.6
Private parties	4.7	2.5	11.8	23.5	11.7

Figure 5.140 PGCV index example calculations.

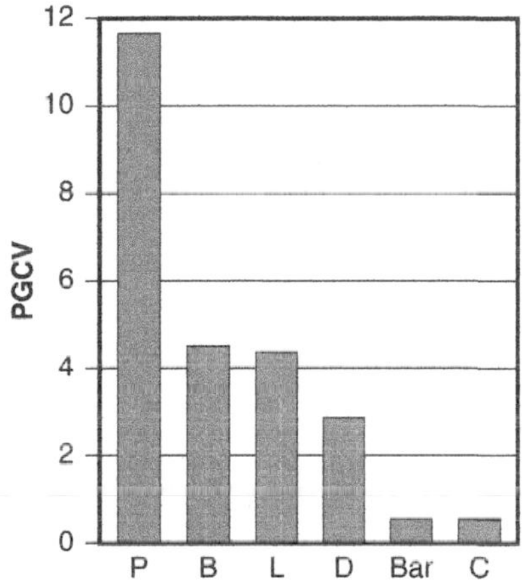

Figure 5.141 PGCV index example.

The PGCV indices are compared in a bar chart in Figure 5.141. Graphs communicate information more quickly than tables, so graphing PGCV results is a good idea.

Private parties, with a PGCV value of 11.7, should be the first priority for improvement. Compare these results to those in the importance–performance analysis example.

Considerations

- In step 1, think about what you really want to do with the information. If you want to improve a particular product or service, collect importance–performance data about features of that product or service. If you want to change the selection of products or services you offer customers, collect importance–performance data about the products or services, not about their features.

- It is possible to collect information about features and then sum the data for each product or service to compare information about the selection of products or services. However, because collecting data is expensive, do this only if you really need both kinds of information.
- Absolute values of PGCV scores are not meaningful. The meaning comes from comparing them to other outputs to determine their relative value to customers.
- Before comparing PGCV scores, be sure that the surveys' numerical scales were the same.
- Because information is grasped more quickly from a graph than a table, graph PGCV scores that you are comparing when you communicate the results. Bar or dot charts are good ways to show this data. If you are comparing several groups, such as different customer segments, grouped bar charts and grouped or two-way dot charts would be useful.
- See *survey* for more information about developing and administering surveys.
- If you want to use PGCV indexes for more detailed analysis, consult a statistician for help.

plan–do–study–act cycle

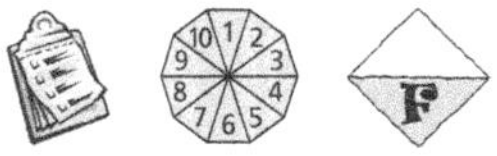

Also called: PDSA, plan–do–check–act (PDCA) cycle, Deming cycle, Shewhart cycle
Variation: standardize–do–study–adjust (SDSA) cycle (also called standardize–do–check–act or SDCA cycle)

Description

The plan–do–study–act cycle (Figure 5.142) is a four-step model for carrying out change. Just as a circle has no end, the PDSA cycle should be repeated again and again for continuous improvement. PDSA is a basic model that can be compared to the ten-step improvement process, but also it can be applied on a small scale, such as within every step of an improvement process.

When to Use

- As a model for continuous improvement, such as:
- When starting a new improvement project, or . . .
- When developing a new or improved design of a process, product, or service, or . . .

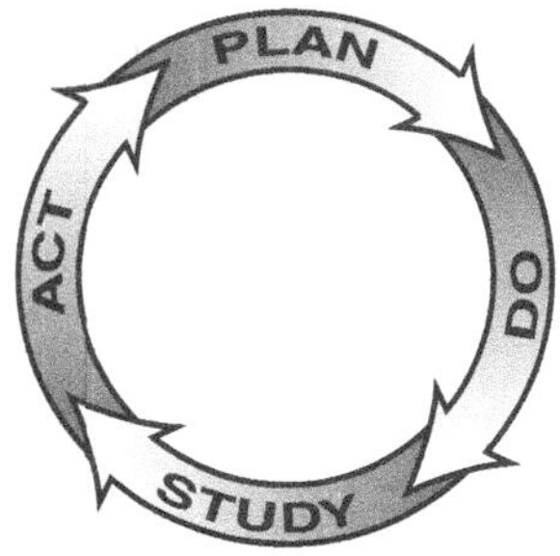

Figure 5.142 Plan–do–study–act cycle.

- When defining a repetitive work process, or . . .
- When planning data collection and analysis in order to verify and prioritize problems or root causes, or . . .
- When implementing any change

Procedure

1. *Plan.* Recognize an opportunity and plan the change.
2. *Do.* Test the change. Carry out a small-scale study.
3. *Study.* Review the test, analyze the results, and identify what you learned.
4. *Act.* Take action based on what you learned in the study step. If the change did not work, go through the cycle again with a different plan. If you were successful, incorporate the learnings from the test into wider changes. Use what you learned to plan new improvements, beginning the cycle again.

Example

Pearl River School District, 2001 recipient of the Baldrige Award, uses the PDSA cycle as a model for defining most of their work processes, from the boardroom to the classroom. PDSA is the basic structure for the organization's overall strategic planning, needs analysis, curriculum design and delivery, staff goal-setting and evaluation, provision of student services and support services, and classroom instruction.

Figure 5.143 shows their A+ Approach to Classroom Success. This is a continuous cycle of designing curriculum and delivering classroom instruction. Improvement is not a separate activity. It is built into the work process.

The A+ Approach begins with a "plan" step called "analyze." In this step, students' needs are analyzed by examining a range of data available in Pearl River's electronic data "warehouse," from grades to nationally standardized tests. Data can be analyzed for individual students as well as stratified by grade, gender, or any other subgroup. Because

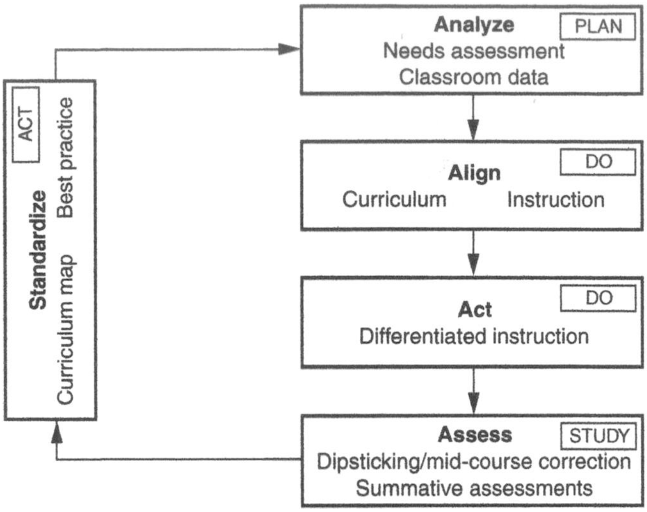

Figure 5.143 Plan–do–study–act example.

PDSA does not specify how to analyze data, a separate data analysis process (Figure 4.14) is used here as well as in other processes throughout the organization.

The A+ Approach continues with two "do" steps. "Align" asks what national and state standards require and how they will be assessed. Teaching staff also plans curriculum by looking at what is taught at earlier and later grade levels and in other disciplines to assure a clear continuity of instruction throughout the student's schooling. Teachers develop individual goals to improve their instruction where the "analyze" step showed any gaps.

The second "do" step is called "act." This is where instruction is actually provided, following the curriculum and teaching goals. Differentiated instruction is used based on each student's learning rates and styles and varying teaching methods.

The "study" step is called "assess." Formal and informal assessments take place continually, from daily teacher "dipstick" assessments to every-six-weeks progress reports to annual standardized tests. Teachers also can access comparative data on the electronic database to identify trends. High-need students are monitored by a special child study team. Throughout the school year, if assessments show students are not learning as expected, mid-course corrections are made such as reinstruction, changing teaching methods, teacher mentoring, and consulting. Assessment data becomes input to the analyze step during the next cycle.

The "act" step is called "standardize." When goals are met, the curriculum design and teaching methods are standardized. Teachers share best practices in formal and informal settings. Results from this cycle become input for the "analyze" phase of the next A+ cycle.

See the Pearl River story on page 65 to learn more about their PDSA-based improvement process.

standardize–do–study–adjust (SDSA) cycle

Also called: standardize–do–check–act (SDCA) cycle

This cycle is used to standardize improvements.

1. *Standardize.* Through training and documentation, a revised process is standardized throughout the organization.
2. *Do.* The process is operated.
3. *Study.* Does output meet customer needs? Are the improvements being retained?
4. *Adjust.* Revise the process, if necessary, to hold the gains or meet customer needs.

Considerations

- The most important aspect of PDSA is that it is a cycle, not a linear process. Feedback from one round of improvement loops into the next round of improvement. The cycle never ends.
- PDSA is a model for implementing change. Many organizations use it as one of their quality improvement processes. However, it does not detail the key steps of understanding customers, collecting and analyzing data, prioritizing opportunities, and identifying root causes. Supplement it with other processes for those key steps, as in the example.
- The ten-step quality improvement process is an expansion of PDCA. Steps 1 to 6 are *plan,* 7 is *do,* 8 is *study,* and 9 and 10 are *act.*
- *Study* is often called *check*, sometimes *evaluate*, and occasionally *assess, learn,* or *review.*

plan–results chart

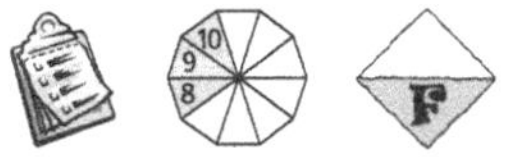

Description

Typically, we look at results on a one-dimensional scale—were we successful or weren't we? It is more useful to look at two dimensions:

- Did we accomplish the plan we developed?
- Did we achieve our desired results?

The plan–results chart, a variation of the generic two-dimensional chart, looks at both dimensions. It is an analytical tool to guide the team's thinking about the outcome of a project and the next steps.

When to Use

- After a solution has been tested, or . . .
- After a project has been implemented, or . . .
- When a project is floundering, about to be or already labeled dead

Procedure

1. Draw a square and divide it into four quadrants, as in Figure 5.144. On the horizontal edge, label the left half "plan not accomplished" and the right half "plan accomplished." On the vertical edge, label the bottom half "results not accomplished" and the top half "results accomplished."
2. Think about the plan, solution, or strategy that your team put in place and carried out. Was it successfully accomplished? Did you do what you had planned to do? Choose "plan accomplished" or "plan not accomplished."
3. Think about the results that your plan was intended to accomplish. Were they successfully accomplished? Choose "results accomplished" or "results not accomplished."
4. The words you have chosen on the edges intersect in one of the four boxes of the chart. The words in the box provide guidance about how to think about your results and your next actions.

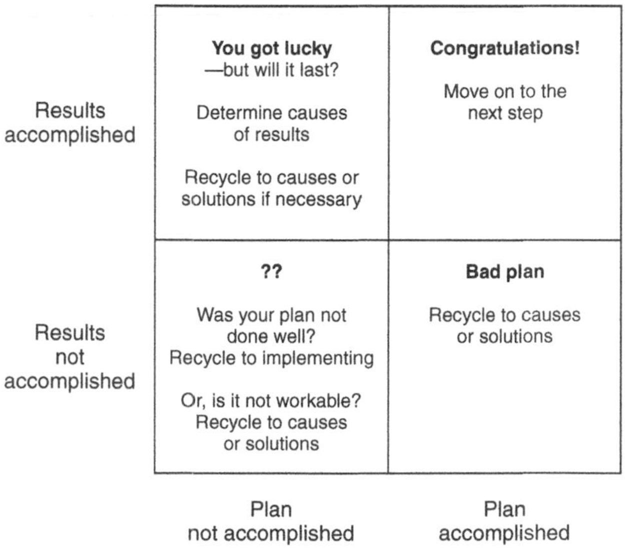

Figure 5.144 Plan–results chart.

- *Plan accomplished, results accomplished.* This is the ideal situation. Everything seems to have worked well. Move on to the next step in your improvement process.
- *Plan accomplished, results not accomplished.* Something was wrong with the plan. It may not have been effective or appropriate. Or it may not be addressing the true root cause of the problem. Do more analysis to learn which is the case and whether you need to recycle to solutions or causes.
- *Plan not accomplished, results not accomplished.* In this situation, you cannot know if the problem lies with the execution of the plan or the plan itself. Usually, you should first go back and carry out the plan correctly. If you recognize that the plan is no good, you should find a better plan or reanalyze the root cause.
- *Plan not accomplished, results accomplished.* Somehow you got lucky. Determine the causes of the results. Analyzing process measures, not just results measures, can help indicate what is going on. If your analysis shows that the results are temporary, you may need to recycle. Return to implementing a better plan, choosing a better solution, or finding the true root cause.

Example

There was a continuing problem with office mail being delivered late or not at all. The Pony Express team developed a new system for mail collection, sorting, and delivery. Team members developed and taught a training class for all mailroom employees to introduce the new system.

Case A. The training classes and rollout of the new system went without a hitch. A month later, data show that mail is still late and lost. *Analysis:* Plan accomplished, results not accomplished. (See Figure 5.145.) Something was wrong with the plan. Maybe the new system doesn't effectively solve the problems, or maybe the team hadn't identified the real problems. More analysis is called for, then recycling.

Case B. Results on a quiz administered after training averaged 58 percent correct, and spot checks show that the new system is not being followed. Yet, to everyone's surprise, results after a month show far fewer lost and late pieces of mail. *Analysis:* Plan not accomplished, results accomplished. Maybe the Hawthorne effect is at work here: much attention was focused on correct mail delivery. But will those results slide after the attention goes away? Analyze process measures and figure out why results have improved.

Case C. All the trainees complained about the new system being confusing and harder to use than the old one. And mail is still late and lost. *Analysis:* Plan not accomplished, results not accomplished. The problem could be execution: poor training and rollout of the new system. Or maybe the new system really isn't better than the old one. The Pony Express team must decide whether they should provide better training or go back to the drawing board.

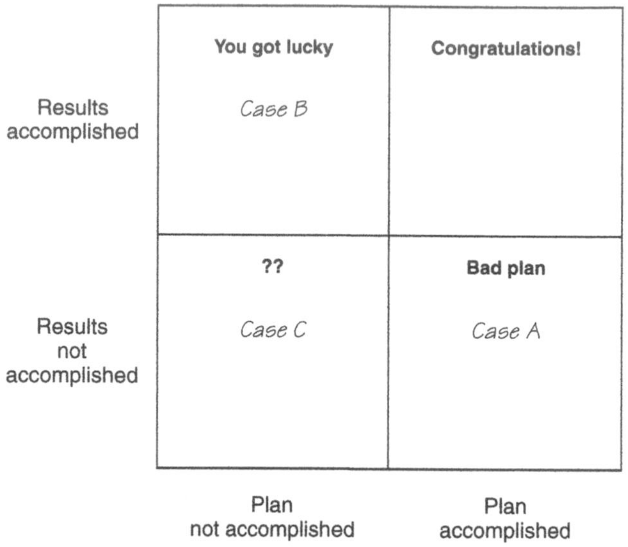

Figure 5.145 Plan–results chart example.

Considerations

- Many groups never review projects or steps of a project. Regular reviews will improve your rate of success. This is the essential *study* step of the PDSA cycle.
- When results are not accomplished, do not get discouraged. Above all, do not quit. Use analytical tools to understand the cause of your difficulty. Cause analysis tools may be especially useful.
- Initial failures can yield valuable lessons that will make your second attempt successful.
- When results are accomplished, but the plan was not: Too often we are tempted to collect our winnings and move on. Before you give in to that temptation, think about what caused the good results. Were they caused by a special cause, a temporary environmental factor, or a Hawthorne effect? Will they last?
- The Hawthorne effect is named after a famous study of the relationship between lighting conditions and productivity. Raising the lighting improved productivity—and so did lowering it! Researchers found that just focusing attention can create positive results. The problem with the Hawthorne effect is that the positive results are temporary. The Hawthorne effect often is alive and well in quality improvement efforts as unusual attention is focused on specific areas.
- Many groups are quick to claim success and move on to the next challenge. Understanding what caused your results is important for the success of the rest of this project and future ones.

PMI

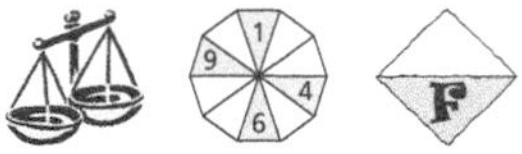

Description

PMI stands for *plus, minus, interesting*. This process structures a discussion to identify the pluses, minuses, and interesting points about an idea.

When to Use

- When evaluating an idea, such as a proposed solution
- Especially when the group is being one-sided in its thinking, aware of only advantages or only disadvantages, or . . .
- When members of the group are polarized and arguing

Procedure

Materials needed: flipchart or whiteboard, marking pen.

1. Review the topic or problem to be discussed. Often it is best phrased as a *why, how,* or *what* question. Make sure the entire team understands the subject of the PMI.
2. Brainstorm pluses—positive aspects of the idea.
3. Brainstorm minuses—negative aspects of the idea.
4. Brainstorm interesting points or implications of the idea. These might be neutral aspects, questions or points to explore, or unusual features.

Example

A work group was looking for ideas to improve safety performance. One day Chris told the subteam about the safety approach her neighbor's company used, which involved safety circles where individuals could confront coworkers' unsafe behavior. The immediate reaction of the team members was skepticism and disinterest, but George suggested that they evaluate the idea before discarding it. They used PMI to explore various aspects of the concept.

Addressing pluses first, they generated the ideas at the top of Figure 5.146. Then they focused on minuses. Finally, they listed interesting points.

At the end of the exercise, the group decided the approach had real possibility. They wanted to pursue it further.

Pluses
More attention to safety
Focus on behavior
Raise awareness of unsafe acts
Up close and personal
Support for confronting unsafe behavior
Creates more teamwork
Learn to be honest with each other
Positive effect in other areas of our lives
Keep each other alert

Minuses
Too personal!
What would we talk about
People wouldn't say what they really think
Too busy to take the time
No one to lead group
Could cause hard feelings
Usual people would dominate
Already have safety training and meetings

Interesting
Peer pressure doesn't affect just kids
Ownership for safety belongs where?
We want to be best
How does it work at XYZ Company?
Different from safety approach we've had
Each person's behavior affects everyone else's safety

Figure 5.146 PMI example.

Considerations

- PMI is similar to the familiar process of listing pros and cons. However, the structure of the exercise and the addition of I—interesting points—makes it more powerful.
- It is important to do one step at a time, in the sequence listed. For example, do not go back and forth between pluses and minuses. By directing attention to one side of an issue at a time, a better view of the entire issue is developed.
- The entire group should look at each side of the issue together. Do not modify the procedure to assign pluses to one subgroup, minuses to another, and interesting points to a third. Thinking together, individuals can drop assumptions and positions to arrive jointly at a better evaluation of the idea.
- If the group is too large to use this tool together, break into smaller groups, each of which generates pluses, minuses, and interesting points. Then join together and share lists.

- Interesting points can lead to modifications and adaptations of an idea that in its present form is not practical.
- This tool can be used by individuals as well as by groups.

potential problem analysis

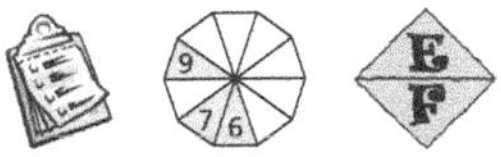

Also called: PPA

Description

Potential problem analysis systematically identifies what might go wrong in a plan under development. Problem causes are rated for their probability of occurrence and how serious their consequences are. Preventive actions are developed as well as contingency plans in case the problem occurs anyway. By using PPA, smooth implementation of a plan is more likely.

When to Use

- Before implementing a plan, and . . .
- When something might go wrong, especially . . .
- When the plan is large and complex, or . . .
- When the plan must be completed on schedule, or . . .
- When the price of failure is high

Procedure

1. Identify broad aspects of the plan that are vulnerable to disruption or failure.
2. For each aspect, identify specific problems that could occur. Write problems in the first column of a table structured like Figure 5.147.
3. For each problem, estimate its risk by rating as high, medium, or low both the probability of the problem occurring and the seriousness of its consequences. Write those assessments under the problem statement. Prioritize the problems by deciding which risks you are willing to accept.

Continue with the following steps for the problems whose risks are unacceptable, recording information on the table. Do steps 4 to 6 for each problem before addressing the next problem.

Chronic Illness Management Program						
Problem	**Causes**	**P**	**S**	**Preventive Action**	**Res. Risk**	**Contingency Plan**
Staff resists role changes Probability H Seriousness H	"Not invented here" syndrome	M	L	Involve staff in role definitions	L	
	Feeling inadequate in new roles	H	H	Training: role playing, what-ifs, IS simulations	M	• Refresher training • Shadow mentor • IS simulations • Responsible: Dr. Peters, CIMP Champion • Trigger: Negative responses to weekly survey
	No perceived advantage	H	M	Visit clinic with CIMP in place	L	
Registry system not ready on time Probability M Seriousness H	Contract negotiations drag	M	H	• Good up-front system definition • Realistic budget	L	
	Last minute changes	H	H	• Good up-front system definition • Extensive software review • Set absolute no-change date	M	• Modify IS module of training • Follow-up training on delayed features • Responsible: Maria Hernandez, IS analyst • Trigger: System shakedown errors

Figure 5.147 Potential problem analysis example.

4. Identify possible causes, where appropriate, and write them in the second column.
5. (Optional) For each cause, rate as high, medium, or low the probability of its occurrence, recorded in the column headed P, and the seriousness of its consequences, recorded in the column headed S.
6. For each cause (or problem, if causes were not detailed), identify preventive actions that would eliminate or reduce its chance of occurring. Write it in the fifth column. Rate the *residual risk*, the probability and seriousness of the cause even with the preventive action in place, and record it in the next column.
7. For each cause whose residual risk is unacceptable, develop a contingency plan to minimize the consequences of the problem if the preventive action fails. Identify what should be done, who is responsible, and what trigger will set the plan into action. Record the plan in the last column.

Example

Figure 5.147 shows a potential problem analysis for a medical group planning to improve the care of patients with chronic illnesses such as diabetes and asthma through a new chronic illness management program (CIMP). The broad aspects of the program vulnerable to disruption were new staff roles, computer system changes, and patient participation. In each of these areas, the planning team identified specific problems that must be considered to ensure success of the program. Two of the potential problems with unacceptable risks are shown on the chart with their analyses.

Of three potential causes of staff resistance, two have low residual risk after preventive actions. The remaining cause has a contingency plan in case the problem occurs despite the preventive action. Similarly, the second problem has only one cause needing a contingency plan.

For the second problem, notice that the preventive action "good up-front definition" will prevent both causes.

This example is also used for the process decision program chart to allow you to compare the two tools.

Considerations

- This is one of several tools that can be used for contingency planning. It analyzes situations more rigorously than PDPC or the contingency diagram. Of the less rigorous tools, PDPC is more systematic than the contingency diagram. FMEA, the most rigorous of them all, is appropriate for new or redesigned processes, products, or services.
- Advantages of this tool are that it specifically addresses causes, rates risk, and separates preventive actions from contingency plans. An advantage of PDPC is that it starts with a tree diagram of the implementation plan to identify problems. Consider combining these tools: start with PDPC to identify problems, then use PPA to analyze them.
- Vulnerable aspects of a plan include:
  - Planned actions that might fail to occur
  - Anything being done for the first time
  - Anything done under a tight deadline
  - Anything done from a distance
  - Anything under the management or responsibility of more than one person
- Ask what? when? where? and how many? or how much? to identify specific problems. These are similar to the questions used in the is–is not matrix. Here is a checklist of other questions that can be used:
  - What inputs must be present? Are there any undesirable inputs linked to the good inputs?
  - What outputs are we expecting? Might others happen as well?

  - What is this supposed to do? Is there something else that it might do instead or in addition?
  - Does this depend on actions, conditions, or events? Are these controllable or uncontrollable?
  - What cannot be changed or is inflexible?
  - Have we allowed any margin for error?
  - What assumptions are we making that could turn out to be wrong?
  - What has been our experience in similar situations in the past?
  - How is this different from before?
  - If we wanted this to fail, how could we accomplish that?
- Identifying every potential problem can overwhelm you, leading to overlooking the most serious problems or abandoning the analysis. In PPA, focus on where you are most vulnerable, on what would hurt the most.
- Identifying potential causes is not always appropriate. For example, if a potential problem is that it might rain on the day of a big event, causes are not useful.
- To rate risk, scales of 1 to 5 or 1 to 10 are sometimes used. If you think you need to assess risk that precisely, consider using FMEA instead.
- Preventive actions may not seem possible. You can't prevent rain. However, you can prevent rain from disrupting an event by holding it under shelter. If preventive actions are truly impossible, contingency plans become your life preserver.
- If a problem is identified for which no preventive or contingent actions can be devised, you must decide whether to accept the risk or to create and select an alternative plan without the risk. Thus, PPA can keep you from embarking on a disastrous course.
- The original version of this tool did not include probability and severity ratings. Adding those steps formalizes the process of focusing on the most significant problems, which otherwise would have to be done intuitively.
- This tool can be transformed into potential opportunity analysis (POA) by brainstorming possible opportunities instead of problems, identifying promoting actions instead of preventive actions, and developing plans to capitalize on the occurrence instead of contingency plans to mitigate the occurrence. Both PPA and POA try to look into the future for unplanned events. PPA is looking for negative events; POA is looking for positive events.
- Potential problem analysis is above all an attitude: we can look into the future, foresee what might happen, and change the future for the better.

presentation

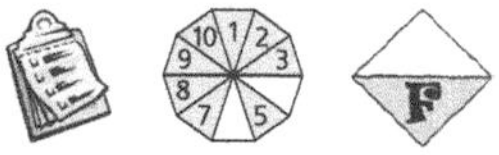

Description

Presentations are formal spoken reports. They communicate to others what you have done or what you plan to do, with the purpose of asking for something (support, approval, resources) or helping the audience do something similar. Usually, visual images or props are used to clarify and enhance the message.

When to Use

- When communicating plans or results to management to gain or confirm their support, or . . .
- When requesting approval, funding, manpower, or other resources, or . . .
- When concluding one phase of a project and requesting approval to move to the next phase, such as at a tollgate review, or . . .
- When explaining your project to customers or others before requesting information from them or after making an improvement to their product or service, or . . .
- Before implementing a change, to inform and get buy-in from those who will be affected, or . . .
- When sharing your best practices or lessons learned with other groups

Procedure

1. Define the subject of the presentation. Determine how much time you will have, where the presentation will be made, and what audiovisual facilities will be available.
2. Identify the audience. Determine who they are: either individual names, for a presentation to a small group of people you know, or demographic or affiliation information, for a larger group. Determine what they want to know about the subject, and write it down.
3. Identify the main ideas you wish to communicate, keeping your audience's needs in mind. Depending on the subject, a model or structure such as your improvement process may be useful as a checklist for ideas. In other situations, brainstorming or nominal group technique may be more beneficial. Summarize the ideas on cards or sticky notes in a few words. These are not notes for speaking, just notes for planning.

4. Determine the best sequence of ideas. Play around with the cards, trying different approaches. An affinity diagram, mind map, storyboard, or flowchart may be useful tools.
5. For every main point, decide how you will explain it. Plan visual aids such as graphs and diagrams that show your message. Decide how you will summarize each point so it communicates something the audience wants to know.
6. Create visuals that support the main ideas. If possible, plan them so that they become the notes to remind you what to say.
7. If you need additional notes, consider using a mind map or flowchart. Never write out your speech and read it.
8. Brainstorm possible questions. Have answers prepared. If necessary, have backup visual aids for anticipated questions. If you list two dozen or more questions, include more information in the presentation.
9. Practice your presentation, first alone and then in front of friendly, honest listeners. Revise the presentation until it feels comfortable, connects with the audience's needs, and fits within the allotted time with allowance for questions.
10. In the 10 or 15 minutes before your presentation, review the list of what your audience wants to know from step 2. Think about the opening few sentences of your presentation. Breathe deeply.
11. When your presentation time comes, stand up. Introduce yourself, if necessary. Speak clearly. Move within the space you have. Tell your story. Stop on time.

Example

The St. Luke's Hospital story on page 74 and the ZZ-400 story on page 88 provide examples of presentations used within improvement processes.

Considerations

Before the Presentation

- Create a checklist of questions about your audience, then get answers. Depending on your situation, items on the checklist might be: their prior knowledge, experience, values, concerns, attitudes, expectations, or restrictions within their situation. These elements may affect what they need to hear and how they will receive your presentation.
- The most basic structure for a presentation is deceptively simple: *opening, body, conclusion.* If you (and your audience) can't clearly identify those three segments, something's wrong with the presentation. A more detailed structure for presentations is Ron Hoff's "All About Them":

- Issue or opportunity the audience cares about
- Your unique point of view on their issue or opportunity
- Evidence to support that point of view
- Your idea or proposal, to resolve the issue or opportunity
- What they should do next

- Lists are memorable. They help an audience hang onto your main points, to see where you're headed and, when the presentation is over, where they've been. (Notice how many lists are in this book? Every tool description is written in the form of a list.) Make lists of your main points, of the steps in getting from A to Z, of what they need to do.

- While planning, think of your presentation as a TV show and each main idea as a scene within the show. Each of these scenes should last less than six minutes. Longer than that and the audience will mentally cut to a "commercial," even if you keep talking! The scenes strung together should tell your entire story. If more than one presenter will tell the story, alternate presenters with every scene.

- Each scene should contain a "flag"—a summarizing statement that drives home the point you are making. The flag is more compelling if you relate it to your audience and what they want to know. For example, suppose the ZZ-400 team is making a presentation about the iron problem described in chapter 4, and they are concluding the scene about root cause analysis:

  Okay flag: "So the point here is that persistence in asking the data 'Why?' finally led us to our root cause."

  Compelling flag when speaking to another improvement team: "So the point here is that you should never stop asking your data 'Why?' until you've reached your root cause.

  Compelling flag when speaking to their management: "So the point here is that using the analysis tools we learned in last spring's training to ask the data 'Why?' finally led us to the root cause of a problem that was wasting 10 percent of our capacity.

- A presentation should not be a mystery story. The audience should know from the beginning your main point and how it relates to them. Summarize at the beginning, as you go through, and at the end. Remember this old saying: "Tell 'em what you're going to tell them; tell 'em; then tell 'em what you told them." By the way, this saying reflects the basic three-point structure mentioned above.

- If you need more notes than your audiovisual slides provide, use only words and brief phrases, written large and dark. Notes are only cues. The idea is to glance at your notes and immediately know what to say next. Index cards, with only a few notes per card, work better than sheets of paper, where it is easy to lose your place. Number the cards sequentially, again in big bold writing.

- Practice your presentation multiple times, first alone until you are comfortable with the material, then with friendly but honest listeners. Start practicing without audiovisual equipment, then add it later. If possible, practice in the room where the real presentation will be made. If that is not possible, try to use a similar room and audiovisual equipment. Videotape your presentation and watch it several times—the first time to get over the shock of seeing yourself on tape and subsequent viewings to identify ways to improve.
- Practice responding to questions, both the ones you've prepared and off-the-wall ones your listeners shoot at you.
- If the presentation will have more than one presenter, such as several members of a team each handling a different section, one person should be overall "director." That person should know the overall purpose of the presentation, understand the audience's needs, know each presenter's strengths to assign parts accordingly, and be willing and able to give constructive criticism to each presenter, including advice about visuals. The presenters and director should rehearse together. Include Q&A practices.

Audio-Visuals

- People have different learning styles. Some receive information best visually, others through hearing, and still others through motion and activity (kinesthetically). The best presentations communicate in multiple ways to reach everyone. Your slides should support your presentation, but they should not *be* the presentation.
- Enliven your presentation with images, graphs, photos, color, humor. Use objects as props. Draw a diagram on a flipchart or whiteboard. If you are using slides, intersperse these other elements to hold the audience's attention.
- It is not necessary—in fact, it is distracting and boring—for the visuals to repeat what you say. On slides with text, aim to have no more than three or four lines, each with no more than three or four words. If you have many slides with just words, ask yourself if they're all necessary. Choose text fonts and colors for readability, not flashiness. Also, graphs and diagrams should convey their messages with a minimum of words on the slide.
- Presentation software, such as Microsoft PowerPoint or Corel Presentations, is widely used because it helps you easily create, change, store, transport, and present audiovisuals. A wide variety of visual effects—images and photos, symbols and graphics, background colors and borders, as well as text—are easily manipulated. The dimensions of sound and motion also can be added to incorporate all the modes of receiving information.

- The animation features available in presentation software add interest and a kinesthetic element. Analyze your audience to determine how much animation is appropriate. Within one slide, use only one type of animation. For example, have all lines of text fly in from the left. If the slide includes a table, graph, or graphic, avoid animation. The slide is already busy enough.
- Presentation software gives you the choice of running the slide show automatically or manually. If your presentation must move quickly, within strict time limits, and without pauses for questions or discussion, choose automatic. Otherwise choose to control the slide progression manually.
- Remember that *you* will make the presentation, not your slides. It is good to have times when no image is visible, especially when you want the audience to focus on you and your message. Use the projector's pause button or project a dark slide during these times, so the audience's attention moves to you.

Handouts

- With presentation software, you can create presentation notes that print beside small images of your slides. Use these notes to include details you want your audience to remember.
- If you will print and photocopy presentation slides and notes for handouts, keep that in mind when choosing slide backgrounds. Dark colors look good when projected, but do not print or copy well.
- Handouts are best distributed at the end of the presentation, unless they contain details that the audience must study immediately to understand your points. In that case, give the audience a moment of silence to look at the written information. If people are studying handouts, they aren't listening to you. Let people know at the beginning that they will receive detailed notes, so they can listen and watch instead of writing.

During the Presentation

Do	Don't
Stand up	Write down then memorize or read a speech
Move around	Lean on a podium, chair, or wall
Keep eye contact with the audience	Keep hands in pockets or fiddle with an object
Read the audience and adjust	Use wordy slides
Let visuals show your points	Read exactly what is written on the slide
Keep the pace brisk	Darken the room to show slides
Stay within your time limit	Apologize for any aspect of your presentation

- As you speak, move around as much as possible within the space you have. You and your audience will feel more comfortable. Point to a feature on a graph, advance the audiovisual machine, write on a flipchart, or simply cross to the other side. Do not stand still gripping a podium, table, or chair.

- Watch your audience's body language and adjust your presentation based on what you see. If they're bored, liven your delivery or have a stretch break. If they're puzzled, ask what is causing confusion and explain a different way.
- Speak to the entire audience when you answer a question. Assume everyone wants to know the answer. Repeat the question if there's a chance the audience couldn't hear it—or if you need to buy time to think.
- You will be nervous. That's good—it gives you energy. If you start getting so nervous that your presentation is in trouble, breathe deeply and concentrate on the audience. What do they need to hear from you? What can you offer them? Nervousness is about you. Great presentations are about them.

prioritization matrix

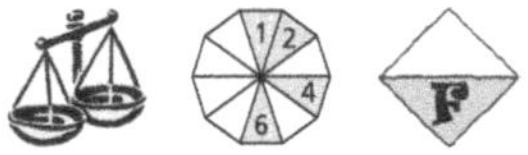

Variations: analytical criteria method (also called full analytical criteria method), consensus criteria method, combination ID/matrix method

Description

A prioritization matrix is an L-shaped matrix that uses pairwise comparisons of a list of options to a set of criteria in order to choose the best option(s). This is the most rigorous, careful, and time-consuming of the decision-making tools described in this book. First, the importance of each criterion is decided. Then each criterion is considered separately, with each option rated for how well it meets the criterion. Finally, all the ratings are combined for a final ranking of options. Numerical calculations ensure a balance between the relative importance of the criteria and the relative merits of the options.

Three different versions of the prioritization matrix are available, depending on the situation: analytical criteria method, consensus criteria method, and combination ID (interrelationship diagram)/matrix method.

When to Use

- When a list of options (initiatives, projects, solutions, major equipment or key personnel selection) must be reduced to one or a few choices, and . . .
- When three or more important criteria must be considered, especially when some are subjective, and . . .

- When the decision is important to the organization's success, with serious consequences for a wrong choice
- Such as: when selecting initiatives, projects, or improvement opportunities; when deciding which opportunities will receive resources and which will not; when choosing the best solution for significant problems; when making major equipment purchase decisions; when making key personnel decisions
- And, when the decision-makers are willing to invest time and effort into the decision-making process
- Also, when prioritizing any list, such as when asking customers to prioritize desired product features

analytical criteria method

When to Use

- When the situations listed for "when to use" any prioritization matrix are true, and . . .
- When the decision is extremely important, with grave consequences for a wrong choice, or . . .
- When all decision-makers must agree with and support the final decision, or . . .
- When individuals have pet projects or hidden agendas, or . . .
- When individuals disagree about the importance of the criteria

Procedure

1. Clarify the goal. Use operational definitions where possible. Make sure everyone understands and agrees on what is to be decided.

Criteria Weighting

2. List criteria that the decision must meet.
3. Create an L-shaped matrix. It should have equal numbers of rows and columns, one more than the number of criteria to allow for labels. List the criteria as row labels and, in the same order, as column headings. The diagonal of cells where the row and column labels are the same will not be used, so block them out, such as by shading them.
4. In the first row under the column headings, compare the first criterion to the second criterion. Which is more important, and how much more? Write in the cell your rating of the relative importance, using this scale:

1 = row is equally important as column

5 = row significantly more important than column

10 = row extremely more important than column

⅕ = row significantly less important than column

⅒ = row extremely less important than column

Always rate the criterion in the *row* label compared to the criterion in the *column* label.

5. Write the inverse of the rating (1, ⅕, or ⅒) in the cell diagonally opposite, the one in the column of the first criterion and the row of the second criterion.
6. Continue to compare each pair of criteria. Always write the inverse of your rating in the cell diagonally opposite.
7. Add the ratings across each row to get a row total. Convert the fractions (⅕, ⅒) to decimals (0.2, 0.1).
8. Add the row totals to get a grand total.
9. Divide each row total by the grand total. These numbers are the relative weights of the criteria. Later, they will be used in the final matrix. If any criteria are much less important than the others, you may choose to eliminate them now to save effort.

Rating Options Against Criteria

10. Create another L-shaped matrix, with equal number of rows and columns, one more than the number of options. Again, mark out the center diagonal cells. Title this matrix with the first criterion.
11. Compare the options to each other pairwise, just as you did the criteria. Using the scale in step 4, ask which option is preferred, thinking about only the first criterion.
12. Write the inverse of the rating in the diagonally opposite cell, just as you did with the criteria.
13. Do the same calculations as in steps 7 through 9. You now have options weighted with respect to the first criterion.
14. Repeat steps 10 through 13 for each of the other criteria. In all, you will make as many matrices as there are criteria.

Combining All Criteria: Summary Matrix

15. Create one more L-shaped matrix. This time, use options as row labels and criteria as column headings. Do not mark out the center diagonal. Write the weight of each criterion from step 9 under each criterion name.

16. In the first column, copy the option weights from the first criterion matrix, step 13. Write "×" and the weight of the first criterion. (The criterion weight is repeated all down the column). In each cell, multiply the two weights and write the result.
17. Repeat for each column, copying weights from the matrix for that criterion.
18. Add the results across each row to get an option score. These scores are relative weights with respect to all criteria. The largest number is the best choice.

Example

Often this method is used with long lists of options or criteria. This concise example makes it easier to see the steps and calculations involved.

A bank was selecting a vendor for a major upgrade of computers throughout the organization. They knew their criteria for selection: low purchase price, technology that anticipated future corporate needs, easy transition for their employees, and good after-sale support.

They had received proposals from three vendors. Vendor A's equipment would have great after-sale support, had a moderate price tag, but had less advanced technology and would entail a tough transition. Vendor B had the most forward-looking technology, but also the highest price and the transition would be difficult, with little support. Vendor C's proposal showed the best price, adequate support, and a very easy transition but less advanced technology. Which was the best choice?

First, the team responsible for the decision identified criteria and compared them in the matrix shown in Figure 5.148. The resulting weights showed that "after-sale support" could be eliminated from the rest of the process.

Then they began comparing vendors relative to the first criterion, low purchase price. Their "low purchase price matrix" is shown in Figure 5.149. They continued to

Criteria	Low purchase price	Anticipate future needs	Easy transition	After-sale support	Sum	Wts
Low purchase price		1/10	5	5	10.1	.22
Anticipate future needs	10		10	10	30.0	.65
Easy transition	1/5	1/10		5	5.3	.12
After-sale support	1/5	1/10	1/5		0.5	.01
					36.1	

Figure 5.148 Analytical criteria method example: criteria matrix.

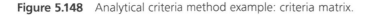

Low Purchase Price	Vendor A	Vendor B	Vendor C	Sum	Wts
Vendor A		5	1/5	5.2	.33
Vendor B	1/5		1/5	0.4	.03
Vendor C	5	5		10	.64
				15.6	

Figure 5.149 Analytical criteria method example: options versus criteria 1 matrix.

Anticipate Future Needs	Vendor A	Vendor B	Vendor C	Sum	Wts
Vendor A		1/5	5	5.2	.25
Vendor B	5		10	15.0	.73
Vendor C	1/5	1/10		0.3	.01
				20.5	

Figure 5.150 Analytical criteria method example: options versus criteria 2 matrix.

Easy Transition	Vendor A	Vendor B	Vendor C	Sum	Wts
Vendor A		5	1/10	5.1	.25
Vendor B	1/5		1/5	0.4	.02
Vendor C	10	5		15.0	.73
				20.5	

Figure 5.151 Analytical criteria method example: options versus criteria 3 matrix.

compare vendors based on the other two criteria. Their matrices are shown in Figure 5.150 and 5.151.

Last, they transferred the criteria and option weights to a summary matrix (Figure 5.152) to determine overall priority. Vendor B was the best choice, for its technology advantage outweighed the other vendors' advantages.

Options vs. All Criteria	Low purchase price 0.22	Anticipate future needs 0.65	Easy transition 0.12	Sum
Vendor A	.33 × .22 .07	.25 × .65 .16	.25 × .12 .03	.26
Vendor B	.03 × .22 .01	.73 × .65 .47	.02 × .12 .00	.48 ⇦
Vendor C	.64 × .22 .14	.01 × .65 .01	.73 × .12 .09	.24

Figure 5.152 Analytical criteria method example: summary matrix.

consensus criteria method

When to Use

- When the situations listed for "when to use" any prioritization matrix are true, and:
- When differences between options are approximately the same
- When it is not appropriate to devote the time required for the analytical criteria method, or . . .
- When a full exchange of views, with the possibility of learning from each other, is not necessary

Procedure

1. Clarify the goal. Use operational definitions where possible. Make sure everyone understands and agrees on what is to be decided.
2. List criteria that the decision must meet.

Criteria Weighting

3. Each person individually prioritizes the criteria by distributing the value of 1.0 among them. In other words, give each criterion a weight from 0 to 1.0, where the weights add up to 1.0.
4. Create an L-shaped matrix with criteria as row labels and group members' names as column headings. Record everybody's weights in their columns. Add across the rows to get a sum for each criterion.

5. If individuals' weights are very different, discuss the reasons for the differences. Repeat steps 3 and 4 if opinions change.

Ranking Options Against Criteria

6. Thinking only about the first criterion, each person individually rank-orders the options, from one to the number of options. The option that meets the criterion best receives the largest number.
7. Create an L-shaped matrix with options as row labels and group members' names as column headings. Title this matrix with the first criterion. Record everybody's rankings in their columns. Add across the rows to get a sum for each option. In the "sum" column, rank the options, from one to the number of options, based on the totals. These are the numbers you will use later.
8. If individuals' rankings are very different, discuss the reasons for the differences. Repeat steps 6 and 7 if opinions change.
9. Repeat steps 6 through 8 for each of the other criteria. In all, you will make as many matrices as there are criteria.

Combining All Criteria: Summary Matrix

10. Create an L-shaped matrix with options as row labels and criteria as column headings. Write the criteria weights under the criteria names.
11. In the first column, copy the option rankings from the first criterion matrix, step 7. Write "×" and the weight of the first criterion. (The same number is repeated all down the column). In each cell, multiply the two weights and write the result.
12. Repeat for each column, copying rankings from the matrix for that criterion.
13. Add the results across each row to get an option score. These scores are relative weights with respect to all criteria. The largest number is the best choice.

Example

Using the same example of a computer upgrade, Figures 5.153 through 5.156 are the matrices used in this method. To make the example simple, we'll assume the decision-making team contained four individuals, Karla, Louis, Marie, and Norm. They did not create a matrix for the first criterion, "low purchase price," since the vendor's bids automatically generated a ranking.

Criteria	K	L	M	N	Sum	Wts
Low purchase price	.15	.15	.25	.30	.85	.21
Anticipate future needs	.70	.65	.55	.60	2.50	.63
Easy transition	.10	.12	.15	.10	.47	.12
After-sale support	.05	.08	.05	.00	.18	.05
					4.00	

Figure 5.153 Consensus criteria method example: criteria matrix.

Anticipate Future Needs	K	L	M	N	Sum	Rank
Vendor A	1	1	2	1	5	0.21 (1)
Vendor B	3	3	3	3	12	0.50 (3)
Vendor C	2	2	1	2	7	0.29 (2)
					24	

Figure 5.154 Consensus criteria method example: criteria versus option 2 matrix.

Easy Transition	K	L	M	N	Sum	Rank
Vendor A	1	2	3	2	8	0.33 (2)
Vendor B	2	1	1	1	5	0.21 (1)
Vendor C	3	3	2	3	11	0.46 (3)
					24	

Figure 5.155 Consensus criteria method example: criteria versus option 3 matrix.

Options vs. All Criteria	Low purchase price 0.21	Anticipate future needs 0.63	Easy transition 0.12	Sum
Vendor A	2 × .21 .42	2 × .63 1.26	2 × .12 .24	1.92
Vendor B	1 × .21 .21	3 × .63 1.89	1 × .12 .12	2.22
Vendor C	3 × .21 .63	1 × .63 .63	3 × .12 .36	1.62

Figure 5.156 Consensus criteria method example: summary matrix.

combination ID/matrix method

When to Use

- When a decision must be made about the best option or set of options, and . . .
- When the issue is complex, with interlocking cause-and-effect relationships, and . . .
- When the decision should be based not on a set of known criteria but on the options that best address root causes or have the broadest impact on the issue, and . . .
- When the decision-makers have good knowledge of the issue and the options
- Such as when choosing which of several related problems to attack first, or when you're not even sure which are the causes and which are the effects

Procedure

Constructing the Matrix

1. Create an L-shaped matrix. It should have five more columns than there are options, to allow for row labels and totals, and one row more than the number of options, to allow for column headings. List the options as row labels and, in the same order, as column headings. (Or number each option and use the numbers as column headings.) The diagonal of cells where the row and column labels are the same will not be used, so block them out, such as by shading them.

2. Consider the first option and the second one. Ask two questions:

   "Does the first option cause or influence the second?" If it does, place an upward-pointing arrow in the box where that row and column intersect.

   "If there is a relationship, how strong is it?" Draw one of these symbols next to the arrow to indicate the strength of the relationship:

   ◎ Strong relationship

   ○ Medium relationship

   △ Weak or possible relationship

3. Compare each option to every other option pairwise.
4. Reflect your marks across the diagonal. For every arrow pointing up, draw an arrow pointing left in the other box where those two options intersect. Also draw the same relationship symbol.

Adding Up the Matrix

5. For each row, count the number of arrows pointing left. Write that number in the first column beyond the options, under "total in."
6. For each row, count the number of arrows pointing up. Write that number in the second column beyond the options, under "total out."
7. Add "total in" and "total out" and write the sums in the next column, under "total in and out."
8. Assign points to the relationship symbols as follows:

   ◎ 9 points   ○ 3 points   △ 1 point

   For each row, add the relationship symbols' values. Write that number in the last column, under "strength."

Analyzing the Matrix

9. Which options rate highest in the "strength" column? Any option chosen should show strong connections to other options. Eliminate any options that have low "strength" numbers.
10. Which options rate highest in the "total in and out" column? The best options have extensive connections to other options. Eliminate options that have a low total in this column.
11. Of the remaining options, which have the highest "total out"? These options affect many others. They are strongly related to the root causes and are often the best ones to address first.

12. Which options have the highest "total in"? These options are affected by many others, so they may serve as indicators or measures of success.

Example

This tool is usually used to sort out a long list of options. This example has only seven to make the procedure easier to see.

Traffic planners are looking at problems in the downtown area. They use a combination ID/matrix diagram, Figure 5.157, to analyze which problem to address first.

"Backed-up traffic"and "short traffic signal cycle" have the highest strength numbers by far. Next is a cluster of options 2, 3, 5, and 6. "Backed up traffic" is a mix of ins and outs, but it has more ins than outs and only one strong out. In contrast, "short traffic signal cycle" has three strong outs, one of them to "backed-up traffic." The first priority should be fixing the short traffic signal.

Considerations

- When you are listing criteria, brainstorming, affinity diagram, and list reduction may be useful. Operational definitions ensure that everyone knows exactly how to apply the criteria.
- When you are listing criteria, be sure the preferred state is clear. For example, instead of "resources," write "low use of resources."
- Do not discuss the relative importance of criteria when first listing them.
- A tree diagram is often used before this tool to generate options.
- Make sure you have calculators or spreadsheet software handy. Check and double-check the math!

	1	2	3	4	5	6	7	Total in	Total out	Total in and out	Strength
1. Narrow streets				↑○				0	1	1	3
2. Pedestrian access problems			←△	←△			←◎	3	0	3	11
3. Interstate off-ramp location		↑△		↑◎				0	2	2	10
4. Backed-up traffic on Main St.	←○	↑△	←◎			↑◎	←◎	3	2	5	31
5. Business park access problems							←◎	1	0	1	9
6. Hard to turn left mid-block				←◎				1	0	1	9
7. Short traffic signal cycle		↑◎		↑◎	↑◎			0	3	3	27

Figure 5.157 Combination ID/matrix method example.

- A very long list of options can first be shortened with a tool such as multivoting or list reduction.
- Before accepting the highest score as the decision, study the results, especially if they are surprising. Graph costs versus benefits. Do a sensitivity analysis: does changing the weights or the scoring system affect the results? If the weights or scoring were controversial, this analysis may help achieve agreement. Look at large differences in scores on important criteria. Compare the most beneficial option with the least expensive. Do these comparisons suggest an improved option?
- The decision matrix is similar to this tool but less rigorous and time-consuming. The effective–achievable matrix is a tool that separates ease-of-accomplishment criteria from effectiveness criteria and visually compares how well various options meet criteria. Paired comparisons can be used for a decision based on individual preferences rather than criteria.

Analytical Criteria Method

- Encourage full discussion whenever you are making a pairwise comparison of criteria in steps 4 and 6 or of options in step 11. Individuals should voice reasons for their opinions. Discuss the comparison until the group reaches consensus on the rating. This is where problems with hidden agendas and pet projects will be resolved. If consensus is reached on each separate comparison, the group automatically has consensus when the final matrix is calculated.
- This method can take a long time, especially if strong differences of opinion exist.
- Sometimes a great deal of conflict is experienced during this process. However, if the decision is critical, bringing that conflict to the surface during the decision-making process is preferable to having the decision sabotaged later by lack of support.
- For complex decisions, this method cuts through the complexity by handling one aspect at a time. Without such a structured tool, discussions can be hard to keep on track and key aspects can be missed.
- This tool is based on the analytical hierarchy process (AHP), a process developed by Thomas L. Saaty. He used a comparison scale from 1 to 9 and more complex mathematics than the analytical criteria method. Other methods for multi-criteria decision analysis also have been developed. Computer software is used for many of these methods because of their mathematical complexity.

Consensus Criteria Method

- It really doesn't matter what value is distributed by each person. Distributing a value of 1.0 is common, but it might be easier to think about distributing 10 points across the criteria. You can work with the resulting higher sums or divide them all by 10.

- If individuals' rankings vary greatly and opinions are not changed by discussion and repeating the ranking, this method does not truly generate consensus. The best solution will not be reached by averaging strong and contrasting opinions to an in-between ranking.
- This method is unable to reflect significant differences between options. For example, suppose that on one criterion, option A is significantly better than option B, which is marginally better than option C. The ranking will be the same as if option A was only marginally better than option B. The final choice will not reflect how outstanding option A is on this criterion. So do not use this method if, for any criterion, differences between options are not approximately consistent.

Combination ID/Matrix Method

- It's okay to start with a mishmash of causes and effects. The tool will sort them out.
- This tool is most useful with a long list of options, 15 to 50.
- Do not ask, "Is this option caused or influenced by the other?" It's less confusing to think only about which options cause others. When you reflect your matrix across the diagonal, the "caused by" question is taken care of.
- Do not draw double-headed arrows. Force yourself to decide which direction the cause or influence works. Similarly, when you reflect your matrix, if you find an arrow already in the second location, you need to rethink that pair of options.
- Do not rely strictly on the math to make your decision. Instead, weigh the indications of the totals with the experience of the group.
- If an option shows approximately equal numbers of in and out arrows, look at the strengths of those arrows. For example, the out arrows may show all weak relationships, but the in arrows strong relationships. This option is an indicator of success, not a cause.
- This tool is based on the relations diagram, also called the interrelationship diagram—thus the "ID" in the name.

process capability study

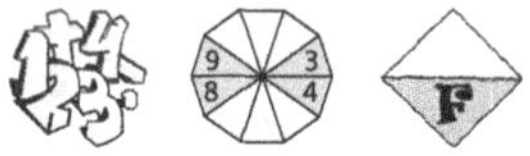

Description

A process capability study analyzes the ability of a stable process to generate output within specification limits for a particular quality characteristic. The study calculates an index, called a process capability index (PCI) which compares process variation to specifications. Several dozen indices have been devised, although only a few are commonly used.

When to Use

- When the process is in statistical control, and . . .
- When the process output forms a normal distribution, and . . .
- When measuring how well a process can meet requirements, such as:
- When monitoring improvement in process performance over time, or . . .
- When comparing two processes to determine which is better able to meet requirements, or . . .
- When communicating with management or a customer about the process's performance

Cpk and Cp, Ppk and Pp

- When minimizing the amount of off-spec product is a more important goal than minimizing variation of the process average from target

Cpm and Cpmk

- When minimizing variation around a target is a more important goal than minimizing off-spec product

Procedure

1. Establish that the process is in statistical control, using a control chart. If not, do not go any further.
2. Establish that the process forms a normal distribution. Use the normal probability plot or a goodness-of-fit test. Statistical software and/or the advice of a statistician would be useful here. If the process is not normal, do not proceed. See "Considerations" for several options.
3. Determine $\hat{\mu}$, the estimated process average, the centerline of your control chart:

   $\hat{\mu} = \bar{\bar{X}}$ if you are using an $\bar{X}$ chart

   $\hat{\mu} = \bar{X}$ if you are using a chart of individuals (X chart)
4. Determine $\hat{\sigma}_X$, the estimated process standard deviation, using the overall standard deviation of sample data.

   Preferred method: Use the root mean square deviation.

$$\hat{\sigma}_X = s = \sqrt{\frac{\sum_{i=1}^{m}(X_i - \hat{\mu})^2}{m-1}}$$

where m = the number of measurements used in the calculation

Statistical software and electronic calculators often use this formula to calculate $\hat{\sigma}_X$. This is sometimes called the overall or long-term deviation.

Alternate method: Use control chart calculations.

From a chart of individuals, use $\hat{\sigma}_X$ from the control chart worksheet (Figure 5.27, page 169).

From an $\bar{X}$ and R chart, calculate:

$$\hat{\sigma}_X = \bar{R} \div d_2$$

where d_2 is read from Table A.2

This is called the within-subgroup or short-term deviation.

Caution: If you use this method, do not use the $\hat{\sigma}_X$ value calculated on the $\bar{X}$ and R control chart worksheet (Figure 5.22, page 162). That is the standard deviation of the subgroup averages, whose spread is narrower than that of the original data. For capability, you must use $\hat{\sigma}_X$ calculated from the original data, which this formula does.

Note: When s is used for $\hat{\sigma}_X$, the resulting values are sometimes called Pp and Ppk instead of Cp and Cpk. For simplicity, this procedure refers only to Cp and Cpk.

5. Determine estimated *process spread* and *process limits:*

$$\text{Estimated process spread} = 6\hat{\sigma}_X$$

$$\text{Estimated upper process limit} = \hat{\mu} + 3\hat{\sigma}_X$$

$$\text{Estimated lower process limit} = \hat{\mu} - 3\hat{\sigma}_X$$

6. Calculate Cp or Cpm. *Note*: Cp or Cpm cannot be calculated if there is only one specification.

$$\text{Cp} = \frac{\text{USL} - \text{LSL}}{6\hat{\sigma}_X} = \frac{\text{Specification spread}}{\text{Estimated process spread}}$$

$$\text{Cpm} = \frac{\text{USL} - \text{LSL}}{6\sqrt{\hat{\sigma}_X^2 + (\hat{\mu} - T)^2}}$$

T = Target value for process average

USL = Upper specification limit

LSL = Lower specification limit

7. Calculate Cpk or Cpmk. Determine the distance from the process average to the closer specification limit. That value goes in the numerator.

$$\text{Cpk} = \frac{\text{smaller}\left\{\text{USL} - \hat{\mu},\ \hat{\mu} - \text{LSL}\right\}}{3\hat{\sigma}_X}$$

$$\text{Cpmk} = \frac{\text{smaller}\left\{\text{USL} - \hat{\mu},\ \hat{\mu} - \text{LSL}\right\}}{3\sqrt{\hat{\sigma}_X^2 + \left(\hat{\mu} - T\right)^2}}$$

8. Larger index values are better. If they are less than the value set by organizational goals or the customer, determine how to improve the process:

   - Cp < 1: If process spread is greater than specification spread, work to reduce variation.
   - Cp ≥ 1 and Cpk < 1: If process spread is less than or equal to specification spread, but one process limit is beyond a specification limit, work to center the process within the specifications.
   - Both Cp and Cpk ≥ 1: If the process limits are within the specification limits, the process is in a good state. Work to reduce variation or improve centering, if either improvement is beneficial for you or your customer.
   - Cpm and Cpmk: Work to bring the process average closer to target or to reduce variation, whichever will provide greater benefit.

Example

Figure 5.158 shows the output from a process with a normal distribution. To make the calculations and relationships easier to understand, it is set up with $\sigma = 2$. The process is in control, so capability limits can be calculated.

The upper process limit falls beyond the upper specification limit, so off-spec product is being made in the range from 16 to 18, as shown by the shaded area.

Specification spread = 16 – 0 = 16

Process spread = 18 – 6 = 12

Cp = specification spread ÷ process spread = 16 ÷ 12 = 1.33

Cpk = $(\text{USL} - \hat{\mu}) \div 3\hat{\sigma}_X = 4 \div 6 = 0.67$

The Cp of 1.33 reflects how the process *could* perform *if* it were centered between the specification limits. Cpk less than 1 indicates that a process limit is outside specifications and the process is currently making significant amounts of off-spec product. Figure 5.159 shows the same process, but centered.

Now suppose that this process drifts over time. Sometimes the mean is centered at 8, sometimes it moves as high as 12, or it could be somewhere in between. But suppose

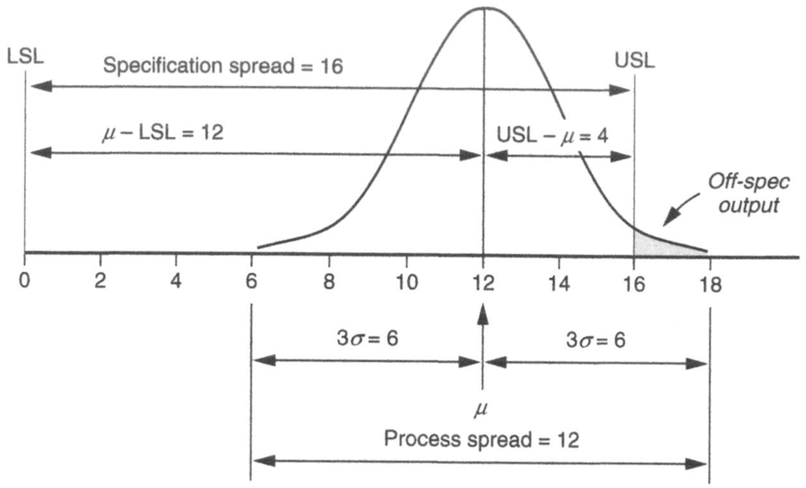

Figure 5.158 Incapable process.

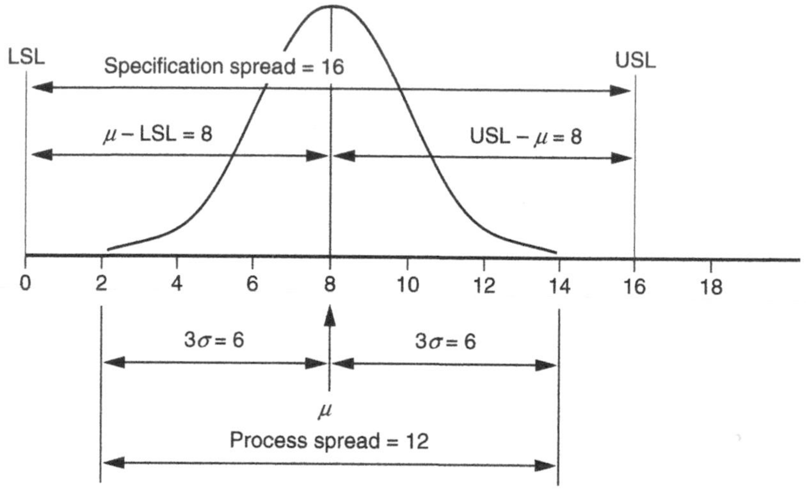

Figure 5.159 Capable process.

also that the variation does not change—a snapshot at any time would show the output as wide as the normal curve in the figures.

The distribution of all the data would be the sum of all the process curves for the entire time period. Suppose that as the process drifts, it spends equal amounts of time at every value between 8 and 12. Then its curve would be wider and taller, as in Figure 5.160.

Since the process is drifting, it is not stable—not in statistical control. It would also not be normal. Under these conditions, it is inappropriate to calculate any capability index. Let's look at what would happen if you did calculate Cp and Cpk for this unstable process, using $\hat{\sigma}_X = \bar{R} \div d_2$ for the standard deviation. This calculation uses averages of

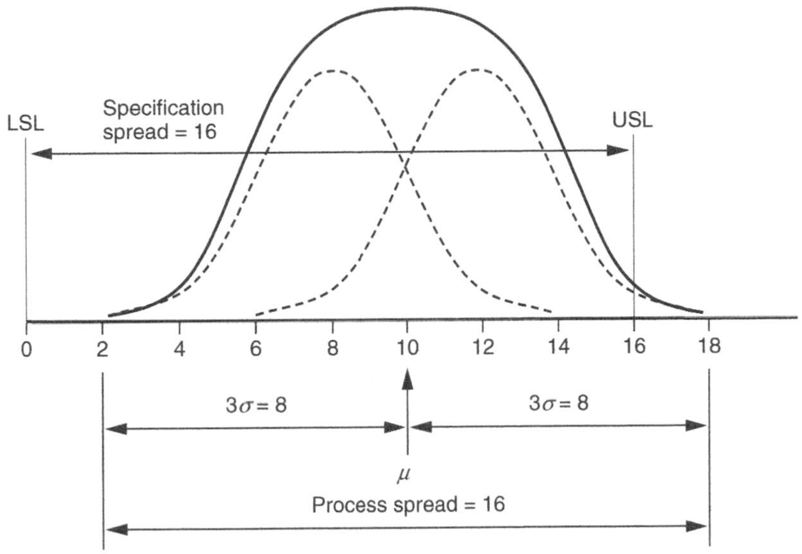

Figure 5.160 Drifting process.

small subgroups, and each subgroup would look like the narrower distributions of Figures 5.158 and 5.159. Thus, $\hat{\sigma}_X$ would be 2, and Cp would be 1.33, as in the previous example. If the process average were 10, then

$$\text{Cpk} = (16 - 10) \div 6 = 1$$

These values of Cp and Cpk are misleading. They indicate that the process has stayed within specification limits, which the diagram clearly shows is not true. Because this process is unpredictable, using s for the standard deviation would also produce misleading results.

Finally, let's look at Cpm and Cpmk for the process in Figure 5.158. Suppose the target for the process is 12, the process average. Then $\bar{X} - T$ in the denominator becomes zero and

$$\text{Cpm} = (16-0) \div 6\sqrt{2^2} = 16 \div 12 = 1.33$$

$$\text{Cpmk} = (16-12) \div 3\sqrt{2^2} = 4 \div 6 = 0.67$$

These results are the same as Cp and Cpk. But suppose the target is 8, the center of the specification spread.

$$\text{Cpm} = (16-0) \div 6\sqrt{2^2 + (12-8)^2} = 16 \div 6\sqrt{4+16} = 0.60$$

$$\text{Cpmk} = (16-12) \div 3\sqrt{2^2 + (12-8)^2} = 4 \div 3\sqrt{4+16} = 0.30$$

Cpm and Cpmk penalize the process severely for being off target.

Considerations

Interpreting the Indices

- Cp measures only variability. Cpk also considers how close the process average is to the specification limits.
- To understand the capability of your process, you must consider both the spread (width) of the process and where the average is positioned. The process in Figure 5.158 is narrow enough to fit the specifications, but because it is off center, it is producing out-of-specification product, represented by the shaded area.
- If the process is perfectly centered, Cp will equal Cpk.
- If Cp is greater than or equal to 1.0 (process spread narrower than specification spread) but Cpk is less than 1 (a process limit outside specifications), the process *would* be capable of meeting specifications *if* it were centered.
- Because a normal distribution has tails extending beyond 3σ on either side, even a perfectly centered distribution with a Cpk of 1.0 will have a small amount of product outside the specifications—0.27%, to be exact. (In the normal distribution, 99.73% of the distribution lies within 3σ of the average.) In addition, processes often drift slightly. Therefore, many customers require their suppliers to demonstrate that their processes have a higher Cpk than 1. With a Cpk of 1.33, the specification lies 1σ beyond the process limit, which provides a comfortable cushion.
- The requirement for six sigma capability has been made famous by Motorola and is the basis for the widely-used Six Sigma process. In this context, "six sigma" means that process variation has been reduced until the distance from either specification limit to the centered process average is at least 6σ. The total specification spread should be 12σ. "Six sigma" also assumes that the process average might drift up to 1.5σ from the centered target. You might want to test your understanding of capability by calculating Cp and Cpk for a "six sigma" process whose specification spread is 12σ and whose process average is 1.5σ from the midpoint. (Answer below.) See *Six Sigma* on page 27 for more information.
- Cpk is highest when the process average is located at the midpoint of the specification spread. If the best target for the process average is not the midpoint, Cpk becomes misleading, and Cpm or Cpmk may be a better index.
- Cpm and Cpmk consider how close the process average is to its target. If the process is perfectly on target, Cpm = Cp and Cpmk = Cpk.
- All the indices are statistical estimates, based on sample data, of unknown true parameters of the process. Therefore, they should be written $\hat{C}p$, $\hat{C}pk$, and so on. However, common usage omits the "hat," and that usage is reflected here.

Problems with Capability Indices

- Use of capability indices is controversial. Too often the number is improperly calculated and is misleading or meaningless. Be sure to follow the rules about statistical control and normal distribution. Reporting process limits and process spread instead of a capability index avoids some of these problems.
- Above all, process capability is a means for communication within and between organizations. Make sure that everyone with whom you are discussing capability—coworkers, managers, customers—agrees on operational definitions of capability, including sampling methods, formulas to be used, and how to handle nonnormal distributions and unstable situations. If software does calculations for you, be sure you know which index is being calculated and what formulas are used, especially for $\hat{\sigma}_X$.
- The purpose of measuring and reporting process capability is to predict future performance of the process. This is possible only if the process is in statistical control. If the process is out of control (unstable), all you can measure and report with any certainty is past performance of the process.
- Do not drop outlying data from the calculation just because you have identified its cause. Include all bad points unless you truly have eliminated forever and always the cause of the outlier.
- A problem with using $\hat{\sigma}_X = \bar{R} \div d_2$ is that the way the data are subgrouped can greatly influence the results.
- Many statisticians consider s the preferable standard deviation to use for calculating capability indices, which are then called Pp and Ppk. Follow the requirements of your organization or your customers (if they require reporting of capability indices) and above all, be sure that reported values are accompanied by explanations of how they were calculated.
- As with any calculated statistic, capability indices have built-in error. Sample size, sampling error, measurement error, distributions that aren't quite normal, especially in the tails—all can cause the calculated capability to differ significantly from the "true" capability. Unfortunately, confidence intervals usually are not reported for capability.
- Neither Cp nor Cpk are directly related to the amount of out-of-spec output produced, although it can be determined from the two together. Two processes with identical Cpk can have different amounts of off-spec output and may need different actions to improve each process.
- If proportion nonconforming or defects are the measurement of interest, it might be more useful to report and monitor them directly. Even then, however, knowledge of the process is essential to know what actions to take to improve the process.

Nonnormal Processes

- The output of many processes do not form normal distributions. (See *histogram* for more information about different kinds of distributions.) The capability indices described here are only meaningful for the normal distribution. For example, we commonly understand Cp = 1 to mean that 99.73% of a stable, centered distribution lies between specifications, but that is true only for a normal distribution. For other distributions, such as the common skewed distribution, Cp and Cpk values calculated by the procedures in this book cannot be compared or interpreted.
- There are other options for calculating a capability index for nonnormal distributions. Whenever you work with a nonnormal distribution, $\bar{R} \div d_2$ cannot be used for $\hat{\sigma}_X$.

  1. Transform the data, including specification limits. For example, a skewed distribution can be transformed into a normal distribution using natural logarithms. Then use the formulas presented here.
  2. Fit a known distribution to the process. Then calculate from that distribution the expected proportion nonconforming. Determine the Cpk that would have the same proportion nonconforming for a normal distribution. Report both the proportion nonconforming and the "equivalent" Cpk.
  3. Do not calculate a capability index. Instead, calculate and report the actual percentage of the process that is outside specifications.

  Consult a statistician for assistance with these methods and for other options.
- Answer to Six Sigma question: Cp = 2.0 and Cpk = 1.5.

process decision program chart

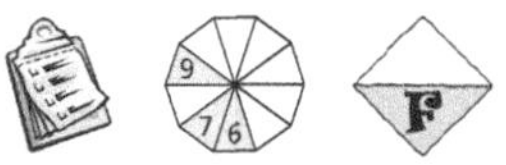

Also called: PDPC

Description

The process decision program chart systematically identifies what might go wrong in a plan under development. Countermeasures are developed to prevent or offset those problems. By using PDPC, you can either revise the plan to avoid the problems or be ready with the best response when a problem occurs. This variation of the tree diagram is one of the seven new QC tools.

When to Use

- Before implementing a plan, especially . . .
- When the plan is large and complex, and . . .
- When the plan must be completed on schedule, or . . .
- When the price of failure is high

Procedure

1. Obtain or develop a tree diagram of the proposed plan. This should be a high-level diagram, showing the objective, a second level of main activities, and a third level of broadly defined tasks to accomplish the main activities.
2. For each task on the third level, brainstorm what could go wrong.
3. Review all the potential problems and eliminate any that are improbable or whose consequences would be insignificant. Show the problems as a fourth level linked to the tasks.
4. For each potential problem, brainstorm possible countermeasures. These might be either actions or changes to the plan that would prevent the problem or actions that would remedy it once it occurred. Show the countermeasures as a fifth level, outlined in clouds or jagged lines.
5. Decide how practical each countermeasure is. Use criteria such as cost, time required, ease of implementation, and effectiveness. Mark impractical countermeasures with an X and practical ones with an O.

Example

A medical group is planning to improve the care of patients with chronic illnesses such as diabetes and asthma through a new chronic illness management program (CIMP). They have defined four main elements and, for each of these elements, key components. The information is laid out in the process decision program chart of Figure 5.161.

Dotted lines represent sections of the chart that have been omitted. Only some of the potential problems and countermeasures identified by the planning team are shown on the chart. Thinking about these helped the team create a better program. For example, one of the possible problems with patients' goal setting is backsliding. The team liked the idea of each patient having a buddy or sponsor and will add that to the program design. Other areas of the chart helped them plan better rollout, such as arranging for all staff to visit a clinic with a CIMP program in place. Still other areas allowed them to plan in advance for problems, such as training the CIMP nurses how to counsel patients who choose inappropriate goals.

This CIMP example is also used for potential problem analysis to allow you to compare the two tools.

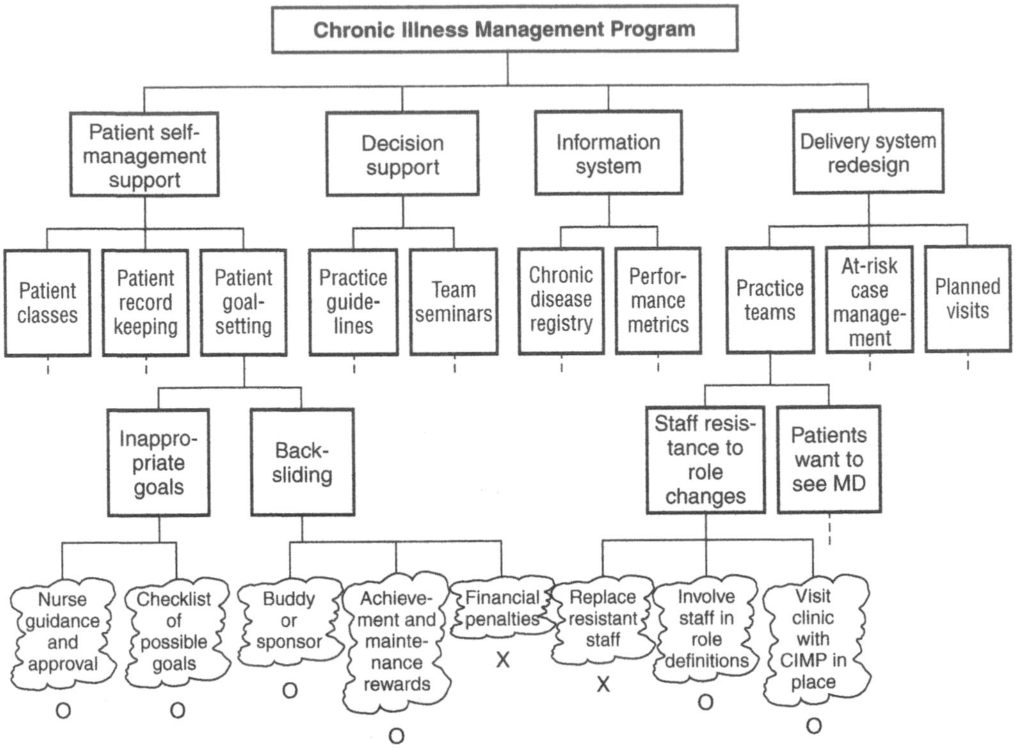

Figure 5.161 Process description program chart example.

Considerations

- This is one of several tools that can be used for contingency planning. PDPC is more systematic than the contingency diagram. Potential problem analysis (PPA) analyzes situations more rigorously. FMEA is most appropriate for new or redesigned processes, products, or services.
- One advantage of this tool is that it starts with a tree diagram of the implementation plan to identify problems. Advantages of PPA are that it specifically addresses causes, rates risk, and separates preventive actions from contingency plans. Consider combining these tools: start with PDPC to identify problems, then use PPA to analyze them.
- Here are some questions that can be used to identify problems:
  - What inputs must be present? Are there any undesirable inputs linked to the good inputs?
  - What outputs are we expecting? Might others happen as well?
  - What is this supposed to do? Is there something else that it might do instead or in addition?

– Does this depend on actions, conditions, or events? Are these controllable or uncontrollable?

– What cannot be changed or is inflexible?

– Have we allowed any margin for error?

– What assumptions are we making that could turn out to be wrong?

– What has been our experience in similar situations in the past?

– How is this different from before?

– If we wanted this to fail, how could we accomplish that?

- Potential problems are sometimes called "what-ifs."
- The tree can be drawn either horizontally or vertically.

project charter

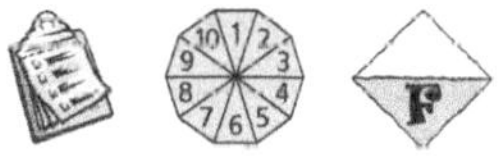

Also called: mission statement

Description

The project charter is a document that authorizes and defines an improvement project.

When to Use

- When a project is started, or . . .
- Whenever someone needs information about the project or clarification of its purpose, scope, timing, or status, or . . .
- At reviews, to ensure that the project is appropriately focused and on time, and to update the charter with current progress, or . . .
- When information learned in the course of the project indicates a need to change the charter

Procedure

1. The project's sponsor (a member of management who initiates and supports the project) and the project's team leader collaborate in writing the charter at the beginning of the project. Many organizations have a standard format. Here are essential elements, with comments or questions that can help fill in the blanks.

Project title	How will this project be named in reports and best-practice databases?
Process name and product or service affected	What functions and outputs of the organization will be the project's focus?
Problem statement	What is wrong with the process, product, or service now?
Objective(s) and/or goal(s)	What improvement or results are expected? State outcomes in terms of process measurements that will be used to measure success.
Scope	Which parts of the process are included and which are not? What are the start and end points of the process?
Business case	What is the importance of the process? What are the current costs of poor quality? What financial impact will the project have: cost savings, increased profit, and cost avoidance? What will the impact be on external customers?
Project start date	This is the date the charter is defined.
Target completion date	Estimate based on project complexity, holidays, workloads, travel or customer contact required, and so on.
Anticipated key milestone dates	Gate reviews, presentations, customer visits, and so on.
People	List names, roles, and contact information: • Project sponsor • Project leader • Team members • Team facilitator • Expert technical resource • Additional process or subject-matter experts • Additional key stakeholders
Signatures and dates	Preparation and approval signatures and dates, revision date.

2. The charter is reviewed and approved by management.
3. As the project becomes better defined during the early steps of defining customer needs, exploring the problem, and establishing measurements, the charter should be reviewed, revised, and reapproved.

Example

Medrad's accounting managers wanted to reduce the work required to process freight invoices in order to free up people to do more value-added work, to eliminate a temporary position, and also to prepare the department to handle future company growth. With the help of a performance improvement facilitator from Medrad's Performance Excellence Center, they defined the project and created a charter, shown in Figure 5.162. In addition to the essential elements listed in the procedure above, the form includes sections for deliverables, support required, and realized financial impact (to be filled in when the project is completed.)

This charter included a general project objective as well as three more specific project goals. Because the project was very unfocused at the beginning, goals and objectives could not be quantified, except for the savings from eliminating the temporary position.

Management and team members realized that major enhancements could be made to the process, but that was not the intent of this project. The charter states that the scope does not include major process improvements.

This revision of the charter was prepared midway through the project, during the improvement phase. Dates had been entered as each of the first three phases were completed. Metrics—baseline and goal—had been added during the measurement phase. The support required had been quantitatively defined during the analysis phase. For this revision, current data was added to the metrics section. (Because of confidentiality, actual numbers for the metrics are not shown, and the names are fictitious.)

See the Medrad freight processing team's story on page 55 for more information about this project.

Considerations

- Use operational definitions to develop statements that are meaningful to everyone involved.
- Three components make up a good improvement objective: direction and magnitude of change, measure of quality, and name of process. String the three together and you have a statement of your objective. For example: "Increase from 93 percent to 98 percent the rate of error-free transactions in the call-handling process." Or, "Reduce by 25 percent the cycle time for packaging parts."
- Use the SMART acronym to evaluate objectives: Are they simple and specific? Measurable? Agreed-upon and accountable? Realistic? Time-related?

Project Charter	
Project Title: Freight Processing Improvement	**Business Unit:** Accounting Operations
Project Leader: John Doe **Team Members & Responsibilities:** Meg O'Toole (Accounting), Deanne Zigler (Supply Chain) **Facilitator:** Steve White **Black Belt:** Steve White **Sponsor:** Maria Martinez **Process Owner:** George Gregory, Red Morris	**Business Unit Description:** Daily processing of accounting transactions: mailing invoices, paying vendors, processing invoices, etc. **Unit Manager:** John Doe

Problem Statement: A large number of hours are spent processing the payment of inbound and outbound freight.

Project Objective: Improve the efficiency of the freight payables process through better utilization of company resources and improvements in the service provided to the department's customers.

Business Case/Financial Impact: The improvements will save $xxM annually in labor. These improvements will enable the process to support future growth of the company.

Project SMART Goals:
- Reduce the labor needed to process freight bills.
- Make processes more robust to handle more capacity.
- Increase awareness of process improvement tools.

	Baseline	Current	Goal
CCID	8–18	10–27	
Defects DPM	xx% xxx,xxx	xx% xx,xxx	x% xx,xxx
Process σ	x.xx	x.xx	x.xx

Project Scope Is: The current process of freight payables, including minor improvements to systems that provide inputs to this process.
Project Scope Is Not: Major system enhancements; other accounting operations functions.

Deliverables:
- Identification of business elements critical in importance to our customers.
- Identification of baseline measurements and performance goals.
- Process plan and budget required for implementation of the improvements.
- Upon approval, successful development, implementation, and control of the improvements.
- List of identified opportunities for future projects.

Support Required: IT—Suzanne Smith to adjust ERP (estimate 8 days). Shipping software consultant to incorporate entry screen changes. Planning—Mark Savoy (4 hours).

Schedule: (key milestones and dates)	Target	Actual	Status
Project Start	7-07-03		
D—Define: Confirm scope and purpose statements. Document findings from site visits.	7-31-03	7-31-03	complete
M—Measure: Determine appropriate measurements. Calculate baseline and performance goals.	8-15-03	8-20-03	complete
A—Analyze: Perform process analysis and identify drivers of baseline performance. Document analysis of opportunity areas. Gain approval for I & C.	9-19-03	10-02-03	complete
I—Innovative improvement: Improve and implement new process.	11-10-03		planning
C—Control: Verify that performance goals were met. Standardize the process and list future improvement opportunities.	12-01-03		

Realized Financial Impact:		**Validated by:**
Prepared by: Steve White	**Date:** 7-7-03	**Revision date:** 10-27-03
Approvals	**Manager:**	**Sponsor:**

Figure 5.162 Project charter example.

- Define a scope that can be achieved in less than six months. Many projects fail because of too ambitious a scope. If a process requires more work than that, attack it in a series of smaller projects.
- See *operational definitions*, *ACORN test,* and *continuum of team goals* to help define the objective or goal.

project charter checklist

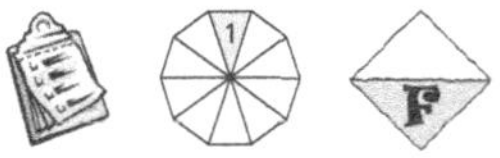

Also called: mission statement checklist

Description

The project charter checklist ensures that the entire team understands what it is supposed to accomplish and how.

When to Use

- During the first or second meeting of a new team
- When a team is given a new or modified charter

Procedure

Discuss the following questions:

1. Is it clear what management expects of us?
2. Is it clear why this project is important to the organization? Is it clear where this project fits into the organization's overall goals and objectives?
3. Does our project cover only part of a larger process? Where do we fit in? Where does our part of the process start and end?
4. Are the boundaries of the project clear? What will be outside our jurisdiction?
5. Are the goals realistic?
6. Will this project work? Does the charter fit in with our knowledge about the process or system?
7. What resources, inside or outside the department, will we need?
8. Do we have the right people on this team to accomplish the mission?
9. What people not on the team will be crucial to our efforts?

10. Who will support our efforts? Who will be opposed? Who will be neutral? How can we reach all of these people?
11. How will opposition be expressed? How can we be sure that any opposition is brought out into the open?
12. How will we know that we have been successful? How will others decide whether we have been successful?

Example

A team in a technical support department was given this objective: "Improve report preparation through review of the steps in the process." As the team discussed the items on the project charter checklist, problems became obvious. Here is some of their discussion.

1. How much improvement do our managers expect? What kind of improvement are they interested in? Are all aspects of report preparation of equal concern?
2. How does report preparation affect the organization? What aspects are causing problems to us or to our customers?
3. How much of the process should we address? Are we talking about only the report-writing process or the entire survey process that culminates in the report?
4. Are we allowed to look at issues within the departments we survey that cause problems for report writing? What about the management review of the reports?
5. How long does management expect this project to last? Will that be enough time, given our other responsibilities?
6. This project could work, depending on the answers to some of our questions.
7. Has management notified the managers of our customer departments that we'll be needing some of their people's time for giving us input? Have they notified our own department, and especially the secretaries?
8. We think we have the right people here.
9. Crucial people: other report writers in our department, the secretaries, and our managers!
10. The report writers aren't going to want to do things differently. The secretaries could be on our side if we find ways to reduce their work.
11. Let's be sure to get input about problems from all the report writers and address them. We could do a barriers-and-benefits exercise with them when we have changes to suggest.
12. How *will* we know we've been successful? How will our managers know? Since we don't know what kind of improvement they're looking for, how will we measure it?

The team should go back to their management to clarify the objective. Unfortunately, this example is fictitious. This was a real objective, as discussed in the final example of the continuum of team goals, but the real team did not use a project charter checklist or the continuum. Months later they thought they were finished; their management said they were not; and the team did not understand why. (See the continuum example on page 153 for more discussion.)

Considerations

- You may choose to divide into smaller groups, each assigned different questions to discuss, then reassemble to share results. The first four questions should be considered by everyone.
- Also see *continuum of team goals* for a tool that can be used with this one.

radar chart

Also called: web chart, spider chart, star chart

Description

The radar chart is a graph that looks like a spiderweb, with spokes radiating from a central point and lines connecting them. It shows measurements where several variables contribute to the overall picture. All variables are considered to be of equal importance on a radar chart.

When to Use

- When tracking or reporting performance or progress, and . . .
- When several variables are being measured to assess overall performance, and . . .
- When it is not necessary to weight the relative importance of the variables

Procedure

1. Identify the variables that will be measured. These may come from customer requirements, key performance indicators, or organizational goals. Other quality tools such as brainstorming or affinity diagrams may be used to develop the variables, or they may have been developed at another point in the quality improvement process.

2. For each variable, determine the measurement scale. It is simplest for each to have the same scale, such as 1 to 5 or a percentage, but different scales can be used if necessary. Determine which end of the scale represents desirable performance.
3. To draw the chart, divide 360 degrees by the number of criteria to determine the angle between spokes. Draw spokes of equal lengths radiating from a central point and spaced evenly around the circle. Label each spoke with its variable. Mark the measurement scale on the spokes, with undesirable performance at the center.
4. For each variable, mark its measurement on the spoke with a large dot. Connect the dots.
5. To show performance at a different time or by another subject, repeat step 4, using different line styles. Add a legend or labels to identify line styles. To show performance of multiple subjects or at multiple times, draw a separate chart for each one.
6. Assess overall performance and determine needed improvement by observing where the "web" lies closest to the center point.

Example

Figure 5.163 is a radar chart of a student's grades over two report periods. The student has improved in four subjects, but seems to be having trouble with art. To avoid confusion as additional report periods are added to the chart, separate sets of spokes could be used for each report period, as in the lower pair of diagrams.

Considerations

- Sometimes the center of the circle is designated as the most desirable end of the measurement scale. In that case, the best performance is on those criteria lying closest to the center. The problem with that approach is that as performance improves, it becomes harder to see the web.
- Sometimes a circle is drawn around the spokes. If the measurement scale is the same for each criterion, smaller concentric circles can be drawn to represent the major scale divisions. For example, if the scale is 1 to 10, circles are drawn 20, 40, 60, and 80 percent of the distance from the center to the outside circle. However, in many cases the rings of circles clutter the graph and make the lines harder to perceive.
- A ruler, compass, and protractor are helpful for drawing the chart by hand.
- Sometimes criteria and their scales are phrased so that a lower number is better. For example, "number of accidents" as a measure of safety would have zero as most desirable. If your criteria mix scales, so that high is good for some and bad

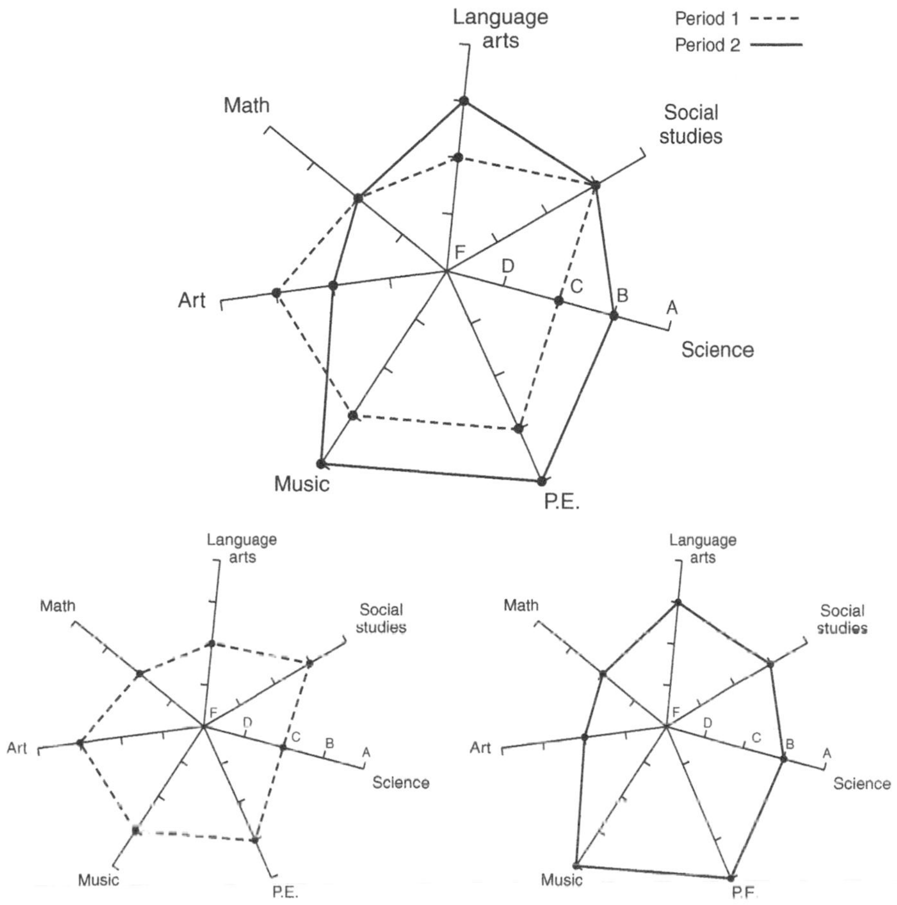

Figure 5.163 Radar chart example.

for others, you have two options. Reword your criteria so that all the scales are directionally consistent. If that isn't possible, be careful to draw your chart so that the preferable end of each scale is at the circumference. Group together spokes whose scales run from high at the center to low at the outside.

- When planning the spokes, group together variables that are related. For example, a series of radar charts comparing various computer systems might show performance-related variables grouped at the top and cost-related variables grouped at the bottom.
- A bar chart or dot chart could be used to show this kind of data. Radar charts are better for showing at a glance which categories need the most improvement or where progress has been made.
- Nominal data are placed on the spokes. The measurement scales along the spokes usually represent ordinal data.

regression analysis

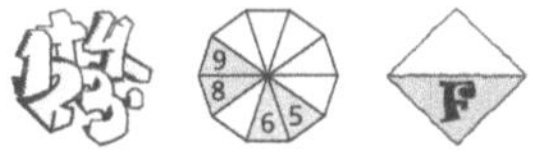

Description

Regression analysis is a statistical tool used to find a model for a relationship between pairs of numerical data. The model is a line or curve that fits the data best. The results of a regression analysis are an equation for that line or curve, a value called r^2 that indicates how good the fit is, and other statistical measures that tell how well the data match the model.

Linear regression identifies the best straight line through a scatter diagram of the data. This type of regression is the most straightforward and is discussed here. *Nonlinear regression* looks for a curve that best fits the data. *Multiple regression* is used when several independent variables affect one dependent variable. Nonlinear and multiple regressions are more complicated. Seek help from a statistician if your data require either one.

When to Use

- When you have paired numerical data, and . . .
- After drawing a scatter diagram of the data, and . . .
- When you want to know how a change in the independent variable affects the dependent variable, or . . .
- When you want to be able to predict the dependent variable if you know the independent variable, or . . .
- When you want a statistical measure of how well a line or curve fits the data

Procedure

While linear regression can be done manually, computer software makes the calculations easier. Follow the instructions accompanying your software. The analysis will generate a graph of the best-fit regression line placed through the data and a table of statistics. These will include:

- *Slope* of the line. The equation for the line has the form $\hat{y} = mx + b$. The slope is the constant m. This tells you that when the independent (x) variable increases by one, the dependent variable ($\hat{y}$) will increase by m. Positive slope means the line slants upward from left to right; negative slope means the line slants downward. (The "hat" on the y reminds us that this is the predicted value of y, not the actual value.)

- *Intercept* of the line. In the line's equation, the intercept is the constant b. This is the value of $\hat{y}$ where the line crosses the y-axis. Knowing the slope and intercept, you can draw the line or predict $\hat{y}$ from a given value of x: $\hat{y} = mx + b$.
- *Coefficient of determination, r^2*. This number, which is between 0 and 1, measures how well the data fits the line. If $r^2 = 1$, the line fits the data perfectly. As r^2 gets smaller, the line's fit becomes poorer, and predictions made from it will be less accurate. You can think of r^2 as the proportion of y's variation that is explained by the regression line. Because most data points don't fall exactly on the line, the rest of the proportion $(1 - r^2)$ is error.
- *Confidence interval,* often 95%. This is a range of values around one or more of the previous statistics. One can be 95% sure that the true value of that statistic lies within the range. A 95% confidence interval for the line is the space in which you can be 95% sure the true regression line will lie.
- The results will include additional values. Consult the software's user guide or help function, a statistics book, or a statistician to learn more about them.

Example

The ZZ-400 manufacturing unit collected data to determine whether product purity and iron contamination were related. This example is part of the ZZ-400 improvement story on page 85. See *scatter diagram* and *stratification* for the scatter diagrams of the data that the team drew first. Then they did a regression analysis.

Figure 5.164 shows the regression line of all the data combined. The value of r^2 is 0.172, which indicates a poor fit.

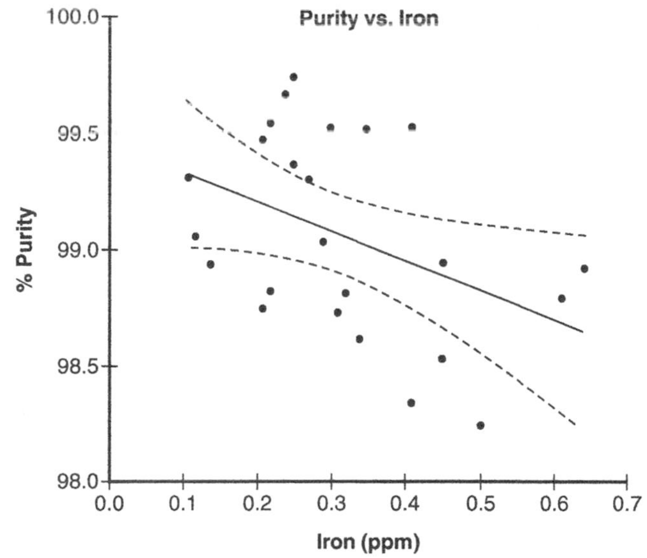

Figure 5.164 Regression analysis example.

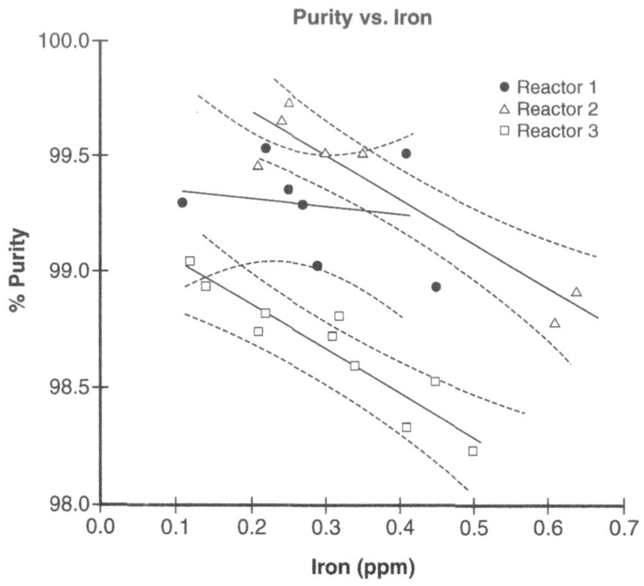

Figure 5.165 Regression analysis example, stratified data.

Table 5.13 Regression analysis example.

	Reactor 1	Reactor 2	Reactor 3
Slope	–0.64	–1.92	–1.83
Intercept	99.47	100.1	99.23
r^2	0.105	0.880	0.858

The data was stratified to show which reactor it came from. Figure 5.165 shows regression lines calculated separately for each reactor's data, and Table 5.13 shows the numerical results.

The fit for the regression lines for reactors 2 and 3 is good. Confidence intervals are shown as dotted lines on either side of those regression lines. Notice how much narrower they are than the confidence intervals for the combined data. The fit is not good for the regression line of reactor 1's data. Confidence intervals are very wide. As the scatter diagram showed, something is different about reactor 1.

Considerations

- Regression analysis models how a dependent variable changes as an independent variable is controlled. It matters which variable is assigned to the x-axis and which to the y-axis, as the result may not be the same when they are reversed. Remember that in regression, the x variable predicts the y variable, and consider carefully how to assign the variables.
- Correlation analysis quantifies the strength of the relationship between two variables, but does not calculate a line to model the data. See *correlation analysis* for more information.
- For a linear regression, an r^2 value of 0 means x and y are not linearly related. The best line is a horizontal line through the data. However, a curve might describe the relationship better. You should always look at a scatter diagram of the data first, then choose a linear or nonlinear regression based on what the data look like.
- Another reason for looking at a scatter diagram first is that very different patterns of data can generate the same statistical results. A visual check of the data can spot outliers and other features of the data that can distort the statistical calculations and might need to be removed.
- Regression analyses usually use a *least squares* method to find the best fit. The vertical distance of every data point from the line, called the *residual*, is calculated; each residual is squared; and all are summed. The line with the smallest sum is the best fit.
- To check the regression analysis, make a scatter diagram of the residuals and their line y values. Most software will do this for you. If the plot is random, everything's fine. If it shows a pattern, the assumptions used by the regression analysis do not match the data. Consult a statistician.
- As with scatter diagrams, a good correlation does not prove that x causes y. Variable y could cause variable x. Or there may be a third variable that influences both.
- As with scatter diagrams, if the regression plot shows no relationship, consider whether the independent (x) variable has been varied widely. Sometimes a relationship is not apparent because the data don't cover a wide enough range. Conversely, be cautious about making predictions outside the range of the data used for the regression. The data may behave differently beyond that range.
- The boundaries of the confidence interval will be curved. This doesn't mean the line could be curved. All the possible regression lines are close to one another near the center of the data and farther apart at extreme values of x.

relations diagram

Also called: interrelationship diagram or digraph, network diagram
Variation: matrix relations diagram

Description

The relations diagram shows cause-and-effect relationships. Just as important, the process of creating a relations diagram helps a group analyze the natural links between different aspects of a complex situation.

When to Use

- When trying to understand links between ideas or cause-and-effect relationships, such as:
- When trying to identify an area of greatest impact for improvement, or . . .
- When a complex issue is being analyzed for causes, or . . .
- When a complex solution is being implemented, or . . .
- After generating an affinity diagram, cause-and-effect diagram, or tree diagram, to more completely explore the relations of ideas

Basic Procedure

Materials needed: sticky notes or cards, large paper surface (newsprint or two flipchart pages taped together), marking pens, tape.

1. Write a statement defining the issue that the relations diagram will explore. Write it on a card or sticky note and place it at the top of the work surface.
2. Brainstorm ideas about the issue and write them on cards or notes. If another tool has preceded this one, take the ideas from the affinity diagram, the most detailed row of the tree diagram, or the final branches on the fishbone diagram. You may want to use these ideas as starting points and brainstorm additional ideas.

3. Place one idea at a time on the work surface and ask, "Is this idea related to any others?" Place ideas that are related near the first. Leave space between cards to allow for drawing arrows later. Repeat until all cards are on the work surface.
4. For each idea, ask "Does this idea cause or influence any other idea?" Draw arrows from each idea to the ones it causes or influences. Repeat the question for every idea.
5. Analyze the diagram:
   - Count the arrows in and out for each idea. Write the counts at the bottom of each box. The ones with the most arrows are the key ideas.
   - Note which ideas have primarily outgoing (from) arrows. These are basic causes.
   - Note which ideas have primarily incoming (to) arrows. These are final effects that also may be critical to address.

   Be sure to check whether ideas with fewer arrows also are key ideas. The number of arrows is only an indicator, not an absolute rule. Draw bold lines around the key ideas.

Example

A computer support group is planning a major project: replacing the mainframe computer. The group drew a relations diagram (Figure 5.166) to sort out a confusing set of elements involved in this project. "Computer replacement project" is the card identifying

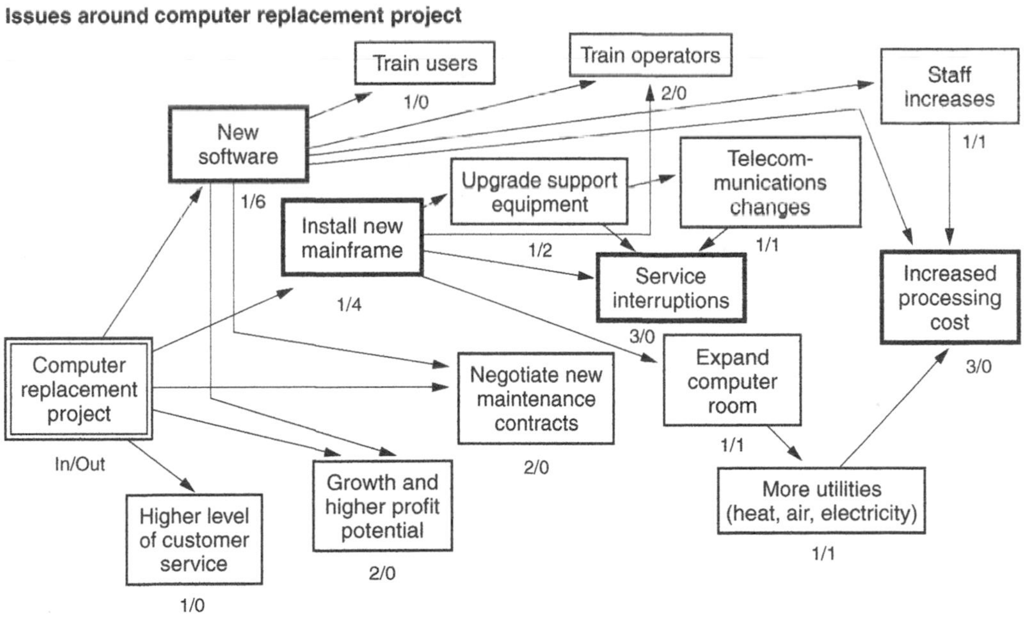

Figure 5.166 Relations diagram example.

the issue. The ideas that were brainstormed were a mixture of action steps, problems, desired results, and less-desirable effects to be handled. All these ideas went onto the diagram together. As the questions were asked about relationships and causes, the mixture of ideas began to sort themselves out.

After all the arrows were drawn, key issues became clear. They are outlined with bold lines. "New software" has one arrow in and six arrows out. "Install new mainframe" has one arrow in and four out. Both ideas are basic causes. Also, "service interruptions" and "increased processing cost" both have three arrows in, and the group identified them as key effects to avoid.

matrix relations diagram

Use a matrix to compare ideas to each other one at a time. This variation is more methodical and neater. It does not visually emphasize relationships as well.

1. Write a statement defining the issue that the relations diagram will explore.
2. Brainstorm ideas about the issue on a flipchart. If another tool has preceded this one, take the ideas from the affinity diagram, the most detailed row of the tree diagram, or the final branches on the fishbone diagram. You may want to use these ideas as starting points and brainstorm additional ideas.
3. Draw an L-shaped matrix with one more row than ideas and four more columns. Write the ideas down the first column as row labels and across the top row as column headings. Or number each idea and use the numbers as column headings. The diagonal of cells where the row and column labels are the same will not be used, so block them out, such as by shading them.
4. Go through the matrix methodically, comparing each idea on the left to each idea across the top. Ask, "Does the first idea cause or influence the second?" In the box where the row and column intersect, draw an arrow pointing up to indicate the idea in the row causing or influencing the idea in the column.
5. Reflect the matrix. Every place there is an arrow pointing up, draw an arrow pointing left in the other box where those two ideas intersect. For example, if row 2, column 3 has an upward arrow, then row 3, column 2 must have a left-pointing arrow.
6. To analyze the matrix, count the "to" arrows (pointing left) and the "from" arrows (pointing up) in each row. Add up the total number of arrows to and from that idea. Write the totals in the far right columns of the matrix.
7. Analyze for key ideas as in step 5 of the basic procedure.

Example

The same project, replacement of the mainframe computer, and the same issues are shown in the matrix relations diagram of Figure 5.167. There's an upward arrow in row 1,

	1	2	3	4	5	6	7	8	9	10	11	12	13	14	15	To	From	Total
1. Computer replacement		↑				↑				↑			↑		↑	0	5	5
2. New software	←		↑	↑	↑					↑		↑	↑			1	6	7
3. Train users		←														1	0	1
4. Train operators		←				←										2	0	2
5. Staff increases		←										↑				1	1	2
6. Install new mainframe	←			↑			↑		↑		↑					1	4	5
7. Upgrade support equipment						←		↑	↑							1	2	3
8. Telecommunications changes							←		↑							1	1	2
9. Service interruptions						←	←	←								3	0	3
10. Negotiate new maintenance contracts	←	←														2	0	2
11. Expand computer room						←								↑		1	1	2
12. Increased processing cost		←			←									←		3	0	3
13. Growth and higher profit potentials	←	←														2	0	2
14. More utilities											←	↑				1	1	2
15. Higher level of customer service	←															1	0	1

Figure 5.167 Matrix relations diagram example.

column 2. Idea 1 influences or causes idea 2. Similarly, the arrow pointing left in row 3, column 2 means idea 2 influences or causes idea 3.

Every upward-pointing arrow is matched with a left-pointing one. For example, row 4 (train operators) has a left-pointing arrow at column 6. Row 6 has an upward-pointing arrow at column 4.

The last three columns sum the arrows pointing left (to), the ones pointing up (from), and the total. These are the same sums the team counted in the previous example to determine key issues.

Considerations

- Fifteen to 50 ideas is the best range of numbers for a relations diagram. Fewer than 15 and you do not need it; more than 50 and the diagram is too complex to handle.
- Do not ask, "Is this idea caused or influenced by any others?" Look for ideas that cause others, and do that for every idea. Then the "caused by" question is unnecessary.
- Do not draw double-headed arrows. Force yourself to decide which cause or influence is strongest.
- Sometimes bold lines are drawn around key effects and double lines are drawn around key causes. Or you may find color-coding more useful.

- It can be useful to share the diagram with others and request their ideas and comments. Do this after step 4 (before the analysis) and when the analysis is completed.
- This tool also has been called the *interrelationship digraph*. The name *relations diagram* is less intimidating and more easily remembered.
- A related tool, the prioritization matrix, variation combination ID/matrix method, starts with a matrix relations diagram and adds assessments of the strength of relationships between ideas. It is used for decision making.

repeatability and reproducibility study

Also called: R&R study, gage R&R, measurement system analysis (MSA)

Description

A repeatability and reproducibility (R&R) study analyzes variation of a measurement system that uses an instrument or gage. This variation is compared to the observed total variation in the process. The main purpose of an R&R study is to ensure that measurement variation is low enough that process measurements truly reflect the process. It is impossible to know whether a product is within specifications or to work on reducing process variation if measurement variation is so great that it masks process variation.

R&R studies focus on two types of variation: repeatability, which is variation obtained by the same instrument in repeated readings, and reproducibility, which is variation that occurs when different people take the measurement.

When to Use

- When a measurement is made using an instrument or device, and . . .
- Before studying process variation or process capability, or . . .
- When selecting a measurement method from several alternatives, or . . .
- When assessing or standardizing measurement methods, procedures, or training, or . . .
- Periodically as an ongoing procedure when maintaining statistical process control of an improved process

Procedure

Planning

1. Determine the part or product, the measurement, and the instrument to be studied.
2. Determine the number of samples to be used and how they will be obtained. Five to ten samples is customary. If each sample may not be consistent throughout, determine procedures to minimize within-sample variation during the study.
3. Determine how many and which operators (people taking measurements) will be used in the study. One to three operators is customary.
4. Determine the number of trials (repeated measurements) each operator will perform. Two or three is customary.
5. Determine calibration procedure, measurement procedure, and analysis procedures.

Testing

6. Calibrate the instrument.
7. Determine a random sequence for the samples. The first operator measures all samples, following the standard operating procedure for the measurement. Record the results.
8. Determine a different random sequence for the samples. The second operator measures all samples, as before. Do not allow operators to see others' results. Repeat until all operators have measured all samples once. This concludes one trial.
9. Repeat steps 7 and 8 until the planned number of trials is completed. Keep operators from seeing sample numbers, previous results, or any other information that might suggest what a measurement "should" be.

Analysis and Improvement

10. Analyze the data. Computer software usually does the math. The most common analysis methods are "range and average" and ANOVA (analysis of variance). Summaries of these methods are given below. The primary results of the analysis are:

    Repeatability (Equipment Variation or EV). This index reports the variability of the instrument when one person repeatedly measures the same item. It is expressed as the width of an interval that contains a certain percentage (often 99%) of the variation due to the instrument. A standard deviation for repeatability may also be given.

    Reproducibility (Appraiser Variation or AV). This index reports the variability between operators when they each measure the same thing. It is also expressed as the width of an interval that contains a certain percentage (often 99%) of the operator variation. A standard deviation for reproducibility may also be given.

Repeatability and Reproducibility (R&R). This index combines the two above to estimate the width of the measurement variation. Again, a standard deviation may also be given. (Note: This index is not a simple sum of the repeatability and reproducibility values, because standard deviations are not additive.)

11. Compare the measurement variation to the total process variation. The simplest way is to calculate the ratio of R&R variation to total process variation. Process variation should be determined from process control charts and expressed in the same terms as R&R variation. If R&R variation is expressed as $\sigma_{R\&R}$, use the process σ for the calculation. If R&R variation is expressed as a 99% interval, use process σ multiplied by 5.15. See "Considerations" for discussion of acceptable measurement variation.

12. If R&R variation is unacceptable, develop action plans for improving it. Further studies or analysis can be done to pinpoint the reason for the variation.

Analysis Methods

Range and Average

This method constructs an $\bar{X}$ and R chart of the study's data. A subgroup is defined as all trials for one sample by one operator. The standard deviation calculated from the overall average range $\bar{R}$ is the standard deviation for repeatability. See *control charts* for more information about $\bar{X}$ and R charts and their formulas. This method can be done without computer software.

1. For each operator, calculate the range R for each sample, the average $\bar{R}_o$ of all those ranges, and the average of all that operator's measurements, $\bar{X}_o$.

2. Calculate the average range $\bar{R}$ of all the operators. Calculate the upper control limit for the ranges.

$$UCL_R = D_4 \times \bar{R}$$

D_4 is the usual R-chart constant from Table A.2, with n = the number of trials.

Plot the ranges and the UCL. If any ranges are higher than the UCL, determine and eliminate the reason for the special cause. Repeat those measurements or, if that is impossible, drop them from the calculation.

(Optional) Plot the average of each operator's repeated measurements on an average chart. Look for indications of differences between operators. Compare the width of the control limits, which represents measurement variation, to the width of the band of plotted points, which is indicative of process variation.

3. Calculate standard deviations for repeatability, reproducibility, and R&R:

$$\sigma_{\text{repeatability}} = \bar{R} \div d_2$$

Table 5.14 Reproducibility constant d_2^*.

Number of Operators	d_2^*
2	1.41
3	1.91
4	2.24

d_2 is the usual *R*-chart constant from Table A.2, with n = the number of trials

$$\sigma_{\text{reproducibility}} = \sqrt{\left(\frac{\bar{X}_{\text{diff}}}{d_2^*}\right)^2 - \frac{\sigma^2_{\text{repeatability}}}{nr}}$$

$\bar{X}_{\text{diff}}$ is the difference between the highest operator's average and the lowest operator's average

d_2^* is from Table 5.14.

n = number of samples

r = number of trials or repeats

$$\sigma_{\text{R\&R}} = \sqrt{\sigma^2_{\text{repeatability}} + \sigma^2_{\text{reproducibility}}}$$

4. To calculate repeatability, reproducibility, and R&R indices, multiply each standard deviation by 5.15 for a 99% interval. For a 95% interval, multiply by 4. For a 99.7% interval, multiply by 6.

ANOVA

ANOVA, or analysis of variance, analyzes the study as a designed experiment. The factors are instrument, operators, and samples. This analysis is done with computer software. The results will include interactions between instrument and operators.

Example

A critical quality characteristic of a manufacturing process involves a laboratory analysis. The characteristic was in statistical control, with a process standard deviation of 13. Customer specifications were much tighter than the process could produce, leading to a Cp of 0.8 and a desire to improve the process. Before looking for process improvements, the team did an R&R study to estimate how much of the observed process variation was actually measurement error.

The study design was ten samples, three analysts, and two trials. Ten samples were saved from the previous weeks of production that represented a range of measurements from 13 to 64. A labeling scheme was devised so that the analysts could not see sample identification when doing the analysis. Three analysts were selected randomly, and their schedules were adjusted so that they would all work the same shift the day of the study. Six random sample sequences were determined, one for each analyst within each trial.

The data and calculations are shown in Table 5.15. The bottom half of Figure 5.168 shows the R-chart of operator ranges. None of the ranges are out of control. Operator C's ranges do show greater variation than the other two operators, which is worth investigating. This means operator C's measurement variation appears to be different from the other operators'. The overall measurement variation is 15% of the total observed process variation, which is acceptable. The team can begin studying ways to reduce process variation, knowing that the process's responses to changes they make will not be swamped by measurement error.

Table 5.15 R&R study example.

Trial / Sample	Operator A			Operator B			Operator C		
	1	2	Range	1	2	Range	1	2	Range
1	45	48	3	46	46	0	49	48	1
2	13	13	0	13	14	1	13	16	3
3	21	19	2	20	17	3	18	22	4
4	64	62	2	63	64	1	61	65	4
5	23	19	4	23	18	5	23	20	3
6	61	60	1	63	61	2	62	64	2
7	34	35	1	33	35	2	36	38	2
8	41	39	2	41	40	1	38	40	2
9	30	30	0	31	32	1	34	35	1
10	52	57	5	53	50	3	53	56	3
$\bar{R}_o$			2.0			1.9			2.5
$\bar{X}_o$	38.3			38.2			39.6		

$\bar{R} = (2.0 + 1.9 + 2.5) \div 3 = 2.13$

$UCL_R = 3.27 \times 2.13 = 7.0$

$\bar{X}_{diff} = 39.6 - 38.2 = 1.4$

Repeatability:

$\sigma_{repeatability} = 2.13 \div 1.13 = 1.9$

99% interval = 9.8

Reproducibility:

$\sigma_{reproducibility} = \sqrt{(1.4 \div 1.91)^2 - (1.9)^2 \div 20} = 0.6$

99% interval = 3.1

R&R:

$\sigma_{R\&R} = \sqrt{(1.9)^2 + (0.6)^2} = 2.0$

99% interval = 10.3

% process variation = $2.0 \div 13 \times 100 = 15\%$

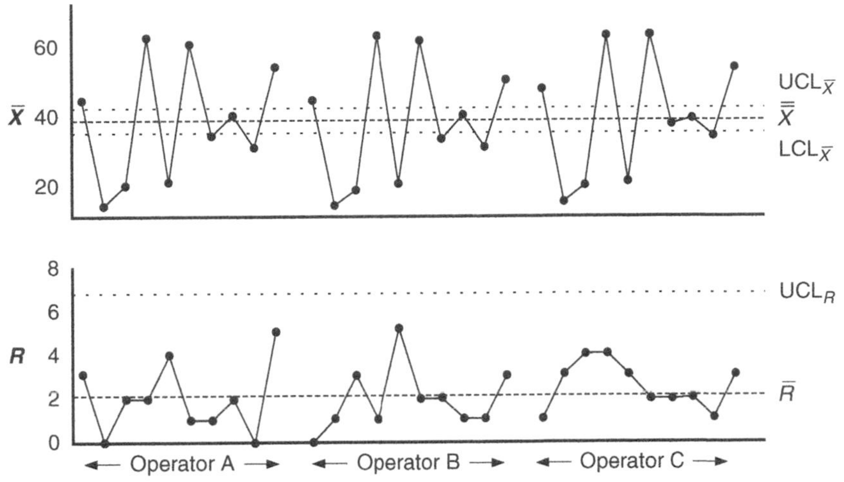

Figure 5.168 R&R study example, average and range charts.

The top half of Figure 5.168 shows the average chart. Although this graph often is not drawn, it can be informative. This average chart does not suggest any major differences between operators. Does it look as though the process is wildly out of control? Actually, this is exactly how you want an average chart to look for an R&R study. Those control limits were determined from the measurement process. Because the control limits are so much narrower than the band swept out by the sample values, it means the measurement system can distinguish between samples and recognize real process changes.

Considerations

- This tool is often called *gage R&R,* where the term gage is expanded to mean any measuring device. In this book, the term "instrument" is used to emphasize that the approach applies to any equipment used to take a measurement.
- The total variation observed in a process includes both true process variation plus variation caused by the act of observing or measuring the process. Even if a process were 100 percent consistent, a control chart of a measured characteristic would probably show variation from the measurement process. In order for you to work on reducing process variation, measurement variation must be small enough, relative to process variation, that it doesn't distort the effects of changes you make to the process.
- The most important sources of measurement variation, and the ones addressed by R&R studies, are *repeatability* and *reproducibility.*

  Repeatability. If the same operator uses the same instrument to measure the same sample repeatedly within a short period of time, how much difference will there be in the results?

Reproducibility. If several different operators use the same instrument on the same sample, how much difference will there be in the results?

Within-sample variation. This is actually part of process variation, but it often affects R&R studies. If the sample itself is not the same from one part to another, such as a part that is out of round or a liquid that is not well-stirred, then repeated measurements may show variation that might be misinterpreted as measurement variation. R&R studies should attempt to eliminate within-sample variation. If your process is likely to have within-sample variation, you can run a study first on special standards that have no within-sample variation. If this is not possible, see Resources for ways to minimize within-sample effects in your study.

- Other sources of measurement error are not included in R&R studies. These are:

  Stability. Does the instrument change over a long period of time, days or weeks? Control charts are used to monitor a measurement process's stability.

  Linearity. Does the instrument measure accurately throughout the entire range of expected measurements?

  Bias or calibration. Does the instrument consistently measure higher or lower than a known true value?

  These types of measurement errors should be studied separately to ensure that your measurement system is accurate.

- Another common cause of measurement problems is *inadequate measurement discrimination.* This means that the measurement units are too large for the amount of variation in the process. For example, a dimension is measured to the nearest centimeter, but the process standard deviation is only half a centimeter. In these situations, the process control chart may indicate lack of control when the process actually is in control. This problem is easily identified with normal process control charts. Count how many values are possible within the limits on the range chart. For a subgroup size of two, three or fewer possible values signals inadequate measurement discrimination. If the subgroup size is greater than two, the signal is four or fewer possible values. The solution is equally simple: use smaller measurement units.

Planning and Running the Study

- If the instrument is automatic or if the same person always operates it, the procedure should be followed with N, the number of operators, equal to one. If several people regularly take measurements, a random sample of operators should be used for the study: at least two, preferably three or four for better statistical results. Using four or more operators makes the study more difficult. Continuing R&R studies should eventually include all operators.

- Do not use untrained or inexperienced operators unless you specifically want to assess training. For overall analysis of the measurement method, the study's operators should have equivalent knowledge of and experience with the measurement procedure.
- Samples should be selected to represent the range of measurements normally encountered in the process. However, if you know that the precision of measurement changes across the measurement range, do separate studies at different measurement levels.
- Generally, R&R studies use ten samples. However, the number of samples times the number of operators should be 16 or greater. If not, consult one of the entries in Resources for modifications to the calculations or be sure your computer software can make the adjustments.
- The design can be modified to study other measurement differences. For example, if you want to compare different measurement methods, "method" could replace the "operator" factor or could be added as an additional factor.
- Take careful notes during the tests of any circumstances that might affect the results.
- If your situation does not seem to fit the standard procedures, get help from Resources, the instrument manufacturer, or an expert.

Analyzing the Results

- Data is usually analyzed using computer spreadsheets or statistical software. However, it is important to know what those programs are doing in order to make appropriate choices when the programs give you options and to interpret and make decisions from the results. If your results show large measurement variability, study the data carefully, such as by making range control charts, to understand what is happening. The entries in Resources describe useful followup studies.
- ANOVA and range methods will usually produce similar although not identical results. The biggest difference is that ANOVA includes interactions between operators and samples. An example of an interaction might be two inexperienced operators obtaining inaccurate measurements at only one end of the measurement range, thus on only three out of ten samples. If ANOVA reveals any significant interactions, they should be investigated and eliminated. A disadvantage of ANOVA methods is that without the range chart, one is less likely to identify inconsistent measurements and remove them from the study. If you are using ANOVA, plot the range chart anyway, to ensure that there are no inconsistencies in your study.
- While plotting range control charts is not essential for the analysis, the chart can make problems and opportunities obvious and also make it easier to communicate the concepts and results of the study to others.

- The average chart can be confusing to those used to process control charts. The plotted points are samples from the process, so they show process variation plus measurement variation. (However, this small selected sample may not be representative of true process variation). The control limits are calculated from repeated measurements of the same sample, so they show only measurement variation. The width of the control limits can be thought of as masking an area where you can't tell what variation comes from the process and what comes from measurement. The narrower the control limits, the more true process variation you can see. That's why you do not want this chart to look as though it is "in control." If the average chart looks "in control," the measurement system is totally unable to distinguish real variation in the process.
- The average and range charts can also indicate differences between operators, such as one operator whose averages are consistently off from the others (biased) or whose ranges are much higher (more variation), as in the example. These indications should be investigated and reasons for differences corrected. See the entries in Resources for more information about follow-up studies.
- If one operator shows consistently lower variation than the others, learn what that person is doing differently and standardize that improved procedure through training and documenting.
- The equations for standard deviation are based on the fundamental principle that their squares are additive. If there are two components, A and B, of total variation:

  $$\sigma^2_{\text{total}} = \sigma^2_{\text{A}} + \sigma^2_{\text{B}}$$

  That is why $\sigma_{\text{repeatability}}$ and $\sigma_{\text{reproducibility}}$ do not add up to $\sigma_{\text{R\&R}}$ and the repeatability and reproducibility indices do not add up to the R&R index.
- Computer software often calculates *part variation.* This represents the amount of variation in the study due to variation from sample to sample. In most cases, this calculation is not meaningful. The sample is small and usually not representative of the entire process. To estimate total process variation, use a standard deviation calculated from process control charts.

Improvement

- Resources should be devoted to improving the measurement system when it cannot detect process variation well enough to determine acceptable product or to identify response to a process change. The comparison of measurement variation to total process variation helps to make this decision.
- A computer software calculation of measurement variation as a percent of total variation should be used with caution. First, the total process variation is usually calculated from the study's samples. As noted above, this is not recommended. Second, the software may have an option to calculate R&R variation as a percent

of process tolerance or specifications. In the past, this was the usual practice, but emphasis on continuous improvement has made that calculation less desirable.

- Common guidelines for R&R as a percent of total process variation are: under 20%, acceptable; 20% to 30%, marginal; over 30%, unacceptable.
- Other ways of comparing measurement variation to process variation have been developed. For example, Wheeler and Lyday (1989) use a discrimination ratio, which reflects how many distinctly different groups of products within the process output could be discriminated by the measurement system. Their guideline for devoting resources to improving the measurement system is a ratio of 4, equivalent to 34% of total product variation. The example above has a discrimination ratio of 9.4.
- A process with a good Cp is probably initially low priority for improvement, even if measurement is a large proportion of its variation. Eventually, however, continuous improvement should address the measurement process so that calculations of Cp better reflect the process itself. See *process capability* for more information about this index.
- If Cp is low, say 0.6 or less, even large measurement variation does not change Cp much. Fixing the measurement system may be necessary to be able to study the process, but measurement improvement alone will not improve the process's capability.
- Periodically repeat R&R studies as you monitor and control the process—that is, forever!

requirements table

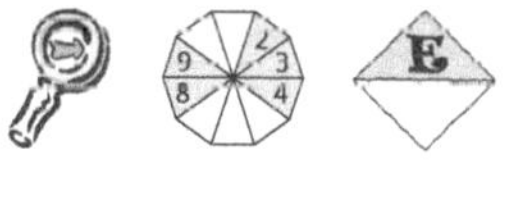

Description

The requirements table is a format for identifying customers and their requirements. It separates customers into four different categories and requirements into two categories. Thinking about the categories leads to a more complete list of customers and requirements.

When to Use

- When developing or working with a list of customers
- When developing or working with a list of customers' requirements

Procedure

1. Define the product or service. Write it at the top of the form.
2. Brainstorm a list of customers. Ask, "Who cares about the quality of what we do or how we do it?" Use these four categories to help develop a complete list:

   External customer. A purchaser or end user of the product or service, or that person's representative, who is outside your company or organization.

   Internal customer. A user of the product or service who belongs to your own company or organization.

   Society. Society has an interest in certain aspects of products, services, and how processes are run. Society's interests usually are represented by agencies such as the EPA, OSHA, certification boards, and so forth. Members of communities in which offices or facilities are located also have interests in aspects of their operation.

   Supplier. Suppliers may have interests in how materials are used or presented or in when or how information is communicated.

3. For each customer, brainstorm requirements and write them in the appropriate columns. Consider the following two types of requirements:

   Product. Requirements or needs about the product or service itself.

   Process. Requirements or needs about how the product or service is prepared or made.

   Table 5.16 shows typical kinds of requirements that concern each type of customer.

Table 5.16 Categories of customers and their requirements.

Those Who Care About Quality	What Is Produced (Product Requirements)	How It Is Produced (Process Requirements)
External customer	Specifications Consistency Packaging Maintainability Durability Safety	Price Time from request to delivery
Internal customer	Internal specifications	Production safety Cost, including labor Schedule
Society	Product liability Product safety	Employee and community health and safety
Suppliers	Appearance and functionality of recognizable supplies or parts	Safe use of supplies When and how information is received

Example

A construction company has contracted to build a home. Figure 5.169 shows the beginning of a requirements table.

The product is the new house. The process is how the house is built—the entire flow of activities of the construction company from signing the contract until the house is turned over to the owners.

Product or service: New house		
Customer	**Product Requirements**	**Process Requirements**
External		
Homeowner	Built to plan Specified materials Durable and strong construction • • •	Finished by September 30 Cost within budget Do not damage trees on site Remove construction debris
Internal		
Rough carpentry crew		Detailed, accurate plans Materials delivered 1 day after order Power available for tools
HVAC crew • • •	Minimum 30" crawl space for ductwork	Schedule before flooring laid 2 weeks notice
Society		
OSHA		Comply with worker safety regulations
Building department	Comply with building codes	Framing inspection before walls closed Electrical inspection . . . Plumbing inspection . . . • • •
Homeowners' association	Meet appearance guidelines	Follow noise regulations
Supplier		
Furnace supplier	Unit sized appropriately for house volume	8 week lead time to factory Install per instructions

Figure 5.169 Requirements table example.

The homeowners are external customers for the house. If the company were building a spec house for unknown future owners, the requirements would be different. Since the homeowners brought a set of plans to the company, their requirements include that the house be built to plan, using materials they specify. They want durable and strong construction. They probably have many other product requirements. Their process requirements are fewer. Users of a product often do not care how something is done as long as it *is* done. The homeowners want the building process to be completed by a certain date, stay within their budget, and not damage the site.

Company members who will be involved in building the house are internal customers, because earlier aspects of the building process affect their work. For example, the rough carpentry crew needs detailed, accurate plans; materials available quickly; and power available for tools. They have no product requirements: they'll build anything. The heating, ventilating, and air conditioning (HVAC) crew wants to be scheduled before flooring is laid, and they require two weeks notice to schedule the job and order supplies. These are process requirements. They also have a product requirement about the physical dimensions of the house necessary to accommodate their work. Other crews should be listed too.

Society's requirements are represented by several groups. OSHA requires the company to comply with worker safety regulations, a process requirement. The local building department requires the house to meet building codes (a product requirement) and has guidelines for when inspections must be done (process). Each of the specific inspection requirements should be listed; the first one, framing, is shown. The local homeowners' association also has product requirements, so that the house will fit into the neighborhood, and process requirements, so that construction activities will not disturb neighbors.

The furnace supplier wants the unit ordered to be appropriately sized for the house volume. If the furnace cannot heat the house, the supplier will receive complaints. Note that here the product requirement is about a portion of the new house. The supplier also has process requirements about order lead time and proper installation of the unit.

This is not a finished example. It shows only a few of the many customers that must be considered and the variety of requirements they may have.

The ZZ-400 story on page 80 describes using a requirements table within the improvement process.

Considerations

- You might not have customers in all four categories. For example, service departments may have no external customers. However, consider each of the four categories to be sure not to omit any customers.
- The reason for categorizing requirements as process or product is to make sure that you think about both types. Not all customers will have both types of requirements.
- Do not waste time and energy debating where to place a customer or requirement. If you're not sure, just put it somewhere and keep going.

- When you are listing requirements for a report or form, include not only what information must be there, but also requirements like arrangement for easy reading or use, formatting as a graph or table, number of copies, legibility, and so on.
- Be as specific as possible when writing requirements. For example, do not say, "adequate lead time." Say instead, "two weeks notice."
- Whenever possible, customers—in all four categories—should be involved in identifying requirements, either through participation on the improvement team or through interviews or surveys. When they are involved, you can know exactly what is important to them and what their specific needs are.

requirements-and-measures tree

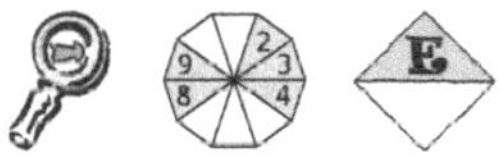

Description

The requirements-and-measures tree organizes customers, their requirements, and related measurements for a product or service. The relationships between all the customers, requirements, and measures become visible.

When to Use

- When developing requirements or measures, or . . .
- To organize a complex set of requirements and/or measures, or . . .
- To visually describe a set of requirements and measures and their relationships

Procedure

Materials needed: flipchart paper, sticky notes, marking pens.

1. Identify one process output. Write it on a sticky note and place at the top of a flipchart page.
2. Identify all customers of that output. Write each one on a sticky note and place on the page under the output.
3. For each customer, identify all requirements. Be as specific as possible, using operational definitions. For example, don't say "timely." Instead say, "received by Friday noon." Write each requirement on a note and place it under the customer's name.

4. At this point, some requirements may be duplicated, or natural groupings may be obvious. Reorganize the requirements if desired. Draw lines to show connections between customers and requirements.
5. For each requirement, brainstorm potential measurements. Follow good brainstorming techniques and try uncritically to generate as many as possible. Then discuss and evaluate the measures. Reduce the list to a manageable number.
6. For each measurement, identify how it will be tracked: what tool, where data are obtained, what frequency, who is responsible.

Example

Figure 5.170 shows the beginning of a requirements-and-measures tree for a construction company. The product is a new house. This is the same situation that was described in the requirements table example. Refer to that example for a discussion of the requirements.

This requirements-and-measures tree shows the branches for the homeowners, the HVAC crew, and the building department. There would be branches for all other customers.

"Cost within budget" could be measured by variance from the budget, which can be monitored by a control chart. "Finished by September 30" could be measured by the number of change requests, monitored on an attribute control chart with control limits based on the company's historical experience, and by the amount the project is ahead or behind schedule, as shown by an activity chart. Scheduling requirements for the work crews and the building department inspections could also be monitored by the same activity chart.

This chart is not complete. Every requirement for every customer should be identified.

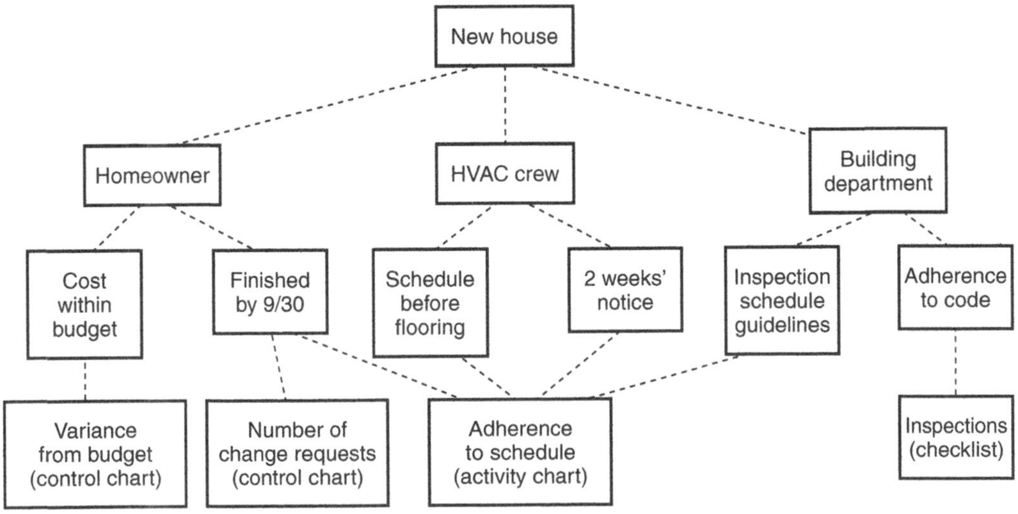

Figure 5.170 Requirements-and-measures tree example.

Considerations

- Do not neglect process requirements—*how* this output is produced. Be sure to include them.
- An affinity diagram can be useful when organizing requirements that have been brainstormed.
- When developing measures, it helps to switch the time frame of your thinking. Think about how to collect measurements of many occurrences over time rather than a single occurrence of the output. For example, if the product is a check request, a requirement might be "contains authorized signature." For one request, that is a yes–no measurement. On a weekly basis, one could record on an attribute control chart the percentage of requests lacking an authorized signature.
- An acronym for remembering the elements of a good measure is RAVE:

  Relevant—measuring what needs to be measured

  Adequate—able to perceive differences

  Valid—accurate and consistent

  Easy to obtain—inexpensive and simple
- When evaluating and deciding on measures, you can use other tools such as the decision matrix, multivoting, and list reduction.
- Using the same tool to measure two or more requirements makes it easier to manage the measurement process. In the example, one activity chart can monitor the scheduling requirements of several customers.
- Think about ways that measurement tools can prevent requirements from being violated, instead of only measuring (inspecting) to detect violations after they have occurred. For example, an activity chart prompts actions at appropriate times, and a control chart chosen for its sensitivity to small changes can alert you when a process begins to drift from historical patterns.

run chart

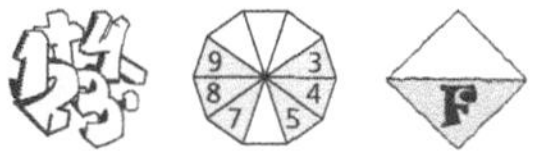

Description

A run chart is a line graph showing a process measurement on the vertical axis and time on the horizontal axis. Often, a reference line shows the average of the data. The run chart reveals patterns in the data over time. Unlike a control chart, it does not show control limits.

When to Use

- When monitoring a continuous variable over time, and . . .
- When looking for patterns, such as cycles, trends, or changes in the average, and . . .
- When you want a quick preliminary analysis to find obvious problems, before the effort of constructing control charts, or . . .
- When insufficient points have been collected to draw a control chart yet

Procedure

1. Decide on the vertical scale, based on the range of measurements you expect to see. Decide on the horizontal time scale, based on the frequency of measurements. Mark and label the scales.
2. If you already have some data, calculate the average:

$$\text{average} = \frac{\text{sum of the values}}{\text{number of values}}$$

   Draw across the chart a reference line showing the average.
3. Plot each measurement in the time order it occurs. Connect points with straight lines.
4. Look for patterns in the data, using these statistical tests:

   Length of Run. Too many consecutive points on one side of the average. A run is 8 points in a row, 10 points on one side out of 11 in a row, 12 points on one side out of 14 in a row, 16 out of 20.

   Number of Runs. The average line is crossed too few or too many times. Use the number of runs table (Table 5.17) for this test. Count the total number of points, skipping any points lying directly on the average line. Determine the number of runs by counting the number of times the line between data points crosses the average line. If the data line touches the average line and then returns to the same side, do not count it as crossing.

   Trend. A series of points heading up or down. If the chart has 20 to 100 points, there should be no more than six in a row steadily increasing or decreasing. For fewer than 20 points, the limit is five. Or use the method described for the scatter diagram to decide if the number of points indicates a trend.

Table 5.17 Number of runs table.

Total Number of Points	OK If Number of Runs Is:	Total Number of Points	OK If Number of Runs Is:
< 10	Too small to tell	36–37	12–25
		38–39	13–26
10–11	3–8	40–41	14–27
12–13	3–10	42–43	15–28
14–15	4–11	44–45	15–30
16–17	5–12	46–47	16–31
18–19	5–14	48–49	17–32
20–21	6–15	50–59	18–33
22–23	7–16	60–69	22–39
24–25	8–17	70–79	26–45
26–27	8–19	80–89	31–50
28–29	9–20	90–99	35–56
30–31	10–21	100–109	39–62
32–33	11–22	110–119	44–67
34–35	11–24	120–129	48–73

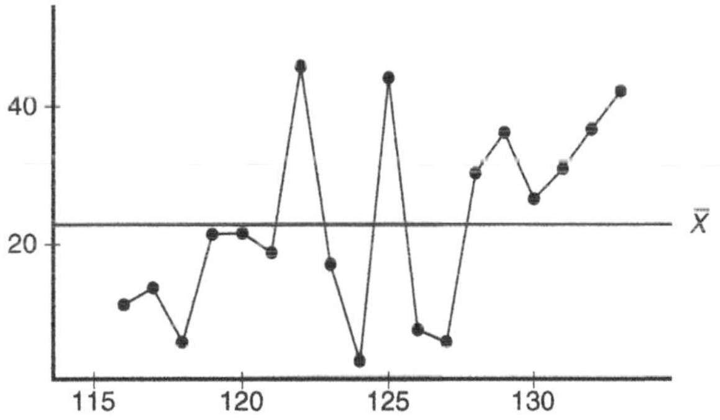

Figure 5.171 Run chart example.

Example

Figure 5.171 shows a typical run chart. The longest run is six points, which is acceptable. There are eighteen total points, and the centerline is crossed five times, which is also within normal range. A trend test shows no abnormal pattern.

However, if the next two points stay above the centerline, then a suspicious pattern would be signaled by both the length-of-run test (8 in a row) and the number-of-runs test (with 20 points, the centerline should be crossed at least six times).

Considerations

- Run charts are not as powerful as control charts for analyzing process data and identifying problems. The out-of-control tests used for control charts cannot be used with run charts because there are no control limits and no calculation of σ. However, run charts provide a quick preliminary check for obvious problems and also are useful if you don't have enough data yet for a control chart.
- As the name "run" indicates, run charts are best for spotting runs of data: points lying on one side of the centerline or a sequence of points headed in one direction.
- These tests for patterns are based on statistics. They test whether the data are truly statistically random. If not, an underlying pattern needs to be identified.
- The tests do not judge whether patterns are desirable or undesirable. For example, a trend may be toward better performance or toward worse performance. It is just as important to do something to understand and maintain good performance as it is to eliminate bad performance.
- See *graph* for more information about graphing techniques. Also see the graph decision tree (Figure 5.68, page 276) in that entry for guidance on when to use run charts and when other graphs may be more useful for your situation.

sampling

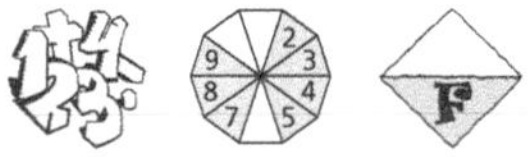

Description

Sampling means selecting a few items out of a larger group in order to question, examine, or test those items and then to draw conclusions about the entire group.

When to Use

- When you need to make conclusions about a large group, and . . .
- When it is expensive, difficult, or would take too long to examine the entire group

For example:

- When monitoring the quality of manufactured items during or after production
- When examining documents during an audit of compliance with procedures
- When gathering employee or customer preferences or feedback
- When testing a new procedure before widespread introduction

Procedure

If sampling is new to you, start by reading the definitions of terms on the next page. These definitions explain the concepts of sampling. Then come back and read the procedure.

1. Determine the population to be studied. Set up a sampling frame.
2. Determine the desired margin of error and confidence level. Those who will use the data should participate in setting these requirements.
3. Select a sampling method and a sample size *n*, balancing cost against required precision. Sample-size calculators for simple random samples are available on the Internet. For other sampling methods, consult a statistician to determine the appropriate sample size.
4. Document the sampling plan. Include how and when to do the sampling. If sampling will be a continuing process, prepare forms such as check sheets to assist those doing the sampling.
5. Select the sample.

Example

Examples are the best way to describe the various methods of sampling. Suppose a company wished to survey its 10,458 employees located around the world. The employee base is the population, and a group had already determined appropriate questions. A group gathered to plan the sampling method.

Simple Random Sample. Gung Ho started the discussion. "I've got it figured out. For a margin of error of 5 and a 95% confidence level, an Internet sample-size calculator showed we need a sample of 371. We'll use a random number generator from the Internet to choose 371 random numbers between 1 and 10,458. Every employee already has a unique employee number. So the employees with those numbers would form our sample."

Systematic Sample. Ima Thinker said, "That employee list is already randomly ordered by location, department, gender—everything except date of employment. Why don't we do this: Randomly pick a starting point, say by throwing a die. If it comes up 4, we start with the fourth person on the list. 10,458 divided by 371 is 28, so we'll sample every 28th person on the list after that."

Convenience Sample. Manny Moneybags said, "Why waste time? There are more than 371 employees here in this building. We'll just survey each one and have our answers by this afternoon." Fortunately, the rest of the group persuaded him that the sample size of 371 was only valid if the selection was totally random. Besides, even if every employee in the building were surveyed, that sample would not be representative of all employees. (Manny Moneybags once conducted a customer survey by questioning the first 100 customers to telephone the call center. In this biased convenience sample, East Coast customers and early risers were overrepresented, and Internet customers were missed entirely.)

Cluster Sample. Vera Practical said, "You don't understand. In order to get more and better information, this survey must be done with face-to-face interviews using trained interviewers. With the methods you're discussing, the travel cost for interviewers will be prohibitive. Here's an idea: Out of our 24 locations, we will randomly select a number of them. Then we'll randomly sample from those "cluster" locations. The sample size will have to be larger, but overall costs will be lower because of reduced travel. A statistician can determine the sample size required."

Proportional Stratified Sample. Will Prevail spoke up. "We must ensure representative response from people with different amounts of experience. A simple random or systematic sample is not the most efficient way to do this. A clustered sample won't do it either. Let's divide all employees into four groups: under 5 years experience, 5 to 10 years experience, 10 to 20 years experience, over 20 years experience. Within each group, we'll select random samples proportional to the number of people in the group. The statistician will determine the sample size. It will be less than 371. And Vera Marie, we can use telephone interviews."

Important Definitions

Population. The entire group of all individuals or items under study. Sometimes called "universe."

Sampling frame. A list, designed to represent the entire population, from which the sample will be drawn.

Sample. (noun) A smaller group of individuals or items selected from the larger population. (verb) To select a group of individuals or items from a population to question or examine.

Random sample. A sample selected by a method in which every possible sample is equally likely.

Representative sample. A sample in which the individuals or items reflect all important characteristics of the population. For example, a customer sample selected from web page visitors would not be representative, for it would not reflect customers who didn't own computers or shop with them.

Sampling error. The difference between the result estimated from the sample and the result one would have gotten if one surveyed the entire population. The result estimated from the sample is rarely *exactly* equal to the unknown, true value. The true sampling error is also unknown, but the maximum sampling error can be calculated.

Margin of error or *confidence interval* or *tolerance specification.* The range of values around the estimate that probably contains the true value. This number is calculated from statistical formulas. For example, suppose a candidate is favored by 38% of poll respondents, with a margin of error of 4%. This means that in the general population, the true percentage of supporters has a high probability of being in the range from 34 to 42 percent. Without more information, however, you don't know how high the probability is. For that, you need the confidence level.

Confidence level. The probability that in repeated sampling, the confidence intervals obtained would contain the true value. In the poll described above, suppose the confidence level were 95%. Then if the poll were repeated many times, in 95% of those polls (or 19 times out of 20), the true value would lie within the confidence interval. Another way of thinking about the confidence level is the *risk* one is willing to take of being wrong, in this case, 5%. When the term "confidence interval" is used to describe the range, the confidence level should be stated with it. In our example: The 4% margin of error means the 95% confidence interval.

Accuracy. How closely the result estimated from the sample compares to the true value.

Precision. How closely the result estimated from the sample compares to the result one would have gotten if one surveyed the entire population.

Bias. A difference between the sample and the population, caused by the sampling method.

Considerations

- One of the advantages of sampling is that the sample results may actually be more accurate than a survey of the entire group. This seems counterintuitive, doesn't it? One is thinking of sampling error when it seems that a 100 percent survey would be more precise. However, other types of errors, such as reporting and nonresponse errors, can cause the result of a 100 percent survey to be different from the *true* value. These errors are often greater than the sampling error and can't be measured or calculated. With a smaller survey, more care can be taken with design, training, and execution to reduce these errors.
- Setting up an appropriate sampling frame can be critical to success of the sampling. The sampling frame ideally would include all the intended population, exclude those not in the intended population, avoid duplication, and be up-to-date and accurate. Suppose the intended population is all voters in an upcoming election. The famous inaccurate poll predicting Dewey beating Truman was due to an inappropriate sampling frame: the telephone directory. In 1948, many rural voters did not have phones. A more appropriate sampling frame might be the official list of registered voters. But what about people who will register between now and the registration deadline? Registered voters who have no intention of voting? Determining an appropriate sampling frame requires the insight of those familiar with the subject of the sampling.
- As sample size is increased, sampling error decreases—at first. Eventually a point is reached where increasing the sample size does not decrease the error. Also, as population increases, the required sample size increases more slowly, becoming a smaller and smaller fraction of population, until with a large population, sample

size does not depend on population size at all. For example, a sample size of 1000 gives a 3.1% margin of error for populations of 50,000 or 200 million. That is why polls of only a few thousand respondents can predict the voting preferences of the entire United States.

- A final note about sample size: It is calculated using the proportion p that respond favorably. Usually, of course, p is unknown ahead of time. Therefore many sample-size calculators assume that $p = 50\%$, which is the worst case and gives the largest sample size. If p can be predicted, use a sample-size calculator that allows you to input p. Alternatively, if after the analysis p turns out to be a small number, you could input the numbers into a confidence interval calculator and recalculate the smaller confidence interval for the data you obtained.
- Avoid selecting random numbers out of your head. The human brain wants to form patterns, so numbers chosen that way often do not pass tests of randomness. Instead, use one of the random number generators available in statistics software or on the Internet.
- As the example implies, convenience sampling is not recommended. If it is unavoidable—say you're testing a new training method and you have to use whoever enrolls for the class—at least do some checks to see if bias exists in your sample.
- Stratified sampling works best when each group is as uniform as possible, with diversity between groups. Cluster sampling is the opposite: it works best when all clusters are as diverse as the total population, so each cluster looks pretty much like the next. In the real world, clusters are often used to avoid problems of geography, with the result that each one is not diverse and looks different from the next. For the same sample size, stratified sampling gives a more precise answer than random sampling, and cluster sampling gives a less precise answer. Therefore, geographic clusters should be avoided. If cost and efficiency dictate using them, a statistician should be consulted to help overcome any nonrandomness. Cluster sampling may be useful when sampling from production lots, where different lots are expected to be alike.
- With systematic sampling, be careful not to introduce bias. For example, if a sample is taken every hour on the hour, shift changes may affect the samples. A better method might be to take one sample per hour, but to throw a die to determine in which 10-minute interval the sample is taken.
- See *hypothesis testing* for other issues related to making decisions based on samples.
- This is an overview of sampling. For more specific details, for assistance with critical projects, or to develop an efficient and effective sampling plan for quality assurance, consult a statistician.

scatter diagram

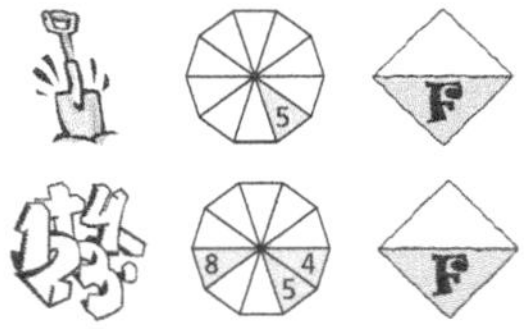

Also called: scatter plot, X–Y graph

Description

The scatter diagram graphs pairs of numerical data, one variable on each axis, to look for a relationship between them. If the variables are correlated, the points will fall along a line or curve. The better the correlation, the tighter the points will hug the line.

When to Use

- When you have paired numerical data, and . . .
- When the dependent variable may have multiple values for each value of the independent variable, and . . .
- When trying to determine whether the two variables are related, such as:
  - When trying to identify potential root causes of problems, or . . .
  - After brainstorming causes and effects using a fishbone diagram, to determine objectively whether a particular cause and effect are related, or . . .
  - When determining whether two effects that appear to be related both occur with the same cause, or . . .
  - When testing for autocorrelation before constructing a control chart

Procedure

1. Collect pairs of data where a relationship is suspected.
2. Draw a graph with the independent variable on the horizontal axis and the dependent variable on the vertical axis. For each pair of data, put a dot or a symbol where the x-axis value intersects the y-axis value. If two dots fall together, put them side by side, touching, so that you can see both.
3. Look at the pattern of points to see if a relationship is obvious. If the data clearly form a line or a curve, you may stop. The variables are correlated. You may wish to use regression or correlation analysis now. Otherwise, complete steps 4 through 7.

4. Divide points on the graph into four quadrants. If there are X points on the graph,
   - Count $X/2$ points from top to bottom and draw a horizontal line.
   - Count $X/2$ points from left to right and draw a vertical line.

   If number of points is odd, draw the line through the middle point.
5. Count the points in each quadrant. Do *not* count points on a line.
6. Add the diagonally opposite quadrants. Find the smaller sum and the total of points in all quadrants.

$$A = \text{points in upper left} + \text{points in lower right}$$

$$B = \text{points in upper right} + \text{points in lower left}$$

$$Q = \text{the smaller of } A \text{ and } B$$

$$N = A + B$$

7. Look up the limit for N on the trend test table (Table 5.18).
   - If Q is less than the limit, the two variables are related.

Table 5.18 Trend test table.

N	Limit	*N*	Limit
1–8	0	51–53	18
9–11	1	54–55	19
12–14	2	56–57	20
15–16	3	58–60	21
17–19	4	61–62	22
20–22	5	63–64	23
23–24	6	65–66	24
25–27	7	67–69	25
28–29	8	70–71	26
30–32	9	72–73	27
33–34	10	74–76	28
35–36	11	77–78	29
37–39	12	79–80	30
40–41	13	81–82	31
42–43	14	83–85	32
44–46	15	86–87	33
47–48	16	88–89	34
49–50	17	90	35

- If Q is greater than or equal to the limit, the pattern could have occurred from random chance.

Example

This example is part of the ZZ-400 improvement story on page 85. The ZZ-400 manufacturing team suspects a relationship between product purity (percent purity) and the amount of iron (measured in parts per million or ppm). Purity and iron are plotted against each other as a scatter diagram in Figure 5.172.

There are 24 data points. Median lines are drawn so that 12 points fall on each side for both percent purity and ppm iron.

To test for a relationship, they calculate:

A = points in upper left + points in lower right = 9 + 9 = 18

B = points in upper right + points in lower left = 3 + 3 = 6

Q = the smaller of A and B = the smaller of 18 and 6 = 6

$N = A + B = 18 + 6 = 24$

Then they look up the limit for N on the trend test table. For $N = 24$, the limit is 6. Q is equal to the limit. Therefore, the pattern could have occurred from random chance, and no relationship is demonstrated.

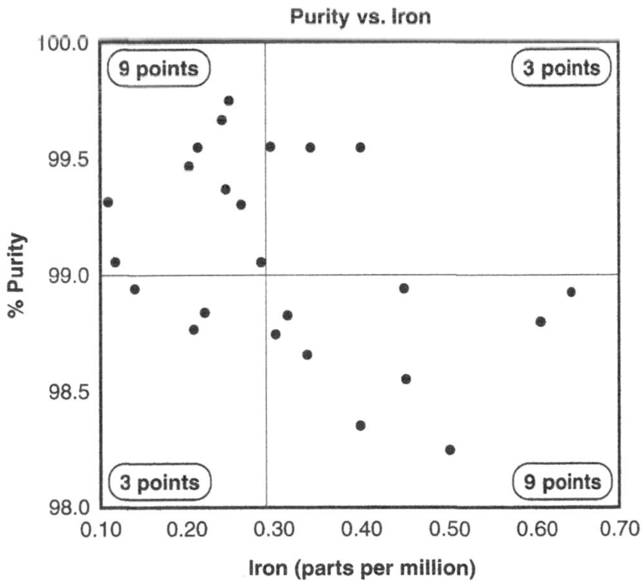

Figure 5.172 Scatter diagram example.

Considerations

- In what kind of situations might you use a scatter diagram? Here are some examples:
  - Variable A is the temperature of a reaction after 15 minutes. Variable B measures the color of the product. You suspect higher temperature makes the product darker. Plot temperature and color on a scatter diagram.
  - Variable A is the number of employees trained on new software, and variable B is the number of calls to the computer help line. You suspect that more training reduces the number of calls. Plot number of people trained versus number of calls.
  - To test for autocorrelation of a measurement being monitored on a control chart, plot this pair of variables: Variable A is the measurement at a given time. Variable B is the same measurement, but at the previous time. If the scatter diagram shows correlation, do another diagram where variable B is the measurement two times previously. Keep increasing the separation between the two times until the scatter diagram shows no correlation.
- Even if the scatter diagram shows a relationship, do not assume that one variable caused the other. Both may be influenced by a third variable.
- When the data are plotted, the more the diagram resembles a straight line, the stronger the relationship. See Figures 5.39 through 5.42, page 198 in the *correlation analysis* entry, for examples of the types of graphs you might see and their interpretations.
- If a line is not clear, statistics (N and Q) determine whether there is reasonable certainty that a relationship exists. If the statistics say that no relationship exists, the pattern could have occurred by random chance.
- If the scatter diagram shows no relationship between the variables, consider whether the data might be stratified. See *stratification* for more details.
- If the diagram shows no relationship, consider whether the independent (x-axis) variable has been varied widely. Sometimes a relationship is not apparent because the data don't cover a wide enough range.
- Think creatively about how to use scatter diagrams to discover a root cause.
- See *graph* for more information about graphing techniques. Also see the graph decision tree (Figure 5.68, page 276) in that section for guidance on when to use scatter diagrams and when other graphs may be more useful for your situation.
- Drawing a scatter diagram is the first step in looking for a relationship between variables. See *correlation analysis* and *regression analysis* for statistical methods you can use.

SIPOC diagram

Description

SIPOC stands for *suppliers–inputs–process–outputs–customers.* A SIPOC diagram shows a high-level flowchart of a process and lists all suppliers, inputs, outputs, and customers. A SIPOC diagram provides a quick, broad view of key elements of a process.

When to Use

- At the beginning of a project, to help define the important elements of the project, or . . .
- When it is not clear what the process inputs are, who supplies them, what the outputs are, or who the customers are, or . . .
- When there are many suppliers, inputs, outputs, and/or customers

Procedure

1. Gather a group of people who are knowledgeable about the process. Identify the process under study.
2. Create a macro or top-down flowchart of the process. Display it where everyone can see it throughout the rest of the procedure. Be sure to include starting and ending points for the process.
3. Identify the outputs of the process. Record all of them on a flipchart, on sticky notes attached to a wall, or on a transparency. Regardless of where they are recorded, have a heading reading "Outputs."
4. Identify the customers who receive the outputs. Record them on a separate flipchart, wall area, or transparency, with the heading "Customers."
5. Identify the inputs that the process needs. Record them separately as before.
6. Identify the inputs' suppliers. Once more, record them separately.
7. Review all your work to find omissions, duplications, unclear phrases, inaccuracies, and so on.
8. Draw a complete SIPOC diagram as shown in the example.

Example

A catalog retailer call center created the SIPOC diagram of their telephone customer service process, shown in Figure 5.173.

Suppliers	Inputs	Process	Outputs	Customers
• Printer • Merchandising dept. • Marketing dept. • Phone company • IS dept. • Training dept.	• Catalog • Telephone lines • Order system • Customer database • Call statistics database • Inventory database • Product training • Customer service training	Customer calls 800 number ↓ Rep answers call ↓ Rep and customer exchange information ↓ Rep enters order into system ↓ Rep and customer conference with product specialist (PS) if required ↓ Call concluded	• Order entry • Payment info to accounting • Product request to warehouse • Inventory database adjustment • Customer database update • Call statistics database update • Credit card transaction data	• Telephone customer • Warehouse dept. • Accounting dept. • Marketing dept. • Credit card processor

Figure 5.173 SIPOC diagram example.

Also see the ZZ-400 story on page 79 and the Medrad story on page 58, which describe using SIPOC diagrams within their improvement processes.

Considerations

- The concepts behind the SIPOC diagram are derived from Juran's triple role concept of customer, processor, and supplier. Juran describes the TRIPROL diagram, which is similar to the SIPOC, commonly used in Six Sigma.
- Sometimes preliminary requirements and measures for inputs and outputs are identified and included on the diagram. If you do so, be sure not to consider output requirements final until you have discussed them with customers.
- See *flowchart* for more information about drawing and analyzing flowcharts.
- See *critical-to-quality analysis* for a tool that examines inputs and outputs in more detail.

stakeholder analysis

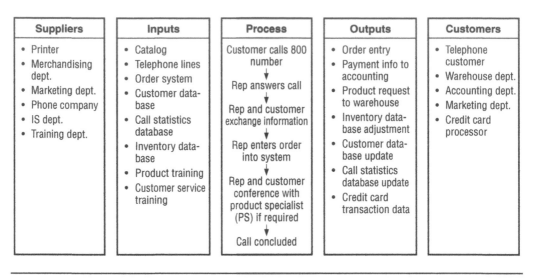

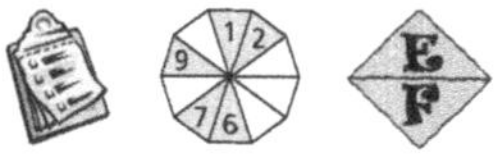

Description

Stakeholder analysis identifies individuals or groups with an interest in an issue. Based on assessments of the interest, influence, and importance of each stakeholder, actions are planned to change those assessments or to work within them to ensure success of a project or plan.

When to Use

- When identifying issues that need to be improved based on stakeholder needs, or . . .
- Before beginning problem analysis, if stakeholders' interests may affect their input, or . . .
- When developing a proposed action, to ensure that its effects on all stakeholders are considered in the design, or . . .
- When planning rollout of a change, to ensure that it will be supported by those who can make it succeed or fail

Procedure

1. Define the issue, problem, proposed action, or plan. Brainstorm the stakeholders of that issue. List all the stakeholders in the left-most column of a table. (Optional) Separate stakeholders into primary and secondary.
2. Identify each stakeholder's interest(s) in the issue. List the interests in the next column.
3. (Optional) Rate the attitude of each stakeholder on a scale of –2 to +2, where –2 is strongly opposed, 0 is neutral or undecided, and +2 is strongly supportive. Write the rating in the third column.
4. Rate the influence of each stakeholder on a scale of 1 to 5 or low, medium, high. Write the rating in the fourth column.
5. Rate the importance of each stakeholder to the project's success on a scale of 1 to 5 or low, medium, high. Write the rating in the fifth column.
6. Construct a two-dimensional chart with influence on one edge and importance on the other. Code each stakeholder with a number or letter. Place each stakeholder's code in the appropriate place on the influence–importance chart.
7. Analyze the influence–importance chart as follows. (See Figure 5.174.)
    - High influence–high importance: Develop good collaboration with these stakeholders.
    - High influence–low importance: These stakeholders could sabotage plans or escalate problems. Work to involve them so their interest increases.
    - Low influence–high importance: Work to give these stakeholders voice and influence. They often need to be defended or protected.

	Low Influence	High Influence
High Importance	Protect and defend: strengthen influence	Collaborate
Low Importance	Do not spend resources	High risk: involve to increase interest

Figure 5.174 Influence–importance chart.

- Low influence–low importance: Perhaps monitor, but do not spend resources here.

8. Develop strategies for each stakeholder. Strategies should aim to increase support and reduce opposition. Write the strategies in the last column of the table.
9. (Optional) Develop a "participation matrix" showing the role each stakeholder should play at each step of the project. List roles—inform, consult, partnership, control—across the top. List stages or steps of the project as row labels down the side. On each row, list stakeholders in appropriate columns. Each stakeholder need not appear in every row, but also cannot appear more than once per row.

Example

A medical group is planning to improve the care of patients with chronic illnesses such as diabetes and asthma through a new chronic illness management program (CIMP). As part of their early planning, they do a stakeholder analysis. The stakeholder analysis table, Figure 5.175, shows five groups of stakeholders: doctors, nurses, administrators, patients, and the information systems department.

Administrators rate high on influence, because they control financing and approval, but low on importance, because they are not personally affected by the program. Patients, on the other hand, have low influence, but extremely high importance, since their health is at stake. Nurses have medium influence and rate high on importance, since they are directly involved in the process. Their attitude is moderately opposed, because they perceive the program as a criticism of their work and because their jobs will change greatly.

Strategies were not developed until the end of the analysis. The influence–importance chart (Figure 5.176) showed that collaboration was needed with doctors and nurses, with efforts to increase the nurses' influence. Patients' rights need to be protected and their influence increased. Administrators pose a high risk; they needed to be involved to increase their interest.

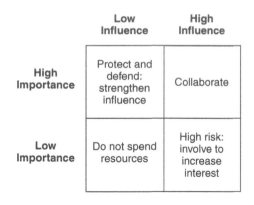

Stakeholder	Interest	Att.	Infl.	Imp.	Strategies
1. Doctors	Improve care	+2	H	M	Consult for suggestions; invite to be on team; MD champion
2. Nurses	Dramatic change in job duties	−1	M	H	Significant team membership; invite RN to lead team
3. Administrators	Reduce costs	+1	H	L	Frequent progress reports; emphasize potential savings
4. Patients	Improved health; changed roles	0	L	H	Publicize effort; seek input during rollout; eventually partner in care
5. Information systems	Must develop new data system	+1	L	M	Keep informed; invite to key meetings

Figure 5.175 Stakeholder analysis table example.

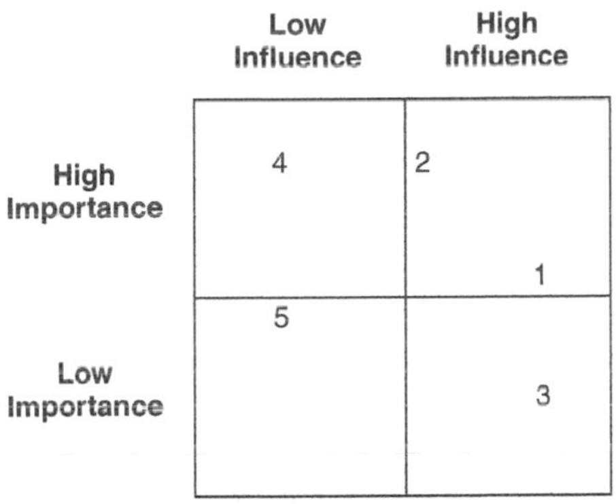

Figure 5.176 Stakeholder analysis example: influence–importance chart.

The strategies on the stakeholder analysis table and the entries in the participation matrix (Figure 5.177) show the result of long discussion about how to structure the project to maximize positive influence and minimize opposition. Notice that the participation matrix shows a gradual change in patients' roles, with the eventual goal of having them partner in care.

Considerations

- A *stakeholder* is anyone with an interest or right in an issue, or anyone who can affect or be affected by an action or change. Stakeholders may be individuals or groups, internal or external to the organization.
- To identify stakeholders and their interests, ask these questions:
  - Who might receive benefits?

	Inform	Consult	Partnership	Control
Identification	Nurses IS	Doctors		
Planning	Patients	Administration IS	Doctors Nurses	CIMP Team
Implementation	Administration	Patients IS	Doctors Nurses	CIMP Team
Monitoring	Administration IS		Nurses Patients	Doctors

Figure 5.177 Stakeholder analysis example: participation matrix.

- Who might experience negative effects?
- Who might be forced to make changes?
- Who might have to change behavior?
- Who has goals that align with these goals?
- Who has goals that conflict with these goals?
- Who has responsibility for action or decision?
- Who has resources or skills that are important to this issue?
- Who has expectations for this issue or action?

- *Primary* stakeholders are those directly affected. *Secondary* stakeholders are intermediaries, or those involved in implementing, funding, monitoring, and so on.
- Consider dividing stakeholder groups based on demographics: gender, occupation, and so on.
- Focus groups and interviews can be useful to get accurate information about stakeholders and to involve them in the analysis process.
- *Influence* is the power to make or affect decisions, to control resources, or to persuade or coerce others who can. This influence may be formal or informal. Those with influence can help or hurt the action or plan, or can worsen or improve a problem.
- *Importance* is the potential to affect or be affected by the issue, problem, or action. Information, rights, and dependencies are often involved.
- Some groups add columns beside Attitude, Influence, and Importance to rate the strength of those assessments. Use ✓ for certain, ? for some uncertainty, ?? for great uncertainty and ??? for simply guessing.

- Some stakeholder analyses do not assess attitude. Decide whether it is appropriate for your situation to do so.
- Attitude should be considered only in light of the stakeholder's influence and importance. For example, neither opposition nor support matter if the stakeholder has little influence and importance.
- Stakeholders with high influence must be involved in some way. If this is not possible with stakeholders who are opposed, their influence must be neutralized.
- On a participation matrix, show plans to increase the involvement of low influence or low importance stakeholders during the course of the project. For example, in the hospital example, the participation matrix shows a desire to actively involve patients and their families as partners in care.
- Other matrices and tables can be used to analyze stakeholders. For example, a table with columns for stakeholder, main objectives, potential positive benefits, potential negative benefits, and overall impact may be useful in complex situations. See the entries in Resources for more analysis ideas.
- Labeling individuals or groups as powerful or powerless, important or unimportant, can ignite strong emotions. Be ready to deal skillfully with sensitive relationships.
- Be careful to use this tool to learn, to build relationships, to seek common ground, to equalize uneven power, and to increase participation and ownership. Be aware that it can be used by some stakeholders to concentrate their power.
- Relations diagrams and mind maps may also be useful to depict stakeholder relationships.

storyboard

Also called: displayed thinking

Description

A storyboard is a visual display of thoughts. It makes all facets of a process, organization, plan, or concept visible at once to anyone. It taps both the creative right brain and the analytical left brain to yield breakthrough thinking.

When to Use

- When developing new ideas, or . . .
- When planning a project, or . . .
- When developing a process flow diagram, or . . .
- When planning a presentation, or . . .
- When exhibiting proposed plans, or . . .
- When exhibiting or documenting project activities or results

Procedure

Materials needed: cards or sticky notes in three different sizes or colors, large piece of newsprint or pinboard, marking pens, images and graphics.

1. Define the topic of the storyboard. Write it on a large card and place it at the top of the work surface (paper or pinboard).
2. Brainstorm important subjects to be considered. Write these on large header cards. The first one should be "Purpose" or "Why work on this topic." Continue to brainstorm headers until the flow of ideas stops.
3. Critically discuss each heading. Identify duplicates and subjects that are less important and should be dropped. Some headings will be a subcategory under another heading and should be set aside. Team members can object to anything on the work surface. The team tries to remove the objection by *plussing* (improving) the idea. Change the card if it is plussed; remove it if the objection cannot be resolved. Continue the discussion until the group has determined the important headers.
4. Choose one header and brainstorm ideas. Write them on small sticky notes and place under the heading. These are called *subbers.* Brainstorm until the group has exhausted all ideas.
5. Critically discuss the subbers, as in step 3. Where choices need to be made, the team may use other tools (such as decision matrix, list reduction, or multivoting) to narrow the list.
6. Continue to generate subbers for each of the headings in turn. Always do a complete brainstorm first before following with a discussion session.
7. Add details to make the story complete.
   - Include copies or thumbnail images of important diagrams or graphs, photos, or sketches.
   - Connect ideas and show relationships with arrows or string.

- Place small sticky notes, called *siders,* next to any card to make comments or amplify ideas.

8. Once the layout is finished, the final storyboard may be refined and polished with computerized word processing or graphics.

Example

This example is part of the ZZ-400 improvement story on page 88. The ZZ-400 purity team made a storyboard (Figure 5.178) to explain to the rest of the plant how they solved a quality problem and the broader learnings from the project. The topic card reads "Iron-Free ZZ-400." The header cards are "What was the problem?" "How did we find the root cause?" "How did we fix it?" and "How will we prevent it from happening again?" Subbers were needed under only the root cause header, to describe the approach.

Small copies of key diagrams, charts, and documents that led them to their solution illustrate the steps. A sider provides contact information.

Considerations

- Storyboards and the process of storyboarding were invented by Walt Disney Studios to develop cartoons and are still used in moviemaking. Keep that in mind as you creatively develop a storyboard that tells the story of your plan or project. What are the main acts, scenes, characters, conflict, and resolution?
- Storyboarding requires a small group. Between 4 and 10 participants is the ideal size.
- Storyboarding alternates between creative brainstorming and critical review. It is essential to keep the two modes separate. Mixing the two is deadly: a group cannot be creative when some members are being critical. It is the facilitator's role to keep the team in the right mode. One helpful technique is to pin a card on the board with "creative" on one side and "critical" on the other, signaling to the team which type of thinking is in play.
- The decision to switch from creative to critical mode or vice versa can be made by the facilitator or suggested by a team member.
- The better defined the topic statement is, the better the final result will be and the easier it will be to get there.
- Common subject headers include purpose, problem or opportunity, future state, who, what, when, how, cost, summary, recommendations, and miscellaneous. The headers you use will depend on your topic.
- To speed the creative process, have participants themselves write their own ideas on the cards.

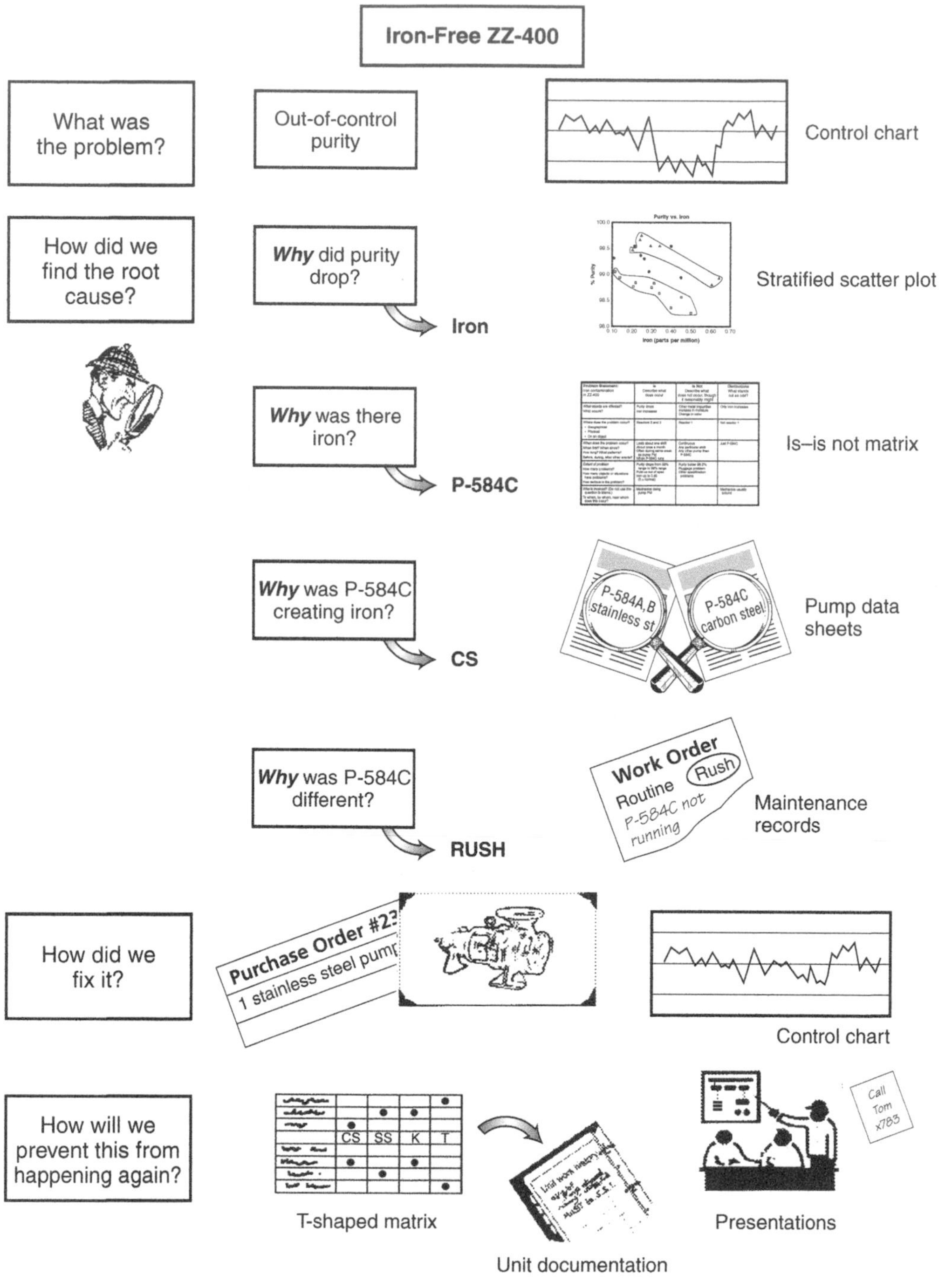

Figure 5.178 Storyboard example.

- Use broad marking pens. With regular pens, it is hard to read ideas from any distance.
- Using cards of different colors for the different header subjects helps the eye and brain sort the different ideas.
- See *brainstorming* for more ideas about that part of the process.
- The storyboard method can be adapted to construct other tools. See *detailed flowchart* and *requirements-and-measures tree* for storyboarding methods for developing those tools.
- See *Mind Maps* for a similar tool for visually communicating plans or ideas.

stratification

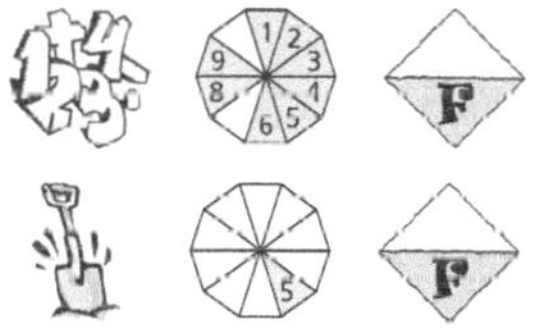

Description

Stratification is a technique used in combination with other data analysis tools. When data from a variety of sources or categories has been lumped together, the meaning of the data can be impossible to see. This technique separates the data so that patterns can be seen. Stratification was one of the original seven QC tools.

When to Use

- Before collecting data, and . . .
- When data come from several sources or conditions, such as shifts, days of the week, suppliers, or population groups, and . . .
- When data analysis may require separating different sources or conditions

Procedure

1. Before collecting data, consider what information about the sources of the data might have an effect on the results. Set up the data collection so that you collect that information also.

2. When plotting or graphing the collected data on a scatter diagram, control chart, histogram, or other analysis tool, use different marks or colors to distinguish data from various sources. Data that are distinguished in this way are said to be *stratified.*

3. Analyze the subsets of stratified data separately. For example, on a scatter diagram where data are stratified into data from source 1 and data from source 2, draw quadrants, count points, and determine the critical value for just the data from source 1, then for just the data from source 2.

Example

This example is part of the ZZ-400 improvement story on page 85. The ZZ-400 manufacturing team drew a scatter diagram (Figure 5.172, page 473) to test whether product purity and iron contamination were related. The plot did not demonstrate a relationship. Then a team member realized that the data came from three different reactors. The team member redrew the diagram, using a different symbol for each reactor's data (Figure 5.179).

Now patterns can be seen. The data from reactor 2 and reactor 3 are circled. Even without doing any calculations, it is clear that for those two reactors, purity decreases as iron increases. However, the data from reactor 1, the solid dots that are not circled, do not show that relationship. Something is different about reactor 1.

See *regression analysis* for the same stratified data analyzed statistically.

See also the bar and dot chart examples in the section on graphs for examples of stratified survey data.

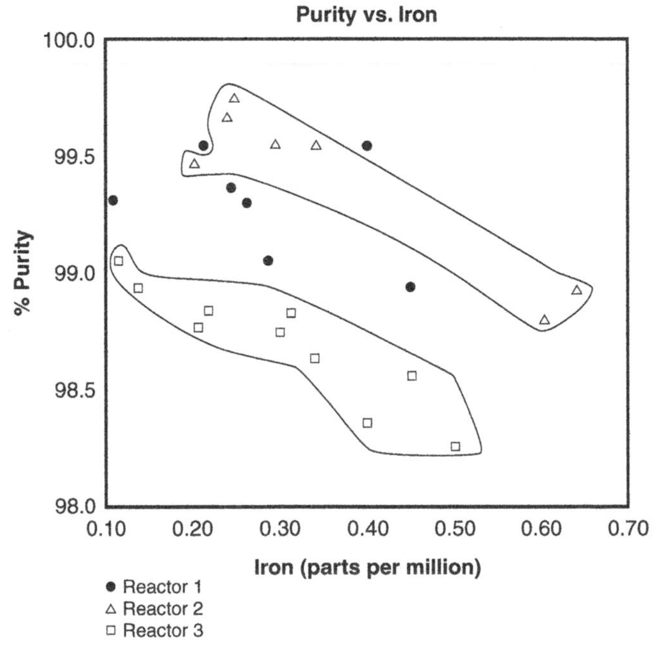

Figure 5.179 Stratification example.

Considerations

- Here are examples of different sources that might require data to be stratified.

Equipment	Suppliers
Shifts	Day of the week
Departments	Time of day
Materials	Products

  Survey data usually benefit from stratification. See *survey* for demographic information that may be needed for analysis.

- Always consider *before* collecting data whether stratification might be needed during analysis. Plan to collect stratification information. After the data is collected it might be too late.
- If you might photocopy your graph, use different marks rather than different colors to stratify the data.
- On your graph or chart, include a legend that identifies the marks or colors used. In Figure 5.179, the legend tells which symbol refers to which reactor.

survey

Variations: questionnaire, e-survey, telephone interview, face to face interview, focus group

Description

Surveys collect data from a targeted group of people about their opinions, behavior, or knowledge. Common types of surveys are written questionnaires, face-to-face or telephone interviews, focus groups, and electronic (e-mail or Web site) surveys. Surveys are commonly used with key stakeholders, especially customers and employees, to discover needs or assess satisfaction.

When to Use

- When identifying customer requirements or preferences, or . . .
- When assessing customer or employee satisfaction

Such as:

- When identifying or prioritizing problems to address, or . . .

- When evaluating proposed changes, or . . .
- When assessing whether a change was successful, or . . .
- Periodically, to monitor changes in customer or employee satisfaction over time

Basic Procedure

Note: It is often worthwhile to have a survey prepared and administered by a research organization. However, you will still need to work with them on the following steps so that the survey will be most useful.

1. Decide what you want to learn from the survey and how you will use the results.
2. Decide who should be surveyed. Identify population groups, and if they are too large to permit surveying everyone, decide how to obtain a sample. (See *sampling* for more information.) Decide what demographic information is needed to analyze and understand the results.
3. Decide on the most appropriate type of survey. See each variation's "when to use" section for guidance.
4. Decide whether the survey's answers will be numerical rating, numerical ranking, yes–no, multiple choice, or open-ended—or a mixture.
5. Brainstorm questions and, for multiple choice, the list of possible answers. Keep in mind what you want to learn and how you will use the results. Narrow down the list of questions to the absolute minimum that you must have to learn what you need to learn.
6. Print the questionnaire or interviewers' question list. (Each variation has additional specific instructions at this step.)
7. Test the survey with a small group. Collect feedback.
   - Which questions were confusing?
   - Were any questions redundant?
   - Were answer choices clear? Were they interpreted as you intended?
   - Did respondents want to give feedback about topics that were not included? (Open-ended questions can be an indicator of this.)
   - On the average, how long did it take for a respondent to complete the survey?
   - For a questionnaire, were there any typos or other printing errors?

   Also test tabulating and analyzing the results. Is it easy? Do you have all the data you need?

8. Revise the survey based on test results.
9. Administer the survey. See the variations for specific details.
10. Tabulate and analyze the data. Decide how you will follow through. Report results and plans to everyone involved. If a sample was involved, also report and explain the margin of error and confidence level.

questionnaire

When to Use

- When possible responses to questions are known and need only be quantified, such as after an initial set of interviews or focus groups has identified key issues, and . . .
- When collecting data from a large group, and . . .
- When data are not needed immediately, and . . .
- When a low response rate can be tolerated, and . . .
- When resources (money, people) for collecting data are limited

Procedure

Follow the basic procedure. Add the following details at steps 6 and 9.

6. Print the survey in a format that is easy to read and easy to fill out. Leave generous space for answers to open-ended questions. Design the form so that tabulating and analyzing data will be easy. At the top, print instructions for completing and returning it.
9. There are three options for administering the questionnaire:
   - *Mail or e-mail the questionnaire with a cover letter.* A cover letter should include the reason for the survey, how recipients were chosen, a deadline and instructions for returning questionnaires, and thanks. If using regular mail, include a stamped and addressed envelope. If using e-mail, design the message so that the respondent can forward it to the address collecting responses, typing answers directly into the message.
   - *Hand-deliver the survey.* Ask the recipient for a 10-minute appointment. At the appointment, explain the purpose of the survey and ask for the recipient's help. This method usually gets a better response rate.

- *Hold a meeting, if all respondents are at the same location.* Send a memo explaining the purpose of the survey and the time required. At the meeting, repeat the purpose before handing out the survey.

For mailed or hand-delivered questionnaires, follow up with nonrespondents to increase the response rate. Send another copy of the survey with a new cover letter, or follow up with a telephone interview.

Example

Pearl River School District, recipient of the Malcolm Baldrige National Quality Award, surveys students, parents, and staff annually using a questionnaire developed and administered by a national research firm. This ensures statistically correct methods, allows them to watch changes over time, and enables comparison with other districts.

The surveys show areas of satisfaction and dissatisfaction, but not reasons. To find causes, they must dig deeper. Recently, the annual survey showed that the high school students ranked the cafeteria service low. This was hardly surprising, since the research firm's database showed that almost all high school students rate their cafeterias low. However, Pearl River wanted to improve this rating.

A committee of student leaders and cafeteria staff prepared a second survey to get more information. Figure 5.180 shows a condensed version of their survey. Demographic information was requested at the top. A short paragraph stated the reason for the survey, some additional information respondents needed to know, and who was eligible to complete the survey. The survey was designed to be filled out quickly with check marks in appropriate boxes. A few open-ended questions were placed at the end. Data from the survey, supplemented with sales data, was analyzed and used to plan menu changes.

Read the Pearl River story on page 65 to learn more about their improvement process. For examples of graphs that can be used to analyze survey responses, see the bar chart and dot chart variations in the *graph* entry.

e-survey

Description

E-surveys are sophisticated Internet-based surveys. They can be structured with interactive features, such as graphics or programming that triggers alternate questions based on previous responses. E-surveys can be more convenient and even enjoyable for respondents, leading to higher response rates. They can also be carried out faster, provide almost instant analysis, cost less than traditional mail or phone surveys, and sometimes allow better targeting of respondents. Respondents can be contacted via e-mail, with a hyperlink to the Web site hosting the survey, or the survey can be located at the organization's Web site, with an invitation to visitors to complete it. To design and carry out an e-survey, get assistance from a research organization.

Pearl River High School Cafeteria Survey

Class: ______________________ Period: ___________

We are seeking your feedback to help us evaluate new offerings and consider other changes. Keep in mind that school cafeterias work under federal nutrition guidelines and that the food service company works under a contract with the school district. This may limit the kinds of items they are allowed to serve (no candy or soda) and the prices they can charge. All 8th, 9th,10th, and 11th graders are invited to take this survey.

1. How many days each week do you typically buy food in the school cafeteria?
   Place a check in the appropriate box beside each type of food.

	0	1	2	3	4	5
Breakfast						
Lunch						
Snacks						

2. Please tell us about which items you buy most often and never buy.
   Place a check in the appropriate box next to each item.

Item	**Buy Most Often**	**Never Buy**
Muffin		
Bagel		
Roll		
Oatmeal		
Deli sandwich		
Hot pretzel		
French fries		

Item	**Buy Most Often**	**Never Buy**
Soup		
Chili		
Salads		
Pasta		
Beverages		
Other (fill in)		
Other (fill in)		

3. Which of the following items, if added to the cafeteria selections, would you purchase?

   ☐ Buffalo wings ☐ Quesadillas ☐ Pancakes

   ☐ Fajitas ☐ Grilled cheese with bacon ☐ Waffles

4. What other suggestions would you make for additions to the cafeteria selections?

5. We welcome any other comments or suggestions you have to increase participation and satisfaction with the cafeteria service:

Thank your for your input. We will report back to you.

Figure 5.180 Survey example.

When to Use

- When all members of the target group have Internet access, and . . .
- When possible responses to questions are known and need only be quantified, such as after an initial set of interviews or focus groups has identified key issues, and . . .
- When collecting data from a large group, and . . .
- When resources (money, people) for collecting data are limited, and . . .
- When data are needed quickly, or . . .
- When a higher response rate than with traditional mail or phone surveys is desired

telephone interview

When to Use

- When most possible answers to the questions are known, and . . .
- When data are needed quickly, and . . .
- When a high response rate is needed, and . . .
- When people resources are available for making calls

Procedure

Follow the basic procedure. Add the following details at steps 6 and 9.

6. Print the interviewers' question list with lots of space to write notes directly under the question. Make one copy for each interview to be conducted. Also prepare a table with columns for name, phone number, busy signals, message left, call-back appointment, and interview completed. (Figure 5.181 shows an example.)

Name	Phone #	Busy	Mess.	Appt.	Done
Lisa Galmon	222-3341	II	I		✓
Sam Hayes	222-3337		III		
Joan McPherson	222-3348	IIII			✓
Albert Stevens	222-3333		I	Fri 9 am	

Figure 5.181 Table for tracking telephone interviews.

9. Fill out the table as calls are made. At the beginning of the phone interview, introduce yourself and explain the survey and its purpose. State how much time will be needed and ask if this is a convenient time. If not, make an appointment to call back. See "Considerations" for interviewing tips.

face-to-face interview

When to Use

- When the group to be surveyed is small, and . . .
- When ample resources (time, people, and possibly travel expenses) are available, and . . .
- When possible answers to the questions are not known, such as when you first begin studying an issue, or . . .
- When the questions to be asked are sensitive, or . . .
- When the people to be surveyed are high-ranking, important, or otherwise deserving of special attention, or . . .
- When close to 100 percent response rate is needed

Procedure

Follow the basic procedure. Add the following details at steps 3, 6, and 9.

3. Almost all the questions should be open-ended. You might start with a few simple yes–no questions.

6. Print the interviewers' question list with lots of space to write notes directly under the question.

9. Make four contacts with each interviewee.

    - *Letter.* Send a letter or e-mail to each interviewee explaining the survey's purpose and stating who will call to make interview arrangements.
    - *Phone.* Call to request an interview. The interviewer should introduce him or herself, describe the survey again briefly, and state how much time is needed—no more than thirty minutes. Set an appointment.
    - *Interview.* At the beginning of the interview, restate the purpose of the survey and the time needed. See "Considerations" for interviewing tips.
    - *Thank-you.* Send a follow-up thank-you letter to the interviewee.

Example

The stories from St. Luke's Hospital (page 72), Medrad (page 59), and ZZ-400 (page 79) all describe using interviews within the improvement process.

focus group

When to Use

- When the subject of the survey is complex or unfocused so that questionnaire or interview questions cannot be written, or . . .
- Before surveying with a questionnaire or interviews, to identify main issues or themes, or . . .
- After a questionnaire, to generate in-depth understanding of the responses

Procedure

A focus group brings together up to a dozen people to discuss their attitudes and concerns about a subject. Leading, documenting, and analyzing a focus group requires an experienced facilitator. Ask for help from your human resources or training departments or from a consultant or research organization.

Considerations

- Conducting a survey creates expectations for change in those asked to answer it. Do not survey if action will not or cannot be taken as a result.
- Satisfaction surveys should be compared to objective indicators of satisfaction, such as buying patterns for customers or attendance for employees, and to objective measures of performance, such as warranty data in manufacturing or readmission rates in hospitals. If survey results do not correlate with the other measures, work to understand whether the survey is unreliable or whether perceptions are being modified, for better or worse, by the organization's actions.
- Surveys of customer and employee satisfaction should be ongoing processes rather than one-time events. Plan–do–study–act is an excellent model for satisfaction surveying. Monitor trends over time with run or line charts.
- Get help from a research organization in preparing, administering, and analyzing major surveys, especially large ones or those whose results will determine significant decisions or expenditures.

- Figure 5.44 on page 204, part of the example for criteria filtering, compares characteristics of different kinds of surveys.

Writing Questions

- Gather input so that the questions will include all topics important to the respondents and to those who will use the data.
- Determine what information is needed before beginning to write questions. Is all this information necessary? Is it sufficient?
- Is some of the information you need available elsewhere? Do not waste survey time gathering information already available.
- Knowing how the data will be analyzed and used affects how the questions are written. For example, will data from students who rarely eat in the cafeteria be analyzed separately from the regulars? If not, will their food preferences bias the results?
- Questions may be about attitude (satisfaction, agreement), behavior (how often, when, how much), or knowledge (true or false: our service department opens at 9 A.M.). Know which type of information you want and phrase your question accordingly.
- Write and rewrite your questions to be short and clear. Try to imagine how the respondent will perceive the question.
- Make your questions as specific as possible. Link questions to particular work processes. Ask yourself, "If this question receives poor responses, will we know what to correct?" If your answer is no, rewrite the question.
- Group the questions by topic and start with easy ones. Put the hardest ones in the middle.
- Remember the "one-time, 30-minute rule."* You have only one time to get information from a respondent and only thirty minutes in which to do it. The number of questions on the survey doesn't matter, but it should take thirty minutes or less to complete. Questions written with forced-choice answers (see below) can be answered faster than open-ended questions.
- When surveying customers about your performance, ask also about the importance of each item. See *importance–performance analysis* for a way to ask and analyze those kinds of questions.
- Avoid jargon, duplicate questions, leading questions, questions phrased in the negative, sensitive or emotional questions, questions that ask two things at once.

---

* Glenda Y. Nogami, "Eight Points for More Useful Surveys," *Quality Progress* (October 1996): 93–96.

- Carefully think through what demographic information you will need to analyze the data. Ask those questions as part of the survey. When you are analyzing results, it will be too late to ask for the information. Demographic information that may be important includes:

Age	Location (zip code)
Gender	Years at location
Marital status	Home ownership
Race or ethnicity	Income level
Religion	Type of job
Education	Years of experience
Hobbies	Size of company

  Often what you ask is a compromise between maintaining the respondents' confidentiality and ensuring necessary analysis later.

- The more time you spend writing the survey, the less time you will spend analyzing it—or repeating it because the information you need isn't there.

Responses

- Questions with forced-choice answers—numerical ratings, rankings, yes–no, or multiple choice—are simple to answer and score, can be analyzed statistically, and can be repeated on periodic surveys to observe trends. Open-ended questions are very difficult to analyze, but they can reveal insights and nuances and tell you things you would never think to ask. A good compromise on a questionnaire or telephone interview is mostly forced-choice questions with a few open-ended ones at the end.

- Types of responses to avoid: Rankings are harder to answer and analyze than ratings. Do not use yes-no questions if a range of responses is possible. Don't use fill-in-the-blank questions if multiple choice would work.

- Check that the choices in a multiple choice answer are mutually exclusive. If not, allow people to select as many as apply.

- The *Likert scale* is the often-used numerical rating of agreement or satisfaction. Use an odd number of choices. Five or seven is best, because the more points on the scale, the harder it is to distinguish between them. On a scale of 1 to 10, what's the difference between a 7 and an 8?

- Write specific descriptions of each rating. For example, instead of "from poor (1) to excellent (5)" write:

  5 = Always exceeds expectations

4 = Sometimes exceeds expectations

3 = Meets expectations

2 = Needs some improvement

1 = Consistently lacking

- Always set up your scale so that a high number means "very satisfied" or "strongly agree." People expect that, so with a scale running the opposite direction, some respondents will answer incorrectly and your data will be unreliable.

Respondents

- See *sampling* for more information about how to select a statistically valid random sample.
- Remember that an average response rate for a blind, random questionnaire is about 35 percent. Be sure to send out enough questionnaires that a 35 percent response rate will generate enough data.
- For product and service planning, survey not only current customers but also potential and past customers.
- Nonresponse, a problem with every questionnaire, can make results unreliable. People who choose to respond often have extremely strong opinions, while nonrespondents have more neutral views. One way of dealing with the issue is to select a random sample of nonrespondents, persistently follow up with that group, and weight their responses. Ask for expert assistance with this.

Interviewing

- If there are several interviewers, they should agree on procedures for recording, summarizing, and analyzing responses. See *operational definitions* for more information.
- When reading questions, speak distinctly.
- Strive for a tone of voice that does not suggest a particular response. Do not react to answers. You should talk as little as possible.
- On open-ended questions, ask follow-up questions to explore the interviewee's thinking. Ask for clarification if you do not fully understand an answer.
- Take complete notes on each question. You will not remember later. It is all right to ask the interviewee to wait a few moments while you write.
- Always ask for permission if you want to make an audio recording of the interview. It is illegal to record a telephone conversation without permission. Take notes to supplement the interview. What if the interviewee's voice cannot be heard on the recording?

- If the interview is not finished within the time promised, ask for permission to continue.
- Always thank the interviewee for his or her time.
- Transcribe your notes as soon as possible after the interview.

Analysis and Reporting

- Methods for analyzing survey data range from simple, such as graphs, to complex, such as correlation or regression analysis and more sophisticated statistics beyond the scope of this book. If your survey calls for complex analysis, especially multiple factors and criteria, get assistance designing and analyzing the survey.
- Results can be biased for many reasons: respondents were chosen nonrandomly, questions were worded poorly, the questionnaire was confusing or poorly organized. If the survey is critical, get expert help.
- Graphs are better than tables for studying and communicating results. Tools that can be useful include dot, bar, and radar charts for reporting Likert scale responses, histograms and other frequency distributions for showing population percentages, importance–performance analysis, matrix diagrams, performance indexing, stratification, and two-dimensional charts.
- For numerical responses, the range of responses is a useful number to calculate, study, and report. Responses that vary from 1 to 5 mean something very different from responses that are all 3s and 4s, although both may average 3.4.
- Response rate must be considered when analyzing data, since nonrespondents often have more moderate views than respondents. If data are stratified, calculate and report response rate for each segment. Consider response rate when interpreting the results.
- Whenever you sample from a group, there will be variation in the results. The data you get may not exactly represent the data of the entire group. The larger your sample, the smaller the variation. You can calculate what the variation is; consult your local statistics expert. See *sampling* for more information.
- To analyze open-ended questions, see *thematic analysis* on page 99.
- If confidentiality was promised to respondents, maintain confidentiality at all costs.
- A complete report of the results and analysis should have an appendix with all the raw data, including responses to open-ended questions. Mask any responses that reveal the respondent's identity or refer to individuals.
- Send everyone who received a survey a summary of the results and what you plan to do. This is not only a courtesy but can improve the response rate the next time you conduct a survey.

table

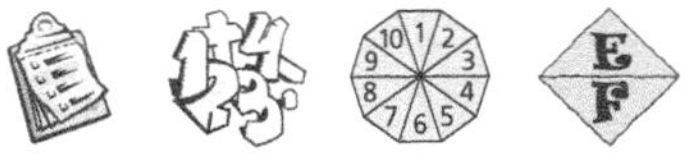

Description

A table is a chart with rows and columns used to organize information. The intersection of a row and column forms a cell. Either words or numbers can be placed in the cells. This is a generic tool that can be adapted for a wide variety of purposes.

When to Use

- When you need to organize data or information, either already collected or that will be collected, and . . .
- When the data or information falls into logical groups or categories

Procedure

1. Decide on the categories into which the data will be organized. If there are two sets of categories, one will be the columns and the other will be the rows. If there is only one set of categories, decide whether it will form the columns or rows. The other will represent each piece of data, data source, time, or a similar indicator of data collection.
2. Create a grid with one more column and row than there are items to go in them. Label the rows in the left column and label the columns in the top row. If required, create columns for totals, averages, and other statistical calculations or for check marks or other notations.
3. Write data in each cell.

Example

A visiting nurse wants to organize information about his patient's medicines. For each prescription, there are six categories of information: drug name, tablet dosage, daily amount, time taken, prescribing doctor, and purpose of drug. That information will be in the columns. Each prescription will form a row.

He creates a grid with seven columns, one more than the number of categories. There are five prescriptions, so he creates six rows. After he fills in data, the table looks like Figure 5.182.

See the St. Luke's Hospital story on page 71 and the Medrad story on pages 60 and 63 for examples of tables used within the improvement process. See also the

Drug	Tablet Dosage	Daily Amount	Time	Doctor	Purpose
Coumadin	5 mg	As directed	Evening	Estrada	Prevent stroke
Klonopin	5 mg	1 tablet	Bedtime	Malone	Control restless leg syndrome
Loratadine	10 mg	1 tablet	Morning	Chin	Prevent allergies
Protonix	40 mg	1 tablet	Morning	Estrada	Prevent heartburn
Verapamil	180 mg	1/2 tablet	Morning	Estrada	Regulate atrial fibrillation

Figure 5.182 Table example.

examples for contingency table, survey tracking, and voice of the customer table. Other examples can be found throughout this book. See the List of Figures and Tables on page xiii.

Considerations

- An L-shaped matrix and a table have the same format. At first glance, they may look identical. However, while an L-shaped matrix is a table, a table is not always an L-shaped matrix. A table may have only one set of labels, for the columns only or the rows only. See the voice of the customer table on page 512 for an example. Also, the purpose of a matrix is to show relationships between elements in the rows and elements in the columns. Tables may list data without showing relationships, as in a contingency table.
- While check sheets can take any form, they are usually constructed as tables.
- If the data will be organized into only one set of categories and the other dimension will represent pieces of data, it is usually better to place the categories in columns and let the rows represent the data, especially if you have not yet collected the data or more may be added. Adding additional rows is usually easier than adding columns.
- Spreadsheet and word processing software can be used to easily construct and modify tables. Use a spreadsheet if you have mostly numerical data and especially if you will do mathematical calculations with the data. Use word processing if your information is mostly words.

tree diagram

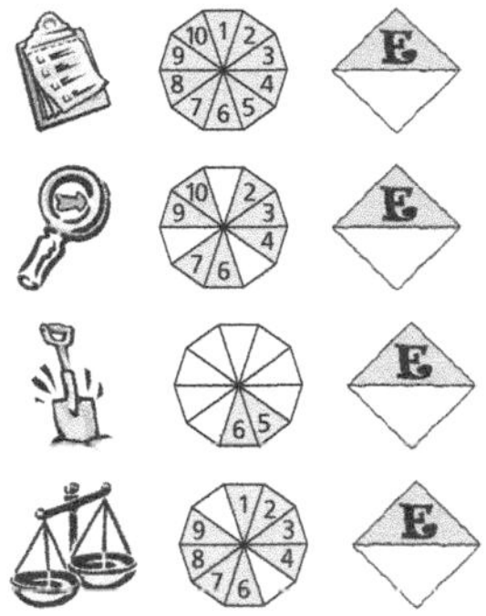

Also called: systematic diagram, tree analysis, analytical tree, hierarchy diagram

Description

The tree diagram starts with one item that branches into two or more, each of which branch into two or more, and so on. It looks like a tree, with trunk and multiple branches. It is used to break down broad categories into finer and finer levels of detail. Developing the tree diagram helps you move your thinking step by step from generalities to specifics. This is a generic tool that can be adapted for a wide variety of purposes.

When to Use

- When an issue is known or being addressed in broad generalities and you must move to specific details, such as:
- When developing logical steps to achieve an objective, or . . .
- When developing actions to carry out a solution or other plan, or . . .
- When analyzing processes in detail, or . . .
- When probing for the root cause of a problem, or . . .
- When evaluating implementation issues for several potential solutions, or . . .
- After an affinity diagram or relations diagram has uncovered key issues, or . . .
- As a communication tool, to explain details to others

The tree diagram is a generic tool type that can be used in many ways. Table 5.19 lists some tree diagram variations that have been named and their uses.

Table 5.19 Tree diagram variations.

Name	Use
Breakdown tree	Show demographic breakdown of a population sample, as in surveys
CTQ (Critical-to-quality) tree (page 211)	Translate customer needs into measurable product and process characteristics
Decision tree or logic diagram (page 224)	Map the thinking process for making a decision
Dendogram	In product design and development, identify product characteristics
Fault tree analysis (page 243)	Identify potential causes of failure
Gozinto chart	In manufacturing, show parts used to assemble a product
How–how diagram	Solve a problem
Job analysis or task analysis	Identify tasks and requirements of a job or task
Organization chart	Identify levels of management and reporting relationships
Process decision program chart (page 428)	Identify potential problems and countermeasures in a complex plan
Requirements-and-measures tree (page 461)	Identify customers, requirements, and measures of a product or service
Why–why diagram or five whys (page 513)	Identify root causes of a problem
Work breakdown structure (WBS)	Identify all aspects of a project, down to work package level

Procedure

1. Develop a statement of the goal, project, plan, problem, or whatever is being studied. Write it at the top (for a vertical tree) or far left (for a horizontal tree) of your work surface.

2. Ask a question that will lead you to the next level of detail. For example:

   - For a goal, action plan, or work breakdown structure: "What tasks must be done to accomplish this?" or "How can this be accomplished?"
   - For root-cause analysis: "What causes this?" or "Why does this happen?"
   - For gozinto chart: "What are the components?"

   Brainstorm all possible answers. If an affinity diagram or relationship diagram has been done previously, ideas may be taken from there. Write each idea in a line below (for a vertical tree) or to the right of (for a horizontal tree) the first statement. Show links between the tiers with arrows.

3. Do a "necessary and sufficient" check. Are all the items at this level *necessary* for the one on the level above? If all the items at this level were present or accomplished, would they be *sufficient* for the one on the level above?

4. Each of the new idea statements now becomes the subject: a goal, objective or problem statement. For each one, ask the question again to uncover the next level of detail. Create another tier of statements and show the relationships to the previous tier of ideas with arrows. Do a "necessary and sufficient check" for each set of items.

5. Continue to turn each new idea into a subject statement and ask the question. Do not stop until you reach fundamental elements: specific actions that can be carried out, components that are not divisible, root causes.

6. Do a "necessary and sufficient" check of the entire diagram. Are all the items *necessary* for the objective? If all the items were present or accomplished, would they be *sufficient* for the objective?

Example

Pearl River School District, 2001 Baldrige Award recipient, uses a tree diagram to communicate how districtwide goals are translated into subgoals and individual projects. They call this connected approach "The Golden Thread."

The district has three fundamental goals. The first, improve academic performance, is partly shown in Figure 5.183. District leaders have identified two strategic objectives that, when accomplished, will lead to improved academic performance. They are academic achievement and college admissions.

Lag indicators are long term and results oriented. The lag indicator for academic achievement is Regents' diploma rate: the percent of students receiving a state diploma by passing eight Regents' exams.

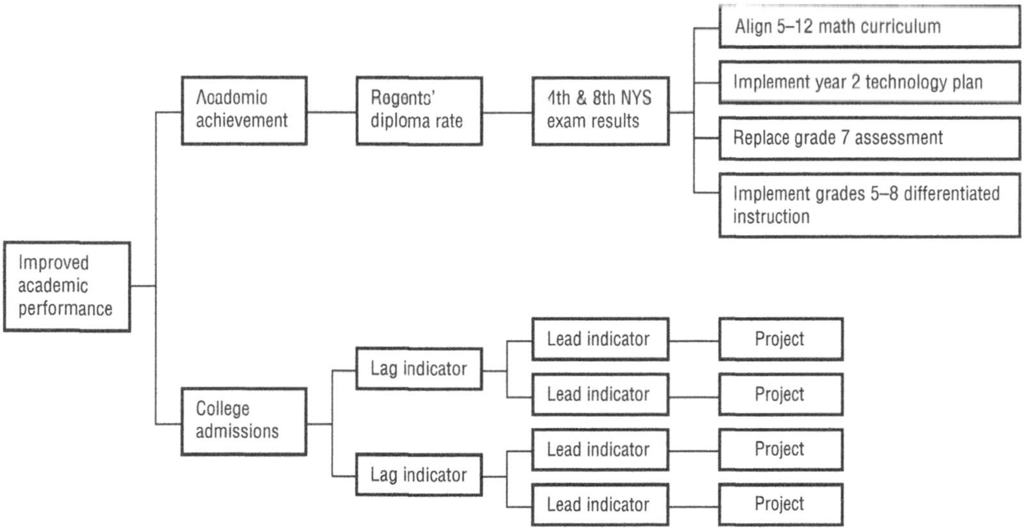

Figure 5.183 Tree diagram example.

Lead indicators are short-term and process-oriented. Starting in 2000, the lead indicator for the Regents' diploma rate was performance on new fourth and eighth grade state tests.

Finally, annual projects are defined, based on cause-and-effect analysis, that will improve performance. In 2000–2001, four projects were accomplished to improve academic achievement. Thus this tree diagram is an interlocking series of goals and indicators, tracing the causes of systemwide academic performance first through high school diploma rates, then through lower grade performance, and back to specific improvement projects. Read the Pearl River story on page 65 to learn more about their improvement process.

Considerations

- If a group is developing the tree diagram, use a large writing surface (newsprint or flipchart paper), sticky notes or cards for the ideas, and marking pens.
- Resist the temptation to get too detailed too quickly. At the first level past the objective statement, the statements should be very broad generalities. The next level can start to break those generalities down to specifics. After that, details will start to emerge. Let the tree structure walk you step by step into succeeding levels of detail.
- Branches of the tree can be different lengths. It may take more levels to work out the details for one element than another.
- With each new idea, check its relevance by asking, "Does this really contribute to the original objective, process, or problem?" If not, the idea does not belong on the diagram.
- You may prefer to follow one branch to its end before addressing other branches.
- Ideas in a later tier may be related to several earlier ideas.
- Sometimes the boxes of each tier are numbered, just like the several-leveled flowchart: 1.0; 1.1, 1.2; 1.1.1, 1.1.2; and so on.
- Review the completed diagram with those who will be involved in the actions, or those involved with the process or problem. Ask for their comments and be willing to make changes based on their suggestions.
- There are two basic types of tree diagrams. The first can be used to break a subject into its basic elements. This component-development type has been called a constituent–component–analysis diagram. The second type shows the means and procedures by which a plan will be carried out. This means-development type has been called a plan-development diagram.

two-dimensional chart

Also called: matrix, perceptual map
See also: effective–achievable chart, importance–performance analysis, plan–results chart

Description

The two-dimensional chart is used to understand the implications of all combinations of two factors or dimensions important in an issue. Intersecting x- and y-axes represent the two dimensions, points on the chart represent possible combinations of the two dimensions, and areas of the chart are assigned interpretations. This tool also can be drawn as a 2 × 2 matrix, where the four boxes of the matrix represent the four quadrants that intersecting x- and y-axes form. This is a generic tool that can be adapted for a wide variety of purposes.

When to Use

- When two different factors are important in evaluating an issue, and . . .
- When the two factors can be rated or scored on a simple scale such as low–high or yes–no, and . . .
- When each combination of possible values of the two factors creates a different interpretation
- Especially when analyzing customer perceptions

Procedure

1. Identify two factors or dimensions that are significant in understanding the issue. Determine a simple scale by which the factors will be valued, such as low–high, yes–no, premium–economy.
2. Draw a graph using two axes intersecting at their midpoints, with one factor on the x-axis and the other on the y-axis. Label the axes with the factor names and indications of the scale. If you prefer, you may draw a box divided into four quadrants, with one factor along the left edge and the other along the bottom or top.
3. Identify the areas on the graph that represent each combination of values of the factors. For each one, decide what interpretation or meaning should be assigned to that area.

4. For each actual situation, determine where on the graph it is placed by its combination of factors. Decide what that means for interpreting or evaluating the situation.

Example

A restaurant team has gathered from customers their views of the importance of various services the restaurant provides and of the restaurant's performance in providing those services. They created a two-dimensional chart (Figure 5.184) with performance on the x-axis and importance on the y-axis.

The team studied the chart to decide what each quadrant meant. The upper right quadrant represents an ideal situation: high-importance services being done well. The upper left quadrant is the most pressing: important services not being done well. The lower left quadrant represents lower priority for attention: services here are not being done well, but they are not important to the customer. Finally, the lower right quadrant represents services that are not important although they are done well. Perhaps too much effort is being put into those services.

Data from the customers is placed on the chart. For example, customers rated breakfast service 4.2 (on a scale where 5 is best) for importance and 3.9 for performance. That resulted in a point in the upper right quadrant. The chart shows that private parties are the highest priority for improvement. Carryout is unimportant and may be using too many resources.

This example is part of the importance–performance analysis, which uses two-dimensional charts. Compare with Figure 5.105 on page 328. For other examples of two-dimensional charts, see *effective–achievable chart, plan–results chart,* and the

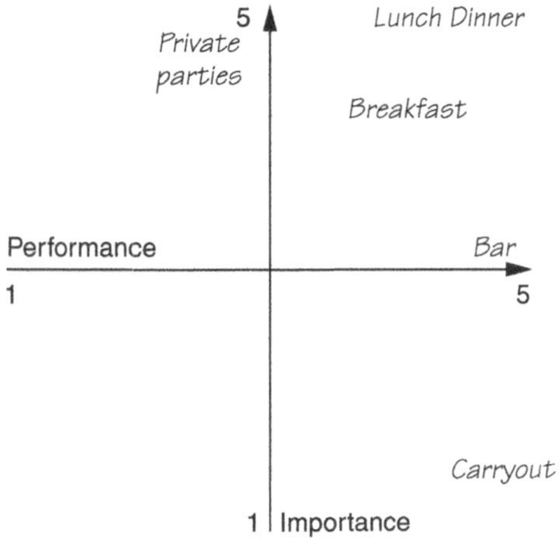

Figure 5.184 Two-dimensional chart example.

influence–importance chart under *stakeholder analysis.* These are variations of the two-dimensional chart. The Kano model on page 18 is also an example.

Considerations

- While it may be possible to measure the factors on a numerical scale, that precision is not necessary. Often lines are drawn with simply "low" at one end and "high" at the other. For example, in the Kano model, numerical precision is possible but not necessary.
- Label the axes consistently and in a way that makes intuitive sense to viewers. Usually, that means low values at the bottom of the y-axis and the left side of the x-axis. If you are using a yes–no scale, place them consistently too.
- The distance between any two points reflects the perceived similarity between them. Points very close together are considered similar.
- Sometimes it is useful to draw arrows (often called vectors) on the chart to indicate magnitude and direction. For example, an arrow might represent how you would like the ratings to change after planned actions or improvements. Or an arrow might show that the farther a point lies in that direction, the greater the customer's perception of a characteristic or the greater the appeal to a market segment.
- Often, variations of this tool have the word "matrix" in their names, such as importance–influence matrix, although they aren't really matrices. This tool is a hybrid between a matrix and a graph, but with its own unique and consistent characteristics.

value-added analysis

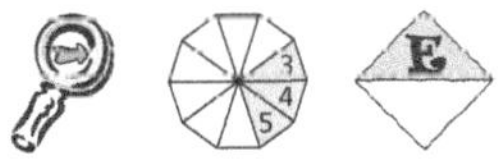

Description

Value-added analysis is a way of studying a process to identify problems. The analysis helps a team look critically at individual steps of a process to differentiate those that truly add value for the customer from those that do not.

When to Use

- When flowcharting a process, to be sure that non-value-added activities are included, or . . .
- When analyzing a flowchart to identify all possible waste in a process

Procedure

1. Obtain or create a detailed flowchart or deployment flowchart of the process.
2. For each step ask the following questions*:
   - Is this activity necessary to produce output?
   - Does it contribute to customer satisfaction?

   If the answer to both questions is yes, then label or color-code (green) the step as *real value-adding (RVA)*.
3. If the answer was no to either question, ask this:
   - Does this activity contribute to the organization's needs?

   If the answer is yes, then label or color-code (yellow) the step as *organizational value-adding (OVA)*.
4. If the answer to all questions is no, then label or color-code (red) the step as *non-value-adding (NVA)*.
5. (Optional) For each step, determine the amount of time required and/or the cost. Determine totals for each category and for the entire process. Determine what fraction or percent of the total time and/or cost is spent in real value-added activities.
6. Study the non-value-added activities to reduce or eliminate them.
7. Study organizational value-adding activities to reduce or eliminate them.

Example

Figure 5.185 shows a value-added analysis for the flowchart for filling an order. (This is the same process as the detailed flowchart example, Figure 5.63, page 261.) Each step has been assigned to one of the three categories: real value-adding, organizational value-adding, or non-value-adding.

Note that looking only at the RVA steps shows a simple, fundamental process that satisfies the customer: order received, receive materials, make product, confirm delivery date with customer, ship product. The steps related to product and materials inventory, product and materials inspection, and the wait steps are NVA. All other steps add value only for the organization and are opportunities for simplification and waste elimination. One could imagine, for example, a pull system to eliminate ordering of materials.

Compare this analysis to the cost-of-poor-quality example on page 201, which uses the same order-filling process. The cost-of-poor-quality analysis identified some of the same steps, but not all. For example, the wait steps were not flagged.

Also see *cycle time chart* for another use of value-added analysis.

---

* Adapted from H. James Harrington, *Business Process Improvement: The Breakthrough Strategy for Total Quality, Productivity, and Competitiveness* (New York: McGraw-Hill, 1991).

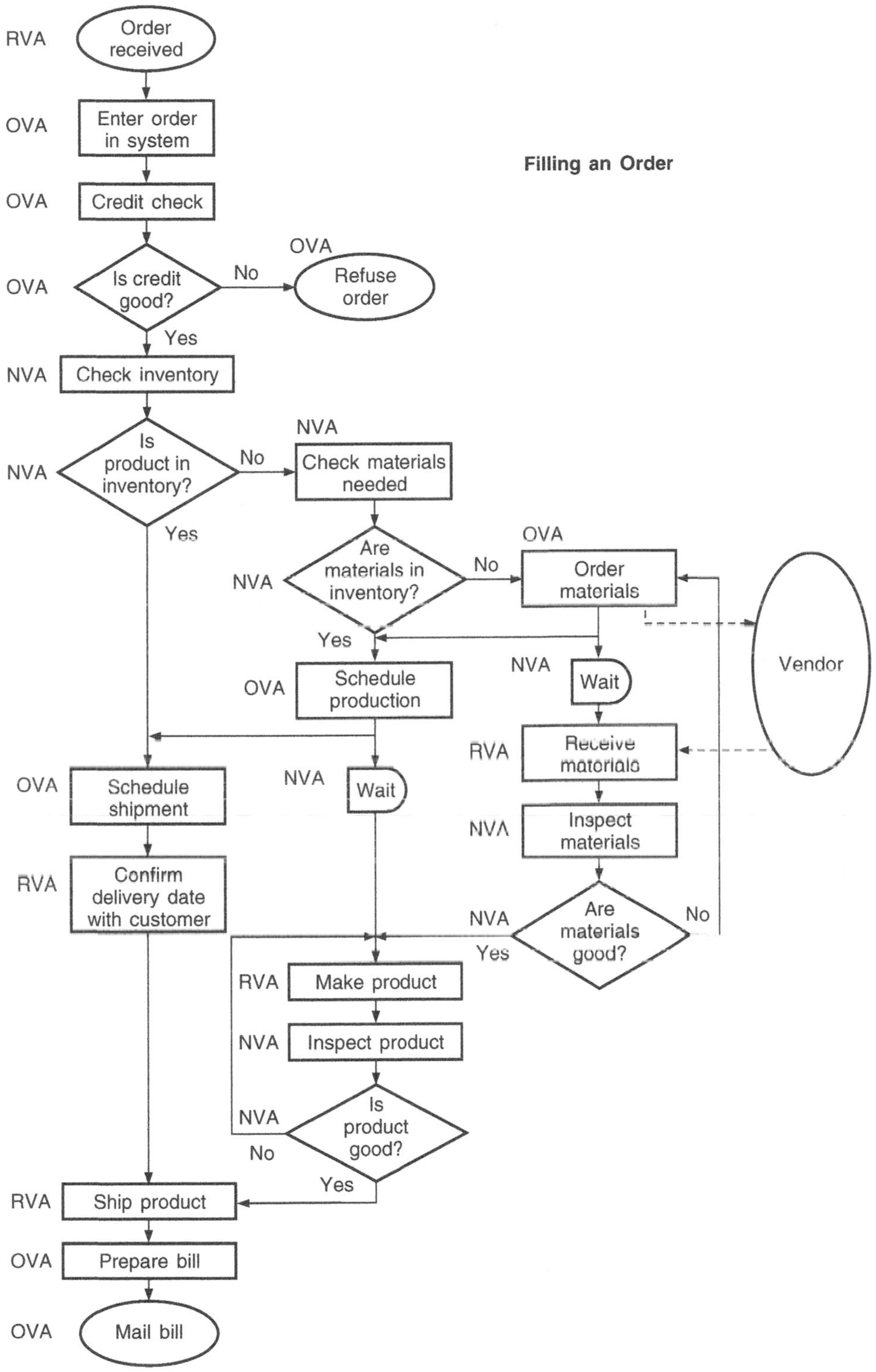

Figure 5.185 Value-added analysis example.

Considerations

- Typical NVA activities are: inspection, rework, inventory, waiting, moving materials. Cost-of-poor-quality activities are usually non-value-added.
- Typical OVA activities are: scheduling, maintenance, information-handling (on paper or in computer systems), accounting, administration.
- Typical RVA activities are: order receipt, manufacturing, delivering product, providing service, communicating information to customer.
- Calculating the percentage of time or cost spent on RVA versus OVA and NVA activities can be eye-opening.
- Try using a cycle time chart to graphically illustrate the frequency and cost of value-added and non-value-added activities.
- For other, complementary ways to analyze a flowchart, see *cost-of-poor-quality analysis* and *critical-to-quality analysis*. While these methods do not identify all the waste that can be found in a value-added analysis, they can be useful starting points if doing a full value-added analysis is overwhelming. They may be more appropriate if the objective is not so broad as identifying all waste.

voice of the customer table

Also called: customer voice table, customer needs table

Description

A voice of the customer (VOC) table captures and organizes information about how the customer uses the product or service. Additional columns on the table translate the customer's voice into meaningful and measurable items for action by the organization. This tool is often used with QFD before building a house of quality.

When to Use

- After obtaining information directly from customers, and . . .
- When organizing and analyzing customer needs and wants, to translate them into organization criteria and actions

Procedure

Part One: Capturing the Customer's Voice

1. Create a table with rows and columns as shown in the example.
2. For each customer response, fill in a row. Under "customer demographic," include sex, age, location, occupation, or anything else considered relevant. Under "voice of the customer," capture as accurately as possible the actual words customers used to describe what they want. Under the remaining "use" columns, answer 5W1H questions to capture details about how customers actually use the product or service.
3. Consider the "voice of the customer" data in light of all the data about how the product or service is used. Under "reworded data," write statements that translate the customer's needs into criteria meaningful to the organization.

Part Two: Translating the Customer's Voice

4. Create another table with rows and columns as shown in the example. Use columns that reflect features meaningful for your product or service.
5. Sort out the data from part one into categories: benefits to the customer and features that provide those benefits. Place under "demanded quality" each piece of reworded data from part one that expresses a qualitative benefit to the customer. Place under the appropriate column each piece of data that expresses a measurable feature.

Example

A catalog clothing retailer has a project to improve their telephone customer service. Customer interviews have provided data about what was important to customers during their calls. Figure 5.186 shows the first entries to a voice of the customer table. The "feature" columns are categories that are meaningful to the customer service department.

Considerations

- Some VOC tables divide each "use" category into two columns. One captures information directly from the customer. The other reflects internally generated ideas about who could use it, what it could be used for, and so on.
- Use feature categories that are meaningful for your organization. For products, categories often used are performance, function, and reliability.
- An affinity diagram or its variation, thematic analysis, can be useful to identify themes from raw VOC data.

Customer Demographic	Voice of the Customer	Use						Reworded Data
		Who	What	When	Where	Why	How	
F, 46, div, urban KY, banker	Should feel like talking to a friend Suggest matching accessories	Self	Business clothes	Weekday evening	From home	No time to shop	Discusses new items matching previous purchases	Friendly rep Rep converses while entering order Smile in rep's voice Rep uses customer's name twice Database lists finishing touches Product specialist can conference
M, 33, mar, rural FL, paramedic	Get the order right Fast-track the order Don't waste my time	Wife (nurse)	Leisure clothing	Weekday morning	From work	Birthday; wife mentioned item	Knows item number Short delivery time Doesn't know size	No errors Rep knows how to deduce correct size Rep knows express shipping methods Short call Instantaneous computer response

Demanded Quality	Features			Other
	Performance	Presence	Knowledge	
Friendly rep	No errors	Rep converses while entering order	Database lists finishing touches	
	Short call	Smile in rep's voice	Product specialist can conference	
	Instantaneous computer response	Rep uses customer's name twice	Rep knows how to deduce correct size	
			Rep knows express shipping methods	

Figure 5.186 Voice of the customer table example.

- This is an overview of a tool that looks simple but is hard to put into practice. Drawing a table is easy; listening hard to customers and translating their words into meaningful internal criteria is not. Consult books and papers about QFD to learn more about this important topic.
- See *house of quality* for more information about translating customer desires into product or service quality characteristics.

why–why diagram

Also called: five whys

Description

The why–why diagram helps to identify the root causes of a problem. In addition, the method helps the team to recognize the broad network of problem causes and the relationship among these causes. It can indicate the best areas to address for short- and long-term solutions.

When to Use

- When the team needs to probe for the root cause of a problem, and . . .
- When the team's analysis of a problem is too superficial, or . . .
- When the many contributing causes to a problem are confusing
- Also, as a graphic communication tool, to explain to others the many causes of a problem

Procedure

Materials needed: cards or sticky notes, newsprint or flipchart paper, marking pens.

1. Develop a statement of the specific problem whose cause you are seeking. Write it on a note and place it at the far left of the work surface.
2. Ask "Why?" this problem does or could occur. List all these causes on notes and place them in a column immediately to the right of the problem.
3. Each of the cause statements now becomes a new problem statement. Again ask "Why?" Sometimes the question needs to be phrased, "Why does this situation cause the problem?" Create another column of cause statements. Show the relationships to the first column of causes with arrows.
4. Continue to turn each cause into a problem and ask "Why?" Do not stop until you reach an answer that is fundamental (company policy or procedure, systems, training needs, and so forth).

Variation

Using a diagram is not essential. Simply ask "Why?" and write down the answer(s). Continue to ask "Why?" for each new statement, until "Why?" has been asked five times or more and a root cause is reached.

Example

Figure 5.187 is a why–why diagram that tries to uncover the reasons that "making travel arrangements is a time-consuming hassle." At the second level, there are two answers to the question "Why?" The secretary interrupts with questions about the travel arrangements, or the tickets arrive with wrong arrangements and must be sent back.

When the question is asked, "Why does the secretary interrupt?" there are six different answers for various types of information the secretary needs. When "Why?" is asked again for each of those, the answers come back to three possibilities: the secretary doesn't know the traveler's requirements, doesn't know company policies, or doesn't know if the trip is firm. Notice that the end of one chain of causes involving the secretary is a very broad systemic problem: no training program exists.

Look at the cause, "Traveler's requirements not known by secretary." One cause of that is "Temporary secretary." It is not useful to ask, "Why is there a temporary secretary?" That question diverts us from our original problem. The appropriate question is, "Why does having a temporary secretary cause this problem?" Then the answer is, "Poor communication of requirements to secretary."

Considerations

- You can construct the diagram directly on flipchart paper or a whiteboard, but using sticky notes makes construction of the diagram easier.
- The team may prefer to follow one cause statement to its end before thinking about other causes.
- Causes in a column on the right may be related to more than one earlier cause.
- If you prefer, place your problem statement at the top of a vertical work surface and list causes in rows under it.
- The variation done without a tree diagram works best in simpler situations when each "Why?" has only one or two answers.
- Don't stop when you reach a "who." Keep asking why. "Whos" are convenient ways to point fingers, but they are not root causes.
- The longer the chain of causes to the end point, the more likely it is that the end point deals with system issues such as management policies. These deeper causes usually lead to a more complete, fundamental solution to a problem. Addressing causes that arise early in the chain often amounts to applying a Band-Aid; addressing deeper causes provides long-term solutions to the problem.
- Tackling deeper causes often solves other related problems. See Figure 5.188. Addressing many problems at the early-cause level is like digging an inch deep. Each solution applies to only one problem. Addressing one problem at a deeper level is like digging a mile deep. You hit bedrock. The deeper cause is likely to involve underlying processes or systems. In digging that deep, you reach a cause

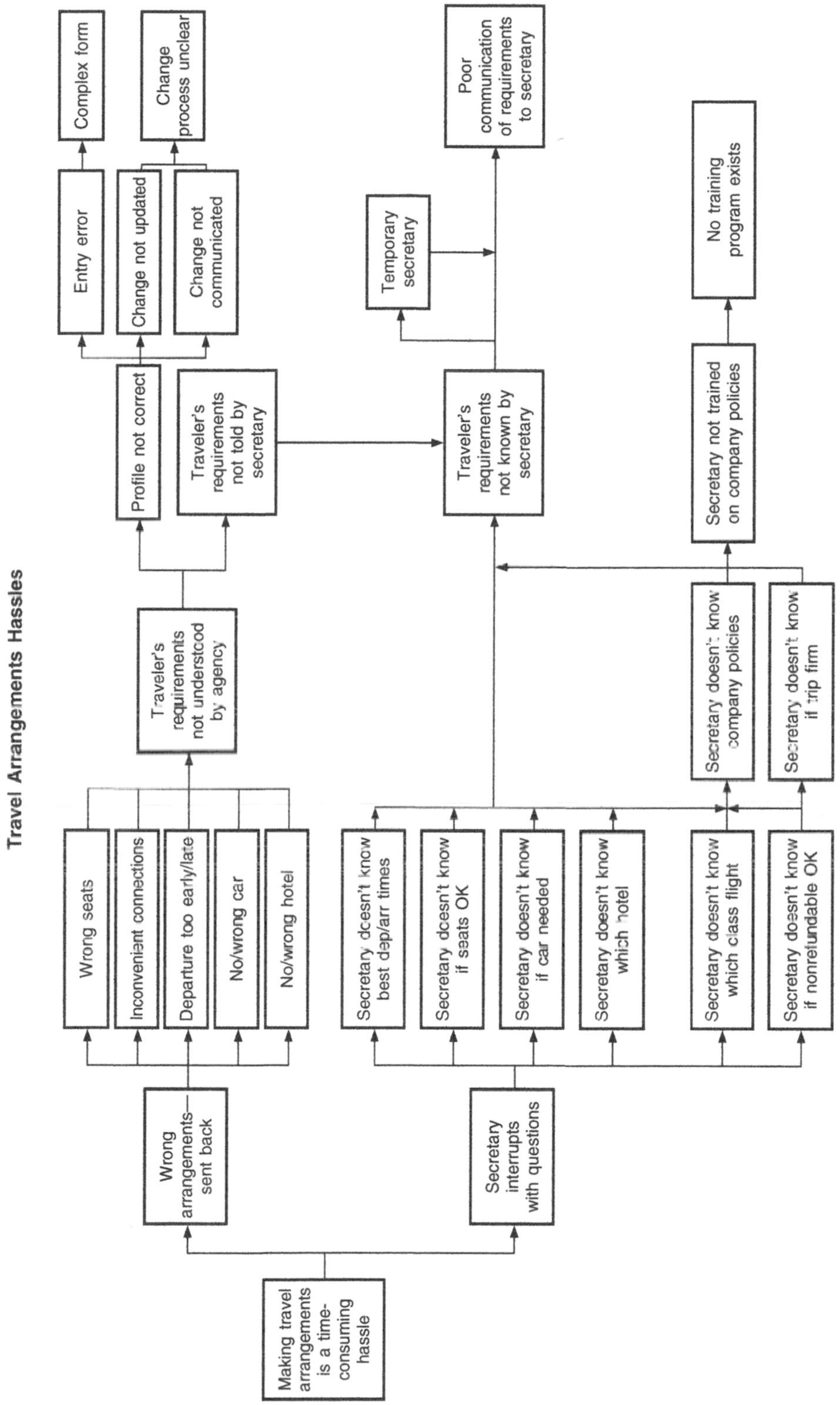

Figure 5.187 Why–why diagram example.

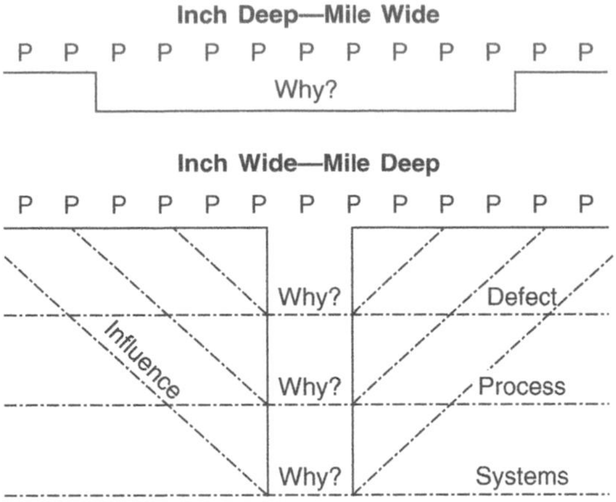

Figure 5.188 Inch deep, mile wide—inch wide, mile deep.

whose solution may influence many other problems. For example, establishing a training program for secretaries on company policies would probably solve other problems in addition to travel arrangement hassles. This principle is sometimes referred to by the phrase "inch deep, mile wide or inch wide, mile deep."

- The why–why diagram is a variation of the tree diagram. See *tree diagram* for more information. Also, see *fault tree analysis* for a similar, more structured tool that generates more information about the relationships between causes and effects at each level.

wordsmithing

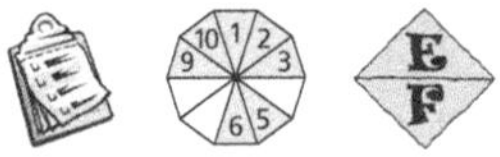

Description

Wordsmithing is a method for writing a statement that includes the ideas of everyone in the group. In a very short period of time, the critical concepts are surfaced and the impossible happens: composition by committee.

When to Use

- When writing a charter or goal statement, or . . .
- When writing any other brief statement, when it is important that the statement represent team consensus

Procedure

Materials needed: flipchart paper and marking pen for each person, tape.

1. Brainstorm thoughts and ideas about your team's issue. Capture the ideas on a flipchart or whiteboard.
2. (Optional) For an objective or goal statement, discuss and decide where on the continuum of team goals the still-unknown statement should be.
3. Each member individually and silently drafts a statement. Give each person a flipchart page and marking pen to write his or her statement—in big letters—then tape each page to the wall.
4. When all the statements are on the wall, everyone should walk around and examine all the statements. All team members underline words or phrases they believe should be in the final version of the statement. If someone has already underlined a phrase you like, underline it again. This is a voting process. Underline words on your own page, too.
5. Identify and discuss the most popular words and phrases. It also can be valuable to discuss minority opinions (single lines). Reach consensus on the key ideas that should be in the final version.
6. At this point, there are two options:
   - Sometimes it is obvious and easy to write the key words or phrases on a clean sheet of flipchart paper, add connecting words and phrases, and *eureka!* the statement is born.
   - If it doesn't seem obvious and easy, one or several volunteers should combine the popular phrases into a draft before the next meeting. The volunteers should take with them all the flipchart pages from today's work and keep in mind the discussions from steps 1, 2, and 5. At the next meeting, the volunteers should present their draft to the entire team for feedback and confirmation or instructions to try again.
7. Assess the statement. Does this statement capture the understanding of the team? For a goal statement, check where the statement is on the continuum of team goals. Is it clear where it is? Is it in the right place?
8. Make the team's agreed-upon modifications to the draft statement and finalize.

Example

The engineering department of a large corporation designs and builds manufacturing facilities for the entire corporation. A team was asked to draft a mission statement for the department. The team brainstormed the following list of ideas:

- Design
- Build/construct
- Plan
- Within budget
- Low cost
- Within schedule
- Better than outsourcing
- Quickly—ASAP
- Best engineering know-how
- Current technology
- Business needs
- Easy to operate
- Flexible
- Safe
- Environmentally responsible
- Customers—plant operations
- Easy to maintain
- Customers—maintenance
- Community
- Manage contractors
- Maintain standards

The group agreed that the final statement belonged at the far left of the continuum of goals, as the statement should encompass all services and activities of the department.

Next came the individual statements and the underlining. Here is what was hanging on the walls after 12 minutes:

> The Engineering Department's mission is to design and construct new manufacturing plants and plant expansions at best total cost and schedule.

> The Engineering Department serves the business units and manufacturing department by designing and building facilities that serve the needs of those customers.

> To ensure that facilities exist to manufacture products for our customers, and that those facilities are built and operate as efficiently and effectively as possible.

> The role of the Engineering Department is to build safe, environmentally responsible, easily maintained and operated facilities according to the best available practices.

> To do the best possible job of building the best possible plants.

Several volunteers took all the flipcharts and over the next week developed the following statement, which was adopted by the team and the entire department:

> The Engineering Department serves the business units, manufacturing department, and our customers by being the best at designing and constructing safe, environmentally responsible, easily maintained and operated facilities—better, faster, and at lower total cost than anyone else.

Considerations

- This process usually takes 30 to 45 minutes up to step 6.
- The more discussion you have in steps 1 and 2, the less you will need in step 5. For most groups, it works best to discuss the ideas thoroughly before trying to generate words.
- However, if there has already been a lot of discussion about the topic, you can minimize or even skip steps 1 and 2. Get your initial ideas out on paper, then discuss more thoroughly in step 5.

work-flow diagram

Also called: spaghetti diagram, work-flow analysis diagram, geographic flowchart, physical layout flowchart

Description

A work-flow diagram is a picture that shows movement through a process. That movement might be of people, materials, paper, or information. The diagram consists of a map (such as a floor plan) of the area where the process takes place and lines showing all movements. The diagram graphically shows inefficiency—unnecessary movement

When to Use

- When the process being studied involves transportation or movement of people, materials, paper, or information, and . . .
- When trying to eliminate waste

Procedure

1. Decide what it is that moves. This may be paper, a file, a person, a piece of information, or materials.
2. Determine the relevant area of movement. Develop or obtain a representation of that area.
3. Develop a list of the process steps, in sequence. The best way to do this is to develop or obtain a detailed flowchart of the process.
4. Draw lines on the layout showing every movement of the item as the process proceeds from step to step. If the movement in a step follows the same path as

the movement in a previous step, draw another line. Different colors or line types can be used to distinguish repeated pathways, different phases of the process, or different people or objects moving.

5. (Optional) Identify the time required for each movement.
6. Analyze the diagram. Look for ways to eliminate or shorten movements.

Example

The paper flow of the process for filling an order is tracked on the work-flow diagram of Figure 5.189. This is the same process flowcharted in Figure 5.63, page 261. The work-flow diagram follows the path of the pieces of paper that are needed to complete the process.

The map of the production office, warehouse, and main offices (located in three separate buildings) was drawn first. Then the team drawing the diagram mentally walked through the process, starting with the first step on the flowchart, "Order received." Each paper movement was drawn with an arrow. If a movement followed the same path as a previous movement, another arrow was drawn.

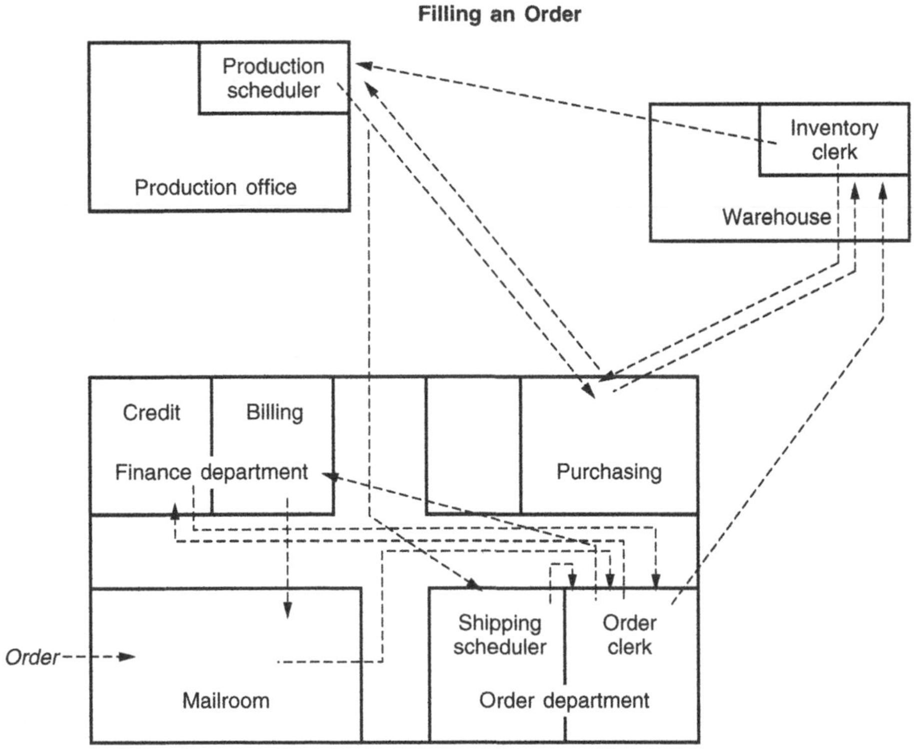

Figure 5.189 Work-flow diagram example.

The paper does a lot of traveling. Notice that credit and billing are located together in the finance department, but they do not interact in this process. Perhaps physical arrangement according to paper flow would be more efficient than the present arrangement by departments.

Considerations

- The relevant area of movement may be an office, a form to be filled out, or five offices located around the world. The representation of that area may be a floor plan, a copy of the form, a map.
- Look for excessive, repetitive, or unnecessary movement.
- To reduce or eliminate movement, you can change either the physical layout or the process, such as changing the sequence of steps.
- If information is passed electronically rather than on paper, the physical layout becomes irrelevant.

Appendix

Table A.1 Area under the normal curve.

Z	0.00	0.01	0.02	0.03	0.04	0.05	0.06	0.07	0.08	0.09
−3.5	.00023	.00022	.00022	.00021	.00020	.00019	.00019	.00018	.00017	.00017
−3.4	.00034	.00033	.00031	.00030	.00029	.00028	.00027	.00026	.00025	.00024
−3.3	.00048	.00047	.00045	.00043	.00042	.00040	.00039	.00038	.00036	.00035
−3.2	.00069	.00066	.00064	.00062	.00060	.00058	.00056	.00054	.00052	.00050
−3.1	.00097	.00094	.00090	.00087	.00085	.00082	.00079	.00076	.00074	.00071
−3.0	.00135	.00131	.00126	.00122	.00118	.00114	.00111	.00107	.00104	.00010
−2.9	.0019	.0018	.0017	.0017	.0016	.0016	.0015	.0015	.0014	.0014
−2.8	.0026	.0025	.0024	.0023	.0023	.0022	.0021	.0021	.0020	.0019
−2.7	.0035	.0034	.0033	.0032	.0031	.0030	.0029	.0028	.0027	.0026
−2.6	.0047	.0045	.0044	.0043	.0041	.0040	.0039	.0038	.0037	.0036
−2.5	.0062	.0060	.0059	.0057	.0055	.0054	.0052	.0051	.0049	.0048
−2.4	.0082	.0080	.0078	.0075	.0073	.0071	.0069	.0068	.0066	.0064
−2.3	.0107	.0104	.0102	.0099	.0096	.0094	.0091	.0089	.0087	.0084
−2.2	.0139	.0136	.0132	.0129	.0125	.0122	.0119	.0116	.0113	.0110
−2.1	.0179	.0174	.0170	.0166	.0162	.0158	.0154	.0150	.0146	.0143
−2.0	.0228	.0222	.0217	.0212	.0207	.0202	.0197	.0192	.0188	.0183
−1.9	.0287	.0281	.0274	.0268	.0262	.0256	.0250	.0244	.0239	.0233
−1.8	.0359	.0351	.0344	.0336	.0329	.0322	.0314	.0307	.0301	.0294
−1.7	.0446	.0436	.0427	.0418	.0409	.0401	.0392	.0384	.0375	.0367
−1.6	.0548	.0537	.0526	.0516	.0505	.0495	.0485	.0475	.0465	.0455
−1.5	.0668	.0655	.0643	.0630	.0618	.0606	.0594	.0582	.0571	.0559
−1.4	.0808	.0793	.0778	.0764	.0749	.0735	.0721	.0708	.0694	.0681
−1.3	.0968	.0951	.0934	.0918	.0901	.0885	.0869	.0853	.0838	.0823
−1.2	.1151	.1131	.1112	.1093	.1075	.1057	.1038	.1020	.1003	.0985
−1.1	.1357	.1335	.1314	.1292	.1271	.1251	.1230	.1210	.1190	.1170
−1.0	.1587	.1562	.1539	.1515	.1492	.1469	.1446	.1423	.1401	.1379
−0.9	.1841	.1814	.1788	.1762	.1736	.1711	.1685	.1660	.1635	.1611
−0.8	.2119	.2090	.2061	.2033	.2005	.1977	.1949	.1922	.1894	.1867
−0.7	.2420	.2389	.2358	.2327	.2297	.2266	.2236	.2207	.2177	.2148
−0.6	.2743	.2709	.2676	.2643	.2611	.2578	.2546	.2514	.2483	.2451
−0.5	.3085	.3050	.3015	.2981	.2946	.2912	.2877	.2843	.2810	.2776
−0.4	.3446	.3409	.3372	.3336	.3300	.3264	.3228	.3192	.3156	.3121
−0.3	.3821	.3783	.3745	.3707	.3669	.3632	.3594	.3557	.3520	.3483
−0.2	.4207	.4168	.4129	.4090	.4052	.4013	.3974	.3936	.3897	.3859
−0.1	.4602	.4562	.4522	.4483	.4443	.4404	.4364	.4325	.4286	.4247
−0	.5000	.4960	.4920	.4880	.4840	.4801	.4761	.4721	.4681	.4641

Continued

Continued

Z	0.00	0.01	0.02	0.03	0.04	0.05	0.06	0.07	0.08	0.09
0	.5000	.5040	.5080	.5120	.5160	.5199	.5239	.5279	.5319	.5359
0.1	.5398	.5438	.5437	.5517	.5557	.5596	.5636	.5675	.5714	.5753
0.2	.5793	.5832	.5871	.5910	.5948	.5987	.6026	.6064	.6103	.6141
0.3	.6179	.6217	.6255	.6293	.6331	.6368	.6406	.6443	.6480	.6517
0.4	.6554	.6591	.6628	.6664	.6700	.6736	.6772	.6808	.6844	.6879
0.5	.6915	.6950	.6985	.7019	.7054	.7088	.7123	.7157	.7190	.7224
0.6	.7257	.7291	.7324	.7357	.7389	.7422	.7454	.7486	.7517	.7549
0.7	.7580	.7611	.7642	.7673	.7704	.7734	.7764	.7794	.7823	.7852
0.8	.7881	.7910	.7939	.7967	.7995	.8023	.8051	.8079	.8106	.8133
0.9	.8159	.8186	.8212	.8238	.8264	.8289	.8315	.8340	.8365	.8389
1.0	.8413	.8438	.8461	.8485	.8508	.8531	.8554	.8577	.8599	.8621
1.1	.8643	.8665	.8686	.8708	.8729	.8749	.8770	.8790	.8810	.8830
1.2	.8849	.8869	.8888	.8907	.8925	.8944	.8962	.8980	.8997	.9015
1.3	.9032	.9049	.9066	.9082	.9099	.9115	.9131	.9147	.9162	.9177
1.4	.9192	.9207	.9222	.9236	.9251	.9265	.9279	.9292	.9306	.9319
1.5	.9332	.9345	.9357	.9370	.9382	.9394	.9406	.9418	.9429	.9441
1.6	.9452	.9463	.9474	.9484	.9495	.9505	.9515	.9525	.9535	.9545
1.7	.9554	.9564	.9573	.9582	.9591	.9599	.9608	.9616	.9625	.9633
1.8	.9641	.9649	.9658	.9664	.9671	.9678	.9686	.9693	.9699	.9706
1.9	.9713	.9719	.9726	.9732	.9738	.9744	.9750	.9756	.9761	.9767
2.0	.9773	.9778	.9783	.9788	.9793	.9798	.9803	.9808	.9812	.9817
2.1	.9821	.9826	.9830	.9834	.9838	.9842	.9846	.9850	.9854	.9857
2.2	.9861	.9864	.9868	.9871	.9875	.9878	.9881	.9884	.9887	.9890
2.3	.9893	.9896	.9898	.9901	.9904	.9906	.9909	.9911	.9913	.9916
2.4	.9918	.9920	.9922	.9925	.9927	.9929	.9931	.9932	.9934	.9936
2.5	.9938	.9940	.9941	.9943	.9945	.9946	.9948	.9949	.9951	.9952
2.6	.9953	.9955	.9956	.9957	.9959	.9960	.9961	.9962	.9963	.9964
2.7	.9965	.9966	.9967	.9968	.9969	.9970	.0971	.9972	.9973	.9974
2.8	.9974	.9975	.9976	.9977	.9977	.9978	.9979	.9979	.9980	.9981
2.9	.9981	.9982	.9983	.9983	.9983	.9984	.9985	.9985	.9986	.9986
3.0	.99865	.99869	.99874	.99878	.99882	.99886	.99889	.99893	.99896	.99900
3.1	.99903	.99906	.99910	.99913	.99915	.99918	.99921	.99924	.99926	.99929
3.2	.99931	.99934	.99936	.99938	.99940	.99942	.99944	.99946	.99948	.99950
3.3	.99952	.99953	.99955	.99957	.99958	.99960	.99961	.99962	.99964	.99965
3.4	.99966	.99967	.99969	.99970	.99971	.99972	.99973	.99974	.99975	.99976
3.5	.99977	.99978	.99978	.99979	.99980	.99981	.99981	.99982	.99983	.99983

Table A.2 Control chart constants.

$\bar{X}$ and R Control Chart				
n	A_2	D_3	D_4	d_2
1	2.660	0	3.267	1.128
2	1.880	0	3.267	1.128
3	1.023	0	2.574	1.693
4	0.729	0	2.282	2.059
5	0.577	0	2.144	2.326
6	0.483	0	2.004	2.534
7	0.419	0.076	1.924	2.704
8	0.373	0.136	1.864	2.847
9	0.337	0.184	1.816	2.970
10	0.308	0.223	1.777	3.078
11	0.285	0.256	1.744	3.173
12	0.266	0.283	1.717	3.258
$\bar{X}$ and s Control Chart				
n	A_3	B_3	B_4	c_4
6	1.287	0.030	1.970	0.9515
7	1.182	0.118	1.882	0.9594
8	1.099	0.185	1.815	0.9650
9	1.032	0.239	1.761	0.9693
10	0.975	0.284	1.716	0.9727
11	0.927	0.321	1.679	0.9754
12	0.886	0.354	1.646	0.9776
13	0.850	0.382	1.618	0.9794
14	0.817	0.406	1.594	0.9810
15	0.789	0.428	1.572	0.9823
16	0.763	0.448	1.552	0.9835
17	0.739	0.466	1.534	0.9845
18	0.718	0.482	1.518	0.9854
19	0.698	0.497	1.503	0.9862
20	0.680	0.510	1.490	0.9869
21	0.663	0.523	1.477	0.9876
22	0.647	0.534	1.466	0.9882
23	0.633	0.545	1.455	0.9887
24	0.619	0.555	1.445	0.9892
25	0.606	0.565	1.435	0.9896

Source: American Society for Testing and Materials. Table adapted from ASTM-STP 15D. Copyright STM. Reprinted with permission.

Resources

This section points you to readily accessible resources where you can learn more about individual tools. In some cases, a reference is the original source for the tool. Other sources provide more detail or depth than could be included in this book. Some of the resources, especially Web sites, have lists of or links to additional references. Many more references to quality tools exist than can be listed here. For example, textbooks on quality control include many tools. Consider these references suggestions and starting points for additional information.

A few tools are not listed; in those cases, neither an original source nor any references that go beyond the information provided in this book could be located. Ironically, some of the most fundamental, widely used tools have the least information available. A half dozen tools were developed by the author or colleagues within the author's organization. For those tools, a note states that this book is the primary reference.

After the individual tool references are two more lists. One shows Web sites that house collections of tools for quality improvement and creativity. The second shows Web sites that provide information on a wide range of statistical topics.

CHAPTER 2: MEGA-TOOLS: QUALITY MANAGEMENT SYSTEMS

Evolution of Quality Control

Juran Institute. www.juran.com.
Skymark Corporation. "Management Thought Leaders." Skymark. www.skymark.com/resources/leaders/biomain.asp.
The W. Edwards Deming Institute. www.deming.org.

Total Quality Management

Imai, Masaaki. *Kaizen: The Key to Japan's Competitive Success.* New York: Random House, 1986.
Ishikawa, Kaoru. *Guide to Quality Control.* Tokyo: Asian Productivity Organization, 1986.
Mizuno, Shigeru (ed.). *Management for Quality Improvement: The 7 New QC Tools.* Cambridge, MA: Productivity Press, 1995.

Quality Function Deployment

Ermer, Donald S., and Mark K. Iniper. "Delighting the Customer: Quality Function Deployment for Quality Service Design." *Total Quality Management* 9, no. 4/5 (July 1998): 86.
Mazur, Glenn. "QFD Case Studies and White Papers." *QFD and Voice of Customer Analysis.* Japan Business Consultants Ltd. www.mazur.net/publishe.htm.
Xie, Min, Kay-Chuan Tan, and Thong Ngee Goh. *Advanced QFD Applications.* Milwaukee: ASQ Quality Press, 2003.

ISO 9000

Bergenhenegouwen, Louise, Annemarie de Jong, and Henk J. de Vries. *100 Frequently Asked Questions on the ISO 9000:2000 Series.* Milwaukee: ASQ Quality Press, 2002.
Cianfrani, Charles A., Joseph J. Tsiakals, and John E. West. *ISO 9001:2000 Explained*, Second Edition. Milwaukee: ASQ Quality Press, 2001.
Seddon, John. "A Brief History of ISO-9000." Lean Service—Systems Thinking for Service Organisations. www.lean-service.com/home.asp.

Malcolm Baldrige National Quality Award

Blazey, Mark L. *Insights to Performance Excellence 2004: An Inside Look at the 2004 Baldrige Award Criteria.* Milwaukee: ASQ Quality Press, 2004.
National Institute of Standards and Technology. "Baldrige National Quality Program." U.S. Commerce Department, Technology Administration, National Institute of Standards and Technology. www.quality.nist.gov.

Benchmarking

American Productivity and Quality Center. www.apqc.org.
Camp, Robert. *Benchmarking: The Search for Industry Best Practices that Lead to Superior Performance.* Milwaukee: ASQ Quality Press, 1989.
Spendolini, Michael J. *The Benchmarking Book.* 2nd ed. New York: AMACOM, 2003.

Six Sigma

Breyfogle, Forrest W., III. *Implementing Six Sigma: Smarter Solutions Using Statistical Methods*. Hoboken, NJ: John Wiley & Sons, 2003.
i-Six Sigma. "New to Six Sigma: A Six Sigma Guide for Both Novice and Experienced Quality Practitioners." i-Six Sigma LLC. www.isixsigma.com/library/content/six-sigma-newbie.asp.
Pyzdek, Thomas. *The Six Sigma Handbook*. New York: McGraw-Hill, 2003.
Wheeler, Donald J. "The Six-Sigma Zone." *SPC Ink*, 2002. SPC Press. www.spcpress.com/ink_pdfs/The%20Final%206%20Sigma%20Zone.pdf.

Lean Manufacturing

AGI-Goldratt Institute. www.goldratt.com.
Goldratt, Eliyahu M., and Jeff Cox. *The Goal: A Process of Ongoing Improvement*. Croton-on-Hudson, NY : North River Press, 1992.
Lean Manufacturing Enterprise Technical Group, Society of Manufacturing Engineers. www.sme.org/cgi-bin/communities.pl?/communities/techgroups/lean_mfg/l_m_e_o.htm
Levinson, William A., and Raymond A. Rerick. *Lean Enterprise: A Synergistic Approach to Minimizing Waste*. Milwaukee, WI: Quality Press, 2002.
Pyzdek, Thomas. *The Six Sigma Handbook*. New York: McGraw-Hill, 2003.

TRIZ

Ideation International. www.ideationtriz.com.
Rantanen, Kalevi, and Ellen Domb. *Simplified TRIZ: New Problem Solving Applications for Engineers & Manufacturing Professionals*. Milwaukee: Quality Press, 2002.
The TRIZ Journal. TRIZ Institute. www.triz-journal.com.

CHAPTER 3: THE QUALITY IMPROVEMENT PROCESS

Kayser, Thomas A. *Mining Group Gold: How to Cash in on the Collaborative Brain Power of a Group*. New York: McGraw-Hill, 1995.
Scholtes, Peter, Brian L. Joiner, and Barbara J. Streibel. *The Team Handbook*. Madison, WI: Joiner/Oriel Inc., 2003.

CHAPTER 4: QUALITY IMPROVEMENT STORIES

Medrad, Inc. www.medrad.com.
Pearl River School District. www.pearlriver.k12.ny.us.
Saint Luke's Hospital of Kansas City. https://www.saintlukeshealthsystem.org/slhs/com/system/baldrige/quality_search.htm.

CHAPTER 5: TOOLS

ACORN Test

Gilbert, Thomas F. *Human Competence: Engineering Worthy Performance*. New York: McGraw-Hill, 1978.

Affinity Diagram

Kabay, M. E. "Computer-Aided Thematic Analysis: Useful Technique for Analyzing Non-Quantitative Data." *Ubiquity* 4: issue 24 (August 6, 2003). Association for Computing Machinery. www.acm.org/ubiquity/views/v4i24_kabay.html.

Mizuno, Shigeru (ed.). *Management for Quality Improvement: The 7 New QC Tools*. Cambridge, MA: Productivity Press, 1995.

Arrow Diagram

Levy, Ferdinand K., Gerald L. Thompson, and Jerome D. West. "The ABCs of the Critical Path Method." In *Managing Projects and Programs*. Harvard Business Review Reprint Series, no. 10811. Boston: Harvard Business School Press, 1989.

Miller, Robert W. "How to Plan and Control with PERT." In *Managing Projects and Programs*. Harvard Business Review Reprint Series, no. 10811. Boston: Harvard Business School Press, 1989.

Mizuno, Shigeru (ed.). *Management for Quality Improvement: The 7 New QC Tools*. Cambridge, MA: Productivity Press, 1995.

Balanced Scorecard

Kaplan, Robert S., and David P. Norton. "The Balanced Scorecard–Measures That Drive Performance." *Harvard Business Review* 70, no. 1 (Jan/Feb92): 71–79.

———. "Using the Balanced Scorecard as a Strategic Management System." *Harvard Business Review* 74, no.1 (Jan/Feb96): 75–85.

———. *The Balanced Scorecard: Translating Strategy into Action*. Boston: Harvard Business School Press, 1996.

Benchmarking

American Productivity and Quality Center. www.apqc.org.

Camp, Robert. *Benchmarking: The Search for Industry Best Practices that Lead to Superior Performance*. Milwaukee: ASQC Quality Press, 1989.

Spendolini, Michael J. *The Benchmarking Book*. 2nd ed. New York: AMACOM, 2003.

Benefits and Barriers

Pritchard, Blanchard. "Quality Management." In *Encyclopedia of Chemical Processing and Design,* vol. 46, exec. editor John J. McKetta. New York: Marcel Dekker, 1994.

Box Plot

Cleveland, William S. *The Elements of Graphing Data.* Murray Hill, NJ: AT&T Bell Laboratories, 1994.

Heyes, Gerald B. "The Box Plot." *Quality Progress* 18, no. 12 (December 1985): 12–17.

———. "The GHOST Box Plot." *Statistics Division Newsletter,* ASQC (summer 1988).

Iglewicz, Boris, and David C. Hoaglin. "Use of Boxplots for Process Evaluation." *Journal of Quality Technology* 19, no. 4 (October 1987): 180–90.

Tukey, John W. *Exploratory Data Analysis.* Reading, MA: Addison-Wesley, 1977.

Brainstorming and Brainwriting

de Bono, Edward. *Serious Creativity: Using the Power of Lateral Thinking to Create New Ideas.* New York: HarperCollins, 1992.

Infinite Innovations Ltd. "Brainstorming.co.uk." www.brainstorming.co.uk.

Iowa State University Extension Service. "Group Decision Making Tool Kit." Iowa State University Extension to Communities. www.extension.iastate.edu/communities/tools/decisions.

Mycoted, Ltd. "Creativity Techniques." *Creativity & Innovation in Science & Technology.* www.mycoted.com/creativity/techniques/index.php.

VanGundy, Arthur B. *Techniques of Structured Problem Solving.* New York: Van Nostrand Reinhold, 1988.

Cause-and-Effect Matrix

Breyfogle, Forrest W., III. *Implementing Six Sigma: Smarter Solutions Using Statistical Methods.* Hoboken, NJ: John Wiley & Sons, 2003.

Checklist

Ishikawa, Kaoru. *Guide to Quality Control.* Tokyo: Asian Productivity Organization, 1986.

Checksheet

Ishikawa, Kaoru. *Guide to Quality Control.* Tokyo: Asian Productivity Organization, 1986.

Juran Institute, Inc. "The Tools of Quality: Part V; Check Sheets." *Quality Progress* 23, no. 10 (October, 1990): 51–56.

Contingency Table

GraphPad Software, Inc. "Creating Contingency Tables." *The Prism Guide to Interpreting Statistical Results.* www.graphpad.com/articles/interpret/contingency/contin_tables.htm.

Ishikawa, Kaoru. *Guide to Quality Control.* Tokyo: Asian Productivity Organization, 1986.

Continuum of Team Goals

This tool was developed by the author in the early 1990s. This book is the primary reference.

Control Charts

Hoyer, Robert W., and Wayne C. Ellis. "A Graphical Exploration of SPC: Part 2." *Quality Progress* 29, no. 6 (June 1996): 57–64.

Ishikawa, Kaoru. *Guide to Quality Control.* Tokyo: Asian Productivity Organization, 1986.

McNees, William H., and Robert A. Klein. *Statistical Methods for the Process Industries.* Milwaukee: ASQC Quality Press, 1991.

Wheeler, Donald, and David Chambers. *Understanding Statistical Process Control.* Knoxville, TN: SPC Press, 1986.

Correlation Analysis

Filliben, James J., and Alan Heckert. "Linear Correlation Plot." In *NIST/SEMATECH e-Handbook of Statistical Methods*, ed. Carroll Croarkin and Paul Tobias, 2004. U.S. Commerce Department, Technology Administration, National Institute of Standards and Technology, Information Technology Laboratory. www.itl.nist.gov/div898/handbook/eda/section3/linecorr.htm.

Ishikawa, Kaoru. *Guide to Quality Control.* Tokyo: Asian Productivity Organization, 1986.

StatSoft, Inc. "Elementary Concepts in Statistics." *Electronic Statistics Textbook.* StatSoft, 2004. www.statsoft.com/textbook/stathome.html.

Cost-of-Poor-Quality Analysis

This tool was developed within Ethyl Corporation in the late 1980s. This book is the primary reference. For information about the concept of cost of poor quality:

Juran, Joseph M., and Frank M. Gryna, Jr. *Quality Planning and Analysis: From Product Development Through Use.* New York: McGraw-Hill, 1980.

Schiffauerova, Andrea, and Vince Thomson. "Cost of Quality: A Survey of Models and Best Practices." Submitted to International Journal of Quality and Reliability Management. Available online at McGill University, Masters in Manufacturing Management. www.mmm.mcgill.ca/documents-frame.html.

Criteria Filtering

VanGundy, Arthur B. *Techniques of Structured Problem Solving.* New York: Van Nostrand Reinhold, 1988.

Critical-to-Quality Analysis and Critical-to-Quality Tree

Berryman, Maurice L. "DFSS and Big Payoffs: Transform Your Organization Into One That's World Class." *Six Sigma Forum Magazine* 2, no. 1 (November, 2002). American Society for Quality. www.asq.org/pub/sixsigma/past/vol2_issue1/berryman.html.

Brown, Tony. "The Critical List." *Quality World,* August, 2002. Institute of Quality Assurance. www.iqa.org/publication/c4-1-71.shtml.

Cycle Time Chart

Harrington, H. James. *Business Process Improvement: The Breakthrough Strategy for Total Quality, Productivity, and Competitiveness.* New York: McGraw-Hill, 1991.

Decision Matrix

Dodgson, John, Michael Spackman, Alan Pearman, and Lawrence Phillips. *Multi-Criteria Analysis Manual,* Jun 25, 2004. UK, Office of the Deputy Prime Minister. www.odpm.gov.uk/stellent/groups/odpm_about/documents/page/odpm_about_608524.hcsp.

Decision Tree

American Association for Artificial Intelligence. "Decision Trees." American Association for Artificial Intelligence. www.aaai.org/AITopics/html/trees.html.

Design of Experiments

Barrentine, Larry. *An Introduction to Design of Experiments: A Simplified Approach.* Milwaukee: ASQ Quality Press, 1999.

Bhote, Keki R. *World Class Quality: Using Design of Experiments to Make It Happen.* New York: AMACOM, 1991.

Wheeler, Donald J. *Understanding Industrial Experimentation.* Knoxville, TN: SPC Press, 1990.

FMEA

FMECA.COM. Haviland Consulting Group. www.fmeca.com.

Stamatis, D. H. *Failure Mode and Effect Analysis: FMEA from Theory to Execution.* Milwaukee: ASQ Quality Press, 2003.

Fault Tree Analysis

Long, Allen. *Fault-Tree.Net.* www.fault-tree.net.

Texas Workers' Compensation Commission. "Fault Tree Analysis." Texas Workers' Compensation Commission, Workers' Health & Safety Division. www.twcc.state.tx.us/information/videoresources/stp_fault_tree.pdf.

U.S. Coast Guard. "Fault Tree Analysis." In *Risk Assessment Tools Reference.* Risk-Based Decision-Making Guidelines, vol. 3. U.S. Department of Homeland Security, U.S. Coast Guard. www.uscg.mil/hq/g-m/risk/E-Guidelines/RBDM/html/vol3/09/v3-09-cont.htm.

Fishbone Diagram

Fukuda, Ryuji. *Managerial Engineering: Techniques for Improving Quality and Productivity in the Workplace.* Cambridge, MA: Productivity Press, 1997.

Ishikawa, Kaoru. *Guide to Quality Control.* Tokyo: Asian Productivity Organization, 1986.

Sarazen, J. Stephen. "The Tools of Quality: Part II; Cause-and-Effect Diagrams." *Quality Progress* 23, no. 7 (July 1990): 59–62.

5W2H

Japan Human Relations Association. *The Service Industry Idea Book : Employee Involvement in Retail and Office Improvement.* Translated by Fredrich Czupryna. Cambridge, MA: Productivity Press, 1990.

Flowchart

Burr, John T. "The Tools of Quality: Part I; Going with the Flow(chart)." *Quality Progress* 23, no. 6 (June 1990): 64–67.

Cram, David M. "Flowcharting Primer." *Training and Development Journal* 34, no. 7 (July, 1980): 64–68.

Force Field Analysis

Lewin, Kurt. *Resolving Social Conflicts and Field Theory in Social Science.* Washington, D.C.: American Psychological Association, 1997.

Stratton, A. Donald. "Solving Problems with CEFFA." *Quality Progress* 19, no. 4 (April 1986): 65–70.

Gantt Chart

Tufte, Edward. "Project Management Graphics (or Gantt Charts)." Online posting, Mar 20, 2002. The Work of Edward Tufte and Graphics Press, Ask E.T. Forum. www.edwardtufte.com/bboard/q-and-a-fetch-msg?msg_id=000076&topic_id=1

Wilson, James M. "Gantt Charts: A Centenary Appreciation." *European Journal of Operational Research* 149, no. 2 (September 2003): 430.

Graph

Chambers, John M., William S. Cleveland, Beat Kleiner, and Paul A. Tukey. *Graphical Methods for Data Analysis.* Boston: Duxbury Press, 1983.

Cleveland, William S. *The Elements of Graphing Data.* Murray Hill, NJ: AT&T Bell Laboratories, 1994.

Filliben, James J., and Alan Heckert. "Exploratory Data Analysis." In *NIST/SEMATECH e-Handbook of Statistical Methods*, ed. Carroll Croarkin and Paul Tobias, 2004. U.S. Commerce Department, Technology Administration, National Institute of Standards and Technology, Information Technology Laboratory. www.itl.nist.gov/div898/handbook/eda/eda.htm.

Rutchik, Robert H. *Guidelines for Statistical Graphs.* U.S. Department of Energy, Energy Information Administration. www.eia.doe.gov/neic/graphs/preface.htm.

Tufte, Edward R. *The Visual Display of Quantitative Information.* Cheshire, CT.: Graphics Press, 2001.

Histogram and other frequency distributions

Chambers, John M., William S. Cleveland, Beat Kleiner, and Paul A. Tukey. *Graphical Methods for Data Analysis.* Boston: Duxbury Press, 1983.

Cleveland, William S. *The Elements of Graphing Data.* Murray Hill, NJ: AT&T Bell Laboratories, 1994.

Filliben, James J., and Alan Heckert. "Exploratory Data Analysis: Histogram." In *NIST/SEMATECH e-Handbook of Statistical Methods*, ed. Carroll Croarkin and Paul Tobias, 2004. U.S. Commerce Department, Technology Administration, National Institute of Standards and Technology, Information Technology Laboratory. www.itl.nist.gov/div898/handbook/eda/section3/histogra.htm.

Gunter, Berton H., "Subversive Data Analysis, Part I: The Stem and Leaf Display." *Quality Progress* 21, no. 9 (September 1988): 88–89.

Ishikawa, Kaoru. *Guide to Quality Control.* Tokyo: Asian Productivity Organization, 1986.

House of Quality

See also the QFD references listed for Chapter 2.

Hauser, John R., and Don Clausing. "The House of Quality." *Harvard Business Review* 66, no. 3 (May–June 1988): 63–73.

Hypothesis Testing

Filliben, James J., and Alan Heckert. "Exploratory Data Analysis: Quantitative Techniques." In *NIST/SEMATECH e-Handbook of Statistical Methods*, ed. Carroll Croarkin and Paul Tobias, 2004. U.S. Commerce Department, Technology Administration, National Institute of Standards and Technology, Information Technology Laboratory. www.itl.nist.gov/div898/handbook/eda/section3/eda35.htm.

Lane, David M. "HyperStat Online Textbook." *Rice Virtual Lab in Statistics,* Rice University. davidmlane.com/hyperstat/index.html.

Importance–Performance Analysis

Andreasen, Alan. *Marketing Social Change: Changing Behavior to Promote Health, Social Development, and the Environment.* San Francisco: Josey-Bass, 1995.

Simply Better! "Voice of the Customer Workbook." U.S. Department of Labor, Employment and Training Administration. www.workforce-excellence.net/pdf/vocwkbk.pdf.

Is–Is Not Matrix

Kepner-Tregoe, Inc. *Analytic Trouble Shooting.* Princeton, NJ: Kepner-Tregoe, 1966.

Kepner, Charles H., and Benjamin B. Tregoe. *The New Rational Manager.* Princeton, NJ: Princeton Research Press, 1981.

Matrix Diagram

Brassard, Michael. *Memory Jogger Plus+: Featuring the Seven Management and Planning Tools.* Methuen, MA: Goal/QPC, 1996.

Mizuno, Shigeru (ed.). *Management for Quality Improvement: The 7 New QC Tools.* Cambridge, Mass: Productivity Press, 1995.

Meeting Evaluation

Scholtes, Peter, Brian L. Joiner, and Barbara J. Streibel. *The Team Handbook.* Madison, WI: Joiner/Oriel Inc., 2003.

Mind Map

Buzan, Tony. Mind Maps®. Buzan Centres Ltd.
www.mind-map.com/EN/mindmaps/definition.html.
www.buzan.org

Mistake Proofing

Chase, Richard B., and Douglas M. Stewart. "Make Your Service Fail-Safe." *Sloan Management Review* 35, no. 3 (Spring 1994): 35–44.

Grout, John. "John Grout's Poka-Yoke Page." Berry College.
www.campbell.berry.edu/faculty/jgrout/pokayoke.shtml.

Nikkan Kogyo Shimbun/Factory Magazine, ed. *Poka-Yoke: Improving Product Quality by Preventing Defects.* Portland, OR: Productivity Press, 1989.

Shingo, Shigeo. *Zero Quality Control: Source Inspection and the Poka-Yoke System.* Translated by Andrew P. Dillon. Portland, OR: Productivity Press, 1986.

Multi-Vari Chart

Bhote, Keki R. *World Class Quality: Using Design of Experiments to Make It Happen.* New York: AMACOM, 1991.

Perez-Wilson, Mario. *Multi-Vari Chart and Analyses.* Scottsdale, AZ: Advanced Systems Consultants, 1993.

Multivoting

Delbecq, Andre L., Andrew H. Van de Ven, and David H. Gustafson. *Group Techniques for Program Planning: A Guide to Nominal Group and Delphi Processes.* Middleton, WI: Green Briar Press, 1986.

Huber, George, and A. L. Delbecq. "Guidelines for Combining the Judgments of Individual Group Members in Decision Conferences." *Academy of Management Journal* 15, no. 2 (June, 1972): 161.

Nominal Group Technique

Delbecq, Andre L., Andrew H. Van de Ven, and David H. Gustafson. *Group Techniques for Program Planning: A Guide to Nominal Group and Delphi Processes.* Middleton, WI: Green Briar Press, 1986.

Dunham, Randall B. "Nominal Group Technique: A User's Guide." *Organizational Behavior Web Course,* University of Wisconsin School of Business. instruction.bus.wisc.edu/obdemo/readings/ngt.html.

Normal Probability Plot

Chambers, John M., William S. Cleveland, Beat Kleiner, and Paul A. Tukey. *Graphical Methods for Data Analysis.* Boston: Duxbury Press, 1983.

Filliben, James J., and Alan Heckert. "Exploratory Data Analysis: Normal Probability Plot." In *NIST/SEMATECH e-Handbook of Statistical Methods*, ed. Carroll Croarkin and Paul Tobias, 2004. U.S. Commerce Department, Technology Administration, National Institute of Standards and Technology, Information Technology Laboratory. www.itl.nist.gov/div898/handbook/eda/section3/normprpl.htm.

Gunter, Bert. "Q–Q Plots." *Quality Progress* 27, no. 2 (February 1994): 81–86.

Operational Definition

Deming, W. Edwards. *Out of the Crisis.* Cambridge, MA.: Massachusetts Institute of Technology, Center for Advanced Engineering Study, 1986.

Scholtes, Peter, Brian L. Joiner, and Barbara J. Streibel. *The Team Handbook.* Madison, WI: Joiner/Oriel Inc., 2003.

Paired Comparison

Manktelow, James, ed. "Paired Comparison." Mind Tools. www.mindtools.com/pages/article/newTED_02.htm.

Peterson, George. "Paired Comparison Tool." U.S. Department of Agriculture, Natural Resources Conservation Service, Social Sciences Institute. www.ssi.nrcs.usda.gov/SSIEnvPsy/nrcs/paircomp.html.

Pareto Chart

Burr, John T. "The Tools of Quality: Part VI; Pareto Charts." *Quality Progress,* 23, no. 11 (November 1990): 59–61.

Juran, Joseph M., and A. Blanton Godfrey. *Juran's Quality Handbook.* New York: McGraw-Hill, 1999.

Ishikawa, Kaoru. *Guide to Quality Control.* Tokyo: Asian Productivity Organization, 1986.

Performance Index

Riggs, James, and Glenn Felix. *Productivity by Objectives.* Englewood Cliffs, NJ: Prentice Hall, 1983.

———. "Productivity Measurement by Objectives." *National Productivity Review* 2, no. 4 (Autumn, 1983): 386–393.

PGCV Index

Hom, Willard C. "Make Customer Service Analyses a Little Easier with the PGCV Index." *Quality Progress* 30, no. 3 (March 1997): 89–93.

Plan–Do–Study–Act Cycle

Deming, W. Edwards. *The New Economics for Industry, Government, Education.* Cambridge, MA: Massachusetts Institute of Technology, Center for Advanced Engineering Study, 1993.

Scholtes, Peter, Brian L. Joiner, and Barbara J. Streibel. *The Team Handbook.* Madison, WI: Joiner/Oriel Inc., 2003.

Plan–Results Chart

This tool was developed within Ethyl Corporation in the late 1980s. This book is the primary reference.

PMI

de Bono, Edward. *Serious Creativity: Using the Power of Lateral Thinking to Create New Ideas.* New York: HarperCollins, 1992.

Potential Problem Analysis

Kepner, Charles H., and Benjamin B. Tregoe. *The New Rational Manager*. Princeton, NJ: Princeton Research Press, 1981.

VanGundy, Arthur B. *Techniques of Structured Problem Solving*. New York: Van Nostrand Reinhold, 1988.

Presentation

Hoff, Ron. *"I Can See You Naked."* Kansas City, MO: Andrews and McNeel, 1992.

Tufte, Edward. *The Cognitive Style of PowerPoint*. Cheshire, CT: Graphics Press, 2003.

Prioritization Matrix

Brassard, Michael. *Memory Jogger Plus+: Featuring the Seven Management and Planning Tools*. Methuen, MA: Goal/QPC, 1996.

Dodgson, John, Michael Spackman, Alan Pearman, and Lawrence Phillips. "The Basic AHP Procedure." In *Multi-Criteria Analysis Manual*, 2004. UK, Office of the Deputy Prime Minister. www.odpm.gov.uk/stellent/groups/odpm_about/documents/page/odpm_about_608524-13.hcsp.

Saaty, Thomas L. *The Analytical Hierarchy Process: Planning, Priority Setting, Resource Allocation*. New York: McGraw-Hill, 1980.

Process Capability Index

Bothe, Davis R. *Measuring Process Capability*. Milwaukee: ASQ Quality Press, 2001.

Gunter, Berton H. "The Use and Abuse of Cpk." *Quality Progress* 22, no. 1 (January 1989): 72–73.

———. "The Use and Abuse of Cpk, Part 2." *Quality Progress* 22, no. 3 (March 1989): 108–9.

———. "The Use and Abuse of Cpk, Part 3." *Quality Progress* 22, no. 5 (May 1989): 70–71.

———. "The Use and Abuse of Cpk, Part 4." *Quality Progress* 22, no. 7 (July 1989): 86–87.

Kotz, Samuel, and Norman L. Johnson. "Process Capability Indices—A Review, 1992–2000 (with Subsequent Discussions and Response)." *Journal of Quality Technology* 34, no. 1 (January 2002): 2–53. American Society for Quality. www.asq.org/pub/jqt/past/vol34_issue1/index.html.

Process Decision Program Chart

Mizuno, Shigeru (ed.). *Management for Quality Improvement: The 7 New QC Tools*. Cambridge, MA: Productivity Press, 1995.

Project Charter

Swinney, Jack. "Project Charter." i-Six Sigma. www.isixsigma.com/library/content/c010218a.asp.
Snee, Ronald D., and Roger W. Hoerl. *Leading Six Sigma: A Step-by-Step Guide Based on Experience with GE and Other Six Sigma Companies.* Upper Saddle River, NJ: Financial Times/Prentice Hall, 2003.

Project Charter Checklist

Scholtes, Peter, Brian L. Joiner, and Barbara J. Streibel. *The Team Handbook.* Madison, WI: Joiner/Oriel Inc., 2003.

Radar Chart

Carangelo, Richard M. "Clearly Illustrate Multiple Measurables with the Web Chart." *Quality Progress* 28, no. 1 (January 1995): 36.
Chambers, John M., William S. Cleveland, Beat Kleiner, and Paul A. Tukey. *Graphical Methods for Data Analysis.* Boston: Duxbury Press, 1983.

Regression Analysis

Chambers, John M., William S. Cleveland, Beat Kleiner, and Paul A. Tukey. *Graphical Methods for Data Analysis.* Boston: Duxbury Press, 1983.
Motulsky, H. J. "Linear Regression." *Curvefit.com. The Complete Guide to Nonlinear Regression.* GraphPad Software, 1999. curvefit.com/linear_regression.htm.

Relations Diagram

Brassard, Michael. *The Memory Jogger Plus+.* Methuen, MA: GOAL/QPC, 1989.
Mizuno, Shigeru (ed.). *Management for Quality Improvement: The 7 New QC Tools.* Cambridge, MA: Productivity Press, 1995.

Repeatability and Reproducibility Study

Barrentine, Larry B. *Concepts for R&R Studies.* Milwaukee: ASQ Quality Press, 2003.
Croarkin, Carroll. "Gauge R&R Studies." In *NIST/SEMATECH e-Handbook of Statistical Methods*, ed. Carroll Croarkin and Paul Tobias, 2004. U.S. Commerce Department, Technology Administration, National Institute of Standards and Technology, Information Technology Laboratory. www.itl.nist.gov/div898/handbook/mpc/section4/mpc4.htm.
Wheeler, Donald J., and Richard W. Lyday. *Evaluating the Measurement Process.* 2d ed. Knoxville, TN: SPC Press, 1989.

Requirements Table

This tool was developed within Ethyl Corporation in the late 1980s. This book is the primary reference.

Requirements-and-Measures Tree

This tool was developed within Ethyl Corporation in the late 1980s. This book is the primary reference.

Run Chart

Pande, Peter S., Roland R. Cavanaugh, and Robert P. Neuman. *The Six Sigma Way Team Fieldbook: An Implementation Guide for Process Improvement Teams.* New York: McGraw-Hill, 2001.

Sampling

Hembree, Barry. "Define the Sampling Plan." In *NIST/SEMATECH e-Handbook of Statistical Methods*, ed. Carroll Croarkin and Paul Tobias, 2004. U.S. Commerce Department, Technology Administration, National Institute of Standards and Technology, Information Technology Laboratory. www.itl.nist.gov/div898/handbook/ppc/section3/ppc33.htm.

Ishikawa, Kaoru. *Guide to Quality Control.* Tokyo: Asian Productivity Organization, 1986.

Prins, Jack. "Test Product for Acceptability." In *NIST/SEMATECH e-Handbook of Statistical Methods*, ed. Carroll Croarkin and Paul Tobias, 2004. U.S. Commerce Department, Technology Administration, National Institute of Standards and Technology, Information Technology Laboratory. www.itl.nist.gov/div898/handbook/pmc/section2/pmc2.htm.

Slonim, Morris James. *Sampling: A Quick Reliable Guide to Practical Statistics.* New York: Simon and Schuster, 1966

Scatter Diagram

Burr, John T. "The Tools of Quality: Part VII; Scatter Diagrams." *Quality Progress* 23, no.12 (December 1990): 87–89.

Chambers, John M., William S. Cleveland, Beat Kleiner, and Paul A. Tukey. *Graphical Methods for Data Analysis.* Boston: Duxbury Press, 1983.

Ishikawa, Kaoru. *Guide to Quality Control.* Tokyo: Asian Productivity Organization, 1986.

SIPOC Diagram

Juran, Joseph M. *Juran on Leadership for Quality.* New York: Free Press, 1989.

Pande, Peter S., Roland R. Cavanaugh, and Robert P. Neuman. *Six Sigma Way Team Fieldbook: An Implementation Guide for Process Improvement Teams.* New York: McGraw-Hill, 2001.

Stakeholder Analysis

Chevalier, Jacques. "Stakeholder Analysis and Natural Resource Management." Carleton University. www.carleton.ca/~jchevali/STAKEH2.html.

Mayers, James. "Stakeholder Power Analysis." Tools for Working on Policies and Institutions. International Institute for Environment and Development, Forestry and Land Use Programme. www.iied.org/forestry/tools/stakeholder.html.

UK, Overseas Development Administration [now Department for International Development], Social Development Department. "Guidance Note on How to Do Stakeholder Analysis of Aid Projects and Programmes." Europe's Forum on International Cooperation. www.euforic.org/gb/stake1.htm.

Storyboard

Forsha, Harry I. *Show Me: The Complete Guide to Storyboarding and Problem Solving.* Milwaukee: ASQ Quality Press, 1995.

Shepard, Dick. "Storyboard: A Creative, Team-Based Approach to Planning and Problem-Solving." *Continuous Journey* 1, no. 2 (December 1992/January 1993): 24–32.

Wallace, Marie. "The Story of Storyboarding—Part 1," Law Library Resource Xchange. (May 22, 1997). www.llrx.com/columns/guide5.htm.

———. "The Story of Storyboarding—Part 2," Law Library Resource Xchange. (Jun 24, 1997). www.llrx.com/columns/guide6.htm.

Stratification

Ishikawa, Kaoru. *Guide to Quality Control.* Tokyo: Asian Productivity Organization, 1986.

Survey

Futrell, David, "Ten Reasons Why Surveys Fail," *Quality Progress* 27, no. 4 (April, 1994): 65–69.

Harrington, H. James. *Business Process Improvement: The Breakthrough Strategy for Total Quality, Productivity, and Competitiveness.* New York: McGraw-Hill, 1991.

Harvey, Jen, ed. "Evaluation Cookbook." Heriot-Watt University, Institute for Computer-Based Learning, Learning Technology Dissemination Initiative. www.icbl.hw.ac.uk/ ltdi/cookbook/contents.html.

Long, Lori. *Surveys from Start to Finish.* Alexandria, VA: American Society for Training and Development, 1986.

Nogami, Glenda Y. "Eight Points for More Useful Surveys." *Quality Progress* 29, no. 10 (October, 1996): 93–96.

Tree Diagram

Mizuno, Shigeru (ed.). *Management for Quality Improvement: The 7 New QC Tools.* Cambridge, MA: Productivity Press, 1995.

Two-Dimensional Chart

This generic tool has been identified and named by the author.

Value-Added Analysis

Harrington, H. James. *Business Process Improvement: The Breakthrough Strategy for Total Quality, Productivity, and Competitiveness.* New York: McGraw-Hill, 1991.

Voice of the Customer Table

Bolt, Andrew, and Glenn H. Mazur. "Jurassic QFD: Integrating Service and Product Quality Function Deployment," from The Eleventh Symposium on Quality Function Deployment, Novi, Michigan, June 1999. QFD and Voice of Customer Analysis. Japan Business Consultants Ltd. www.mazur.net/works/jurassic_qfd.pdf.
Terninko, John. *Step-by-Step QFD: Customer-Driven Product Design.* Boca Raton, FL: CRC Press, 1997.

Why–Why Diagram

Bailie, Howard H. "Organize Your Thinking with a Why–Why Diagram." *Quality Progress* 18, no. 12 (December, 1985): 22–24.

Wordsmithing

This tool was developed within Ethyl Corporation in the late 1980s. This book is the primary reference.

Work-Flow Diagram

Harrington, H. James. *Business Process Improvement: The Breakthrough Strategy for Total Quality, Productivity, and Competitiveness.* New York: McGraw-Hill, 1991.

TOOL COLLECTIONS ONLINE

Boeing Company. "Advanced Quality System Tools." www.boeing.com/companyoffices/doingbiz/supplier/d1-9000-1.pdf.
Center for Innovation in Engineering Education. "Review of Japanese and Other Quality Tools." Vanderbilt University, Engineering Science 130: Introduction to Engineering. www.vanderbilt.edu/Engineering/CIS/Sloan/web/es130/quality/qtooltoc.htm.
Manktelow, James. "Mind Tools." www.mindtools.com.
Mycoted, Ltd. "Creativity Techniques." Creativity & Innovation in Science & Technology. www.mycoted.com/creativity/techniques/index.php.

SkyMark Corp. "Classic Tools." SkyMark. www.skymark.com/resources/tools/management_tools.asp.

Sytsma, Sid, and Katherine Manley. "Quality Tools Cookbook." www.sytsma.com/tqmtools/tqmtoolmenu.html.

Team Creations. "Team Tool Box." The Team Resource Center. www.team-creations.com/toolbox.htm.

Tidd, Joe, John Bessant, and Keith Pavitt. "Innovation Management Toolbox." Wiley College Textbooks. www.wiley.co.uk/wileychi/innovate/website/index.htm.

STATISTICS REFERENCES ONLINE

Croarkin, Carroll, and Paul Tobias, ed. "NIST/SEMATECH e-Handbook of Statistical Methods." U.S. Commerce Department, Technology Administration, National Institute of Standards and Technology, Information Technology Laboratory. www.itl.nist.gov/div898/handbook/index.htm.

Lane, David M. "HyperStat Online Textbook." Rice Virtual Lab in Statistics, Rice University. davidmlane.com/hyperstat/index.html.

Pezullo, John C. "Interactive Statistical Calculation Pages." members.aol.com/johnp71/javastat.html.

StatSoft, Inc. "Electronic Statistics Textbook." StatSoft. www.statsoft.com/textbook/stathome.html.

University of California at Los Angeles, Department of Statistics. "Statistics Calculators." calculators.stat.ucla.edu.

Young, Anne, and Bob Gibberd. "SurfStat Australia: An Online Text in Introductory Statistics." www.anu.edu.au/nceph/surfstat/surfstat-home/surfstat.html.

Index

C

D

E

F

J

K

L

M

N

Q

T

X

Y

Z